Emerging Drug Delivery and Biomedical Engineering Technologies: Transforming Therapy

This book details the advances in drug discovery, manufacturing, and delivery of molecules, and the present need for emerging technologies. Throughout the text, emerging technologies are described, including methods such as additive manufacturing hot-melt extrusion, microfluidics, which have the potential to produce drug delivery systems that were not possible a few years ago, but also biosensors and engineering approaches for diagnosis. This book is of great use to both entry-level and experienced researchers in the field of emerging technologies for drug delivery and biomedical engineering applciations.

Features:

- Describes technologies that are significantly enhancing the delivery of drugs and monitoring patients.
- Presents new data on mobile and wearable point-of-care testing systems.
- Features hot topics such as bioprinting, cold plasma, microfluidics and microneedles.
- Will appeal to experienced researchers and those considering entering the field of emerging technologies.

Drugs and the Pharmaceutical Sciences
A Series of Textbooks and Monographs

Series Editor
Anthony J. Hickey
RTI International, Research Triangle Park, USA

The Drugs and Pharmaceutical Sciences series is designed to enable the pharmaceutical scientist to stay abreast of the changing trends, advances and innovations associated with therapeutic drugs and that area of expertise and interest that has come to be known as the pharmaceutical sciences. The body of knowledge that those working in the pharmaceutical environment have to work with, and master, has been, and continues, to expand at a rapid pace as new scientific approaches, technologies, instrumentations, clinical advances, economic factors and social needs arise and influence the discovery, development, manufacture, commercialization and clinical use of new agents and devices.

RECENT TITLES IN SERIES

Pharmaceutical Inhalation Aerosol Technology, Third Edition,
Anthony J. Hickey and Sandro R. da Rocha

Good Manufacturing Practices for Pharmaceuticals, Seventh Edition,
Graham P. Bunn

Pharmaceutical Extrusion Technology, Second Edition,
Isaac Ghebre-Sellassie, Charles E. Martin, Feng Zhang, and James Dinunzio

Biosimilar Drug Product Development,
Laszlo Endrenyi, Paul Declerck, and Shein-Chung Chow

High Throughput Screening in Drug Discovery,
Amancio Carnero

Generic Drug Product Development: International Regulatory Requirements for Bioequivalence, Second Edition,
Isadore Kanfer and Leon Shargel

Aqueous Polymeric Coatings for Pharmaceutical Dosage Forms, Fourth Edition,
Linda A. Felton

Good Design Practices for GMP Pharmaceutical Facilities, Second Edition,
Terry Jacobs and Andrew A. Signore

Handbook of Bioequivalence Testing, Second Edition,
Sarfaraz K. Niazi

FDA Good Laboratory Practice Requirements, First Edition,
Graham Bunn

Continuous Pharmaceutical Processing and Process Analytical Technology,
Ajit Narang and Atul Dubey

Project Management for Drug Developers,
Joseph P. Stalder

Emerging Drug Delivery and Biomedical Engineering Technologies: Transforming Therapy,
Dimitrios Lamprou

For more information about this series, please visit: www.crcpress.com/Drugs-and-the-Pharmaceutical-Sciences/book-series/IHCDRUPHASCI

Emerging Drug Delivery and Biomedical Engineering Technologies
Transforming Therapy

Edited By

Dimitrios Lamprou
Queen's University Belfast, School of Pharmacy

CRC Press
Taylor & Francis Group
Boca Raton London New York

CRC Press is an imprint of the
Taylor & Francis Group, an **Informa** business

First edition published 2023
by CRC Press
6000 Broken Sound Parkway NW, Suite 300, Boca Raton, FL 33487-2742

and by CRC Press
4 Park Square, Milton Park, Abingdon, Oxon, OX14 4RN

Library of Congress Cataloging-in-Publication Data
Names: Lamprou, Dimitrios, editor.
Title: Emerging drug delivery and biomedical engineering technologies :
transforming therapy / edited by Dimitrios Lamprou, Queens University
Belfast, School of Pharmacy.
Description: First edition. | Boca Raton : CRC Press, 2023. |
Series: Drugs and the pharmaceutical sciences |
Includes bibliographical references and index. |
Identifiers: LCCN 2022047398 (print) | LCCN 2022047399 (ebook) |
ISBN 9781032122717 (hardback) | ISBN 9781003224464 (paperback) |
ISBN 9781032124223 (ebook)
Subjects: LCSH: Pharmaceutical technology. | Drug delivery systems. | Drugs—Design.
Classification: LCC RS192 .E44 2023 (print) | LCC RS192 (ebook) |
DDC 615.1/9—dc23/eng/20230125
LC record available at https://lccn.loc.gov/2022047398
LC ebook record available at https://lccn.loc.gov/2022047399

ISBN: 978-1-032-12271-7 (HB)
ISBN: 978-1-032-12422-3 (PB)
ISBN: 978-1-003-22446-4 (EB)

DOI: 10.1201/9781003224464

Typeset in Times
by codeMantra

Contents

Preface

The strategies and manufacturing processes for the production of dosage forms and medical devices are experiencing significant advancement. Trends in technological advances have given the potential to new technologies that can transform therapy. Emerging technologies are shaping the future of manufacturing and delivering traditional and new medicines, and monitoring diseases. These technologies are ranging from apps to bioprinting to microneedles to cold plasma. This book covers a variety of emerging systems such as drug-loaded nanofibers or particles that can be incorporated into more complex objects, the importance of printing in personalised medicine, microfluidics and biosensors for drug delivery applications and disease monitoring, cellular therapy, cold-plasma applications in tissue engineering and devices and apps used to improve patient compliance.

Key Features

- Comprehensive coverage of emerging technologies used in development and manufacturing.
- It discusses the advantages of emerging technologies.
- It presents new opportunities available and future trends.
- This book has been compiled by highly experienced contributors in their fields.

Readership: Researchers in industry and academia working in drug delivery and biomedical engineering applications using traditional and emerging technologies. Undergraduate and postgraduate students in Pharmacy, Pharmaceutical Sciences, Biotechnology, and Bioengineering will find this book very interesting. The book also covers the aspects of regulatory and continuous manufacturing, aspects that are very important for scientists in pharma, biopharma and biofabrication industries. Clinicians and regulatory experts will also benefit from this book.

Editor

Dimitrios Lamprou (PhD, MBA) is a Professor (Chair) of Biofabrication and Advanced Manufacturing and a well-known expert in emerging technologies for drug delivery and medical implants. He is currently the author of over 140 peer-reviewed publications and of over 350 conference abstracts and has over 140 oral talks in institutions and conferences across the world. His research and academic leadership have recognised with a range of awards, including the Royal Pharmaceutical Society of Great Britain Science Award and the Scottish Universities Life Sciences Alliance Leaders Scheme Award.

Contributors

Prashant Agrawal
Department of Mathematics, Physics and Electrical
    Engineering
Northumbria University
London, United Kingdom

Gozal Ahmadova
Institute for Innovation and Knowledge Management
ESADE Business School
Barcelona, Spain

Manal Eid Alkahtani
UCL School of Pharmacy
University College London
London, United Kingdom

Brayan J. Anaya
Department of Pharmaceutics and Food Science
School of Pharmacy, Complutense
    University of Madrid
Madrid, Spain

G. P. Andrews
Pharmaceutical Engineering Group
School of Pharmacy, Queen's University
Belfast, United Kingdom

Heather E. Barry
School of Pharmacy
Queen's University Belfast
Belfast, United Kingdom

Madalina M. Barsan
Laboratory of Functional Nanostructures
National Institute of Materials Physics
Măgurele, Romania

Abdul W. Basit
Department of Pharmaceutics
UCL School of Pharmacy
University College London
London, United Kingdom

John Batani
Faculty of Engineering and Technology
Botho University
Maseru, Lesotho

Luís B. Bebiano
i3S – Instituto de Investigação e Inovação em Saúde
Universidade do Porto
Porto, Portugal
and
INEB – Instituto Nacional de Engenharia
BiomédicaUniversidade do Porto
Porto, Portugal
and
ISEP – Instituto Superior de Engenharia do Porto
Politécnico do Porto
Porto, Portugal

Flávia Castro
i3S – Instituto de Investigação e Inovação em Saúde
Universidade do Porto
Porto, Portugal
and
INEB – Instituto Nacional de Engenharia Biomédica
Universidade do Porto
Porto, Portugal

Jose R. Cerda
Department of Pharmaceutics and Food Science
School of Pharmacy, Complutense University of Madrid
Madrid, Spain

Tiffany G. Chan
Department of Bioengineering
Imperial College London
London, United Kingdom

Enrica Chiesa
Department of Drug Sciences
University of Pavia
Pavia, Italy

Innocent Chingombe
ICAP
Columbia University
Harare, Zimbabwe

Itai Chitungo
Faculty of Medicine
College of Medicine and Health Sciences, University of
    Zimbabwe
Harare, Zimbabwe

Bice Conti
Department of Drug Sciences
University of Pavia
Pavia, Italy

Dan J. Corbett
School of Pharmacy
Queen's University Belfast
Belfast, United Kingdom

Javier Cudeiro-Blanco
Department of Bioengineering
Imperial College London
London, United Kingdom

Diganta B. Das
Department of Chemical Engineering
Loughborough University
Loughborough, United Kingdom

Victor C. Diculescu
Laboratory of Functional
    Nanostructures
National Institute of Materials Physics
Măgurele, Romania

Rossella Dorati
Department of Drug Sciences
University of Pavia
Pavia, Italy

Tafadzwa Dzinamarira
ICAP
Columbia University
Harare, Zimbabwe
and
School of Health Systems & Public Health
University of Pretoria
Pretoria, South Africa

Sotirios I. Ekonomou
Faculty of Health and Applied
    Sciences (HAS)
University of the West of England
Bristol, United Kingdom

Moe Elbadawi
UCL School of Pharmacy
University College London
London, United Kingdom

Teodor A. Enache
Laboratory of Functional Nanostructures
National Institute of Materials Physics
Măgurele, Romania

YongQing Fu
Department of Mathematics, Physics and Electrical
    Engineering
Northumbria University
London, United Kingdom

Simon Gaisford
Department of Pharmaceutics
UCL School of Pharmacy
University College London
London, United Kingdom

Ida Genta
Department of Drug Sciences
University of Pavia
Pavia, Italy

Charlotte R. Haigh
Department of Chemical Engineering
Loughborough University
Loughborough, United Kingdom

Maurice Hall
School of Pharmacy
Queen's University Belfast
Belfast, United Kingdom

Lezley-Anne Hanna
School of Pharmacy
Queen's University Belfast

D. S. Jones
Pharmaceutical Engineering Group
School of Pharmacy, Queen's University
Belfast, United Kingdom

Aytug Kara
Department of Pharmaceutics and Food Science
School of Pharmacy, Complutense University of Madrid
Madrid, Spain

Dimitrios A. Lamprou
School of Pharmacy
Queen's University Belfast
Belfast, United Kingdom

S. Li
Pharmaceutical Engineering Group
School of Pharmacy, Queen's University
Belfast, United Kingdom

Yi-Chen Ethan Li
Department of Chemical Engineering
Feng Chia University
Taichung, Taiwan

D. Liu
Pharmaceutical Engineering Group
School of Pharmacy, Queen's University
Belfast, United Kingdom

Francis C. Luciano
Department of Pharmaceutics and
    Food Science
School of Pharmacy, Complutense
    University of Madrid
Madrid, Spain

Christopher Markwell
Department of Mathematics, Physics and Electrical
    Engineering
Northumbria University
London, United Kingdom

Elliot Mbunge
Faculty of Science and Engineering, Department of
    Computer Science
University of Eswatini
Kwaluseni Campus, Eswatini

Sophie V. Morse
Department of Bioengineering
Imperial College London
London, United Kingdom

Sterghios A. Moschos
Department of Applied Sciences
Northumbria University
London, United Kingdom

Benhildah Muchemwa
Faculty of Science and
    Engineering, Department of
    Computer Science
University of Eswatini
Kwaluseni Campus, Eswatini

Godfrey Musuka
ICAP
Columbia University
Harare, Zimbabwe

Mine Orlu
UCL School of Pharmacy
University College London
London, United Kingdom

Maryam Parhizkar
School of Pharmacy
University College London
London, United Kingdom

Bruno Pereira
i3S – Instituto de Investigação e
    Inovação em Saúde
Universidade do Porto
Porto, Portugal
and
IPATIMUP – - Instituto de Patologia e Imunologia
    Molecular da
Universidade do Porto
Porto, Portugal

Rúben F. Pereira
i3S – Instituto de Investigação e
    Inovação em Saúde
Universidade do Porto
Porto, Portugal
and
INEB – Instituto Nacional de Engenharia
BiomédicaUniversidade do Porto
Porto, Portugal
and
ICBAS – Instituto de Ciências Biomédicas Abel Salazar
Universidade do Porto
Porto, Portugal

Silvia Pisani
Department of Otorhinolaryngology
IRCCS Foundation Policlinico S.Matteo
Pavi, Italy

Antonios N. Pouliopoulos
Department of Surgical and
    Interventional Engineering
King's College London
London, United Kingdom

Laia Pujol Priego
Institute for Innovation and Knowledge Management
ESADE Business School
Barcelona, Spain

Angelo Kenneth Romasanta
Institute for Innovation and Knowledge Management
ESADE Business School
Barcelona, Spain

Saliha Saad
Faculty of Health and Applied Sciences (HAS)
University of the West of England
Bristol, United Kingdom

Corinna Schlosser
UCL School of Pharmacy
University College London
London, United Kingdom

István Sebe
University Pharmacy Department of Pharmacy
  Administration
Semmelweis University
Budapest, Hungary
and
Research & Development Directorate
Egis Pharmaceuticals PLC
Budapest, Hungary

Dolores R. Serrano
Department of Pharmaceutics and Food Science
School of Pharmacy, Complutense University of Madrid
Madrid, Spain

Alexandros Ch. Stratakos
Faculty of Health and Applied Sciences (HAS)
University of the West of England
Bristol, United Kingdom

Edina Szabó
Department of Organic Chemistry and Technology
Budapest University of Technology and Economics
Budapest, Hungary

Stephen Todryk
Department of Applied Sciences
Northumbria University
London, United Kingdom

Hamdi Torun
Department of Mathematics, Physics and Electrical
  Engineering
Northumbria University
London, United Kingdom

Dimitrios Tsaoulidis
Department of Chemical and Process Engineering
University of Surrey
Guildford, United Kingdom

Aniko Varadi
Faculty of Health and Applied Sciences (HAS)
University of the West of England
Bristol, United Kingdom

Jonathan Wareham
Institute for Innovation and Knowledge Management
ESADE Business School
Barcelona, Spain

Edward Weaver
School of Pharmacy
Queen's University Belfast
Belfast, United Kingdom

Prateek R. Yadav
Department of Chemical Engineering
Loughborough University
Loughborough, United Kingdom

Iván Yuste
Department of Pharmaceutics and Food Science
School of Pharmacy, Complutense University of Madrid
Madrid, Spain

Romána Zelkó
University Pharmacy Department of Pharmacy
  Administration
Semmelweis University
Budapest, Hungary

1 Advances in Drug Delivery and the Need for Emerging Technologies

Edward Weaver and Dimitrios A. Lamprou
Queen's University Belfast

CONTENTS

1.1 INTRODUCTION

The world of pharmaceutics and healthcare is a constantly developing one, with new demands and pressures being exerted constantly. To maintain the balance and to strive deeper into the capacity of treating diseases, much of the responsibility falls upon novel emerging technologies. As capabilities within human medical expertise have developed, the demand by patients for a higher level of healthcare has increased [1], which is a large stimulant in the drive for discovering new technologies. The term "emerging technology" may not necessarily relate to being a completely novel process, but it can apply to a technology that is beginning to show great promise in recent years, whether that be due to a new application of it or whether it be related to progress obtained in related fields.

Progress is coming both from the areas of novel manufactured therapeutics as well as the invention of diagnostic and assisting aids. While many emerging technologies show great promise, it's also important to consider the economic viability of the processes, as one of the major barriers to development is funding. The benefit of researching these new technologies is that new methods of manufacturing or synthesis can be discovered that may reduce the overall cost of a process. For example, when the concept of 3D printing (3DP) was established in the 1980s, the printers were viable only for industrial use due to their price; however, as the level of R&D has increased within the field and more manufacturers can supply the technology, the price of the printers has reduced significantly, meaning that now

3D printers are also available for individual use. This in turn means that more processes can be explored using the technology, leading to exponential growth in the knowledge obtained. Moreover, as the vast majority of the population now has access to a mobile phone [2], technologies involving the use of apps to support healthcare are a hugely developing market.

The desire to progress in a research field is inevitable, and this review will explore certain areas that are showing great progress to achieve the task of implementing these emerging technologies on a wide scale in the near future.

1.2 EMERGING MANUFACTURED THERAPEUTICS

1.2.1 REGENERATIVE MEDICINE

Regenerative medicine is a multi-faceted discipline that is based around the replacement of damaged cells and tissues from sources such as stem cells, donor tissues or printing apparatus. The concept of organ and tissue transplantation has been widely used as a means of life preservation and elongation for decades; however, there have always been issues ensuring a steady demand for transplantable organs. This explains why other methods are being researched to provide alternative approaches to the renewal of proper body functionality via the use of regenerative medicine.

Wound healing is also an area being thoroughly explored, owing to the skin's importance to a human's survival. Being the largest organ in the human body, severe

DOI: 10.1201/9781003224464-1

1

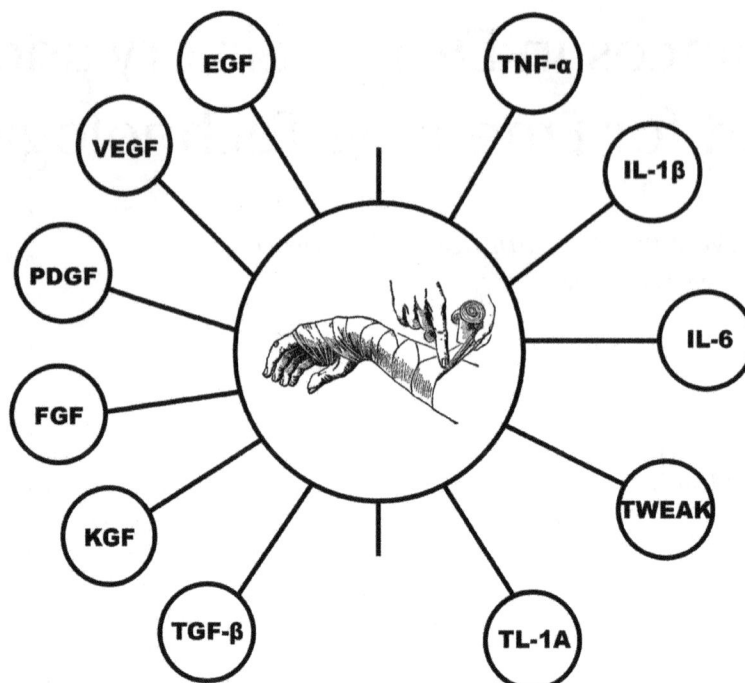

FIGURE 1.1　A list of many of the integral growth factors and cytokines involved in wound repair. All acronyms are as follows: EGF (epidermal growth factor), VEGF (vascular endothelial growth factor), PDGF (platelet-derived growth factor), FGF (fibroblast growth factor), KGF (keratinocyte growth factor), TGF-β (transforming growth factor beta), TNF-α (tumour necrosis factor alpha), IL-1β (interleukin 1 beta), IL-6 (Interleukin 6), TWEAK (tumour necrosis factor-like weak inducer of apoptosis), TL-1A (tumour necrosis factor-like cytokine 1A). More information about the factors can be found here [6,7].

damage to the skin via routes, such as impact trauma or burning, can readily cause life-threatening issues. Novel regenerative wound therapy is being proposed to act alongside conventional skin therapies including debridement [3]; therefore, a localised skin graft would not be considered as a regenerative therapy. Due to its complexity, skin regeneration is a complex and expensive process to perform artificially, with a constant need to assess a cost-benefit analysis before proceeding with the procedure. A vast multitude of factors are triggered upon a trauma to the skin, including the coagulation cascade, growth factors (epidermal growth factor, transforming growth factor beta, etc.) and a vast array of inflammatory mediators [4]. To complicate issues further, active pharmaceutical ingredients (APIs), such as the growth factors mentioned above, that are delivered to the trauma site are often cleared quickly and have low stability. The natural mediators released upon trauma have also been shown to interfere with the therapy provided [4]. Hence there must be put in place an effective drug delivery system (DDS) to combat this challenge.

A few growth factor-based medicines have already been approved for clinical use, including Regranex®, which exploits platelet-derived growth factor (PDGF) in a topically applied setting to assist with the treatment of diabetic foot ulcers alongside conventional treatment. This artificially produced PDGF promotes the chemotactic recruitment and proliferation of cells involved in wound repair, allowing the formulation to provide approximately twice the standard of care for diabetic foot ulcer (DFU) compared

to monotherapy alone [5]. A more comprehensive display of some of the growth factors and cytokines involved in wound healing can be found in Figure 1.1.

The use of stem cells for tissue regeneration has been a focus for the emerging drug market for decades; however, it is constantly being met with contradiction for its therapeutic use, whether that be from an efficacy standpoint or an ethical one. Stem cells act as multipotent bases to differentiate into theoretically any desired cell within the body (depending upon their source), which is obvious why they're an attractive target for emerging medical biotechnologies. Sources of multipotent stem cells include, but are not limited to, bone marrow, peripheral blood, adipose, placenta and foetal tissue. Stem cells can be artificially triggered via environmental stimuli to differentiate and proliferate into a desired cell type. For example, using embryonic stem cells, retinal degenerative disease has seen an improved prognosis [8], which is a huge step considering that previous therapies offered no regenerative effect at all. Other examples of the use of stem cells have seen improved treatments for diabetic neuropathy [9], Parkinson's disease [10] and Alzheimer's disease [11]. While the concept of the use of stem cells is relatively old, what classes it as an emerging technology is the huge jump in viability for the process that has been gained in recent times. Coupling the use of stem cells alongside technologies like cellular scaffolding from electrospinning or microRNA co-administration has meant that the implementation of stem cells for their regenerative properties has become far more successful. It is safe to say that the widespread use of stem cell therapy is not yet ready, but with the amount of research currently being undertaken in the area and with the advancement of technology alongside, it's only a matter of time. The ethical opposition to its use has also dissipated slightly, owing to the easing access to renewable sources of stem cells, such as bone marrow, rather than using more controversial sources like embryos.

Cold plasma, also known as non-thermal plasma, is an innovative technology often associated with wound healing and tissue regeneration. Discovered approximately 25 years ago [12], cold plasma is a partially ionised gas consisting of reactive species including electrons, ultraviolet photons and radicals. Its reactive species are what give rise to its therapeutic effect within the field of medicine. What distinguishes cold plasma as a promising emerging technology is its safety profile upon administration, as cold plasma appears to be completely biocompatible with tissues in the body [13]. The main beneficial species within cold plasma for skin regeneration come in the form of the reactive nitrogen and oxygen species present, as these are responsible for the expression of a number of growth factors and cytokines associated with tissue repair e.g. vascular endothelial growth factor (VEGF) [14]. While cold plasma is yet to be used clinically, the initial efficacy and safety studies performed display great promise around tissue regeneration. For example, a study performed on sheep in the area of wound closing presented a complete wound closure between 28 and 42 days depending upon the level of cold plasma treatment, compared to unsuccessful closure within 42 days for the control group of no treatment [13]. Another example displayed a greater acceleration of wound healing in murine subjects compared to no treatment [15]. As mentioned, the safety profile of cold plasma appears to be one of the most beneficial factors of the therapy, as it doesn't affect the number of proliferative and apoptotic cells, as well as not affecting the DNA within cells [16].

1.2.2 3D Printing

The area of 3DP, also known as additive manufacturing (AM), has made huge leaps in its progress in recent years, with new uses being discovered frequently. When stripped to its basics, there exist nine different forms of 3DP; however, not all may be suitable for printing for medical purposes, owing to the conditions experienced during printing, whether that be temperature, reagent or environmentally based [17]. 3DP is already an extremely accomplished discipline, with higher-quality printers being developed steadily. The decrease in cost for 3D printers has dropped dramatically over the decades, making them much more accessible to the masses [18]. 3D printers that are frequently used for medical applications are fused deposition modelling (FDM), selective laser sintering (SLS) and stereolithography (SLA). The research being undertaken in the area of 3DP is extremely extensive and owing to the wide scope of printing variables available; a wide range

of materials can be used to produce various products and devices. As mentioned though, it is essential to choose the correct printer depending on the material being used, as well as the properties of the printed object. For example, FDM printers generally have a maximum heating temperature of 300°C, meaning that only materials with melting temperatures less than this may be suitable for the use of this printer type. This bracket of materials would include many polymers and some polymer ceramic composites such as polylactic acid (PLA) and aluminium (III) oxide composites, respectively [19]. The resolutions of each printer will vary upon individual materials being used and what conformation they're being printed in, so it's difficult to say which type of printer will provide the highest quality of results. Table 1.1 displays the advantages and disadvantages of each type of commonly used printer to allow a comparison between the characteristics and usages for each device type.

3D printed tablets are an emerging technology being researched heavily within the field of 3DP. The aim of 3DP tablet is to fully customise and optimise the properties that a tablet possesses, for example decreasing the disintegration time or tablet friability. The designs and constructs obtained via 3DP are far more complex than what could be achieved using a standard mould/press procedure. Designs proposed include a multi-channelled dosage form to aid tablet disintegration and dissolution [37], tablets with braille/patterns for the visually impaired [38] and complex geometrical shapes [39], which can be seen in Figure 1.2.

3DP has also been employed in the fabrication of reservoir microdevices to assist in the prolonged delivery of APIs via the favoured oral route [40]. The printed devices are approximately 500 μm in diameter, printed from a sacrificial material like polyvinyl alcohol (PVA) and filled with a desired compatible API. The concept of the device is to allow APIs, which may experience high rates of clearance, to experience a prolonged release due to the device having mucoadhesive properties within the intestine. This new technology also allows enhanced mucoadhesion compared to previously proposed manufacturing processes [40], owing to its composition and the ability to precisely control the shape and orientation of the materials within the device. Once the devices adhere within the intestine, the sacrificial material gradually dissolves, releasing the entrapped API in a controlled manner. Contained within a gastro-resistant capsule, the microdevices can be delivered orally and released throughout the intestine to achieve their goal of prolonged drug release.

3DP has also being used to assist with the planning and proceeding of surgical operations via the use of 3D-printed replica models [41]. The knowledge of an individual's anatomy pre-surgery can be an extremely helpful tool in decreasing the risk of performing the surgery, as well as decreasing the time of the procedure. This is particularly important when sensitive areas such as the base of the skull or cranial areas are involved [42]. The models act as a detailed guide for surgery and allow

TABLE 1.1

Summary of 3D Printers and Their Usages

Type of Printer	Main Compatible Materials	Advantages	Disadvantages	Common Usages
Selective Laser Sintering (SLS)	• Thermoplastic polymers [20] • Laser-insensitive APIs	• Reduced energy consumption • Wide range of usage • Starting materials don't require pre-processing [20]	• Potential for material degradation during printing due to laser • Printed objects may require post-treatment	• Medical implants and surgical tools • Tissue Engineering • Prosthetic manufacture • Tablet printing
Stereolithography (SLA)	• Acrylate-based resins [21] • Epoxy resins • Ceramics	• High resolution [21]	• High cost • Additives affect the printing process dramatically [21] • Low printing velocities • Compatible with only one material at a time	• Microfluidic-chip printing • Robotics and aerospace • Manufacturing implant devices
Fused deposition modelling (FDM)	• Thermoplastic polymers [22] • Polymer-matrix composites • Fibre-reinforced composites [19]	• Cost effective • High-resolution prints • Integrates simply with computer-aiding design software [19]	• Materials with melting points >300°C may be unsuitable • Varying quality depending upon parameters used [19]	• Microfluidic device casting • Tools and equipment manufacturing [23]
Digital light processing (DLP)	• Photocurable resins [24] • Elastomers [25] • Bioinks [26]	• High resolution • High printing speed [27]	• Only compatible with photo-sensitive materials	• Bioprinting [26,28] • Microneedle printing [29]
Multi-jet fusion (MJF)	• Polyamide 12 [30]	• Faster than SLS [30]	• New technology so gaps in research	• Scaffolds for biomedical applications [31]
Direct metal laser sintering (DMLS)	• Metal • Alloyed powders • Metal matrix composites [32]	• Starting materials don't require pre-processing [20]	• Limited process parameters have been explored • Materials can be expensive [32]	• Biomedical applications and implants [33,34] • Microwave components [35]
Electron beam melting (EBM)	• Metal • Alloyed Powders [20]	• Higher temperatures than SLS can be achieved, up to 1100°C [36], meaning some materials can be used for EBM which can be processed with SLS	• Restricted to conductive materials • Requires vacuum conditions [36]	• Dental prosthesis • Artificial joints • Aeronautic applications [36]

custom 3D-printed operative tools and guides to be fabricated in case they are required for surgery [43]. Images acquired from computerised tomography (CT) and magnetic resonance imagining (MRI) can be transferred and translated into a computerised design ready for printing, which will be individualised for each patient. This will act as a tangible aid for the surgical team to prepare for the operation. The models can also be used as an educative tool as opposed to dissected specimens, to train the future medical personnel. Owing to the customisable nature of 3DP, the printed models can also be coloured or accentuated, to highlight areas within the model and ameliorate learning.

In circumstances as experienced with the covid-19 pandemic, 3DP provides a viable means to create objects without the necessity for a supply chain. As the presence of 3D printers increases in locations like hospitals, the specialised tools and devices are able to be printed on site within a short amount of time which could help prevent disruption to the provision of healthcare [44].

1.2.3 3D AND 4D BIOPRINTING

As mentioned in Section 1.2.1, 3D bioprinting (3DBP) is a revolutionary concept that has recently been developed to aid the artificial production of tissues and organs, ready for transplantation. With high waiting times for life-critical transplants of various organs including kidneys and livers, it's essential that an alternative means is optimised for ensuring a readily available source of organs and tissues. Recent statistics suggest that the number of live-donor registrants has decreased over the last 5 years [45] meaning less availability of potentially higher-quality organs for transplantation.

3DBP incorporates printing bioinks, containing materials such as cells or stem cells, alongside other biocompatible

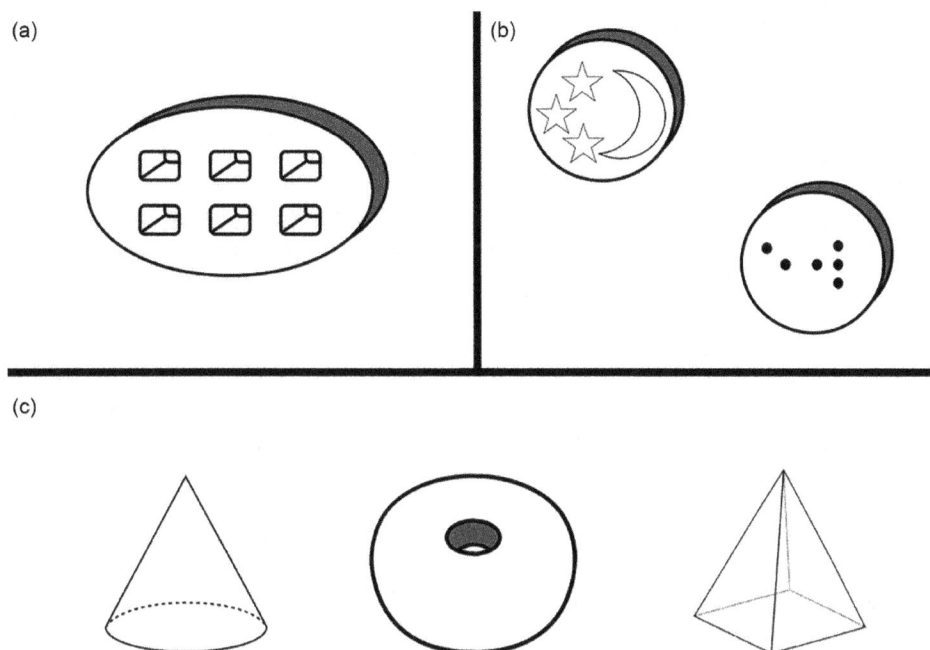

FIGURE 1.2 (a–c) Proposed designs for 3D-printed tablets of various designs.

materials to produce an artificial organ/tissue to mimic its naturally occurring equivalent. While this concept appears extremely complex, having to imitate the intricacies of human tissue into a layer-by-layer printed model, recent advances in CT and MRI have allowed relatively simple transference of specific areas required for production [46], as a computerised design must be conceived prior to the printing process. This will also mean that the provision of patient-specific care is becoming far more of a reality. Previous methods used before 3DBP relied on the use of an internal central scaffolding within the printed organ to support the attachment and proliferation of cells, using both naturally occurring materials including gelatin and collagen, or using synthetic polymers including PLA and poly-(lactic-co-glycolic acid) (PLGA) [47]. The cell/scaffolding complex would then be left to develop under precisely controlled conditions to ensure tissue-organ development. This method could however lack the ability to produce complex microstructures within the tissues, which is why the 3DBP method could be favourable, owing to its capacity to place individual cells exactly in their predetermined places. The tissue/organ is constructed fully, layer-by-layer, using the bioprinter, ready for use for its intended purpose.

It should also be noted that there is an alternative method for 3DBP, other than the layer-by-layer AM process. The method is referred to as indirect 3DBP [48] and revolves around the production of sacrificial moulds, which are then coupled with the desired biomaterial, after which the mould is selectively removed. This method is generally deemed as a less efficient version of 3DBP as it involves more steps, often can't produce the same resolution as direct 3DBP and is potentially a more wasteful process [49]. After allograft transplantation, patients are usually required to take lifelong

immunosuppressants to minimise the chance of organ rejection, which is obviously not an ideal situation given the prevalence of infectious diseases worldwide, including the recent covid-19 virus pandemic. A recent study incorporating regulatory T-cells within a bioprinted hydrogel has shown that the need for such high levels of immunosuppression post-transplantation has been reduced [50].

Bone grafting has also been shown to be successful using 3DBP, for those individuals who may require it following a lack of capacity to naturally regrow or replace their own bone structures. Using an extrusion-based 3D printer, a chitosan-based hydrogel scaffold has been developed to act as the basis for a successful artificial bone graft surgery, which will couple with calcium carbonate composites to simulate bone [51]. One of the main functions of bone is obviously to provide structural integrity within the skeleton, hence precise placement must be involved during the manufacturing of the scaffold to provide the optimum basis for the bone graft. The 3DP method is perfect for this role owing to its ability to place materials in specific predetermined places.

The bioprinted organs not only act as a viable alternative for transplantation, but they can also be used as templates for diagnostic tests like high-throughput screening for APIs or medicines. The production of a 3D-simulated environment to investigate the action of an API in a particular tissue is a highly sought-after process as it will aid the accuracy to which data can be obtained in pre-clinical trials. 2D models have been used previously for high-throughput screening; however, it has often been noted that the 2D models don't accurately represent the environment of complex tissues and organs, which is why the implementation of a 3D model would help mimic truly the extracellular matrix that would be experienced *in vivo* [52]. As the cost of producing these artificial organs is steadily decreasing [53], it is becoming gradually more viable to produce these models for commercial use, especially if only a small area of tissue is required for production. Printers that cost as low as €150 have been modified minorly to produce viable tissues that would be viable for high-throughput screening.

4D bioprinting (4DBP) has also been given rise thanks to advancements in 3DBP. 4DBP differs from 3DBP in the way that the printed 3D lattice structure is able to respond effectively to external stimuli, as would be seen in native tissues. This is achieved via the use of responsive biomaterials and bioinks, including thermal-, humidity-, electrical- and magnetic responsive materials [54]. The dynamic capacity of these 4D-printed structures provides them with greater efficacy of mimicking the native tissue which in turn increases their usefulness upon transplant. While 3D-printed structures can reach a level of dynamic capacity once *in situ*, there is a chance that the tissue can take time to reach this state; whereas 4D-printed tissues will immediately be established within the environment [54]. Thermally responsive materials are among the most studied materials used for 4DBP, such as the responsive material poly(N-isopropylacrylamide) (PNIPPAAm) [55]. These materials are able to change shape reversibly depending on the temperature that they're exposed to, whether that be to fold, shrink or swell. This property is especially useful for tissues like muscles and skin, which are constantly changing dimensions due to these kinds of stimuli.

1.2.4 ELECTROSPINNING

Electrospinning is a process whereby nano-scale fibres can be continuously manufactured from polymer materials, with the use of a high-voltage power supply. Compared to other fibre manufacturing methods, electrospun (e-spun) fibres often boast favourable characteristics including a high surface area to volume ratio, high porosity and high tensile strengths [56], as well as being a continuous process instead of batch. The produced fibres have multiple applications both in and out of the medical field. To name a few uses outside the area of focus, e-spun fibres have been used for air filtration, environmental protection, insulation materials and energy storage [57]. Within medicine, there are some exciting emerging applications that e-spun products are being used for, which will be discussed below. The examples discussed in this review will focus on solution electrospinning, although melt electrospinning is also a prominent area of emerging technologies [58,59].

Owing to their excellent biocompatibility, a common usage for e-spun fibres is to be applied as a scaffolding material for tissue and cell culture, which ties it into the field of regenerative medicine mentioned previously. Multiple research studies have investigated the optimal choice of polymer to use as a scaffold, depending upon the specific environment required. Amongst many examples, poly(glycerol sebacate) (PGS) has shown promise as a scaffold for nerve tissue engineering [60], while a collagen/chitosan/hyaluronic acid polymer solution displayed ameliorated qualities for bone regeneration. The scaffolding mimics partially the extracellular matrix (ECM) of the tissues due to its nano-fibrous nature, which explains its importance when choosing the correct polymers. The naturally occurring ECM consists of a gel-nanofiber matrix, consisting over polymer structural assistants like collagen and proteins [61]. The fibres within the ECM have diameters of approximately 100 nm [62], providing a target size for the artificial fibres. Electrospinning has become the most common method for the production of scaffolding for tissue engineering [63], as its relatively simple concept and its scalability lend itself towards the process. The ability of the fibres to form a 3D matrix compared to 2D is also an excellent feature as mimics the ECM to a closer extent.

E-spun fibres can act as excellent structure materials to facilitate drug delivery via various delivery routes. Their high surface area to volume ratio allows a very high loading capacity of medicines within the fibres, and once again owing to their adaptability, different polymers can be used to carry a wide range of medicines. Drug-loaded e-spun fibres provide viable routes of administration across transdermal, topical and oral routes, as well as minorly invasive injections offering parenteral delivery [64]. These locally injected implants provide excellent control over release kinetics, whether it be first order or biphasic, etc., depending upon the choice of e-spun material used. The biomedical applications for the injectable e-spun fibre scaffolds are very wide and have already been shown to have efficacy in treatments including sexually transmitted disease prevention, neuron regeneration, spinal fusion and prolonged hormone delivery, amongst many more [65–68]. The e-spun fibres are so versatile that they can be adapted to each individual task depending on what is required for the DDS, once again pointing towards a potential goal of patient-specific treatment. The microstructure, which refers to the individual fibres within the e-spun material, can be specifically orientated within the macrostructure by using different methods of electrospinning such as emulsion or co-axial electrospinning [69,70]. This allows a higher level of customisation within the final product material, which can boast more beneficial characteristics like increased hydrophilicity or high surface area to volume ratios.

A particularly poignant development for electrospinning is the application of e-spun technology for chemotherapeutic usage. E-spun implantable fibres bearing both singular and multi APIs have been used as an effective means to deliver sustained targeted release of therapy to cancerous tumours [71]. The fibres can act as carriers both directly for the API and also for APIs encapsulated within nanocarriers, for example nanoparticles (NPs) such as liposomes [72]. This characteristic is possible due to the choice of fibre material acting as either a lipophilic or hydrophilic material meaning that the fibres can form complexes with the material to provide either a sustained release or fast-acting localised delivery. The NPs themselves can help providing sustained release by augmenting the release profiles, and the nanofibers will serve as a biocompatible stable scaffold that offers targeted delivery as well as increasing the stability of the loaded micelles [73]. This is potentially a powerful complex of actions formed, especially for diseases like cancer given that even once a tumour is surgically removed, there is still the capacity for metastatic cancer cells to proliferate and spread throughout the body. Multiple prominent chemotherapeutic APIs including doxorubicin, paclitaxel, 5-fluorouracil (5-FU) and cisplatin have been demonstrated to be compatible with e-spun nanofibers, meaning that their usage for a wide range of cancers may be possible. Topical delivery of 5-FU via an electrospun patch has also been proven to be an effective means of drug delivery [74], as it improves patient compliance, compared to the traditional delivery method of a twice-daily application of a 5-FU cream. The patch was shown to deliver the API in a controlled manner over 28 days meaning that the patient needs only to replace it after this amount of time.

There are multiple other applications for utilising e-spun fibres as a scaffold for supporting the delivery of APIs both with and without nanocarrier involvement, including the delivery of curcumin, ciprofloxacin and chloramphenicol to name a few [75–77]. However, the use of e-spun fibres as bandages could be a concept that would gain wide-scale attention owing to its simplicity and its scope of application. As mentioned, the e-spun fibres act as a perfect environment to support the association of APIs, hence the potential to create a medicated plaster or bandage via the usage of electrospinning is incredibly viable. The medicated bandages offer a scaffold to provide the controlled release of the API to the affected area, which can contribute towards antimicrobial control or wound healing [78,79]. A combination of effects is also possible, as seen in studies performed by Yan et al., which integrated silver nanoparticles and a cellulose scaffold upon a PLA e-spun fibre bandage [78]. This combination therapy truly represents the possibility that this emerging technology can provide in this area, as not only harmful threats from microbes are mitigated, but also cell proliferation and repair are enhanced.

1.3 MEDICAL DEVICES AND ASSISTANTS

1.3.1 Mobile Device Applications

Given the dramatic rise in ownership of mobile phone devices, particularly smart devices, the area of phone applications (apps) to assist with the promotion of healthcare is one of extreme interest and development. Worldwide, over half the population owns a smartphone [2]. This statistic is higher in developed areas, for example, Europe has a smartphone ownership rate of 80% [2]. Mobile health (mHealth) would be most useful for the elderly population, and within technologically developed countries such as South Korea, 73.6% of adults aged in their 60's owned a smart device [80], highlighting the scope of who this technology would be available for [81]. Another study performed in 2017,

however, highlighted that the main users of mHealth apps were the younger generation who are generally more familiar with the use of apps [82]. This could highlight that mHealth apps may need to be created to be more user friendly for the elderly population.

The ability to portray information across a vast connected network simultaneously is very potent, especially given the importance of phones in modern-day life. mHealth apps have a wide array of functionality, from monitoring disease progression to assisting with initial diagnosis, to customising life regimens based upon genetics. It must be noted that many of the mHealth apps available currently aren't based upon evidence-based medicine, which could be a factor hindering its expansion [81]. Popular apps including Generis™ portray medical information to its users; however, not all of it has sufficient scientific backing.

That being said, there are plenty of apps available for patients that will provide users with a useful service that can help improve the quality of life. For example, mySugr®, which is an app to assist patients monitor their diabetic conditions, or MDacne, which provides users with Food and Drug Administration (FDA) approved treatment plans for acne virtually, from registered dermatologists. Apps such as these can also help users place trust in these emerging mHealth apps, as full confidentiality of sensitive material is provided by the app. Studies into data collection from mHealth apps have indicated a degree of uncertainty from users to willingly give their data on their health onto the app [83], which is another barrier to their wide-scale use.

Within the UK, there is currently a large push to change medical prescriptions from physical copies to an online format, also known as the electronic prescription service (EPS). This has many benefits for those who have access to this service, as requesting prescriptions is made far simpler, as well as allowing the patient to track their order and medical history. For those patients who have poor mobility, e.g., the elderly or disabled, repeat prescriptions can be requested without the need to be physically present. Delivery can also be arranged to further facilitate the patient's care. The EPS system also allows prescriptions to be sent on transferred electronically anywhere within the UK, which has been especially important during the covid-19 pandemic for those isolating away from their regular pharmacies.

1.3.2 Microneedles

Microneedles (MNs) are a novel method to facilitate the delivery of APIs via the transdermal route. Transdermal delivery of APIs is limited, without providing an alternate formulation for delivery, and is generally restricted to APIs with suitable lipophilicity (log P within 2–3 [84]), low melting points and molecular weights <500 Da [85]. MNs can appear in many different forms such as hollow and dissolvable, though they comprise the same base design. The needles measure in size within the micron range, 10–2000 μm in height and 10–50 μm in width [86], which will bypass the stratum corneum and allow delivery of APIs to the viable epidermis (VE) [87]. This action is essential for improving the bioavailability of the API, whether the MNs deliver the drug directly or if the MNs act solely as a pore creator for facilitated delivery. One of the biggest attractions to MNs is that owing to their size, MNs avoid complications arising from patients experiencing pain, irritation or phobia for the delivery device. Many of the nerve fibres associated with pain of injection lie below the VE [88], meaning that traditional injections can act as stimuli for these nerves, while MNs avoid this. Obviously, this factor will improve patient compliance with these formulations. Outlined in this section will be the various forms of MNs available and how they are developing MNs as a prominent emerging technology; namely, these are solid, hollow, coated, dissolving and hydrogel microneedles.

The simplest form of MN is a solid MN, which serves no other purpose than to act as a pre-treatment for other formulations, by creating micro-size pores into the VE; onto which another medicine e.g., a cream, can be placed. The pores increase the amount of API that is able to be absorbed by the body, as the relatively impermeable stratum corneum is essentially bypassed. For example, a 6.74-fold increase of absorption using the antiepileptic drug tiagabine hydrochloride was observed, compared to solely transdermal delivery [89]. Solid MNs are amongst the simplest to manufacture [89] but also require a second formulation to have a therapeutic effect. All the following microneedles mentioned are coupled intrinsically with an API, meaning that they alone can have a therapeutic benefit. Solid and coated MNs will generally have the highest mechanical strengths within the MN category, meaning that the chances of breakage upon insertion are the lowest. This is an important factor to consider as these MNs may pose the greatest threat of causing an immune response, should breakage occur within the skin.

Hollow MNs employ a drug reservoir within the MN base to provide medicinal action, as the MNs are inserted into the target tissue and the reservoir it triggered to release the payload through micro-sized channels directly into the VE. Hollow MN are often harder to manufacture due to their fragile structure [84]; however, they can provide excellent transdermal delivery of various medicines. Hollow MNs have already shown promise in the area of biologics [90] and nanoparticles [91], amongst others. The flow of medicine through the hollowed needles from the reservoir is pressure driven upon insertion, and this form of MN is able to deliver a higher drug load as compared to other types [84].

Dissolving MNs are fabricated from water-soluble materials, including maltose, dextran, hyaluronic acid (HA), albumin and polyvinylpyrrolidone, amongst others [92]. The API is contained within the needles themselves, which is then released upon dissolution. The main benefit of this form of MN is that they offer high levels of biocompatibility, which is incredibly useful in case the needle breaks upon insertion, leaving small remnants of the MN left in the skin. This has been an issue that has been related to solid MNs in the past [92]. A limitation of this technology is

that owing to space limitation, a limited amount of API can be delivered at once. However, this does allow an emerging application of MNs, which is vaccine administration. Vaccines are potent medicines, which require only small quantities to attain a therapeutic effect [93], which makes them perfect candidates for delivery from DDSs with limited drug doses available.

Coated MNs consist of a solid MN core, with a therapeutically active coating. The duration of needle insertion depends on the diffusion coefficient presented by the coating material [94] and has seen a variance of 1–15 min and beyond in terms of insertion times [94,95]. The active ingredients within the DDS must be potent as again space to implant the drug is limited.

Lastly, hydrogel MNs represent a promising new area of medical development in that they possess multiple actions. They are the most recently developed form of MN and boast favourable properties such as excellent biocompatibility [96]. Hydrogel MNs can be used for multiple purposes, whether that be for diagnostic applications or for the delivery of APIs. The mechanism of action of the hydrogel MNs arises from the imbibement of interstitial fluid (ISF) from the skin [97], causing the needles to swell and alter shape. ISF is quickly becoming a well-recognised powerful source of biomarkers which can be used for the diagnosis of various disease states including inflammation (e.g. in sepsis) or cancer [98]. Hydrogel MNs can also effectively deliver APIs in a controlled fashion via diffusion from APIs contained in a base reservoir. The rate of diffusion can be controlled effectively via alterations of the polymer contents of the MNs, both in the choice of material and the density of MN formed [96].

There are many ways of creating MNs, such as micromolding or photolithography, although the main emerging technology to fabricate MNs is AM, as discussed earlier. This is mainly due to the improved printing quality and availability of this technology. AM allows the rapid fabrication of high-quality MN devices of various types, which is advancing research within MN technology exponentially. Many different printers can be used depending on the material being printed to form the MNs; for example, digital light processing (DLP) is commonly used to print hydrogel MNs [29,99], while printing solid MNs may be more suitable using SLA [100]. 3D-printed MNs are commonly found being used for performing cutting-edge research globally, and it is fast becoming the most favourable way of producing MNs.

1.3.3 MICROFLUIDICS (MFs)

The concept of MFs is also not a new one; however, the realisation for many novel applications for the system would deem MFs as a widespread emerging technology. MFs has been shown to have competency both in therapeutics and diagnostics [101]. When combined, this is commonly referred to as theranostics. Given the propensity for MFs to assist with nanocarrier formulation, many of its uses for therapeutic applications involve this. Multiple nanocarriers including liposomes, solid lipid nanoparticles (SLNs), chitosan and gold NPs, amongst others [102–105], have been successfully formulated using an MF apparatus. Given its timely relevance, the covid-19 vaccines operate via an mRNA-loaded liposomal carrier; hence it is theorised that MFs could provide an effective method for future vaccine formulation that operate in this way, as MFs have already been demonstrated to be compatible with the formulation of RNA-loaded liposomes [106]. This use highlights the potential of MFs for the prevention of covid-19; however, it is also possible that MFs could be applied towards a diagnostic goal. MFs has become well known as a diagnostic tool for biomarker detection, as it can provide a platform for detecting disease/pathogen biomarkers, even using a minimal concentration of target receptors in the test sample. Covid-19 detections via these means have been achieved in multiple cases already, which also produce a very high-quality level of avoiding false positive/false negative results [44,107]. An exciting novel use of MFs is that of acting as a portable point-of-care (PoC) diagnostic device, which has already allowed diagnosis of covid-19 via a paper-based MF device [108].

MFs has recently become an area of focus around organ-on-a-chip (OOAC) technology, as the chips are highly customisable to a high variety of conditions depending upon what is required for assaying. In 2020, OOAC technology is listed in the top 10 emerging technologies worldwide [109], and upon its discovery, 28 OOAC companies started business within 7 years [110]. OOAC is usually used for high-throughput drug screening and pre-clinical evaluations to provide accurate simulated data on the activity of a desired API within a specific environment. This is possible by tailoring the chips to simulate the specific physiological environments and then regulating various parameters throughout sampling. The technology has evolved to the point where factors including fluid shear force, concentration gradients, mechanical stress and cell patterning can all be adapted throughout the process of sampling [109]. This technology is often favourable during early studies as it can help avoid performing a number of unnecessary animal studies, meaning that it's also considered to be very ethical. Species variation is often an issue during animal testing, although this is not an issue for OOAC, as the environment can be precisely engineered to mimic the target patient's physiology. There also exist 2D cell culture experiments that assist with predicting the efficacy of an API; however, these also will not provide as accurate results *in vitro* as will be obtained from OOAC [109]. Livers, kidneys, hearts, muscle and lungs amongst others are examples of organs that have been simulated on an MF chip, as well as multiple "organs" within a singular MF system to simulate a multifaceted delivery [111].

1.3.4 BLOCKCHAIN

Blockchain is an emerging technology relating to the prevention of counterfeit medicines making it to the pharmaceutical supply chain. As noted in some papers, this factor can be more of a problem in certain areas of the world due to lower quality checks and security [112]; however, it is a very prominent issue that must be controlled everywhere. To provide examples of this, in the Philippines, 30% of medicines surveyed in drug stores were identified to be falsified counterfeit [113]. In addition, a pharmacy in Peru was found to possess 56% of its pharmaceutical products being falsified medicines [114]. Blockchain is a technology that acts as a secure database to monitor the quality of items within the distribution supply chain. The security of blockchain arises from data being stored on independent devices throughout the system, with a high level of transparency for any amendments that are made to the data. Any amendments must also be accepted unanimously by all users within the chain, meaning that additions from rogue individuals are not possible. Blockchain is not limited to pharmaceutical usage as well, as it is also being implemented in areas such as finance and e-finance. To summarise blockchain in the pharmaceutical field, the blockchain network consists of a ledger, a contract repository, an initial document repository and the drug distribution history [113]. With these components combined, it is possible to tell exactly what level of distribution is planned within a supply chain and what level is achieved. Individual shipments can be registered on the network with both their origin and destination, allowing the receiver to accept the medicines with confidence that they are genuine; or if anything counterfeit is suspected, the blockchain can be checked and the medicines can be validated/rejected. A visual representation of this can be seen in Figure 1.3.

Counterfeit medicines within the supply chain have a huge impact upon a community, whether that be from a healthcare perspective or from an economic one. Poor quality medicines may have a negative impact upon a patient's quality of life due to possessing potentially a lower efficacy. This may be caused by the diminished presence (if any) of active ingredient, poor packaging or the use of expired stock. The economic burden is caused by the extensive checking that must be performed upon the medicines as well as the cost of containment should a falsified medicine be identified [115]. It was reported in 2018, according to the World Health Organization (WHO), that the total global sales of counterfeit medicines were between $200 and 431 billion [116], which will mostly be collected by criminal organisations rather than acting as a taxable commodity for governments.

1.4 CONCLUSIONS AND FUTURE PROSPECTIVE

In conclusion to this chapter, the authors discuss their opinions on the longer-term viability of the technologies discussed throughout the review. The research into emerging therapeutics including regenerative medicine and 3DBP is showing great promise, particularly given the various routes that are available to carry out each process. 3DBP especially is a contemporary technology that is developing rapidly due

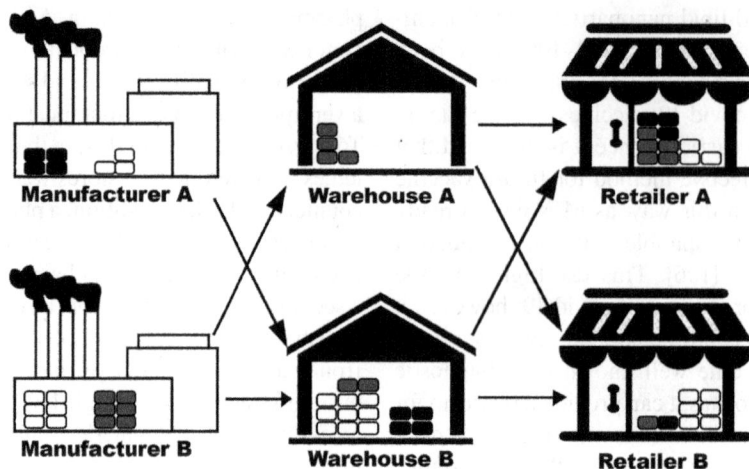

FIGURE 1.3 A representation of a standardised supply chain; each step of which can be monitored and tracked using blockchain to ensure the correct medicine has been transferred along the chain successfully to its end destination.

to the interest and funding that is being focussed towards it. We predict that in the future, it is likely that 3DBP becomes common practice to act as a source of organs and tissue for allograft transplantation. The capacity for AM in general to interact with other technologies, such as electrospinning and microfluidics, makes the technology extremely attractive to invest in, as it can influence the progression of many other processes.

As discussed, owing to the rising proportion of the population with access to smartphones, the capacity for apps to influence healthcare is growing dramatically. The technology still requires further development in terms of evidence-based medicine being portrayed; however, this too is likely to increase in usage and impact in the near future. mHealth app development will also help empower people to take more charge of their own healthcare.

MN technology is strongly developing, although there are still many areas that require evolution, as their usage within clinical practice is still very limited [117]. There are promising signs of their applications, but limitations such as mechanical stability and low dosage capacity are currently hindering them from expanding further. MFs are showing encouraging signs of growth from both a therapeutic and diagnostic viewpoint. It is likely that it will grow to be a commonly used technology across vast fields of healthcare, especially given the growth of AM methods for production. Lastly, blockchain is a new technology that will grow exponentially and will become a widely employed method of ensuring that only suitable quality medicines will come to market and reach the patient.

REFERENCES

1. Feehan, M., et al., Patient preferences for healthcare delivery through community pharmacy settings in the USA: A discrete choice study. *Journal of Clinical Pharmacy and Therapeutics*, 2017. **42**(6): pp. 738–749.
2. Olson, J.A., et al., Smartphone addiction is increasing across the world: A meta-analysis of 24 countries. *Computers in Human Behavior*, 2020. **129**: p. 107138.
3. Tottoli, E.M., et al., Skin wound healing process and new emerging technologies for skin wound care and regeneration. *Pharmaceutics*, 2020. **12**(8): p. 735.
4. Park, J., S. Hwang, and I.-S. Yoon, Advanced growth factor delivery systems in wound management and skin regeneration. *Molecules*, 2017. **22**(8): p. 1259.
5. McLaughlin, P.J., et al., Topical naltrexone is a safe and effective alternative to standard treatment of diabetic wounds. *Advances in Wound Care*, 2017. **6**(9): pp. 279–288.
6. Yamakawa, S. and K. Hayashida, Advances in surgical applications of growth factors for wound healing. *Burns & Trauma*, 2019. **7**.
7. Mirza, R.E. and T.J. Koh, Contributions of cell subsets to cytokine production during normal and impaired wound healing. *Cytokine*, 2015. **71**(2): pp. 409–412.
8. Tang, Z., et al., Progress of stem/progenitor cell-based therapy for retinal degeneration. *Journal of Translational Medicine*, 2017. **15**(1): pp. 1–3.
9. Evangelista, A.F., et al., Bone marrow-derived mesenchymal stem/stromal cells reverse the sensorial diabetic neuropathy via modulation of spinal neuroinflammatory cascades. *Journal of Neuroinflammation*, 2018. **15**(1): pp. 1–17.
10. Mendes Filho, D., et al., Therapy with Mesenchymal stem cells in Parkinson disease: History and perspectives. *The Neurologist*, 2018. **23**(4): pp. 141–147.
11. Han, L., et al., MicroRNA Let-7f-5p promotes bone marrow Mesenchymal stem cells survival by targeting caspase-3 in Alzheimer disease model. *Frontiers in Neuroscience*, 2018. **12**: p. 333.
12. Laroussi, M., Cold plasma in medicine and healthcare: The new frontier in low temperature plasma applications. *Frontiers in Physics*, 2020. **8**: p. 74.
13. Melotti, L., et al., Could cold plasma act synergistically with allogeneic mesenchymal stem cells to improve wound skin regeneration in a large size animal model? *Research in Veterinary Science*, 2021. **136**: pp. 97–110.
14. Dunnill, C., et al., Reactive oxygen species (ROS) and wound healing: The functional role of ROS and emerging ROS-modulating technologies for augmentation of the healing process. *International Wound Journal*, 2017. **14**(1): pp. 89–96.
15. Schmidt, A., et al., A cold plasma jet accelerates wound healing in a murine model of full-thickness skin wounds. *Experimental Dermatology*, 2017. **26**(2): pp. 156–162.

16. Dijksteel, G.S., et al., Safety and bactericidal efficacy of cold atmospheric plasma generated by a flexible surface dielectric barrier discharge device against *Pseudomonas aeruginosa* in vitro and in vivo. *Annals of Clinical Microbiology and Antimicrobials*, 2020. **19**(1): p. 37.

17. Yan, Q., et al., A review of 3D printing technology for medical applications. *Engineering*, 2018. **4**(5): pp. 729–742.

18. Li, Y., et al., Cost, sustainability and surface roughness quality – A comprehensive analysis of products made with personal 3D printers. *CIRP Journal of Manufacturing Science and Technology*, 2017. **16**: pp. 1–11.

19. Mohan, N., et al., A review on composite materials and process parameters optimisation for the fused deposition modelling process. *Virtual and Physical Prototyping*, 2017. **12**(1): pp. 47–59.

20. Awad, A., et al., 3D printing: Principles and pharmaceutical applications of selective laser sintering. *International Journal of Pharmaceutics*, 2020. **586**: p. 119594.

21. Schmidleithner, C. and D.M. Kalaskar, *Stereolithography*. 2018, InTech.

22. Wang, J., et al., A novel approach to improve mechanical properties of parts fabricated by fused deposition modeling. *Materials & Design*, 2016. **105**: pp. 152–159.

23. Salentijn, G.I., et al., Fused deposition modeling 3D printing for (bio)analytical device fabrication: Procedures, materials, and applications. *Analytical Chemistry*, 2017. **89**(13): pp. 7053–7061.

24. Mu, Q., et al., Digital light processing 3D printing of conductive complex structures. *Additive Manufacturing*, 2017. **18**: pp. 74–83.

25. Zhao, T., et al., Superstretchable and processable silicone elastomers by digital light processing 3D printing. *ACS Applied Materials & Interfaces*, 2019. **11**(15): pp. 14391–14398.

26. Kim, S.H., et al., Precisely printable and biocompatible silk fibroin bioink for digital light processing 3D printing. *Nature Communications*, 2018. **9**(1): pp. 1–14.

27. Patel, D.K., et al., Highly stretchable and UV curable elastomers for digital light processing based 3D printing. *Advanced Materials*, 2017. **29**(15): p. 1606000.

28. Liu, Z., et al., Additive manufacturing of hydroxyapatite bone scaffolds via digital light processing and in vitro compatibility. *Ceramics International*, 2019. **45**(8): pp. 11079–11086.

29. Yao, W., et al., 3D printed multi-functional hydrogel microneedles based on high-precision digital light processing. *Micromachines*, 2020. **11**(1): p. 17.

30. O'Connor, H.J., A.N. Dickson, and D.P. Dowling, Evaluation of the mechanical performance of polymer parts fabricated using a production scale multi jet fusion printing process. *Additive Manufacturing*, 2018. **22**: pp. 381–387.

31. Habib, F.N., et al., Fabrication of polymeric lattice structures for optimum energy absorption using multi jet fusion technology. *Materials & Design*, 2018. **155**: pp. 86–98.

32. Nandy, J., H. Sarangi, and S. Sahoo, A review on direct metal laser sintering: Process features and microstructure modeling. *Lasers in Manufacturing and Materials Processing*, 2019. **6**(3): pp. 280–316.

33. Girardin, E., et al., Biomedical Co-Cr-Mo components produced by direct metal laser sintering. *Materials Today: Proceedings*, 2016. **3**(3): pp. 889–897.

34. Brezinová, J., et al., Direct metal laser sintering of Ti6Al4V for biomedical applications: Microstructure, corrosion properties, and mechanical treatment of implants. *Metals*, 2016. **6**(7): p. 171.

35. Chio, T.-H., G.-L. Huang, and S.-G. Zhou, Application of direct metal laser sintering to waveguide-based passive microwave components, antennas, and antenna arrays. *Proceedings of the IEEE*, 2017. **105**(4): pp. 632–644.

36. Körner, C., Additive manufacturing of metallic components by selective electron beam melting—A review. *International Materials Reviews*, 2016. **61**(5): pp. 361–377.

37. Sadia, M., et al., Channelled tablets: An innovative approach to accelerating drug release from 3D printed tablets. *Journal of Controlled Release*, 2018. **269**: pp. 355–363.

38. Awad, A., et al., 3D printed tablets (printlets) with braille and moon patterns for visually impaired patients. *Pharmaceutics*, 2020. **12**(2): p. 172.

39. Goyanes, A., et al., Effect of geometry on drug release from 3D printed tablets. *International Journal of Pharmaceutics*, 2015. **494**(2): pp. 657–663.

40. Vaut, L., et al., 3D printing of reservoir devices for oral drug delivery: From concept to functionality through design improvement for enhanced mucoadhesion. *ACS Biomaterials Science & Engineering*, 2020. **6**(4): pp. 2478–2486.

41. Atalay, H.A., et al., Impact of personalized three-dimensional (3D) printed pelvicalyceal system models on patient information in percutaneous nephrolithotripsy surgery: A pilot study. *International braz j urol*, 2017. **43**(3): pp. 470–475.

42. Barber, S.R., et al., Virtual functional endoscopic sinus surgery simulation with 3D-printed models for mixed-reality nasal endoscopy. *Otolaryngology–Head and Neck Surgery*, 2018. **159**(5): pp. 933–937.

43. Aimar, A., A. Palermo, and B. Innocenti, The role of 3D printing in medical applications: A state of the art. *Journal of Healthcare Engineering*, 2019. **2019**: p. 5340616.

44. Lamprou, D.A., Emerging technologies for diagnostics and drug delivery in the fight against COVID-19 and other pandemics. *Expert Review of Medical Devices*, 2020. **17**(10): pp. 1007–1012.

45. Kirchner, V.A., et al., Current status of liver transplantation in North America. *International Journal of Surgery*, 2020. **82**: pp. 9–13.

46. Wang, X., Advanced polymers for three-dimensional (3D) organ bioprinting. *Micromachines*, 2019. **10**(12): p. 814.

47. Zhang, Y.S., et al., 3D bioprinting for tissue and organ fabrication. *Annals of Biomedical Engineering*, 2017. **45**(1): pp. 148–163.

48. Naghieh, S., et al., Indirect 3D bioprinting and characterization of alginate scaffolds for potential nerve tissue engineering applications. *Journal of the Mechanical Behavior of Biomedical Materials*, 2019. **93**: pp. 183–193.

49. Zhang, Y., et al., Recent advances in 3D bioprinting of vascularized tissues. *Materials & Design*, 2020. **199**: p. 109398.

50. Kim, J., et al., Encapsulation of human natural and induced regulatory T-cells in IL-2 and CCL1 supplemented alginate-GelMA hydrogel for 3D bioprinting. *Advanced Functional Materials*, 2020. **30**(15): p. 2000544.

51. Kurian, M., R. Stevens, and K. McGrath, Towards the development of artificial bone grafts: Combining synthetic biomineralisation with 3D printing. *Journal of Functional Biomaterials*, 2019. **10**(1): p. 12.

52. Mazzocchi, A., S. Soker, and A. Skardal, 3D bioprinting for high-throughput screening: Drug screening, disease modeling, and precision medicine applications. *Applied Physics Reviews*, 2019. **6**(1): p. 011302.

53. Kahl, M., et al., Ultra-low-cost 3D bioprinting: Modification and application of an off-the-shelf desktop 3D-printer for biofabrication. *Frontiers in Bioengineering and Biotechnology*, 2019. **7**: 184.

54. Li, Y.-C., et al., 4D bioprinting: The next-generation technology for biofabrication enabled by stimuli-responsive materials. *Biofabrication*, 2016. **9**(1): p. 012001.

55. Gao, B., et al., 4D bioprinting for biomedical applications. *Trends in Biotechnology*, 2016. **34**(9): pp. 746–756.

56. Hong, J., et al., Cell-electrospinning and its application for tissue engineering. *International Journal of Molecular Sciences*, 2019. **20**(24): p. 6208.

57. Xue, J., et al., Electrospun nanofibers: New concepts, materials, and applications. *Accounts of Chemical Research*, 2017. **50**(8): pp. 1976–1987.

58. Robinson, T.M., D.W. Hutmacher, and P.D. Dalton, The next frontier in melt electrospinning: Taming the jet. *Advanced Functional Materials*, 2019. **29**(44): p. 1904664.

59. Wunner, F.M., et al., Melt electrospinning writing of highly ordered large volume scaffold architectures. *Advanced Materials*, 2018. **30**(20): p. 1706570.

60. Hu, J., et al., Electrospinning of poly(glycerol sebacate)-based nanofibers for nerve tissue engineering. *Materials Science and Engineering: C*, 2017. **70**: pp. 1089–1094.

61. Xie, X., et al., Electrospinning nanofiber scaffolds for soft and hard tissue regeneration. *Journal of Materials Science & Technology*, 2020. **59**: pp. 243–261.

62. Kim, M.-C., et al., Cell invasion dynamics into a three dimensional extracellular matrix fibre network. *PLOS Computational Biology*, 2015. **11**(10): p. e1004535.

63. Hanumantharao, S. and S. Rao, Multi-functional electrospun nanofibers from polymer blends for scaffold tissue engineering. *Fibers*, 2019. **7**(7): p. 66.

64. Chen, S., et al., Electrospinning: An enabling nanotechnology platform for drug delivery and regenerative medicine. *Advanced Drug Delivery Reviews*, 2018. **132**: pp. 188–213.

65. Blakney, A.K., Y. Jiang, and K.A. Woodrow, Application of electrospun fibers for female reproductive health. *Drug Delivery and Translational Research*, 2017. **7**(6): pp. 796–804.

66. Johnson, C.D., et al., Injectable, magnetically orienting electrospun fiber conduits for neuron guidance. *ACS Applied Materials & Interfaces*, 2018. **11**(1): pp. 356–372.

67. Qu, Y., et al., Injectable and thermosensitive hydrogel and PDLLA electrospun nanofiber membrane composites for guided spinal fusion. *ACS Applied Materials & Interfaces*, 2018. **10**(5): pp. 4462–4470.

68. Mofidfar, M. and M.R. Prausnitz, Electrospun transdermal patch for contraceptive hormone delivery. *Current Drug Delivery*, 2019. **16**(6): pp. 577–583.

69. Zhang, C., F. Feng, and H. Zhang, Emulsion electrospinning: Fundamentals, food applications and prospects. *Trends in Food Science & Technology*, 2018. **80**: pp. 175–186.

70. Shahriar, S., et al., Electrospinning nanofibers for therapeutics delivery. *Nanomaterials*, 2019. **9**(4): p. 532.

71. Chen, S., et al., Emerging roles of electrospun nanofibers in cancer research. *Advanced Healthcare Materials*, 2018. **7**(6): p. 1701024.

72. Chandrawati, R., et al., Enzyme prodrug therapy engineered into electrospun fibers with embedded liposomes for controlled, localized synthesis of therapeutics. *Advanced Healthcare Materials*, 2017. **6**(17): p. 1700385.

73. Khodadadi, M., et al., Recent advances in electrospun nanofiber-mediated drugdelivery strategies for localized cancer chemotherapy. *Journal of Biomedical Materials Research Part A*, 2020. **108**(7): pp. 1444–1458.

74. Jun, E., et al., Synergistic effect of a drug loaded electrospun patch and systemic chemotherapy in pancreatic cancer xenograft. *Scientific Reports*, 2017. **7**(1): p. 12381.

75. Puiggalí-Jou, A., et al., Smart drug delivery from electrospun fibers through electroresponsive polymeric nanoparticles. *ACS Applied Bio Materials*, 2018. **1**(5): pp. 1594–1605.

76. Günday, C., et al., Ciprofloxacin-loaded polymeric nanoparticles incorporated electrospun fibers for drug delivery in tissue engineering applications. *Drug Delivery and Translational Research*, 2020. **10**(3): pp. 706–720.

77. Zelkó, R., D.A. Lamprou, and I. Sebe, *Recent Development of Electrospinning for Drug Delivery*. 2020, Multidisciplinary Digital Publishing Institute.

78. Yan, D., et al., Surface modified electrospun poly(lactic acid) fibrous scaffold with cellulose nanofibrils and Ag nanoparticles for ocular cell proliferation and antimicrobial application. *Materials Science and Engineering: C*, 2020. **111**: p. 110767.

79. Memic, A., et al., Latest progress in electrospun nanofibers for wound healing applications. ACS Applied Bio Materials, 2019. **2**(3): pp. 952–969.

80. Lee, D. and S. Han, Reliability and validity of knee joint angles of the elderly measured using smartphones. *Journal of International Academy of Physical Therapy Research*, 2020. **11**(3): pp. 2107–2112.

81. Ramey, L., et al., Apps and mobile health technology in rehabilitation: The good, the bad, and the unknown. *Physical Medicine and Rehabilitation Clinics of North America*, 2019. **30**(2): pp. 485–497.

82. Carroll, J.K., et al., Who uses mobile phone health apps and does use matter? A secondary data analytics approach. *Journal of Medical Internet Research*, 2017. **19**(4): p. e125.

83. Di Matteo, D., et al., Patient willingness to consent to mobile phone data collection for mental health apps: Structured questionnaire. *JMIR Mental Health*, 2018. **5**(3): p. e56.

84. Nagarkar, R., et al., A review of recent advances in microneedle technology for transdermal drug delivery. *Journal of Drug Delivery Science and Technology*, 2020. **59**: p. 101923.

85. Wang, Y., et al., Influencing factors and drug application of iontophoresis in transdermal drug delivery: An overview of recent progress. *Drug Delivery and Translational Research*, 2022. **12**(1): pp. 15–26.

86. Hao, Y., et al., Microneedles-based transdermal drug delivery systems: A review. *Journal of Biomedical Nanotechnology*, 2017. **13**(12): pp. 1581–1597.

87. Dharadhar, S., et al., Microneedles for transdermal drug delivery: A systematic review. *Drug Development and Industrial Pharmacy*, 2019. **45**(2): pp. 188–201.

88. Ali, R., et al., Transdermal microneedles—A materials perspective. *Aaps Pharmscitech*, 2020. **21**(1): pp. 1–14.

89. Nguyen, J., et al., The influence of solid microneedles on the transdermal delivery of selected antiepileptic drugs. *Pharmaceutics*, 2016. **8**(4): p. 33.

90. Dul, M., et al., Hydrodynamic gene delivery in human skin using a hollow microneedle device. *Journal of Controlled Release*, 2017. **265**: pp. 120–131.

91. Niu, L., et al., Intradermal delivery of vaccine nanoparticles using hollow microneedle array generates enhanced and balanced immune response. *Journal of Controlled Release*, 2019. **294**: pp. 268–278.

92. Ita, K., Dissolving microneedles for transdermal drug delivery: Advances and challenges. *Biomedicine & Pharmacotherapy*, 2017. **93**: pp. 1116–1127.

93. Ripolin, A., et al., Successful application of large microneedle patches by human volunteers. *International Journal of Pharmaceutics*, 2017. **521**(1): pp. 92–101.

94. Al Sulaiman, D., et al., Hydrogel-coated microneedle arrays for minimally invasive sampling and sensing of specific circulating nucleic acids from skin interstitial fluid. *ACS Nano*, 2019. **13**(8): pp. 9620–9628.

95. Gill, H.S. and M.R. Prausnitz, Coated microneedles for transdermal delivery. *Journal of Controlled Release: Official Journal of the Controlled Release Society*, 2007. **117**(2): pp. 227–237.

96. Dreiss, C.A., Hydrogel design strategies for drug delivery. *Current Opinion in Colloid & Interface Science*, 2020. **48**: pp. 1–17.

97. Sivaraman, A. and A.K. Banga, Novel in situ forming hydrogel microneedles for transdermal drug delivery. *Drug Delivery and Translational Research*, 2017. **7**(1): pp. 16–26.

98. Zhang, X., et al., Encoded microneedle arrays for detection of skin interstitial fluid biomarkers. *Advanced Materials*, 2019. **31**(37): p. 1902825.

99. Kundu, A., et al., DLP 3D printed "Intelligent" Microneedle Array (iµNA) for stimuli responsive release of drugs and its in vitro and ex vivo characterization. Journal of Microelectromechanical Systems, 2020. **29**(5): pp. 685–691.

100. Pere, C.P.P., et al., 3D printed microneedles for insulin skin delivery. *International Journal of Pharmaceutics*, 2018. **544**(2): pp. 425–432.

101. Weaver, E., et al., The present and future role of microfluidics for protein and peptide-based therapeutics and diagnostics. *Applied Sciences*, 2021. **11**(9): p. 4109.

102. Moradikhah, F., et al., Microfluidic fabrication of alendronate-loaded chitosan nanoparticles for enhanced osteogenic differentiation of stem cells. *Life Sciences*, 2020. **254**: p. 117768.

103. Bressan, L.P., et al., 3D-printed microfluidic device for the synthesis of silver and gold nanoparticles. *Microchemical Journal*, 2019. **146**: pp. 1083–1089.

104. Sedighi, M., et al., Rapid optimization of liposome characteristics using a combined microfluidics and design-of-experiment approach. *Drug Delivery and Translational Research*, 2019. **9**(1): pp. 404–413.

105. Arduino, I., et al., Preparation of cetyl palmitate-based PEGylated solid lipid nanoparticles by microfluidic technique. *Acta Biomaterialia*, 2021. **121**: pp. 566–578.

106. Filipczak, N., et al., Recent advancements in liposome technology. *Advanced Drug Delivery Reviews*, 2020. **156**: pp. 4–22.

107. Dong, X., et al., Rapid PCR powered by microfluidics: A quick review under the background of COVID-19 pandemic. *TrAC Trends in Analytical Chemistry*, 2021. **143**: p. 116377.

108. Gong, F., et al., Pulling-force spinning top for serum separation combined with paper-based microfluidic devices in COVID-19 ELISA diagnosis. *ACS Sensors*, 2021. **6**(7): pp. 2709–2719.

109. Wu, Q., et al., Organ-on-a-chip: Recent breakthroughs and future prospects. *BioMedical Engineering OnLine*, 2020. **19**(1): pp. 1–9.

110. Zhang, B., et al., Advances in organ-on-a-chip engineering. *Nature Reviews Materials*, 2018. **3**(8): pp. 257–278.

111. Low, L.A., et al., Organs-on-a-chip. *Advances in Experimental Medicine and Biology*, 2020. **1230**: pp. 27–42.

112. Bryatov, S. and A. Borodinov. Blockchain technology in the pharmaceutical supply chain: Researching a business model based on Hyperledger Fabric. in *Proceedings of the International Conference on Information Technology and Nanotechnology (ITNT)*, Samara, Russia. 2019.

113. Sylim, P., et al., Blockchain technology for detecting falsified and substandard drugs in distribution: Pharmaceutical supply chain intervention. *JMIR Research Protocols*, 2018. **7**(9): p. e10163.

114. Medina, E., E. Bel, and J.M. Suñé, Counterfeit medicines in Peru: A retrospective review (1997–2014). *BMJ Open*, 2016. **6**(4): p. e010387.

115. Van Baelen, M., et al., *Fighting Counterfeit Medicines in Europe: The Effect on Access to Medicines*. 2017, SAGE Publications Sage UK: London, England.

116. Miller, H.I. and W. Winegarden, *Fraud in Your Pill Bottle the Unacceptable Cost of Counterfeit Medicines*. 2020, Pacific Research Institute.

117. Jeong, S.-Y., et al., The current status of clinical research involving microneedles: A systematic review. *Pharmaceutics*, 2020. 12(11): p. 1113.

2 Hot-Melt Extrusion

An Emerging Manufacturing Technology for Drug Products

S. Li, D. Liu, D. S. Jones, and G. P. Andrews
Queen's University of Belfast

CONTENTS

2.1 GENERAL INTRODUCTION

Hot-melt extrusion (HME) has been widely utilised as a 'continuous manufacturing (CM) –ready' technique owing to its versatility and unique characteristics. Extrusion has been used in the plastics, rubber and food industries since the early 1930s whilst later (1970s) expanding its application to pharmaceutical products and, since, helping drive a paradigm shift in drug product development and manufacturing (Patil et al., 2016). Pharmaceutical HME processing was first used to produce a drug-containing polymeric matrix using poly (vinyl acetate-co-methacrylic acid) and epoxy resin (El-Egakey et al., 1971). To date, HME has been extended to address a plethora of pharmaceutical challenges with over 500 papers published in the scientific literature over the last decade.

HME technology is the processing of active pharmaceutical ingredients (APIs) with thermoplastic excipients in a non-ambient state, through controlled thermal and mixing profiles, prior to pushing the final blended matrix through an exit orifice of defined geometry (Jana & Miloslava, 2012; Simões et al., 2019). Melt extrusion is usually carried out at temperatures above the glass transition temperature (T_g) or the melting temperature (T_m) of the main matrix component, utilising a rotating screw(s) to drive product along the barrel chamber, taking advantage of both thermal and mechanical energy to mix input materials. It is both time- and cost-efficient, solvent-free and environmentally friendly, and requires a narrow footprint relative to batch manufacturing procedures (Simões et al., 2021). However, the large energy consumption required for downstream processing, the limited supply of thermoplastic excipients, the metastable nature

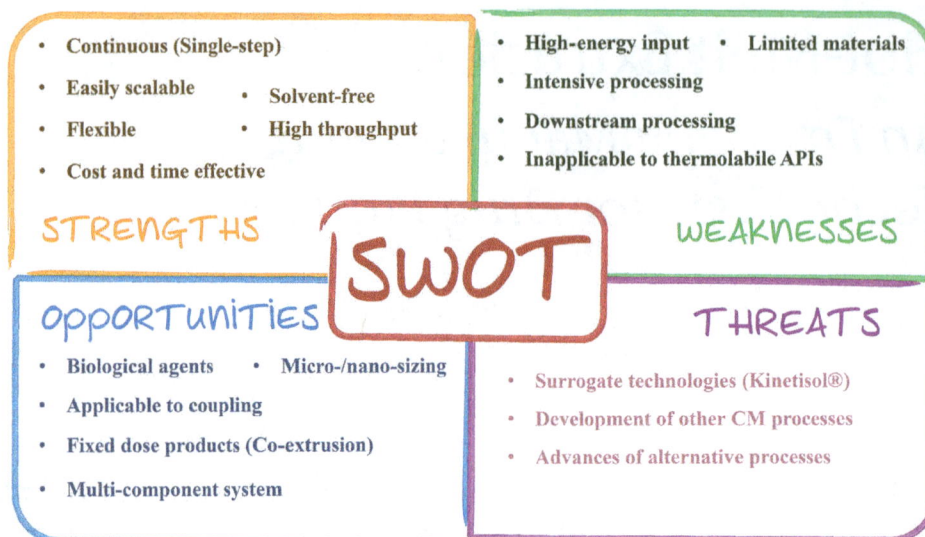

FIGURE 2.1 A SWOT analysis of pharmaceutical hot-melt extrusion technology.

of some amorphous-based end-products and the lack of suitability for some thermolabile active compounds, such as bio-compounds, are significant limitations hindering HME achieving its full pharmaceutical potential (Lang et al., 2014). A simple SWOT analysis of HME is presented in Figure 2.1 to help demonstrate the potential of HME technology as a pharmaceutical product manufacturing technique.

2.2 A BRIEF OVERVIEW OF PHARMACEUTICAL HME PROCESSING

A typical HME process consists of pumping drug compound(s) and excipients into a feed hopper of an extruder, followed by melting, compaction and mixing of the input materials under a pre-set temperature/screw profile, ending with the final homogeneous extrudate being forced through the exit orifice (die) with a pre-defined output geometry. Due to the stringent requirements on product quality within the pharmaceutical industry, twin-screw extruders (TSEs), particularly intermeshing co-rotating TSEs, have been more widely studied for their applicability in processing drug-containing matrices. The melt flow patterns within an intermeshing co-rotating TSE are generally more complex when compared to a ram or a single-screw extruder. There is also enhanced axial melt flow movement if compared with a counter-rotating or a non-intermeshing TSE. Such complex and thorough melt flow patterns result in excellent mixing within the extruder barrel driving better control of drug content homogeneity.

Within a modular TSE, the screws are often assembled using a profile of varying elements intended for different functionalities namely material forwarding, compression (usually facilitated with elevated temperature to soften or melt the fed mass) and metering. Screw elements designed for these functions are broadly categorised as conveying elements, mixing/kneading elements and zoning elements, respectively. A schematic of typical conveying and mixing elements along with their key features is presented in Figure 2.2. In general,

conveying elements are designed with helix flights. By changing the helix angle and the flight depth, the conveying elements can be used in the feeding zone, discharging zone or as an intermittent step along prolonged mixing zones to facilitate the forward transport of material. Typical mixing elements, on the other hand, are blocks of kneading discs assembled at defined offset angles. It is widely accepted that smaller offset angles between kneading discs provide increased dispersive mixing, whilst larger angles (up to 90°) tend to result in distributive mixing. Where functioning zones need to be isolated from an adjacent zone, zoning elements may be used to create sufficient distance for effective isolation. These elements usually have a helix geometry similar to conveying elements but exhibiting wider axial widths between flights and greater flight depth, to reduce the impact of the previous zone on the melt flow properties.

As shown in Figure 2.3, in a typical pharmaceutical HME process, a physical blend of the raw materials is fed into the feed hopper and taken through a series of functioning zones along the length of the extruder barrel. Upon exiting, the extruded intermediate drug product often requires immediate auxiliary processing, such as pelletisation, encapsulation, pulverisation, granulation, tableting and/or calendaring, to achieve the final dosage form. A plethora of controllable processing parameters make HME a versatile processing tool with a considerable degree of customisation and flexibility, allowing for the bespoke manufacture of a diverse range of drug products, leading to widespread application and opportunity.

2.3 EXAMPLE APPLICATIONS OF PHARMACEUTICAL HME

With significant advancement in polymer engineering and an increasing demand for drug-enabling technologies, HME has found a prominent role as a scale-ready, manufacturing technique, with a steadily increasing number of innovative HME drug products, both reported in the literature and on

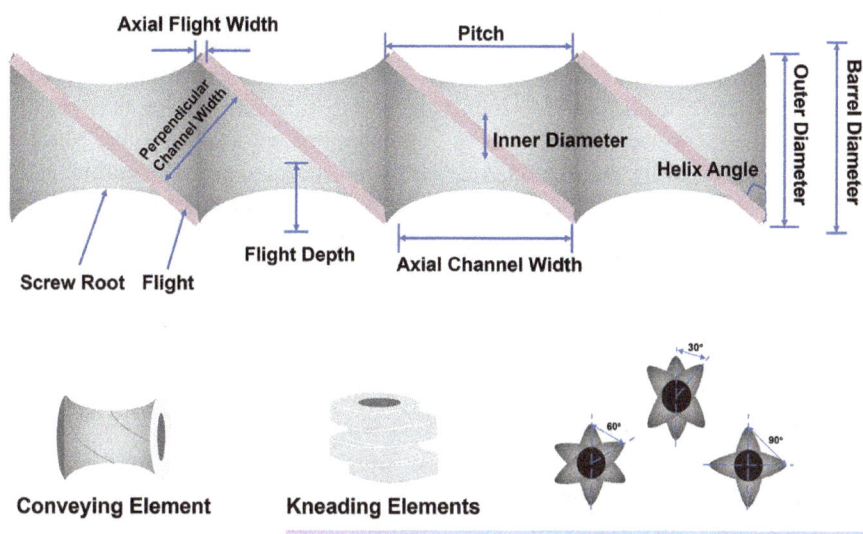

FIGURE 2.2 General characteristics of extrusion screws and depiction of element type.

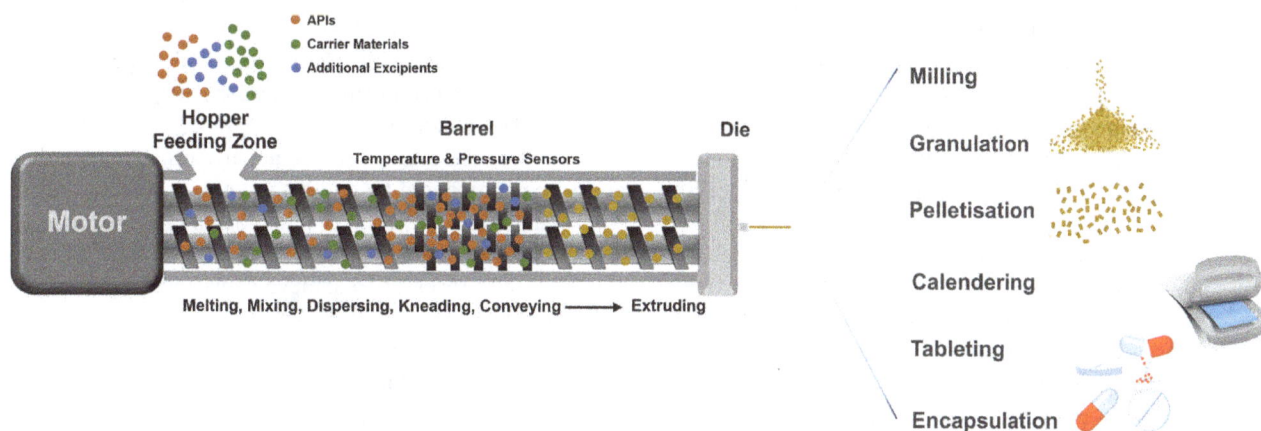

FIGURE 2.3 Overview of typical pharmaceutical HME processing.

the market. Many major players within the pharmaceutical industry are moving towards HME production with several commercial products already on the market (Matić et al., 2020). This has been due to the significant success in developing a variety of drug delivery strategies with ease of scale-up. Commonly, the utilisation of HME includes drug enablement, controlled delivery, taste-masking, personalised medicine (via multilayer co-extrusion or fused-deposition modelling additive manufacturing), abuse deterrent/tamper resistant, solvent-less granulation and self-emulsifying systems. HME is a versatile technology as is regarded as an emerging, scalable process for the manufacturing of drug products. Examples of the diversity of drug products that have been manufactured using HME are detailed in the following section.

2.3.1 DRUG-ENABLED FORMULATIONS

The pharmaceutical industry has witnessed ongoing and increasing formulation challenges as numerous new

chemical entities (NCEs) exhibit poor aqueous solubility, hence resulting in limited therapeutic potential. Arising primarily from the implementation of combinatorial chemistry and high-throughput screening (HTS) in drug discovery processes, poor aqueous solubility now affects up to 40% of approved drug products and 70%–90% of drug molecules within the development pipelines (classified as biopharmaceutical classification system (BCS) class II or class IV (Kalepu & Nekkanti, 2015)). Enhancing drug dissolution, thus facilitating drug absorption and bioavailability, by modification of the physical form of the active ingredient has led to increased interest and exploration in advanced manufacturing techniques such as HME (Burcham et al., 2018).

2.3.1.1 Drug-Polymer Amorphous Solid Dispersions (ASDs)

Improving drug solubility and bioavailability via the production of drug-polymer amorphous solid dispersions

(ASDs) is one of the most reported applications of pharmaceutical HME. Solubility enhancement via amorphisation is achieved by disruption or elimination of the drug crystal lattice, reducing the energy required to facilitate dissolution. Amorphous drug compounds, however, possess inherent stability challenges, both during drug dissolution and throughout their shelf life, owing to their high energy state. Drug-polymer ASDs predominantly produce a 'spring and parachute' release profile in which supersaturation is observed on release of amorphous drug from the polymer matrix. The initial 'spring' is driven by the inherent characteristics of the amorphous drug and the hydrophilicity and swelling behaviour of the carrier. Polymeric carriers also promote solubility enhancement by exerting an inhibitory effect against recrystallisation of supersaturated API, forming a 'parachute', a slow recrystallisation of drug to equilibrium solubility. Precipitation inhibition is related to the level of interaction between drug and polymer in solution and therefore may be affected by polymer structure, rate of permeation in vivo, as well as level of supersaturation (Bevernage et al., 2012; Warren et al., 2010).

At present, it is still difficult to reliably predict how a polymer may affect the rate of drug recrystallisation, and there are often performance disparities exhibited by structurally related polymers. With that said, however, there have been considerable insights provided within the literature on how polymer excipients can be used to retard the crystal growth of poorly soluble drugs. Interaction between polymer excipients, in this case, between HPMCAS and felodipine, under controlled recrystallisation conditions revealed the importance of polymer pKa (Schram et al., 2015). The authors reported: at pH values below the pKa of the polymer, drug recrystallisation occurred more rapidly and the polymer was observed to form coiled globules on the crystal surface. Conversely, when above the pKa, the polymer excipient was more uniformly distributed over the crystal surface reducing the area available for crystal growth.

Another common consideration during the screening of ASD polymer candidates is drug-polymer (D-P) miscibility. It is widely recognised that defining limits of D-P miscibility can help identify stable drug concentrations within polymer matrices. In this respect, the Hansen solubility parameter (δ) calculated using the group contribution method has been reported as a theoretical guide for early-stage polymer screening in ASD development. For a drug-polymer pair, it is recommended that a $\Delta\delta_t < 7.0\,MPa^{1/2}$ indicates significant miscibility and high probability of forming glass solutions during melt extrusion, whereas a $\Delta\delta_t > 10.0\,MPa^{1/2}$ suggests a lack of miscibility and high probability of phase separation (Forster et al., 2001).

More recent work has further explored the use of thermodynamic models such as the Flory–Huggins (F–H) interaction theory to comprehensively understand the theoretical miscibility in a drug-polymer binary system as a function of varying temperature, particularly temperatures that are practically difficult to attain (Li et al., 2016a; Lin & Huang, 2010; Marsac et al., 2009; Tian et al., 2013; Zhao

et al., 2011). The application of F-H modelling to facilitate understanding of the D-P miscibility is of particular relevance in the context of HME, since the method enables an approximate correlation between the HME processing temperature and an 'allowable' drug loading that would generate amorphous drug. Additionally, the construction of such phase diagrams permits an indication of the stability of candidate formulations and hence comparison across relevant temperatures and drug loadings.

To date, many solubility-enhanced ASD products, manufactured by HME, have been marketed and used in multiple disease treatments, such as Technivie®, Mavyret®, Viekira pak® for hepatitis C, Venclexta® for leukaemia, Onmel® and Noxafi® for skin diseases, Belsomra® for insomnia, Cesamet® as antiemetic, Fenoglide® and Rezulin® for cardiovascular diseases and Kaletra® and Norvir® for HIV, to name just a few (Kallakunta et al., 2019b).

2.3.1.2 Nano-Particulate Delivery Systems

Over the last 5 years, there has been a noticeable emergence of nanomedicine and nanotechnology academic research in pharmaceutical technology. Nano-sized drug delivery platforms possess improved therapeutic efficacy due to enhanced drug dissolution and improved cellular uptake (Khan et al., 2022). The commercialisation of drug products containing nanoparticles, however, has always been hindered by challenges associated with large-scale manufacture. Existing technologies typically used to produce lab-scale nano-pharmaceuticals usually require sophisticated processes to achieve consistent large-volume manufacturing.

Of recent years, HME has been investigated for its suitability as a continuous manufacturing (CM) process for production of drug-loaded nanoparticles. Nano-suspensions of phenytoin have been fed into polymer melts using a melt extruder equipped with a side feeder (Baumgartner et al., 2014). The concept was centred around rapid devolatisation of water from the nano-suspension using the high-pressure, high-temperature HME environment, whereas the shearing forces generated by the rotating screws could disperse nano-crystals within the molten polymer. Common polymers, such as Soluplus®, Kollidon® VA64, Eudragit® EPO, HPMCAS and PEG 20000, which are frequently used in the development of oral dosage forms, were investigated. It was shown that the presence of polymeric matrices during devolatisation helped 'de-aggregate' the nano-crystals and enhance the dissolution performance of the drug, regardless of the polymer types. A similar study utilising a combination of HME and high-pressure homogenisation (Ye et al., 2016) was used to homogenise efavirenz nano-suspension in a Soluplus® matrix using HME, followed by water removal. The study investigated two different drug loadings, 1.02% and 2.06%, respectively, which both showed good stability, over a 6-month timeframe under the ICH long-term storage conditions.

More sophisticated HME-based platforms for nanoparticle fabrication have utilised a bottom-up route that incorporates an auxiliary device for further size reduction and homogenisation. In this sequential process, the barrel of the extruder provides an emulsification site, whereas the auxiliary process further reduces and regulates the final particle size. Patil et al. (2014) reported the manufacture of solid lipid nanoparticles (SLN), involving the use of HME and high-pressure homogenisation (HPH) in a sequential manner. In this study, a surfactant-containing aqueous phase was injected into a hot melt extruder and blended with gravimetrically fed Compritol® 888 ATO within the extruder barrel. The 'hot pre-emulsion' resulting from HME processing was then introduced to a HPH process. It has been shown that lower screw speeds and non-conveying screw configuration both resulted in a decrease in the particle size of the final SLN. The authors attributed such impact to prolonged 'contact' (residence) time between the molten lipid and the aqueous surfactant solution.

More recently, a nanostructured lipid carrier (NLC) formulation for lidocaine has been successfully prepared. This study resulted in an HME NLC with an average particle size of less than 50 nm, polydispersity index <0.3 and an encapsulation efficiency of 73.9%, by combining HME with an auxiliary probe sonicator (Bhagurkar et al., 2017). It has been reported that HME screw speed had a significant impact upon the stability of the NLCs, with high screw speeds resulting in increased particle size after storage. In contrast, the duration of probe sonication was found to have little influence on the quality of the NLCs. The same sequential processing principle has also been successfully applied to other systems with positive outcomes (Muhindo et al., 2021; Shadambikar et al., 2021).

The preparation of nanoparticles has also been achieved using HME via a 'top-down' production approach, in which coarse particles were broken down to reduced sizes via the application of aggressive shearing. For example, to enhance therapeutic activity of Iron (II) sulphate, HME has been used to reduce particle size from a micron-scale down to approximately 350–400 nm (after dispersing at 5–50 mg/mL) by crushing coarse particles in the presence of surfactant and polymer excipients, followed by pulverisation using a grinder (Koo et al., 2019).

2.3.1.3 Cocrystals and Salts

As previously discussed, HME is a non-ambient shear process, and there is the potential to use high levels of thermal and mechanical energy to drive interactions between active ingredients and functional excipients. In this case, reactive extrusion can be used to produce molecular complexes such as cocrystals or salts, with an increasing number of articles published in this area over recent years.

Cocrystals are crystalline single-phase materials in which two or more distinct molecules interact, at defined stoichiometry, via non-covalent interactions. Cocrystallisation during HME processing has been thought to be driven by thermo-mechanochemistry, wherein parent reagents are subjected to both high shear forces and elevated temperatures within the extruder barrel.

Early work in this area reported the use of HME as a mechanochemical tool to produce ibuprofen-nicotinamide cocrystal agglomerates at an equal molar ratio (Dhumal et al., 2010). It was suggested that the successful extrusion of cocrystal product required a processing temperature above the eutectic point of the two parent compounds, but very importantly below the melting point of the resultant cocrystal. Within this defined processing window, it was shown that higher extrusion temperatures resulted in enhanced cocrystal conversion. This work was later extended to other cocrystal systems, including for example, caffeine-oxalic acid, nicotinamide-trans cinnamic acid, carbamazepine-saccharin and theophylline-citric acid, via liquid-assisted extrusion. Importantly, the addition of benign solvents, such as water and ethanol, in small 'catalytic' amounts, could serve as a process 'plasticiser' helping to drive cocrystallisation at significantly lower processing temperatures (Daurio et al., 2011). However, in some cases, the use of solvents also resulted in the formation of cocrystal hydrates. It was shown that hydrates were typically formed when extrusion was performed at temperatures lower than required to evaporate the solvent. A further report by the same team demonstrated successful extrusion cocrystallisation of a model system containing AMG 517 and sorbic acid. Of note was the superior downstream processibility of extruded cocrystals relative to those produced using conventional solution-based synthesis (Daurio et al., 2014).

With a clear demonstration that HME could be used to bulk produce pharmaceutical cocrystals, albeit with and without solvent, further explorative studies examined the feasibility of matrix-assisted extrusion cocrystallisation. In this work, the concept focused on driving cocrystalisation reactions whilst concurrently formulating with an active ingredient (cocrystal) within an excipient matrix, in a single-step extrusion process.

Early work by Boksa et al. (2014) reported a carbamazepine-nicotinamide cocrystal embedded in a Soluplus® matrix. Further work by Li et al. (2016b) reported for the first time the *in situ* formation of an equimolar ibuprofen-isonicotinamide cocrystal within a sugar alcohol (xylitol) matrix. Moreover, this team identified selection criteria for defining acceptable matrix excipients and the use of Flory–Huggins phase diagrams to help define an HME processing space to enhance cocrystal yield (Li et al., 2018). It has been proposed that during matrix-assisted extrusion cocrystallisation, the matrix carrier plays a dual role in acting as a molten 'solvent' facilitating the cocrystal reaction and as a functional component of the formulation, allowing for the formation of an acceptable solid product after exit from the extrusion dies. Moreover, the physicochemical properties of the matrix excipients may impart additional functionality and be used to modify drug release behaviour and/or mechanical properties (Boksa et al., 2014).

Salt formation is one of the most common techniques used to modify drug properties (e.g., solubility) and accounts for ~40% of all marketed pharmaceutical formulations (Bharate, 2021). Cocrystals differ from salts wherein no proton transfer occurs. Drug and its respective salts are distinct APIs as per FDA (The United States Food and Drug Administration) guidelines and enjoy patent protection individually, whilst cocrystals are regarded as intermediate products or in-process materials and are treated the same way as new polymorphs by the regulatory authorities (Food and Drug Administration, 2018). Similar to cocrystal synthesis, conventional preparation of pharmaceutical salts relies heavily on the use of organic solvents (Hossain et al., 2021). It would be of significant relevance to the industry if more economical and environmentally friendly salt production was available. In this context, a number of recent studies have examined the feasibility of producing pharmaceutical salts using HME. The impact of extrusion parameters on reaction yield has been investigated. Processing at temperatures close to or above the melting point of the salt negatively impacted yield, whilst increasing the length of the mixing zone along the extrusion barrel had a positive impact upon salt formation (Lee et al., 2017). Additional studies further demonstrated the suitability of HME for the preparation of salts containing an equimolar mixture of indomethacin and tromethamine (Bookwala et al., 2018).

2.3.2 CONTROLLED RELEASE DRUG DELIVERY PLATFORMS

2.3.2.1 Sustained Delivery of Small Molecule Drugs

Drug-eluting medical devices, such as intravaginal rings, surgical sutures, subcutaneous implants and catheters, are local drug delivery systems, which are capable of releasing APIs to surrounding tissues in a sustained and regulated manner for achieving optimised therapeutic effect. A recent example demonstrates the versatility of HME for these applications, with a biphasic release profile being achieved with a hot melt-extruded drug-eluting device (Deng et al., 2021a).

Drug-loaded intravaginal rings (IVRs) are a promising female-controlled strategy mainly used for contraceptive purposes and prevention of sexually transmitted diseases. APIs are usually extruded into nonbiodegradable polymers followed by a shaping process such as injection moulding and/or 3D printing. Polymers commonly used in traditional IVR formulations include polyurethane (Gupta et al., 2008), silicone (Woolfson et al., 2003) and poly-ethylvinyl acetate (pEVA) (Abdelkader et al., 2021), where the embedded drug is regulated by the released via diffusion, governed by the inherent characteristics of the matrix and the API.

Typical fabrication processes for **drug-loaded surgical sutures** include electrospinning, melt-spinning (modified HME) and coating, and have been commonly employed for the manufacture of absorbable drug-eluting sutures (Deng, et al., 2021b). The use of HME in suture production is somewhat limited to the generation of monofilaments where drugs are added during manufacture and homogeneously dispersed in the carrier material. This can be distinguished from electrospinning that is more flexible; being able to synthesise either coated or homogeneous

sutures. Electrospun fibres often demonstrate low mechanical strength (Brennan et al., 2018; Hu et al., 2010), whereas extruded suture threads are usually tenacious without voids (Tummalapalli et al., 2016). It has been shown that inclusion drug can further impact mechanical strength, as in such cases, the manufacture of high drug-loaded sutures may not be possible using electrospinning

Early attempts have focused on the use of HME to fabricate immunostimulatory sutures in which cytosine–phosphonothioate–guanine oligonucleotides were dispersed in a matrix of poly (lactic-co-glycolic acid) (PLGA) (Intra et al., 2011). Extruded sutures had diameters ranging from 50 to 300 μm and showed sustained release for up to 35 days. The antimicrobial properties of extruded sutures were significantly higher than the commercially available sutures used for wound closure. Other relevant work in this area focused on the use of poly(caprolactone) (PCL) to produce antimicrobial chlorhexidine-PCL sutures. Chlorhexidine live toxicity was significantly reduced after incorporation into a PCL monofilament, and drug release was reasonably short, extending to 6 days, making this product clinically sub-optimal for this application (Scaffaro et al., 2013). The feasibility of delivering nitric oxide in a solid state by incorporating it into a PCL-based suture through melt extrusion was demonstrated by Lowe et al. (2014). The nitric oxide-PCL sutures had high tensile strength and performed effectively in an antithrombotic and antimicrobial capacity, and had a therapeutic efficacy at lower drug loads, opening the possibility of sustained drug release for these agents.

In general, thermolabile compounds are often immediately discounted as being available to HME processing. However, it has been shown that thermosensitive APIs can be successfully processed when incorporated in the lamellae structure of hydrotalcite (Catanzano et al., 2014). PCL was used to form hydrotalcite-like compounds containing diclofenac via HME. The extruded fibres exhibited a prolonged drug release profile, extending beyond 70 days, with an adjustable release rate, tailored through adjustment of the diclofenac and hydrotalcite-like compounds. This extrusion strategy has been further examined in a recent study using the same API, and a unique polymer matrix comprising PCL, polyethylene glycol (PEG) and chitosan-keratin. The addition of PEG and chitosan-keratin significantly impacted the tensile properties of the suture, with optimum tensile strength obtained at a defined ratio of 80:19:1 (PCL/PEG/chitosan-keratin).

In addition to the manufacture of surgical sutures, recent work has demonstrated the combined use of solvent casting and HME for the manufacture of an **antimicrobial catheter**. Niclosamide and polyurethane were hot-melt extruded to fabricate a drug-eluting catheter, which exhibited both *in vitro* and *in vivo* antibacterial activity and reduced bacterial colonisation. All niclosamide-loaded catheters displayed a biphasic drug release profile with burst release exhibited in the first 24 hours and sustained release over the following 27-day period (Vazquez-Rodriguez et al., 2022).

In addition, drug-eluting implants have been proposed as an alternative route to oral administration for APIs with poor bioavailability from the gastrointestinal tract (GIT). Drug-eluting implants have been successfully manufactured using HME for osteoporosis treatment. A bisphosphonate-loaded polymer implant containing poly (lactic acid) (PLA) and PLGA as implant matrices were manufactured and assessed principally for drug release behaviour. These implants offered a way to enhance bioavailability and reduce the GIT irritation, which is often observed when bisphosphonates are administered orally. Moroever, drug release behaviour was dependent upon polymer excipient type (Dharmayanti et al., 2020). Further studies have also confirmed the use of PCL/PEG1500-based subdermal implants containing raloxifene hydrochloride, showing that implants with high drug loads exhibited a comparatively slower release of drug (Muhindo et al., 2022).

2.3.2.2 Implantable Devices for Delivery of Biologics

Biologics, i.e. therapeutic protein products (TPPs), represent an emerging sector of the pharmaceutical industry. Their scaled production to commercialization, however, are still limited by a number of well-recognised development challenges including limited half-life when administered systematically, poor patient compliance as a result of frequent injection or infusion, fragile folded structures that are prone to denaturation or unfolding upon exposure to external stimuli such as heat, mechanical stress, moisture and the susceptibility to enzymatic denaturing (Mitragotri et al., 2014; Pérez et al., 2010; Vaishya et al., 2015; Zheng & Pokorski, 2021).

Historically, HME has been overlooked as a viable production method for the manufacture of TPPs, quite possibly on the assumption that the fragile structure of the biologics would not survive the hostile environment within an extruder barrel. In the last 10 years, protein entrapment within polymeric matrices produced via HME have been reported. Early studies, however, mostly showed an attempt at the general applicability of the technology using biodegradable polymers, with little effort to understand the impact of the numerous process variables on product quality (Ghalanbor et al., 2013; Stanković et al., 2013). Polymers that facilitate low-temperature extrusion such as PLGA, a polyester approved by the FDA for use in implantable formulations, or multiblock poly(ε-caprolactone)-b-polyethylene glycol-b-poly(ε-caprolactone) (PCL-PEG-PCL), were amongst the first reported polymeric matrix carriers used for protein extrusion.

The feasibility of extruding bovine serum albumin (BSA) within PLGA matrices and the influence on protein release has been previously reported (Cossé et al., 2017). Confocal laser scanning microscopy and Raman spectroscopy were used to determine that BSA was homogenously distributed throughout the PLGA matrix following extrusion. Raman data also suggested a conformational change of the BSA structure, which may account for incomplete protein release. Furthermore, storage stability studies revealed the PLGA susceptibility to moisture uptake and an impact on protein release.

Other studies have attempted alternative ways to maintain protein stability during melt processing including the use of covalently attaching PEG (PEGylation). The effect of PEGylation and its effect on improving the thermal stability of lysozyme during HME in a PLGA matrix was reported by Lee et al. (2015). In their work, a single 20 K Da PEG chain covalently attached (monoPEGylation) to lysozyme was shown to significantly improve thermal stability of the protein. The authors postulated that a longer PEG chain would have increased efficacy in shielding protein from thermal exposure. However, it was noted that conjugates with multiple PEG chains (5–10 K Da) attachments could result in poor *in vivo* activity.

In related work, Teekamp et al. (2015) stabilised alkaline phosphatese (AP), prior to HME processing using spray drying, with insulin in a 1:10 protein:insulin weight ratio. The collected AP-insulin powders were extruded, in parallel to pure AP powders, using six different biodegradable polymers, namely PCL, PLGA (two different molecular weights) and the PEG-copolymers (hydrophilic). Data confirmed that the use of insulin was beneficial in protecting protein, at low (55°C) and intermediate (90°C) temperatures. The inclusion of hydrophilic co-polymer blocks, on the other hand, was seen to further improve protein stabilities during HME, particularly at intermediate temperatures.

2.3.3 CONTINUOUS PROCESSING – EXTRUSION GRANULATION

Granulation is a common unit operation used throughout the pharmaceutical industry. Conventionally, this is conducted in batch processes such as wet granulation, fluidised bed granulation, melt granulation and roller compaction, with a view to improve powder handling or compression properties of a formulation (Ennis & Litster, 1997). In recent years, TSE has received increased attention as a promising alternative for the production of drug-containing granules.

Unlike conventional methods, TSE granulation processes are considered to be 'regime separated' in that wetting and nucleation steps are separate to consolidation and growth. This allows for better control of granule density and size. Moreover, twin-screw granulation has compatibility with in-line/online monitoring, which provides more precise and real-time informative data to assess uniformity and decomposition of active drug ingredients (Sarabu et al., 2019; Tambe et al., 2021). To date, TSE granulation is broadly classified into twin-screw melt granulation (TSMG) and twin-screw dry granulation (TSDG), depending on the binder used and the binding mechanism driving granulation.

TSMG uses a meltable binder and processing temperatures where only the binder is melted and uniformly distributed in the system, achieving agglomeration of particles and ultimately formation of granules. Several studies have confirmed successful TSMG using a range of different binders, including hydroxypropylcellulose (Klucel™ EF), lipids (Comprital® 888 ATO, Precerol® ATO5, Geleol™) and polyethylene oxide (PEO). There have also been commercial products developed using TSGM including Covera-HS® (Verpamil hydrochloride), Nurofen Meltlets Lemon® (ibuprofen) and Eucreas™ (Metformin hydrochloride) (Kallakunta et al., 2018; Dhaval et al., 2022).

Batra et al. (2017) screened different polymeric binders, including PVP, HPMC, HPC methacrylate polymers, Poloxamers® and Soluplus®, for their appropriateness for use in TSMG containing acetaminophen and metformin (Batra et al., 2017). In their work, it was found that with careful selection of processing conditions and optimisation, all investigated polymers were shown to function as suitable binders, producing pharmaceutically acceptable granules. The quality of granules and subsequently compressed tablets were somewhat compromised when using low melting binders. 'Sticky' agglomerated material was problematic, particularly in downstream sieving processes, required for size reduction. Moreover, when 10% w/w binder loading was used, some polymers (HMPC and poloxamer) were not able to produce tablets with sufficient tensile strength.

Twin-screw dry granulation differs from TSMG in the sense that TSDG is usually carried out at significantly lower extrusion temperatures, making it more suitable for thermally labile materials. The kneading effect of screw elements and softening of ingredients primarily under frictional heat facilitate a uniform distribution of the binder and the subsequent generation of granules. Successful applications of TSDG using sildenafil citrate and ondansetron, respectively, incorporating cellulose derivatives including HPC, HEC and EC as binders were first reported by Repka et al. (2017). In all processing, a temperature of 65°C was used, which was below the Tg of all binder polymers; screw speeds ranging between 100 and 200 rpm were used; and a range of screw configurations were investigated. In the same year, Richter suggested low binder contents (~10% w/w) may be preferential in TSDG, which may help achieve higher drug loadings (Richter, 2017).

More recently, the feasibility of TSDG using three different drug compounds, theophylline, APAP and lidocaine HCl, respectively, with a number of polymeric binders namely HPC, EC, Klucel® EF, Klucel® MF and Eudragit® RSPO has been investigated (Kallakunta et al., 2019a). The authors reported that the crystallinity of the drug impacted drug loading. Moreover, the viscosity of the polymer binder had an influence upon both processing and quality of the granules, with higher polymer viscosity resulting in enhanced granulation and larger and harder granules.

2.3.4 PATIENT-CENTRED DRUG PRODUCTS

Combination pharmacotherapies are common in chronic conditions such as HIV, diabetes and cardiovascular disease; consequently, an impetus has been established for novel polytherapy products. Several advantages are associated with the use of fixed-dose combination (FDC) products such as improved concordance through reduced tablet burden and the simplification of drug regimens. Moreover, complimentary FDCs enable synergistic effects without dose escalation of a single therapy, reducing the risk of side effects.

2.3.4.1 Fixed-Dose Combinations (FDCs) via Co-extrusion

Formulation and process development of FDCs are more challenging than those for single drug products due to the complexity in dosing ratio, pill size and additionally the requirement for different drug release profiles (Andrews et al., 2019b). In light of this, hot-melt co-extrusion where multiple strands of molten matrices are manufactured in layered structures has been presented as a potential solution.

One of the major advantages conferred by multi-layered extrudates is the potential to continuously manufacture complex drug formulations in a single step, where independent modulation of drug release of each individual drug can be made achievable. Moreover, separation of non-compatible drug combinations permits flexibility when designing a dosage form. Although the principles that underpin extrusion apply in the manufacture of dosage forms by co-extrusion, other additional considerations exist, such as the interaction between the adjacent drug layers and the possibility of inter-diffusion between layers. It is therefore important to study the API and carrier physicochemical properties to determine the appropriate of polymers to be used for this purpose. Greater control over processing parameters must also be exerted to modulate the melt viscosity and extrusion velocity in order to promote adhesion and uniformity between the adjoining layers (Vynckier, et al., 2014b).

Commercial products manufactured using co-extrusion are mostly combination products where an active drug or a drug combination is incorporated into a medical device. Contraceptive devices, including Implanon® (implantable rod of 4 cm in length and 2 cm in diameter containing etonogestrel, aimed for contraceptive efficacy for up to 3 years) and NuvaRing® (a 21-day release intravaginal ring incorporated with a FDC of etonogestrel and ethinyl estradiol), are amongst the most recognised medicinal products by co-extrusion. In both devices, a reservoir structure is designed where the active ingredients are embedded in an inner ethylene vinyl acetate (EVA) matrix which is enveloped by an outer EVA membrane to regulate drug release rate.

Currently, there are no marketed oral solid dosage forms produced via co-extrusion; however, promising research into the area suggests this as a viable formulation platform. A sustained release bi-layer unit with the model drug theophylline incorporated in an inner PEG core has been investigated by Quintavalle et al. (2008). In this work, it was shown that the drug release mechanism was regulated by the outer lipophilic shell, manufactured from microcrystalline wax. Additional studies by Dierickx et al. (2012) manufactured FDC mini-matrices, containing metoprolol tartrate and hydrochlorothiazide (HCTZ), using bi-layer co-extrusion. In this work, PCL was used for the core whilst PEO for the coat. The authors reported good adhesion between the two layers, even though there were significant inherent differences between the polymers. Further work was conducted by the same group co-extruding a lipophilic core (PCL or EC) with a hydrophilic coat (PEO or Soluplus®, respectively), whilst both layers contained diclofenac sodium (Dierickx et al., 2013) and reported the development of a metoprolol tartrate and HCTZ FDC via co-extrusion using EC as the core matrix carrier with PEO as the coat (Vynckier et al., 2014a). More recently, Andrews et al. (2019b) co-extruded a bi-layer system aimed to deliver biphasic release of a simvastatin and aspirin in a FDC for improved chronic management of cardiovascular diseases. In this work, aspirin was designed to be delivered using an immediate release coat layer based on Kollidon® VA64, whilst simvastatin was embedded in a core matrix consisting of a blend of anionic polymers to attain site-specific drug release targeting the distal regions of the small intestine.

2.3.4.2 Fused Deposition Modelling 3D Printing

3D printing is a novel manufacturing method which offers unparalleled flexibility in oral solid dosage form preparation. Fused deposition modelling 3D printing (FDM 3DP), or sometimes referred to as "additive manufacturing", is an emerging pharmaceutical technology that can be directly coupled with HME. The technique operates with a layer-by-layer process in which extruded molten mass is deposited to build in the x, y and z planes.

Over the last decade, there has been a significant uplift in published articles within the academic literature showcasing HME FDM 3DP dosage units prepared for various drug compounds. For example, FDM 3DP controlled release tablets were reported for acetaminophen in HPMC E5 (Zhang et al., 2017a,b), glipizide (Li et al., 2017) and budesonide (Goyanes et al., 2015), respectively, in polyvinyl alcohol, theophylline and metformin hydrochloride in thermoplastic polyurethanes (Verstraete et al., 2018). Likewise, pantoprazole sodium immediate release dosage forms were reported to have been printed with Povidone K12, PEG 6000, PEG 20000, Kollidon® VA 64 and poloxamer 407, respectively (Kempin et al., 2018). There have also been reports describing 3D-printed gastroretentive tablets where buoyancy was achieved by adjusting the infill density setting during printing (Chai et al., 2017). More recently, patient-centric amorphous aripiprazole containing orodispersible films have been prepared using HME FDM 3D printing (Lee et al., 2022). It was demonstrated that the technology could achieve fine-tuning of the puncture strength of the printed films, facilitate water penetration and accelerate disintegration of the films, when compared to conventional solvent casting film preparation. The application of HME FDM 3D printing has also been expanded to a variety of shaped dosage forms for personalised medicine, such as solid implants and vaginal rings (Dumpa et al., 2021).

To date, reported studies utilising FDM 3D printing have been primarily executed as batch, without full integration of a continuous extrusion line. In keeping with the paradigm shift towards continuous process, a recent article by Tan et al. (2018) reviewed the use of HME directly linked to FDM 3D printing. Bandari et al. (2021) further expanded this topic. In their work, the limited progress in HME-based FDM 3DP drug products was linked to the disconnected unit operations and the lack of academic research that truly assessed the continuous nature of an integrated HME-3DP process. Specifically, several processing aspects were discussed, including in-process quality attributes (material-, equipment- and process-specific parameters), in-line analytical tool selection/alignment, and synchronisation of HME throughput to 3D printing speed that are critical to the successful implementation of HME FDM 3DP.

2.3.5 Abuse-Deterrent Formulations

The abuse of prescription drugs, such as opioids, anti-anxiety medications, sedatives and stimulants, has become a rising issue. Whilst active measures are being taken worldwide for early identification and intervention of prescription drug abuse to prevent such problems to progress to addiction, the development of abuse-deterrent dosage forms that can resist or reduce intentional tampering has also emerged as a new formulation focus.

It was found in a rehabilitation programme with drug abusers that 80% illegally used opioid tablets that were tampered through crushing or chewing, in order to achieve accelerated drug release, whilst another commonly used opioid abuse method was to take excessive quantities of intact tablets (Passik et al., 2006). There is currently no 'ceiling-dose' for the majority of opioids due to the phenomenon of tolerance (Chou et al., 2009). It is, therefore, considered more reasonable to deter tampering via physical, rather than pharmacological, intervention (Stanos et al., 2012). Amongst the several strategies that have been used to deter drug abuse, matrix tablets manufactured using high molecular weight polymers via thermal processing, such as melt extrusion, are a good example of an effective abuse-deterrent formulation (ADF).

Xu and co-workers investigated the impact of different HME post-processing on the abuse-deterrent efficiency of PEO-based ADF tablets (Xu et al., 2019). Tablets compressed using cryo-milled extrudates exhibited more susceptibility to physical manipulation and had a high degree of extractability in the presence of alcohol. The study also revealed that the heat exposure of PEO during the HME processing resulted in decreased solution viscosity of the formulation and reduced forces required for both syringing and injection.

Extruded AD pellets have been reported by Maddineni et al. (2014) with formulations optimised using a Design of Experiments (DoE) approach. In their work, lidocaine hydrochloride was embedded in polymeric matrices consisting of three polymer excipients, namely PolyOx™ WSR301 (PEO), Benecel™ K15M (HPMC) and Carbopol® 71G, at varying ratios. The tamper resistant attributes were evaluated based on post-milling particle size, gelling and percentage of drug extraction in both aqueous and alcoholic media. Their results suggested a particle size requirement of greater than 150 µM to prevent snorting, and good gelling ability offered by the combination of HPMC and Carbopol in both water and alcohol which would prevent drug content extraction.

Some studies presented the combination of HME and a 'forming unit' as a sufficient process to produce ADFs. For instance, a series of related patents describing AD pills produced by HME-calendaring were shown to meet regulatory guidelines for extended-release formulations whilst also exhibiting tamper resistance (Maddineni et al., 2014). Tablets included a matrix agent with an average molecular weight between 50 and 350 K Da, and the extruded formulation was directly formed into a tablet via calendaring, without further processing. The extruded matrix was shown to exhibit increased resistance to physical or mechanical stress such as pulverising or grinding, rendering the dosage form less amenable to abuse. It was also stated that the matrix agent might exert additional abuse-deterrent effects 'by acting as a gelling or viscosity increasing agent'.

Tamper-resistant formulations have also been reported using co-extrusion. A co-extruded, multilayer, multiparticulate, dosage unit containing the opioid hydromorphone and naltrexone, an opioid antagonist, for the treatment of opioid dependence was reported by Flath and Masselink (2005). The dosage form was co-extruded with three distinct layers, namely a core containing naltrexone, which was encapsulated up by a hydrophobic middle sheath, and an exterior shell with hydromorphone. If the dosage unit was tampered with, for example by chewing, or dissolved, then the entrapped naltrexone would be exposed from the core and start to exert its antagonising effect against hydromorphone, thus preventing abuse of the opioid active.

2.4 HME AND CONTINUOUS MANUFACTURING

2.4.1 Principle of Quality by Design (QbD)

The International Conference on Harmonization (ICH) guidelines Q8(R2) (ICH, 2009) advocates that 'quality cannot be tested into a product but must be incorporated by design', along with ICH Q9, Q10 and Q11 emphasising quality by design (QbD) to accelerate the industrial shift to CM. With the advent of process analytical technology (PAT) tools, real-time data on critical quality attributes and their relationship with processing parameters can be collected during manufacturing, which in turn allows online adjustment of production conditions for precise quality control (Figure 2.4).

2.4.2 Process Analytical Technology (PAT) Tools

The USFDA introduced the concept of PAT at the beginning of the 21st century, through publication of 'PAT-A

Framework for Innovative Pharmaceutical Manufacturing and Quality Assurance', with the intention to guide fundamental understanding of the design of CM processes and to encourage more effective quality control of drug products. The PAT concept is the implementation of in-line or online monitoring and data analysis tools for controlling and measuring the process performance and ultimately ensuring the quality of products. A wide range of fast and non-destructive PAT tools have been reported to support interfacing with HME processes, to help reduce batch variability and losses and to meet specifications. A few examples are provided in the sections below.

2.4.2.1 Near- and Mid-Infrared Spectroscopy

(NIR and MIR, respectively) are amongst the most well-established PAT tools for in-line quality monitoring. Islam et al. (2015) implemented NIR, coupled with a transmission probe and a reflectance probe, as a PAT tool to monitor drug (indomethacin) transformation within hydrophilic polymeric matrices during HME processing. In their study, the reflectance probe was positioned in the first mixing zone, whilst the transmission probe was positioned at the die zone immediately prior to the exit orifice. The assumption within this work was that, within the first mixing zone, the fed indomethacin would not have converted to the amorphous form and therefore should remain reflective allowing collection of reflectance spectra. When approaching the die, on the other hand, optimised HME conditions should have rendered all indomethacin content amorphous and dispersed in a transparent amorphous matrix; therefore, the detection of transmission signals would be more appropriate. The measured amorphous conversion yields under various processing conditions using the two NIR probes were ascertained using offline confocal Raman spectroscopy. It was also found that the screw speed had an impact upon the recorded spectra, but not the homogeneity of the dispersed drug molecules in the final extrudates.

2.4.2.2 Raman Spectroscopy

Raman is a spectroscopic method that is more applicable for wet granulation processing in HME, attributed to its lower susceptibility to water interference compared with NIR. However, the Raman effect is inherently weak and might be subject to interference caused by heating and photodecomposition during HME, leading to the disruption of spectra acquisition. As a result, in-line HME product characterisation using Raman spectroscopy requires rigorous design and has been underutilised. More recently, there has been an increased interest in this spectroscopic technique for in-line, as an alternative to in-line NIR or MIR, for content-based real-time quality control. Andrews et al. (2019a) built an in-line Raman-PLS model for the detection of Ramipril thermal degradation and for the prediction of optimum HME processing space for this thermally labile active compound. Latter work by this group also demonstrated the application of this non-destructive in-line tool for simultaneous quality control for two active compounds

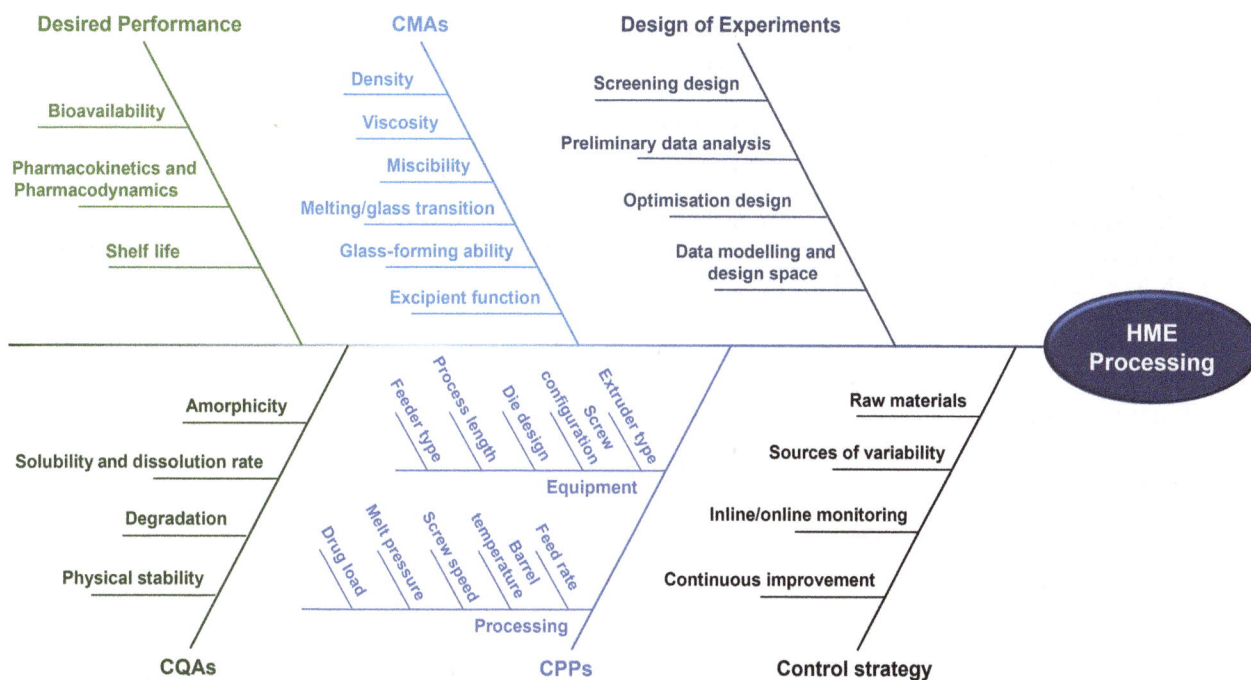

FIGURE 2.4 HME product development roadmap adapting to a QbD embedded CM Processing concept.

in a FDC formulation (Dadou et al., 2020, 2021), bringing further advances in the use of chemometric methods in the quality analysis/control of pharmaceutical product manufacturing using continuous processes.

2.4.2.3 In-Line UV/Visible Spectroscopy

The ultraviolet-visible spectroscopy (UV/VIS) provides real-time analysis on colour, concentration, solid-state and particle size, when combined with HME. Wesholowski et al. (2018) investigated the use of in-line UV quantification of the drug load using two model drugs, carbamazepine and theophylline, respectively, within a copovidone matrix manufactured via HME. The authors concluded that this PAT tool required only minimal effort for data evaluation and was sufficient for real-time data acquisition at a measurement frequency of 1 Hz. The use of in-line UV-Vis as a fast-working PAT tool for melt-extruded piroxicam (PRX)/Kollidon® VA64 ASDs has been reported by Schlindwein et al. (2018). The authors selected absorbance of the extrudates as the critical quality attributes (CQAs) and investigated the impact of HME processing conditions as well drug concentration on response signals.

2.4.2.4 In-Process Rheometry

Over the last five years, in-process rheology has been reported as another promising PAT tool to help build improved understanding and control of the overall HME process. The main principle behind the use of rheology as a PAT tool is based on the hypothesis that material changes, including amorphisation, melting and formation of interactions, within the extruder barrel should be reflected in changes in the melt flow properties, most notably the viscosity of the melt and back pressure at the extrusion exit.

Kelly et al. (2018) inserted a rheological slit die in the final zone of their extruder and measured the pressure changes within the die across a range of extrusion throughputs. The measured pressure changes were used to calculate important rheological parameters including melt exit pressure and shear viscosity and suggested that these parameters were sensitive to API loading in this case. Additionally, several other widely used characteristic techniques, such as differential scanning calorimetry (DSC), X-ray diffraction (XRD) and thermogravimetric analysis (TGA), have also have been suggested as potential PAT tools in conjunction with extruders (Ren et al., 2019).

2.5 SUMMARY AND FUTURE PERSPECTIVES

The undesirable physicochemical properties of newly discovered drug compounds, the quest for low-cost, high-effective drug products and the rising demand on medical services and healthcare pose unprecedented challenges for drug product development. The proliferation of intellectual properties derived from HME and the marketed products produced via HME are indicative of the importance of this technology, underpinning its versatility and relatively untapped potential. Although HME has been routinely reported for the manufacture of ADSs, drug-loaded sustained delivery devices, we have reported not only on those applications but extended our discussion to other less well-described applications. For example, HME has been used for its ability to combine both thermal and mechanical energies in reactive HME to produce pharmaceutical multicomponent systems. With the aid of equipment modifications and careful selection of ingredients, HME has demonstrated practicality in offering an alternative to complex, lengthy and costly batch granulation processes conventionally used for particle size enlargement and regulation. The advancement in equipment integration allowing explorations of end-to-end manufacturing models based on HME and auxiliary shaping/processing units will help further define the potential of HME for the manufacture of products with complex structures. Moreover, the chapter also introduces CM using HME within a QbD framework, presenting case studies where laboratory-scale CM trials have been established and verified using a range of real-time quality measurement tools.

Understandably, the pharmaceutical industry's uptake of new technologies, even those identified as having significant advantages over traditional methods, is painfully slow. Nonetheless, with HME expanding its application scope into many advanced formulation and production fields, it is now regarded as an advantageous manufacturing platform for drug products. With further technical progress, regulation updates and exploratory research, in the near future, HME will no longer work as a substitute, but potentially become a mainstay in pharmaceutical manufacturing.

REFERENCES

Abdelkader, H., Fathalla, Z., Seyfoddin, A., Farahani, M., Thrimawithana, T., Allahham, A., Alani, A. W. G., Al-Kinani, A. A., & Alany, R. G. (2021). Polymeric long-acting drug delivery systems (LADDS) for treatment of chronic diseases: Inserts, patches, wafers, and implants. *Advanced Drug Delivery Reviews*, *177*, 113957).

Andrews, G. P., Jones, D. S., Senta-Loys, Z., Almajaan, A., Li, S., Chevallier, O., Elliot, C., Healy, A. M., Kelleher, J. F., Madi, A. M., Gilvary, G. C., & Tian, Y. (2019a). The development of an inline Raman spectroscopic analysis method as a quality control tool for hot melt extruded ramipril fixed-dose combination products. *International Journal of Pharmaceutics*, *566*, 476–487.

Andrews, G. P., Li, S., Almajaan, A., Yu, T., Martini, L., Healy, A., & Jones, D. S. (2019b). Fixed dose combination formulations: Multilayered platforms designed for the management of cardiovascular disease. *Molecular Pharmaceutics*, *16*(5), 1827–1838.

Bandari, S., Nyavanandi, D., Dumpa, N., & Repka, M. A. (2021). Coupling hot melt extrusion and fused deposition modeling: Critical properties for successful performance. *Advanced Drug Delivery Reviews*, *172*, 52–63.

Batra, A., Desai, D., & Serajuddin, A. T. M. (2017). Investigating the use of polymeric binders in twin screw melt granulation process for improving compactibility of drugs. *Journal of Pharmaceutical Sciences*, *106*(1), 140–150.

Baumgartner, R., Eitzlmayr, A., Matsko, N., Tetyczka, C., Khinast, J., & Roblegg, E. (2014). Nano-extrusion: a promising tool for continuous manufacturing of solid nano-formulations. *International Journal of Pharmaceutics*, *477*(1–2), 1–11.

Bevernage, J., Brouwers, J., Annaert, P., & Augustijns, P. (2012). Drug precipitation-permeation interplay: Supersaturation in an absorptive environment. *European Journal of Pharmaceutics and Biopharmaceutics*, *82*(2), 424–428.

Bhagurkar, A. M., Repka, M. A., & Murthy, S. N. (2017). A novel approach for the development of a nanostructured lipid carrier formulation by hot-melt extrusion technology. *Journal of Pharmaceutical Sciences*, *106*(4), 1085–1091.

Bharate, S. S. (2021). Recent developments in pharmaceutical salts: FDA approvals from 2015 to 2019. *Drug Discovery Today*, *26*(2), 384–398.

Boksa, K., Otte, A., & Pinal, R. (2014). Matrix-Assisted Cocrystallization (MAC) simultaneous production and formulation of pharmaceutical cocrystals by hot-melt extrusion. *Journal of Pharmaceutical Sciences*, *103*(9), 2904–2910.

Bookwala, M., Thipsay, P., Ross, S., Zhang, F., Bandari, S., & Repka, M. A. (2018). Preparation of a crystalline salt of indomethacin and tromethamine by hot melt extrusion technology. *European Journal of Pharmaceutics and Biopharmaceutics*, *131*, 109–119.

Brennan, D. A., Conte, A. A., Kanski, G., Turkula, S., Hu, X., Kleiner, M. T., & Beachley, V. (2018). Mechanical considerations for electrospun nanofibers in tendon and ligament repair. *Advanced Healthcare Materials*, *7*(12), 1701277.

Burcham, C. L., Florence, A. J., & Johnson, M. D. (2018). Continuous manufacturing in pharmaceutical process development and manufacturing. *Annual Review of Chemical and Biomolecular Engineering*, *9*, 253–281.

Catanzano, O., Acierno, S., Russo, P., Cervasio, M., del Basso De Caro, M., Bolognese, A., Sammartino, G., Califano, L., Marenzi, G., Calignano, A., Acierno, D., & Quaglia, F. (2014). Melt-spun bioactive sutures containing nanohybrids for local delivery of anti-inflammatory drugs. *Materials Science and Engineering: C*, *43*, 300–309.

Chai, X., Chai, H., Wang, X., Yang, J., Li, J., Zhao, Y., Cai, W., Tao, T., & Xiang, X. (2017). Fused deposition modeling (FDM) 3D printed tablets for intragastric floating delivery of domperidone. *Scientific Reports*, *7*(1), 1–9.

Chou, R., Fanciullo, G. J., Fine, P. G., Adler, J. A., Ballantyne, J. C., Davies, P., Donovan, M. I., Fishbain, D. A., Foley, K. M., Fudin, J., Gilson, A. M., Kelter, A., Mauskop, A., O'Connor, P. G., Passik, S. D., Pasternak, G. W., Portenoy, R. K., Rich, B. A., Roberts, R. G., … Miaskowski, C. (2009). Clinical guidelines for the use of chronic opioid therapy in chronic noncancer pain. *Journal of Pain*, *10*(2), 113–130.

Cossé, A., König, C., Lamprecht, A., & Wagner, K. G. (2017). Hot melt extrusion for sustained protein release: Matrix erosion and in vitro release of PLGA-based implants. *AAPS PharmSciTech*, *18*(1), 15–26.

Dadou, S. M., Senta-Loys, Z., Almajaan, A., Li, S., Jones, D. S., Healy, A. M., Tian, Y., & Andrews, G. P. (2020). The development and validation of a quality by design based process analytical tool for the inline quantification of Ramipril during hot-melt extrusion. *International Journal of Pharmaceutics*, *584*, 119382.

Dadou, S. M., Tian, Y., Li, S., Jones, D. S., & Andrews, G. P. (2021). The optimization of process analytical technology for the inline quantification of multiple drugs in fixed dose combinations during continuous processing. *International Journal of Pharmaceutics*, *592*, 120024.

Daurio, D., Medina, C., Saw, R., Nagapudi, K., & Alvarez-Núñez, F. (2011). Application of twin screw extrusion in the manufacture of cocrystals, part I: Four case studies. *Pharmaceutics*, *3*(3), 582–600.

Daurio, D., Nagapudi, K., Li, L., Quan, P., & Nunez, F. A. (2014). Application of twin screw extrusion to the manufacture of cocrystals: Scale-up of AMG 517-sorbic acid cocrystal production. *Faraday Discussions*, *170*, 235–249.

Deng, X., Gould, M., & Ali, M. A. (2021a). Fabrication and characterisation of melt-extruded chitosan/keratin/PCL/PEG drug-eluting sutures designed for wound healing. *Materials Science and Engineering: C*, *120*, 111696.

Deng, X., Qasim, M., & Ali, A. (2021b). Engineering and polymeric composition of drug-eluting suture: A review. *Journal of Biomedical Materials Research - Part A*, *109*(10), 2065–2081.

Dharmayanti, C., Gillam, T. A., Williams, D. B., & Blencowe, A. (2020). Drug-eluting biodegradable implants for the sustained release of bisphosphonates. *Polymers*, *12*(12), 2930.

Dhaval, M., Sharma, S., Dudhat, K., & Chavda, J. (2022). Twin-screw extruder in pharmaceutical industry: History, working principle, applications, and marketed products: An in-depth review. *Journal of Pharmaceutical Innovation*, *17*, 294–318.

Dhumal, R. S., Kelly, A. L., York, P., Coates, P. D., & Paradkar, A. (2010). Cocrystalization and simultaneous agglomeration using hot melt extrusion. *Pharmaceutical Research*, *27*(12), 2725–2733.

Dierickx, L., Remon, J. P., & Vervaet, C. (2013). Co-extrusion as manufacturing technique for multilayer mini-matrices with dual drug release. *European Journal of Pharmaceutics and Biopharmaceutics*, *85*(3 PART B), 1157–1163.

Dierickx, L., Saerens, L., Almeida, A., de Beer, T., Remon, J. P., & Vervaet, C. (2012). Co-extrusion as manufacturing technique for fixed-dose combination mini-matrices. *European Journal of Pharmaceutics and Biopharmaceutics*, *81*(3), 683–689.

Dumpa, N., Butreddy, A., Wang, H., Komanduri, N., Bandari, S., & Repka, M. A. (2021). 3D printing in personalized drug delivery: An overview of hot-melt extrusion-based fused deposition modeling. *International Journal of Pharmaceutics*, *600*, 120501.

El-Egakey, M. A., Soliva, M., & Speiser, P. (1971). Hot extruded dosage forms. I. Technology and dissolution kinetics of polymeric matrices. *Pharmaceutica Acta Helvetiae*, *46*(1), 31–52. http://www.ncbi.nlm.nih.gov/pubmed/5542801.

Ennis, J. B., & J.D. Litster. (1997). Principles of size enlargement. In R. Perry & D. Green (Eds.), *Perry's Chemical Engineers' Handbook* (pp. 20–56). New York: McGraw-Hill.

Flath, R. P., & Masselink, J. K. (2005). *Tamper resistant co-extruded dosage form containing an active agent and an adverse agent and process of making same.* Google Patents.

Food and Drug Administration. (2018). *Regulatory classification of pharmaceutical co-crystals, guidance for industry.* U.S. Department of Health and Human Services, February.

Forster, A., Hempenstall, J., Tucker, I., & Rades, T. (2001). Selection of excipients for melt extrusion with two poorly water-soluble drugs by solubility parameter calculation and thermal analysis. *International Journal of Pharmaceutics*, *226*(1–2), 147–161.

Ghalanbor, Z., Körber, M., & Bodmeier, R. (2013). Interdependency of protein-release completeness and polymer degradation in PLGA-based implants. *European Journal of Pharmaceutics and Biopharmaceutics*, *85*(3 PART A), 624–630.

Goyanes, A., Chang, H., Sedough, D., Hatton, G. B., Wang, J., Buanz, A., Gaisford, S., & Basit, A. W. (2015). Fabrication of controlled-release budesonide tablets via desktop (FDM) 3D printing. *International Journal of Pharmaceutics*, *496*(-2), 414–420.

Gupta, K. M., Pearce, S. M., Poursaid, A. E., Aliyar, H. A., Tresco, P. A., Mitchnik, M. A., & Kiser, P. F. (2008). Polyurethane intravaginal ring for controlled delivery of dapivirine, a non-nucleoside reverse transcriptase inhibitor of HIV-1. *Journal of Pharmaceutical Sciences*, *97*(10), 4228–4239.

Hossain Mithu, M. S., Economidou, S., Trivedi, V., Bhatt, S., & Douroumis, D. (2021). Advanced methodologies for pharmaceutical salt synthesis. *Crystal Growth and Design*, *21*(-2), 1358–1374.

Hu, W., Huang, Z.-M., & Liu, X.-Y. (2010). Development of braided drug-loaded nanofiber sutures. *Nanotechnology*, *21*(31), 315104.

ICH. (2009). International conference on harmonisation of technical requirements for registration of pharmaceuticals for human use. ICH harmonised tripartite guideline pharmaceutical development Q8(R2). *International Conference on Harmonisation*, *3*(June), 12–24.

Intra, J., Zhang, X-Q., Williams, R. L., Zhu, X., Sandler, A. D., & Salem, A. K. (2011). Immunostimulatory sutures that treat local disease recurrence following primary tumor resection. *Biomedical Materials*, *6*(1), 011001.

Islam, M. T., Scoutaris, N., Maniruzzaman, M., Moradiya, H. G., Halsey, S. A., Bradley, M. S. A., Chowdhry, B. Z., Snowden, M. J., & Douroumis, D. (2015). Implementation of transmission NIR as a PAT tool for monitoring drug transformation during HME processing. *European Journal of Pharmaceutics and Biopharmaceutics*, *96*, 106–116.

Jana, S., & Miloslava, R. (2012). Hot-melt extrusion. *Ceska a Slovenska Farmacie: Casopis Ceske Farmaceuticke Spolecnosti a Slovenske Farmaceuticke Spolecnosti*, *61*(3), 87–92.

Kalepu, S., & Nekkanti, V. (2015). Insoluble drug delivery strategies: review of recent advances and business prospects. *Acta Pharmaceutica Sinica B*, *5*(5), 442–453.

Kallakunta, V. R., Patil, H., Tiwari, R., Ye, X., Upadhye, S., Vladyka, R. S., Sarabu, S., Kim, D. W., Bandari, S., & Repka, M. A. (2019a). Exploratory studies in heat-assisted continuous twin-screw dry granulation: A novel alternative technique to conventional dry granulation. *International Journal of Pharmaceutics*, *555*, 380–393.

Kallakunta, V. R., Sarabu, S., Bandari, S., Tiwari, R., Patil, H., & Repka, M. A. (2019b). An update on the contribution of hot-melt extrusion technology to novel drug delivery in the twenty-first century: part I. *Expert Opinion on Drug Delivery*, *16*(5), 539–550.

Kallakunta, V. R., Tiwari, R., Sarabu, S., Bandari, S., & Repka, M. A. (2018). Effect of formulation and process variables on lipid based sustained release tablets via continuous twin screw granulation: A comparative study. *European Journal of Pharmaceutical Sciences*, *121*, 126–138.

Kelly, A. L., Gough, T., Isreb, M., Dhumal, R., Jones, J. W., Nicholson, S., Dennis, A. B., & Paradkar, A. (2018). In-process rheometry as a PAT tool for hot melt extrusion. *Drug Development and Industrial Pharmacy*, *44*(4), 670–676.

Kempin, W., Domsta, V., Grathoff, G., Brecht, I., Semmling, B., Tillmann, S., Weitschies, W., & Seidlitz, A. (2018). Immediate release 3D-printed tablets produced via fused deposition modeling of a thermo-sensitive drug. *Pharmaceutical Research*, *35*(6), 1–2.

Khan, K. U., Minhas, M. U., Badshah, S. F., Suhail, M., Ahmad, A., & Ijaz, S. (2022). Overview of nanoparticulate strategies for solubility enhancement of poorly soluble drugs. *Life Sciences*, *291*, 120301.

Koo, J. S., Lee, S. Y., Azad, M. O. K., Kim, M., Hwang, S. J., Nam, S., Kim, S., Chae, B.-J., Kang, W.-S., & Cho, H.-J. (2019). Development of iron(II) sulfate nanoparticles produced by hot-melt extrusion and their therapeutic potentials for colon cancer. *International Journal of Pharmaceutics*, *558*, 388–395.

Lang, B., McGinity, J. W., & Williams, R. O. (2014). Hot-melt extrusion – basic principles and pharmaceutical applications. *Drug Development and Industrial Pharmacy*, *40*(9), 1133–1155.

Lee, H. L., Vasoya, J. M., Cirqueira, M. de, L., Yeh, K. L., Lee, T., & Serajuddin, A. T. M. (2017). Continuous preparation of 1:1 haloperidol–maleic acid salt by a novel solvent-free method using a twin screw melt extruder. *Molecular Pharmaceutics, 14*(4), 1278–1291.

Lee, J.-H., Park, C., Song, I.-O., Lee, B.-J., Kang, C.-Y., & Park, J.-B. (2022). Investigation of patient-centric 3D-printed orodispersible films containing amorphous aripiprazole. *Pharmaceuticals, 15*(7), 895.

Lee, P. W., Towslee, J., Maia, J., & Pokorski, J. (2015). PEGylation to improve protein stability during melt processing. *Macromolecular Bioscience, 15*(10), 1332–1337.

Li, Q., Wen, H., Jia, D., Guan, X., Pan, H., Yang, Y., Yu, S., Zhu, Z., Xiang, R., & Pan, W. (2017). Preparation and investigation of controlled-release glipizide novel oral device with three-dimensional printing. *International Journal of Pharmaceutics, 525*(1), 5–11.

Li, S., Tian, Y., Jones, D. S., & Andrews, G. P. (2016a). Optimising drug solubilisation in amorphous polymer dispersions: Rational selection of hot-melt extrusion processing parameters. *AAPS PharmSciTech, 17*(1), 200–213.

Li, S., Yu, T., Tian, Y., Lagan, C., Jones, D. S., & Andrews, G. P. (2018). Mechanochemical synthesis of pharmaceutical cocrystal suspensions via hot melt extrusion: Enhancing cocrystal yield. *Molecular Pharmaceutics, 15*(9), 3741–3754.

Li, S., Yu, T., Tian, Y., McCoy, C. P., Jones, D. S., & Andrews, G. P. (2016b). Mechanochemical synthesis of pharmaceutical cocrystal suspensions via hot melt extrusion: Feasibility studies and physicochemical characterization. *Molecular Pharmaceutics, 13*(9), 3054–3068.

Lin, D., & Huang, Y. (2010). A thermal analysis method to predict the complete phase diagram of drug-polymer solid dispersions. *International Journal of Pharmaceutics, 399*(1–2), 109–115.

Lowe, A., Deng, W., Smith, D. W., & Balkus, K. J. (2014). Coated melt-spun acrylonitrile-based suture for delayed release of nitric oxide. *Materials Letters, 125*, 221–223.

Maddineni, S., Battu, S. K., Morott, J., Soumyajit, M., & Repka, M. A. (2014). Formulation optimization of hot-melt extruded abuse deterrent pellet dosage form utilizing design of experiments. *Journal of Pharmacy and Pharmacology, 66*(2), 309–322.

Marsac, P. J., Li, T., & Taylor, L. S. (2009). Estimation of drug-polymer miscibility and solubility in amorphous solid dispersions using experimentally determined interaction parameters. *Pharmaceutical Research, 26*(1), 139–151.

Matić, J., Paudel, A., Bauer, H., Garcia, R. A. L., Biedrzycka, K., & Khinast, J. G. (2020). Developing HME-based drug products using emerging science: a fast-track roadmap from concept to clinical batch. *AAPS PharmSciTech, 21*(5), 1–8.

Mitragotri, S., Burke, P. A., & Langer, R. (2014). Overcoming the challenges in administering biopharmaceuticals: Formulation and delivery strategies. *Nature Reviews Drug Discovery, 13*(9), 655–672.

Muhindo, D., Ashour, E. A., Almutairi, M., Joshi, P. H., & Repka, M. A. (2021). Continuous production of raloxifene hydrochloride loaded nanostructured lipid carriers using hot-melt extrusion technology. *Journal of Drug Delivery Science and Technology, 65*, 102673.

Muhindo, D., Ashour, E. A., Almutairi, M., & Repka, M. A. (2022). Development and evaluation of raloxifene hydrochloride-loaded subdermal implants using hot-melt extrusion technology. *International Journal of Pharmaceutics, 622*, 121834.

Passik, S. D., Hays, L., Eisner, N., & Kirsh, K. L. (2006). Psychiatric and pain characteristics of prescription drug abusers entering drug rehabilitation. *Journal of Pain and Palliative Care Pharmacotherapy, 20*(2), 5–13.

Patil, H., Kulkarni, V., Majumdar, S., & Repka, M. A. (2014). Continuous manufacturing of solid lipid nanoparticles by hot melt extrusion. *International Journal of Pharmaceutics, 471*(1–2), 153–156.

Patil, H., Tiwari, R. V., & Repka, M. A. (2016). Hot-melt extrusion: from theory to application in pharmaceutical formulation. *AAPS PharmSciTech, 17*(1), 20–42.

Pérez, C., Castellanos, I. J., Costantino, H. R., Al-Azzam, W., & Griebenow, K. (2010). Recent trends in stabilizing protein structure upon encapsulation and release from bioerodible polymers. *Journal of Pharmacy and Pharmacology, 54*(3), 301–313.

Quintavalle, U., Voinovich, D., Perissutti, B., Serdoz, F., Grassi, G., Dal Col, A., & Grassi, M. (2008). Preparation of sustained release co-extrudates by hot-melt extrusion and mathematical modelling of in vitro/in vivo drug release profiles. *European Journal of Pharmaceutical Sciences, 33*(3), 282–293.

Ren, Y., Mei, L., Zhou, L., & Guo, G. (2019). Recent perspectives in hot melt extrusion-based polymeric formulations for drug delivery: Applications and innovations. *AAPS PharmSciTech, 20*(3), 92.

Repka, M. A., Park, J.-B., Tiwari, R. V., Patil, H. G., Jr, Morott, J. T, Lu, W., Upadhye, S. B., & Vladyka, R. S. (2017). *Twin-screw dry granulation for producing solid formulations* (Issue WO2017185040A1). https://patents.google.com/patent/WO2017185040A1/en.

Richter, M. (2017). *Dry granulation as a twin-screw process in pharmaceutical applications.* Thermo Fisher Scientific, Karlsruhe, Germany. files/1525/LR79-dry-granulation-twin-screw-process-pharmaceutical-applications.pdf.

Sarabu, S., Bandari, S., Kallakunta, V. R., Tiwari, R., Patil, H., & Repka, M. A. (2019). An update on the contribution of hot-melt extrusion technology to novel drug delivery in the twenty-first century: part II. *Expert Opinion on Drug Delivery, 16*(6), 567–582.

Scaffaro, R., Botta, L., Sanfilippo, M., Gallo, G., Palazzolo, G., & Puglia, A. M. (2013). Combining in the melt physical and biological properties of poly(caprolactone) and chlorhexidine to obtain antimicrobial surgical monofilaments. *Applied Microbiology and Biotechnology, 97*(1), 99–109.

Schlindwein, W., Bezerra, M., Almeida, J., Berghaus, A., Owen, M., & Muirhead, G. (2018). In-line UV-Vis spectroscopy as a fast-working process analytical technology (PAT) during early phase product development using hot melt extrusion (HME). *Pharmaceutics, 10*(4), 166.

Schram, C. J., Beaudoin, S. P., & Taylor, L. S. (2015). Impact of polymer conformation on the crystal growth inhibition of a poorly water-soluble drug in aqueous solution. *Langmuir, 31*(1), 171–179.

Shadambikar, G., Marathe, S., Ji, N., Almutairi, M., Bandari, S., Zhang, F., Chougule, M., & Repka, M. (2021). Formulation development of itraconazole PEGylated nano-lipid carriers for pulmonary aspergillosis using hot-melt extrusion technology. *International Journal of Pharmaceutics: X, 3*, 100074.

Simões, M. F., Pinto, R. M. A., & Simões, S. (2019). Hot-melt extrusion in the pharmaceutical industry: toward filing a new drug application. *Drug Discovery Today, 24*(9), 1749–1768.

Simões, M. F., Pinto, R. M. A., & Simões, S. (2021). Hot-melt extrusion: a roadmap for product development. *AAPS PharmSciTech*, 22(5), 184.

Stanković, M., de Waard, H., Steendam, R., Hiemstra, C., Zuidema, J., Frijlink, H. W., & Hinrichs, W. L. J. (2013). Low temperature extruded implants based on novel hydrophilic multiblock copolymer for long-term protein delivery. *European Journal of Pharmaceutical Sciences*, 49(4), 578–587.

Stanos, S. P., Bruckenthal, P., & Barkin, R. L. (2012). Strategies to reduce the tampering and subsequent abuse of long-acting opioids: Potential risks and benefits of formulations with physical or pharmacologic deterrents to tampering. *Mayo Clinic Proceedings*, 87(7), 683–694.

Tambe, S., Jain, D., Agarwal, Y., & Amin, P. (2021). Hot-melt extrusion: Highlighting recent advances in pharmaceutical applications. *Journal of Drug Delivery Science and Technology*, 63, 102452.

Tan, D. K., Maniruzzaman, M., & Nokhodchi, A. (2018). Advanced pharmaceutical applications of hot-melt extrusion coupled with fused deposition modelling (FDM) 3D printing for personalised drug delivery. *Pharmaceutics*, 10(4), 203.

Teekamp, N., Olinga, P., Frijlink, H. W., & Hinrichs, W. (2015). Protein stability during hot melt extrusion: The effect of extrusion temperature, hydrophilicity of polymers and sugar glass pre-stabilization. *Uniwersytet Śląski*, 343–354.

Tian, Y., Booth, J., Meehan, E., Jones, D. S., Li, S., & Andrews, G. P. (2013). Construction of drug-polymer thermodynamic phase diagrams using flory-huggins interaction theory: Identifying the relevance of temperature and drug weight fraction to phase separation within solid dispersions. *Molecular Pharmaceutics*, 10(1), 236–248.

Tummalapalli, M., Anjum, S., Kumari, S., & Gupta, B. (2016). Antimicrobial surgical sutures: Recent developments and strategies. *Polymer Reviews*, 56(4), 607–630.

Vaishya, R., Khurana, V., Patel, S., & Mitra, A. K. (2015). Long-term delivery of protein therapeutics. *Expert Opinion on Drug Delivery*, 12(3), 415–440.

Vazquez-Rodriguez, J. A., Shaqour, B., Guarch-Pérez, C., Choińska, E., Riool, M., Verleije, B., Beyers, K., Costantini, V. J. A., Święszkowski, W., Zaat, S. A. J., Cos, P., Felici, A., & Ferrari, L. (2022). A Niclosamide-releasing hot-melt extruded catheter prevents Staphylococcus aureus experimental biomaterial-associated infection. *Scientific Reports*, 12(1), 12329.

Verstraete, G., Samaro, A., Grymonpré, W., Vanhoorne, V., van Snick, B., Boone, M. N., Hellemans, T., van Hoorebeke, L., Remon, J. P., & Vervaet, C. (2018). 3D printing of high drug loaded dosage forms using thermoplastic polyurethanes. *International Journal of Pharmaceutics*, 536(1), 318–325.

Vynckier, A.-K., Dierickx, L., Saerens, L., Voorspoels, J., Gonnissen, Y., de Beer, T., Vervaet, C., & Remon, J. P. (2014a). Hot-melt co-extrusion for the production of fixed-dose combination products with a controlled release ethylcellulose matrix core. *International Journal of Pharmaceutics*, 464(1–2), 65–74.

Vynckier, A.-K., Dierickx, L., Voorspoels, J., Gonnissen, Y., Remon, J. P., & Vervaet, C. (2014b). Hot-melt co-extrusion: requirements, challenges and opportunities for pharmaceutical applications. *The Journal of Pharmacy and Pharmacology*, 66(2), 167–179.

Warren, D. B., Benameur, H., Porter, C. J. H., & Pouton, C. W. (2010). Using polymeric precipitation inhibitors to improve the absorption of poorly water-soluble drugs: A mechanistic basis for utility. *Journal of Drug Targeting*, 18(10), 704–731.

Wesholowski, J., Prill, S., Berghaus, A., & Thommes, M. (2018). Inline UV/Vis spectroscopy as PAT tool for hot-melt extrusion. *Drug Delivery and Translational Research*, 8(6), 1595–1603.

Woolfson, A. D., Malcolm, R. K., & Gallagher, R. J. (2003). Design of a silicone reservoir intravaginal ring for the delivery of oxybutynin. *Journal of Controlled Release*, 91(3), 465–476.

Xu, X., Siddiqui, A., Srinivasan, C., Mohammad, A., Rahman, Z., Korang-Yeboah, M., Feng, X., Khan, M., & Ashraf, M. (2019). Evaluation of Abuse-Deterrent Characteristics of Tablets Prepared via Hot-Melt Extrusion. *AAPS PharmSciTech*, 20(6), 1–11.

Ye, X., Patil, H., Feng, X., Tiwari, R. V., Lu, J., Gryczke, A., Kolter, K., Langley, N., Majumdar, S., Neupane, D., Mishra, S. R., & Repka, M. A. (2016). Conjugation of Hot-Melt Extrusion with High-Pressure Homogenization: a Novel Method of Continuously Preparing Nanocrystal Solid Dispersions. *AAPS PharmSciTech*, 17(1), 78–88.

Zhang, J., Feng, X., Patil, H., Tiwari, R. V., & Repka, M. A. (2017a). Coupling 3D printing with hot-melt extrusion to produce controlled-release tablets. *International Journal of Pharmaceutics*, 519(1–2), 186–197.

Zhang, J., Yang, W., Vo, A. Q., Feng, X., Ye, X., Kim, D. W., & Repka, M. A. (2017b). Hydroxypropyl methylcellulose-based controlled release dosage by melt extrusion and 3D printing: Structure and drug release correlation. *Carbohydrate Polymers*, 177, 49–57.

Zhao, Y., Inbar, P., Chokshi, H. P., Malick, A. W., & Choi, D. S. (2011). Prediction of the thermal phase diagram of amorphous solid dispersions by flory-huggins theory. *Journal of Pharmaceutical Sciences*, 100(8), 3196–3207.

Zheng, Y., & Pokorski, J. K. (2021). Hot melt extrusion: An emerging manufacturing method for slow and sustained protein delivery. *WIREs Nanomedicine and Nanobiotechnology*, 13(5), e1712.

3 3D Printing Technologies for Personalized Drug Delivery

Aytug Kara, Jose R. Cerda, Iván Yuste, Francis C. Luciano,
Brayan J. Anaya, and Dolores R. Serrano
Complutense University of Madrid

CONTENTS

3.1 INTRODUCTION: UNLOCKING THE POTENTIAL OF 3D PRINTING TECHNOLOGIES IN THE MANUFACTURING OF PERSONALIZED DRUG DELIVERY SYSTEMS

The current society is moving from the pharmacological concept of "One-size-fits-all" toward personalized therapies adapted to patients' needs. Science is moving forward, and along with new knowledge, it has appeared as the need to titrate the doses to find the optimal balance between efficacy and toxicity. This is especially important in pediatric and geriatric populations where the physiological needs are distinctively different from the dose administered to adults (Figure 3.1). Also, innovative medicines are needed to target diseases that before did not have a clinical solution [1,2].

3D printing technologies have emerged in the last decade as a promising approach to manufacturing personalized drug delivery systems. A large window has been opened in this sense along with exponential growth of interest in this field. For example, the number of articles published in PUBMED in 2012 regarding 3D printing of medicines was just thirteen while last year, in 2021, over 1500 manuscripts were published on this topic. Nevertheless, the implementation of 3D-printed medicines in hospitals and community pharmacies is slowly growing. The main reason behind this resistance is the lack of standardization, the need for specialized trained healthcare professionals along with the necessity for more user-friendly 3D printers and software interfaces.

There is a wide range of 3D printing technologies available; however, there are just a handful of techniques that have shown greater feasibility in the manufacturing of medicines. In this chapter, two of the main technologies that have shown the greatest applicability will be discussed in more depth to understand which are the main requirements to implement these techniques in clinical practice, which are the main limitations, and how these challenges can be overcome. These techniques can be grouped as (1) semisolid extrusion-based 3D printing techniques including pressure-assisted microsyringes (PAM) and fused deposition modeling (FDM) and (2) laser-based systems including selective laser sintering (SLS) and stereolithography (SLA) [3]. The application of these 3D printing technologies in the development of oral, parenteral, topical, and transdermal dosage forms will be discussed. Also, briefly, it will illustrate the potential of these technologies to engineer microfluidic chips with multiple applications in pharmacy and medicine such as nanomedicine manufacturing or organ-on-chip.

DOI: 10.1201/9781003224464-3

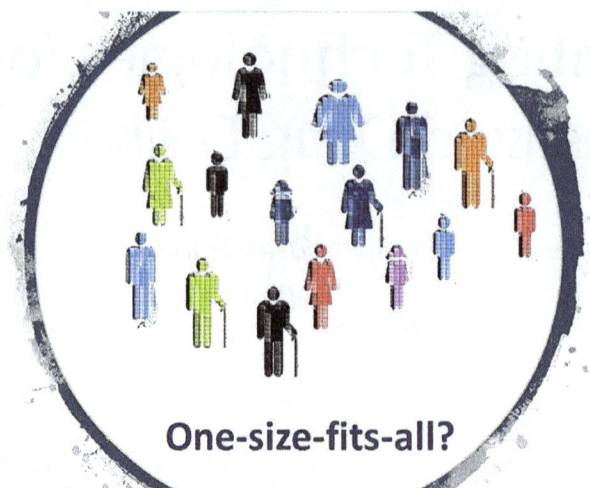

FIGURE 3.1   Does one-size-pill fit all the requirements of the population?

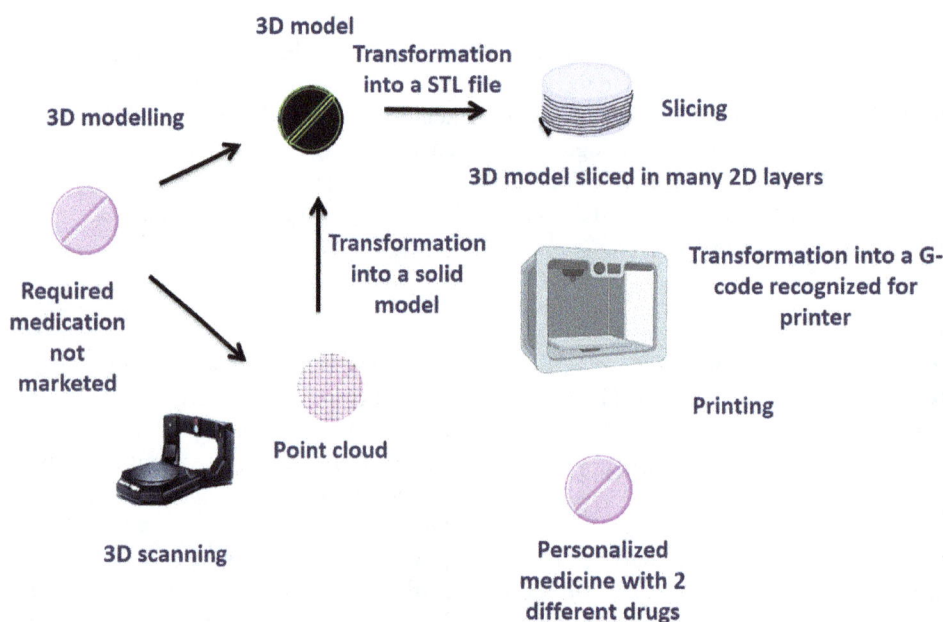

FIGURE 3.2   Schematic representation of the 3D printing process of a tablet.

3.2   UNDERSTANDING HOW 3D PRINTING TECHNOLOGIES WORK WHEN FABRICATING PERSONALIZED MEDICINES AND WHICH CHALLENGES SHOULD BE OVERCOME FOR THEIR IMPLEMENTATION IN CLINICAL PRACTICE

Before the printing process itself, it is necessary to bear in mind a few steps such as the geometrical design of the 3D printed solid dosage form, the slicing, and finally, the creation of the g-code file recognized for the printer (Figure 3.2). The geometrical design is key as we need to ensure that all the required dose is contained within the volume of the dosage form we have created. This is simple at a first sight but it can be an extra challenge to overcome

when implementing it into clinical practice. 3D printing is a volume-centered manufacturing technique. The dose that a medicine contains has to be transformed in volume taking into account the density of the materials we are working with. Based on this calculation and using the appropriate software, the exact dimensions of the solid dosage form should be engineered. There is much software available to create geometrical structures that are useful for pharmacists such as Tinkercad, AutoCAD, Solidwork, Mesh mixer, Autofusion, Rhinoceros, and many others. Also, 3D scanners can be used to extract the dimensions of commercially available solid dosage forms leading to a cloud point that allows its transformation into a solid model. Upon the 3D model has been created, most printers have their own slicing software that allows to input the instructions required during printing, such as height of each layer, printing and travel speed, type of infilling, the temperature of the process in the semisolid extrusion-based systems or potency of the laser in the laser-based 3D printing techniques. Also, there is free software to perform the slicing process, such as Cura, one of the most commonly utilized ones. After the slicing, a G-code file is generated and contains all the information necessary for the printing of the drug delivery system.

3.2.1 Semisolid-Based 3D Printing Technologies

Among the semisolid-based 3D printing techniques, FDM is one of the most commonly utilized systems (Figure 3.3a). Before printing, a filament with an adequate diameter (usually 1.75 or 3 mm) should be manufactured. The filament contains a mixture of the drug with suitable excipients that confer the filament with the physicochemical characteristics required for printing such as enough tensile strength and hardness. The filament is fed up through a nozzle head that heats the material above its glass transition to make it extrudable through the orifice of the nozzle. The material is then deposited on the platform at the exact coordinates according to the G-code file developed previously, and the extruded material rapidly cools down maintaining its structure upon deposition on the platform. Several layers are deposited sequentially, and the 3D printed dosage form is created. It can be considered a green technology as is solvent-free, and the waste generated is minimal [4]. However, the major two challenges to overcome are the manufacturing of suitable filaments for printing and the applicability of this technique with thermolabile drugs.

When fabricating filaments for FDM, the drug is usually mixed with cellulose-based polymers such as hydroxypropyl methylcellulose (HPMC), hydroxypropyl cellulose (HPC), ethylcellulose (EC), and other biodegradable polymers such as polyvinylpyrrolidone (PVP), polycaprolactone (PCL), polyvinyl alcohol (PVA), polylactic acid (PLA) and copolymers such as Soluplus (polyethylene glycol-PCL-PVA), Kollidon (PVP-PVA), and Resomer (Poly(lactic-co-glycolic acid)), among others. The addition of polyethylene glycol (PEG) as a plasticizer (normally between 5% and 20% weight ratio) is highly recommended to make the filament flexible to some extent. Also, PEGs have a low glass transition temperature (well below 0°C) bringing down the glass transition of the mixture, and hence, the printing temperature can be also reduced. The fabrication of filaments can be performed on hot-melt extruders. The main inconvenience is the large batch size required (at least 50 g of powder mixture) which limits its use with expensive drugs and also, the complexity of producing high-quality filaments as they are often too brittle causing blockages on the printer head. To overcome this issue, novel 3D printers have been engineered by incorporating a small hot-melt extruder device before the nozzle head extruder, which allows us to load in the printer directly the powder mixture that we want to print rather than necessitating the fabrication of a filament in a previous step. This process can be denominated as direct powder extrusion, and it has been successfully implemented at a laboratory scale [5,6]. 3D printers that allow direct powder extrusion are more costly than basic FDM printers. The price range varies from a few thousand euros (eg. Hyrel 3D printers) to more innovative and costly printers that allow working under Good Manufacturing Practice (GMP) protocols, such as M3DIMAKER [7].

The PAM technique is also commonly employed in the manufacturing of drug delivery systems (Figure 3.3b). This technique necessitates a semisolid formulation as starting material, and hence, a solid filament is not required making

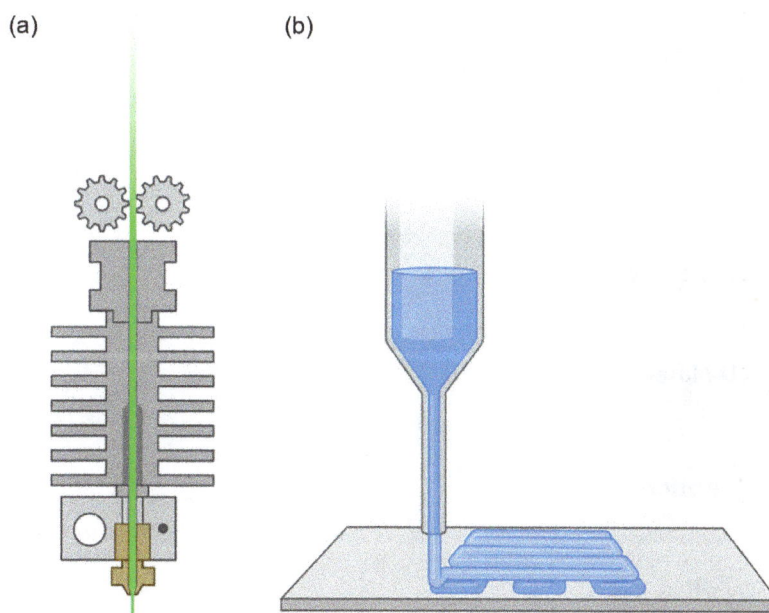

FIGURE 3.3 Semisolid-based 3D printing techniques. (a) FDM, (b) PAM.

the process easier. This semisolid formulation is extruded by pressurized air through the nozzle of the printer by a gear system like in the FDM technique that guides and forces the filament to enter the extruder nozzle head. The material is already in the semisolid state, and hence, heating will be minimal as pressure only needs to be applied to deform the material plastically; hence, this technique becomes more useful for thermolabile drugs. Also, high drug content formulations can be easily printed which is interesting, especially for those medicines that require a high drug dose to be administered. The viscosity is the key factor to consider when printing with this technique. Too high viscosity will lead to nozzle clogging, while too low viscosity will result in a 3D-printed structure unable to maintain its integrity and its 3D structure. Unlike FDM, most formulations require the use of solvents. For this reason, PAM is not considered green technology and makes necessary a drying step post-printing which may last up to 24–48h making the overall process extremely slow. Also, according to the European Pharmacopoeia, the residual solvent content must be determined as left-over solvent is especially critical for the pediatric population [8].

3.2.2 Laser-Based 3D Printing Technologies

Regarding laser-based 3D printing technologies, both SLA and SLS have gained importance over the last few years which is related to their greatest resolution achieved during printing reaching up to 10–25 μm layer height and around 0.5% error on dimension accuracy. In the case of bottom-up SLA printers, the build platform is positioned in the tank that contains the photopolymerizable resin mixed with the drug at a distance equivalent to one layer height for the surface of the liquid. The UV laser selectively cures the resin in those coordinates executed in the G-code file and is located under the resin tank which has a silicon coating layer allowing the laser to pass through but avoiding the cured resin from sticking to it. The light of the UV laser activates the monomer carbon chains of the liquid resin creating strong bonds between each other which leads to the solidification of the resin. The laser beam is focused on a predetermined path using a set of mirrors. After solidification of each layer, the build platform moves up the distance corresponding to the next layer height, and the cured resin is detached from the bottom of the tank as the build platform moves upward, known as the peeling step. The process is repeated till the whole 3D structure is printed (Figure 3.4a). In the top-down SLA printer, the UV light is located above the tank, and the build platform moves downward after every layer is cured.

A post-printing step is commonly required, consisting in washing with usually isopropanol alcohol and curing the 3D

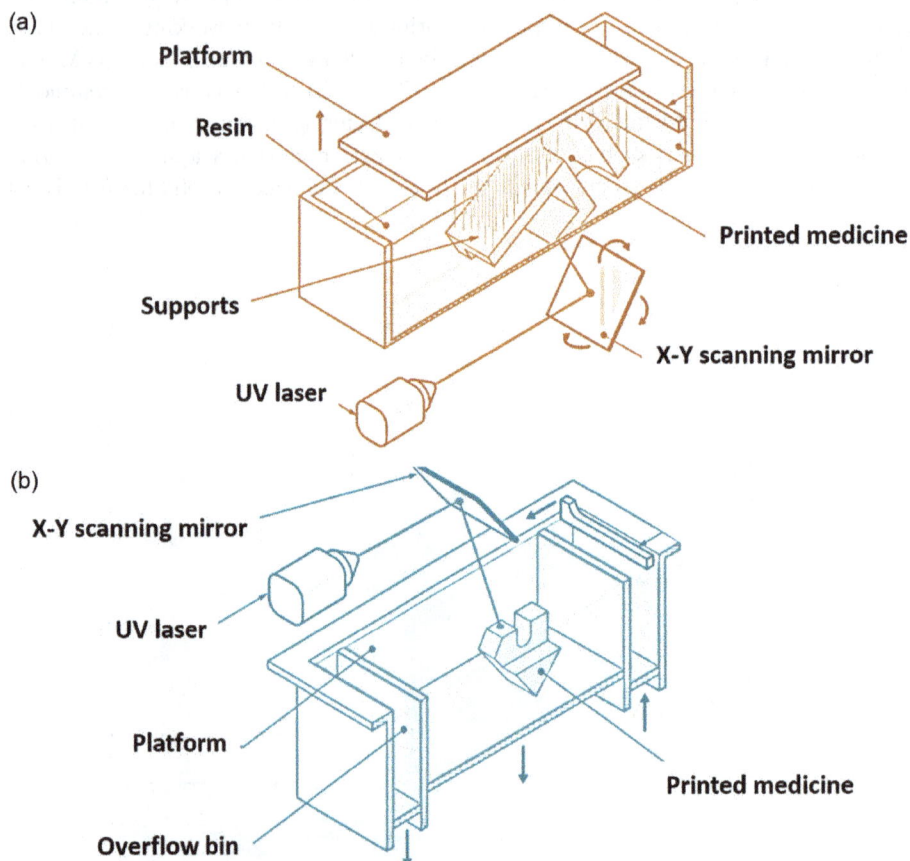

FIGURE 3.4 Laser-based 3D printing techniques. (a) SLA, (b) SLS.

object to remove any residue of unpolymerized resin which limits its pharmaceutical application [1,9]. The printing speed of SLA printers has improved with the development of digital light processing (DLP). The UV light source in the SLA printers is based on a point laser, while the source of light is a digital projector in the DLP printers which significantly reduces the time of printing as all at once is printed the entire layer [10]. Apart from the high resolution, SLA and DLP technologies can print medicines at low temperatures being suitable with thermolabile drugs, but certain drugs can degrade with UV light; this is why the potency of the laser and time exposure are critical factors during the process. However, the main challenge to overcome with this technology is the fact that most resins are made of photocrosslinkable polymers that are cytotoxic. Even though less toxic photocrosslinkable resins have been developed, such as polyethylene glycol diacrylate (PEGDA), further toxicological studies have to be implemented to ensure its safety in humans, especially after chronic administration [10].

One challenge to overcome when using SLA printers is the curling that occurs during the curing process of the resin that slightly shrinks upon exposure to UV light. This can provoke internal stress between the next layer and the previously solidified material resulting in the part curling [11]. Similarly, warping can occur during FDM printing because when the extruded material cools down, its dimensions can decrease. Differential cooling causes internal stress in the structure that pulls the underlying layer upward, causing it to wrap [12].

The second laser-based technology commonly employed for manufacturing drug delivery systems is known as SLS, which uses a powder bed to build up the 3D-printed medicine. Unlike powder bed printers in which a spray solution is utilized to bind particles together, SLS makes use of a laser directed to draw a specific pattern onto the surface of the powder bed that contains our drug along with excipients (Figure 3.4b). After the first layer is completed, a roller distributes a new layer of powder on top of the previous one. This process is repeated several times while the 3D object is printed. SLS is also a green technology as is a solvent-free technique. In terms of speed, it is a faster technique than powder bed or SLA as no post-printing step is required. There is no need for the solvent to be evaporated or for the resin to be washed out and cured. SLS also has a good resolution as SLA. However, the harsh printing conditions can degrade drugs, and low potency SLS printers have been developed to allow their implementation in the pharmaceutical field [13].

3.3 APPLICATION OF 3D PRINTING TECHNOLOGIES IN THE DEVELOPMENT OF ORAL SOLID DOSAGE FORMS: POLYPILLS

The first marketed drug that used 3D technology was Spritam® in 2016. It contains levetiracetam, an antiepileptic that is administered under the tongue, and in less than 4 s is dispersed in the mouth and starts to get absorbed, allowing to stop epileptic seizures before they become severe. It contains one gram of active ingredient, which is very high. This medicine is only able to be manufactured using a 3D technology patented system by the Apprecia Pharmaceuticals laboratory called ZipDose, which is based on binder injection. First, a layer of the powder mixture is deposited, which contains the active ingredient and the excipients, and then a binding liquid is added so that the powder molecules bind together and prepare to adhere to the next layer; This process is repeated several times, obtaining a solid form with high porosity, which allows being an orodispersible tablet and in a matter of seconds getting disintegrated and available for absorption.

3D printing techniques have been extensively used for the manufacture of solid dosage forms. Many innovations have been introduced in this field such as geometrical shapes that cannot be manufactured using conventional industrial techniques such as channeled mini-tablets to enhance drug release [14], a combination of immediate and sustained drug release profiles within the same dosage form [15], solid dosage forms transformed into 3D-printed gummies to facilitate the swallowing process in children and elderly population [16], 3D-printed tablets with Braile and Moon patterns for visually impaired patients [17] and QR-encoded solid dosage forms to avoid counterfeiting [18], among others.

3D printing of polypills has gained importance over the last years taking into account the rising numbers of polymedicated patients. Low patient compliance is related to the difficulty of following complex therapeutic regimens that can involve the intake of more than 5–6 pills daily. This is especially important in the treatment of the metabolic triad: obesity, hypertension, and diabetes, but also in the management of infectious diseases and mental health issues.

When 3D printing solid dosage forms, two different approaches can be utilized [19]. The first approach is called the sandwich strategy or in layers. It is characterized by printing a consecutive number of layers, and for each one, a different drug-polymer mixture is incorporated. The upper and bottom layers will be more exposed to the physiological fluids of the gastrointestinal tract, and hence, it is recommended to position those drugs in which an immediate-release profile is desired. However, the central layers are reserved for delayed or sustained-release drugs. This technique is easier to implement when FDM and PAM printers possess just a single nozzle head in which the pharmaceutical ink is changed after every layer has been printed. Similarly, this technique has been applied in SLA printers, in which the resin of the tank is replaced once each layer is printed. The second 3D printing strategy is denominated as the segmented approach. In this case, different drugs and excipients are combined in each layer modulating the final release profile. This strategy is slightly more difficult to implement, but a successful performance has been achieved with printers that possess two nozzle head extruders that can print simultaneously in the same plane. A schematic representation of both strategies is illustrated in Figure 3.5.

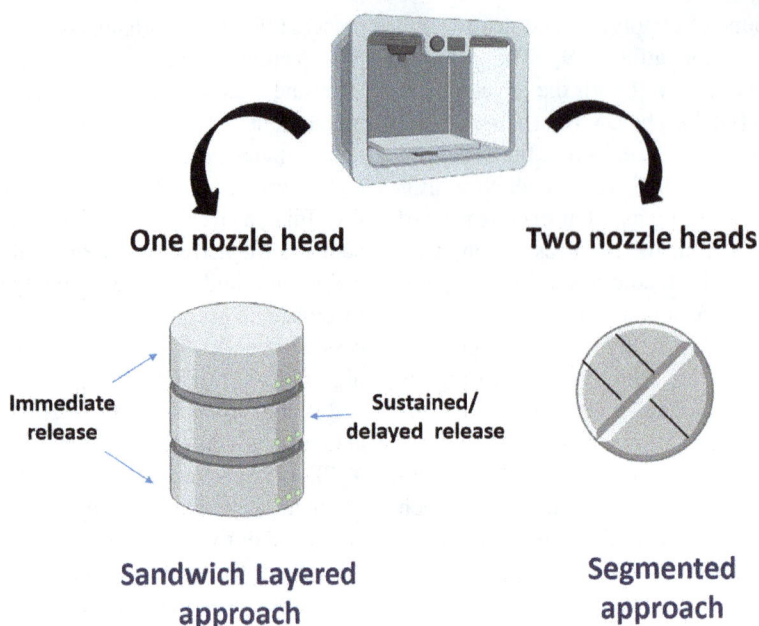

FIGURE 3.5   Different approaches for polypill 3D printing.

In Table 3.1, examples of different 3D printed polypills are illustrated. A three-drug polypill has been manufactured using PAM for patients with type 2 diabetes and high blood pressure. This polypill is made up of captopril (angiotensin-converting enzyme, ACE inhibitor) with a zero-order prolonged release due to the incorporation of mannitol as an osmotic agent, glipizide, a hypoglycemic drug, and nifedipine, a calcium antagonist [20]. The zero order is very useful for controlling blood pressure levels over long periods. Glipizide and nifedipine are embedded in a hydrophilic matrix of hydroxypropyl methylcellulose (HPMC) that allows first-order release by diffusion, unlike captopril. A separating layer between both compartments with different kinetic releases was made of croscarmellose sodium and sodium starch glycolate as disintegrants, polyvinylpyrrolidone K30 as a binder, and mannitol as a diluent. This allowed that upon oral administration each compartment behaves as an individual tablet. The main drawback of this polypill is that DMSO (dimethyl sulfoxide) was used as a solvent, to avoid blockage of the nozzles in the printing process. Even though DMSO is not found in toxic concentrations, its chronic administration could potentially result in toxicity to the patient's health.

Similarly, a five-in-one-dose combination polypill has been successfully printed using PAM [15]. A combined sandwich and segmented approach was followed as a four-nozzle head printer was used. The tablet was consisting of two layers: the immediate and the sustained release. The immediate-release layer contained aspirin as an antiplatelet and hydrochlorothiazide as a diuretic. Sodium starch glycolate and polyvinylpyrrolidone K30 were used as a disintegrant and a binder, respectively, to form the paste necessary for PAM extrusion. The

sustained-release layer was physically separated from the previously mentioned layer by a hydrophobic cellulose acetate shell acting as a permeable barrier including also mannitol as a filler and PEG as a plasticizer. This layer contained atenolol as a beta-blocker, ramipril as an ACE inhibitor, and pravastatin as a 3-hydroxy-3-methylglutaryl–coenzyme A (HMG-CoA) reductase inhibitor to reduce cholesterol and triglycerides blood levels mixed with HPMC 2208 to form a hydrophilic matrix and lactose as a filler.

Pluronic F127, a thermoreversible gel agent, has been used to manufacture polypills for patients with hyperglycemia. A gel matrix was created including metformin, acarbose, and glyburide also known as glibenclamide, combined with Pluronic F127 [21]. The drugs were first dissolved in deionized water at 4°C and then mixed with Pluronic F127 at a concentration of 30%. Then, solutions were sonicated to remove bubbles and transferred to the PAM printer to induce gelation at room temperature before printing at room temperature. Drugs were printed in a cylindrical shape by utilizing a core-shell, multilayer, or gradient distribution. Based on the printed structure, the release profile can be controlled. The core-shell structure resulted in a delayed release of the drug located in the inner section of the tablet. The multilayer system resulted in a pulsed release over time depending on the drug that was exposed to the media while the gradient structure allowed a sustained release over a 5 h period.

Apart from medicated polypills, poly-vitamin supplements have also been successfully printed using a PAM printer [22]. The polypill had two different compartments, a sustained-release one containing caffeine encapsulated within an outer shell compartment made up of an

TABLE 3.1
3D Printed Polypills

3D Printing Technique	Disease	Drug	Excipients	Printing Characteristics	References
PAM	Hypertension	Captopril Nifedipine Glipizide	HPMC CA PEG 6000	Regen HU printer with 4 nozzle heads 400 μm print tip Vacuum dryer at 40°C for 24 h for complete drying 25°C for printing	[20]
	Hypertension Dyslipidemia Thrombosis	Aspirin Hydrochlorothiazide Pravastatin Atenolol Ramipril	HPMC CA PEG 6000 Mannitol SG PVPK30 Lactose	Regen HU printer with four-nozzle heads 400 μm print tip Vacuum dryer at 40°C for 24 h for complete drying 25°C for printing	[15]
	Hyperglycemia	Metformin Acarbose Glibenclamide	Pluronic® F-127	Dual extrusion F5200N.1 PAM printer Printing pressure of 12 psi and 38–46 psi for gradient pills Build time: 90 min. Vacuum and UV light for 12 h to enhance consistency	[21]
	Lack of vitamins	Caffeine Vitamin B1 Vitamin B3 Vitamin B6	Craft Blend R30M Craft Blend R4H	BioX from Cellink Commercial resins to achieve immediate or sustained release 8 h post-processing to remove solvents	[22]
FDM	Hypertension Dyslipidemia	Lisinopril Indapamida Rosuvastatina Amlodipino	PVA Sorbitol	Marketbot 2 nozzle heads 100% infill Layer thickness of 166 μm 150°C for printing	[24]
	HIV	Ritonavir Lopinavir	HPMCAS PEG MgSt	Direct powder extrusion (Hyrel printer) 80°C for printing Layer thickness of 100 μm 1000 μm nozzle head	[5]
	Parkinson	Levodopa Benserazide Pramipexole	PVA Mannitol EVA PVP-VA FS	Hot-melt extrusion followed by FDM Prusa i3 MK3 printer with a single nozzle 185°C for PVA printing 220°C for EVA and PVP-VA printing. Sustained release over 24 h	[23]
SLA	Inflammation	Paracetamol Acetyl salicylic acid Naproxen Cloramphenicol Caffeine Prednisolone	PEGDa PEG 300 TPO	Form 1+ SLA 3D printer 25°C for printing 405 nm laser Layer thickness of 300 μm	[25]

Key: HPMC, Hydroxypropyl methylcellulose; CA, Acetate cellulose; PEG, polyethylene glycol; PVA, polyvinyl alcohol; PEGDa, polyethylene glycol diacrylate; PVPK30, polyvinylpyrrolidone K30; SG, Sodium starch glycolate; HPMCAS, Hydroxypropyl methylcellulose acetate succinate; MgSt, Magnesium stearate; TPO, thermoplastic polyolefin used as photoinitiator; EVA, ethylene-vinyl acetate copolymer (82:18, w:w); PVP-VA, vinylpyrrolidone-vinyl acetate copolymer 60:40; FS, fumed silica.

immediate-release formulation of vitamins B1, B3, and B6. The immediate-release compartment was created by mixing the vitamins with a commercially available gel (Craft Bland R30M) that consists of a mixture of disintegrants and binders leading to a 30 min dissolution time. The sustained-release compartment was formulated with Craft Blend R4H forming a gel matrix that maintained integrity over a 4 h period. The disadvantage of this process was that an 8 h post-processing step was required to remove the solvents included in the commercial inks.

Other 3D-printed polypills manufactured with FDM extrusion contained lisinopril and amlodipine for hypertension, indapamide as a diuretic, and rosuvastatin for dyslipidemia. Filaments were manufactured using a hot-melt extruder using PVA and sorbitol as excipients. A novel approach was utilized consisting of using distilled water as a temporary co-plasticizer to reduce the extrusion and printing process from 170°C to 90°C and 210°C to 150°C, respectively. The printing was used with a Marketbot 2 printer with two printer heads allowing the consecutive printing of the first two layers, followed by a thorough cleaning before the printing of the third and fourth layers started.

Mini-tablets have been designed to ensure ease of swallowing for children containing ritonavir and lopinavir for the treatment of HIV [5]. The printing technique used was direct powder extrusion using an FDM printer which has a directly connected mini-extruder in which the powder mixture consisting of active ingredients and excipients was loaded without the need to previously manufacture a filament. These tablets were previously fabricated by hot-melt extrusion followed by a conventional FDM printer. However, a significant drug degradation (> 30%) was observed at the temperature required for extrusion at 120°C. Direct powder extrusion allowed a successful printing for both drugs as the printing temperature was reduced to 80°C. Hydroxypropyl methylcellulose acetate succinate (HMPCAS) combined with PEG 4000 as a plasticizer and magnesium stearate as a lubricant were utilized to create a sustained zero-order drug release matrix over a 24 h period.

Polypills for Parkinson's disease are also been developed using FDM [23]. Three drugs are incorporated: levodopa in combination with benserazide, a dopa decarboxylase inhibitor, supplemented with pramipexole, a dopamine agonist. Hot-melt extrusion was used to manufacture two different composition filaments. Pramipexole was incorporated in a rapid release matrix that contained polyvinyl alcohol, mannitol as plasticizer, and fume silica as glidant while levodopa and benserazide were extruded along with 34% ethylene-vinyl acetate copolymer (82:18, w:w), 15% vinylpyrrolidone-vinyl acetate copolymer 60:40 (PVP-VA), and 0.5% fume silica for prolonged drug release. Levodopa is absorbed in the upper gastrointestinal tract. For this reason, the polypill was designed as mini-tablets and mini-hollow cylinders to float in the stomach for prolonged periods resulting in a sustained release over 24 h.

Using the stereolithography (SLA) technique, the manufacture of polypills has also been feasible. For example, a polypill was printed containing six different drugs: paracetamol (antipyretic and analgesic), caffeine, naproxen (non-steroidal anti-inflammatory, NSAID), chloramphenicol (broad-spectrum antibiotic), prednisolone (anti-inflammatory corticosteroid), and aspirin (NSAID). The combination of chloramphenicol and prednisolone is commonly used in inflammatory infections of the conjunctiva and cornea such as conjunctivitis, keratoconjunctivitis, or blepharoconjunctivitis. The rest of the components of the polypill were added to increase the efficacy of the anti-inflammatory action [25]. PEGDa was chosen as a photopolymerizable monomer and thermoplastic polyolefin (TPO) as the photoinitiator. The formulations were prepared by dissolving the drug and photoinitiator in liquid PEGDa and PEG 300 Da. Once the dissolution was complete, the mixture was poured into the platform or tray of resin for printing. Three different types of polypills were manufactured: (I) cylindrical, (II) ring shape, and (III) ring shape with PEG300 filler.

To be able to print with multiple drugs sequentially, an SLA printer is required to replace the resin contained in the tank after the layer containing a specific drug has been printed. In this way, multiple pills can be manufactured in about 30 min by making changes to the print settings. In this work, the effect of geometry was evaluated using release tests for the three types of manufactured polypills. In the case of the cylinder-shaped type I polypill, none of the drugs reached 100% release at 20 h. Therefore, to increase the dissolution rate, the type II polypill was designed in the shape of a ring and thus with increased surface area exposed to the dissolution medium. The results obtained showed an increase in the release rate of water-soluble drugs, while drugs with low water solubility located did not show significant changes compared to type I [8]. In this respect, the type III geometry was designed to understand the effect of a soluble filler such as PEG300. With this design, a substantial release increase was observed compared to type I. Paracetamol, caffeine, and aspirin reached 100% release at 20 h, and chloramphenicol increased to more than 80% while Prednisolone did not increase to more than 45% [25]. This demonstrates the complexity of being able to adjust the desired release profile of oral solid dosage forms using 3D printing techniques.

3.4 APPLICATION OF 3D PRINTING TECHNOLOGIES IN THE DEVELOPMENT OF PARENTERAL DOSAGE FORMS

3D printing has also been applied in the fabrication of parenteral applications, and although several studies have been performed over the last few years, there is a lack of literature concerning this topic [26]. Parenteral dosage forms can be defined as sterile drug products, which can be presented as solutions, dispersions, suspensions, emulsions, reconstituted lyophilized powder, implants (including nano and microparticles), and a combination of drug device. Most common administration routes can be subcutaneous, intramuscular, or intravenous, although intrathecal, intracisternal, intraspinal, intraepidural, and intradermal can also be used [27,28].

The above-described 3D printing technologies can be used for the development of parenteral dosage forms but depending on the objectives and specific requirements several considerations must be taken into account for selecting the proper one [26]. Additionally, metal printers have been extensively used to manufacture prostheses and dental implants adapted to patients' needs [29]. The major benefit of 3D printing technology for parenteral dosage forms is the possibility of creating personalized products with precise control over dimension and microstructure, being able to apply them for different purposes, such as tissues and organs [26]. 3D-printed microfluid chips to mimic the *in vivo* tissue microenvironment have also been developed, but this topic is out of the scope of this chapter [30].

Several considerations should be born in mind when choosing the printing material for parenteral constructs: (1) being biocompatible for minimizing immune response in the body; (2) providing suitable mechanical properties when required such as prosthesis; (3) ability to promote cell adhesion and proliferation (bone and tissue scaffolds); and (4) sterility.

The most commonly utilized polymers for 3D printing parenteral implants can be divided into the following categories: those used for semisolid extrusion or those utilized for stereolitography (Figure 3.6). Polylactic acid (PLA), polyglycolic acid (PGA), polyvinyl alcohol (PVA), polycaprolactone (PCL), and their copolymers are the most well-known synthetic polymers for such applications. They have the advantage compared to natural polymers in that their degradation time in the body is much longer, lasting up to 36 months. For example, in the PLGA copolymer, depending on the ratio of polylactic and polyglycolic groups, the degradation time is very different from 1 to 2 months when the ratio is 50:50 or above a year in the extremes (with a higher percentage of PLA or a higher percentage of PGA). In general, the tensile strength of these polymers is optimal for FDM printing, while natural polymers tend to be utilized in PAM printers resulting in 3D-printed materials with less rigidity and a much faster degradation time, normally below a month [26]. Several photocross-linkable synthetic polymers are also used in

SLA printers for parenteral applications, highlighting those derived from methacrylates (PEGDA, PEGDMA, and GELMA) and popypropylene fumarate as they have shown reduced toxicity compared to other synthetic materials. It is important in the post-processing step, in which the 3D structure is washed and cured for several hours to remove any trace of unreacted resin that triggers toxicity in the human body.

3D printing has been successfully used to manufacture parenteral dosage forms such as long-acting implants and biomedical devices such as stents. In Table 3.2, the most innovative applications are summarized.

3D-printed long-acting implants have been developed to precisely control drug release possessing a morphology and microstructure well-adapted to patients' needs. A wide range of drugs such as hormones, cytostatics, anesthetics, and antimicrobials have been evaluated [31–34]. 3D-printed rod-shaped implants made of PVA and PLA were fabricated with FDM followed by dip coating with a PCL polymer mixture. The rods were designed with gaps or windows in their surface able to module drug release. The smaller the number and the window size, the slower the drug release which lasted up to 300 days [34]. Personalized vaginal rings were 3D printed with FDM with different morphological shapes, O, Y and M, to minimize pelvic inflammatory diseases and uterine perforations. Rings were made of PLA: PCL (8:2 w:w) and Tween 80 loaded with progesterone, and release was prolonged for up to 7 days [35]. 3D-printed scaffolds for breast cancer have been fabricated using PLGA in

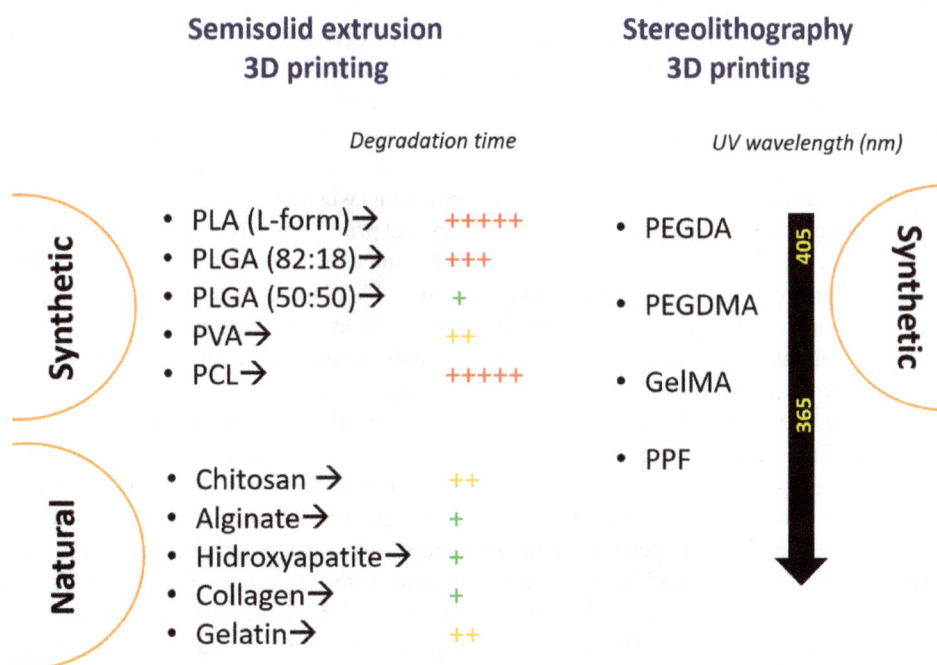

FIGURE 3.6 Most commonly used polymers in the fabrication of parenteral solid dosage forms. Key: PLA, poly(lactic acid); PLGA, poly(lactic-co-glycolic acid); PCL, polycaprolactone; PVA, poly(vinyl alcohol); CS, chitosan; HA, hyaluronic acid, PEGDA, poly(ethylene glycol) diacrylate; PEGDMA, poly(ethylene glycol) methacrylate; GelMA, gelatin methacrylate; PPF, polypropylene fumarate. Degradation time is depicted with a + symbol.

TABLE 3.2

Examples of 3D-Printed Parenteral Implants and Stents

Parenteral Dosage Form	Polymer /Drug	3D Printing Technique	Applications	References
Long-acting implants	PVA PLA PCL	FDM	Rod-shaped implant with different windows size on the surface able to modulate drug release up to 300 days	[34]
	PLA PCL Tween 80 Progesterone	FDM	Vaginal rings with different shapes to minimize pelvic inflammation and control progesterone release for up to 7 days	[35]
	PLGA Doxorubicin Cisplatin	Customized E-jet printer	Scaffolds with 200 μm allowing drug release over 30 days and enhancing antitumoral efficacy	[36]
	Alginate Dexamethasone PCL Bevacizumab	PAM	Rod-shaped implant with a core made of alginate and dexamethasone release in 7 days and a shell made of PCL and bevacizuman release in 60 days. Angiogenesis was suppressed in one month	[37]
Stents	PLA PCL	FDM	PLA core and PCL shell 1400 MPa Young's modulus 3% degradation over 6 weeks	[39]
	PCL Amoxicillin Cefotaxime	FDM	Stent for salivary duct obstruction made of PCL and loaded with amoxicillin and cefotaxime which were released within 28 and 3 days, respectively, upon implantation	[40]

Key: PLA, poly(lactic acid); PLGA, poly(lactic-co-glycolic acid); PCL, polycaprolactone; PVA, poly(vinyl alcohol).

combination with doxorubicin and cisplatin with a pore size above 200 μm allowing a sustained release of both cytostatic drugs over 30 days and eliciting a superior efficacy to conventional intravenous treatment [36]. 3D-printed core-shell rods have been developed for the treatment of retinal vascular diseases. Rods consisted of a core made of alginate and dexamethasone coated by a shell of PCL and bevacizumab. Dexamethasone was released over 1 week while bevacizumab showed a sustained release over 60 days. This drug combination showed angiogenesis suppression in an *in vivo* model over 1 month period [37].

The feasibility of 3D-printed stents has been widely demonstrated providing advantages over conventional methods such as laser-cutting manufacturing. The latter can damage the microstructure of the stent provoking microcracks due to high temperatures required which can affect the overall functionality of the device [38]. Biodegradable stents have been fabricated using FDM in combination with a composite PLA-PCL material. Each of the materials on their own does not possess suitable properties for stent applications. However, a stent made of a PLA core and a PCL shell resulted in Young's modulus of 1400 MPa and a 3% degradation over 6 weeks [39]. Additionally, stents can be loaded or coated with antimicrobials to prevent infection and anti-inflammatory drugs to reduce post-surgical side effects [40]. To treat recurrent obstructive salivary glands, a stent was printed mimicking the salivary ducts using PCL loaded with amoxicillin and cefotaxime with a sustained release over 28 and 3 days, respectively, to avoid infection post-implantation.

Also, 3D printing techniques have been successfully used in regenerative medicine. These advances are highlighted for the treatment of bone defects as 3D printing has made feasible the fabrication of personalized scaffolds with fine control of shape, porosity, and mechanical properties with the ability to release antimicrobial drugs to prevent infections when required [41,42]. In the field of ophthalmology, 3D printing is moving forward. It is worth highlighting 3D-printed retinas and corneas fabricated using natural biocompatible polymers such as collagen, alginate, and human stem cells and 3D-printed intraocular lenses that required an optimal tune of the refractive index of the printed layers [43,44].

3D printing is becoming an interesting and highly applicable technique in the development of novel parenteral dosage forms, with a wide variety of applications. However, there is an unmet clinical need to develop biocompatible novel materials that mimic the properties of natural materials but are able to provide the physicochemical properties adapted to patients' needs. Also, advanced printers are required that facilitate the clinical translation of these novel materials allowing them to work under Good Manufacturing Conditions (GMP). Additionally, the combination of 3D-printed materials with biomedical electronics will facilitate monitoring physiological processes allowing for a fine adjustment of drug release [26,45,46].

3.5 APPLICATION OF 3D PRINTING TECHNOLOGIES IN THE DEVELOPMENT OF TOPICAL AND TRANSDERMAL DOSAGE FORMS

3D printing has also shown relevant applications in the field of topical and transdermal drug delivery. Several examples will be discussed including wound dressings, microneedles, and facial masks. Wound dressings made with 3D printing techniques have been shown to possess optimal flexibility and adhesive strength properties comparable to commercial dressings for wound-healing applications. Wound healing is a complex process as several factors interplay a role in the process including hemostasis, inflammation, proliferation, and remodeling.

Hence, 3D printing is required to obtain precise control over the spatial distribution of the biological components and materials to result in an enhancement of cell proliferation (e.g., angiogenesis and reepithelialization) and healing rather than inflammation and scarring [47,48]. In Table 3.3, 3D-printed wound dressing compositions have been summarized.

Antibacterial wound dressings have been fabricated using a novel bioink, consisting of GElMA and xanthan gum with excellent printability and swelling properties. N-halamines were incorporated into the matrix due to their strong biocidal properties, and titanium dioxide nanoparticles were added to protect the N-Cl bonds of the N-halamines from degradation during the UV crosslinking process. A customized direct ink writing 3D printer

TABLE 3.3
3D-Printed Wound Dressings, Microneedles, and Facial Masks

Topical Dosage Form	Polymer /Drug	3D Printing Technique	Applications	References
Wound dressing	GElMA Irgacure Xanthan gum N-halamines TiO$_2$ nanoparticles	PAM-SLA	Antibacterial biofilm growth Enhancement of wound healing in *in vivo* models	[49]
	Chitosan Pectin Lidocaine	PAM	Absorption of exudates while maintaining a moist wound-healing environment	[50]
	N,O-carboxymethyl chitosan Potato starch Mupirocin	Inkredible+ cellink printer (PAM+SLA)	Sustained release over 8 h and good *in vitro* antibacterial efficacy	[51]
	PCL Copper Zinc Silver	FDM	Antibacterial dressings adapted to complex anatomical parts of the body	[52]
Microneedle	Cisplatin	SLA	1000 μm in height, 1000 μm in width 80% capacity penetration and 1 h drug release	[55]
	Biocompatible Class I resin (Dental SG) Rifampicin	SLA	1150 μm in height, 950 μm in width Hollow microneedle with 360 μL reservoir Transdermal delivery	[56]
	Biocompatible Class I resin (Dental SG) Insulin Xylitol	SLA	Microneedles coated by inkjet printing Dimensions: 1000×1000×1000 μm Transdermal delivery equivalent to subcutaneous administration	[57]
	SI500 photoresin Human cells Alginate	SLA	600 μm in height, 1000 μm in width Hollow microneedle loaded with human cells encapsulated in alginate beads. Good cell viability upon extrusion	[58]
Mask	Flex-EcoPLA or PCL Salicylic acid	FDM	Hot-melt extruded filaments High percentage of drug degradation	[59]
	PEGDA PEG Salicylic acid	SLA	Minimal drug degradation Suitable mask for anti-acne purposes. Sustained drug release	[59]

Key: GelMA, gelatin methacrylate; TiO$_2$, titanium dioxide; PCL, polycaprolactone; PLA, poly(lactic acid); PEGDA, poly(ethylene glycol) diacrylate.

was employed. A pressure-assisted syringe was used to extrude the gel mixture through a needle with a tip diameter of 300 μm under 0.3–0.5 MPa. After ink deposition, a 365 nm UV light was applied to photo-crosslink the dressing in which Irgacure 2959 was used as photoinitiator. Wound dressings were lyophilized to obtain a porous matrix. Dressings were able to inhibit the bacterial biofilm and *in vivo* significantly accelerate wound healing in a mouse model [49]. Wound dressings of chitosan-pectin hydrogels containing a local anesthetic drug, lidocaine, have been successfully manufactured using PAM followed by lyophilization. The combination of chitosan and pectin allowed the exudates to be absorbed while maintaining a moist wound-healing environment [50]. Similarly, modified chitosan, N, O-carboxymethyl chitosan, was used to develop a novel bioink for wound dressing applications in combination with potato starch and mupirocin leading to a sustained release over 8 h [51]. Wound dressings loaded with metals, such as copper, zinc, or silver, have been fabricated by hot-melt extrusion using PCL. A 3D scanner was employed to capture images of different parts of the body, and consequently, printed dressings were adapted to the specific anatomy of the patient exhibiting a sustained release of metal ions for 72 h being effective against *S. aureus* [52].

Regarding the microneedle technology, it has shown great potential in controlled drug delivery as can pierce through the stratum corneum layer of the skin into the epidermis evading interaction with nerve fibers and hence, being painless and patient friendly [53]. This is achieved due to their height, between 25 and 2000 μm combined with a diameter usually between 50–250 μm in the base and 1–25 μm in the tip. There exits several techniques to manufacture microneedles such as droplet-borne air-blowing method, lithography, and molding; the latter is the most commonly employed method but is highly time-consuming, and its scale-up is challenging [54]. 3D printing has emerged as a new technology for rapid prototyping of microneedles. In Table 3.3, several examples are illustrated.

3D-printed polymeric microneedles (1000 μm in height, 1000 μm in width) have been fabricated by SLA using commercial resins followed by the coating of cisplatin formulations to enhance drug delivery in epidermoid skin tumors. Microneedles showed an optimal piercing capacity with an 80% penetration rate releasing the cisplatin within one hour upon skin application [55]. 3D-printed hollow microneedles were fabricated using SLA resulting in an efficient transdermal delivery of rifampicin. A biocompatible Class I resin (Dental SG) was used to manufacture microneedles of 1150 μm in height and 950 μm in base diameter. Each microneedle has a circular opening at the top of the reservoir with a 360 μL of capacity to fill the drug solution. A cap was also designed to prevent drug leakage after loading. Upon administration, rifampicin was detected in the bloodstream for 24 h [56].

3D-printed microneedles have also been developed for insulin skin delivery. Microneedle base dimensions were 1×1 mm with a 1 mm in length. The surface of the microneedles was coated with insulin through inkjet printing. Different insulin mixtures with xylitol, mannitol, and trehalose were tested. The insulin structure, α-helix and β-sheet, was preserved showing the optimum performance with the xylitol mixture. The *in vivo* administration of the microneedle showed rapid low glucose levels along with a longer duration equivalent to insulin subcutaneous administration [57]. Also, 3D-printed hollow microneedles have been successfully developed for microencapsulation cell extrusion. The dimensions of the cones were 1000 μm in base diameter, 600 μm in height, and 400 μm the tip diameter. Cells were encapsulated in atomized alginate capsules (3.5 w/v) crosslinked with $CaCl_2$ (1.5%), followed by filling through extrusion inside the microneedle reservoir. The viability of the cells was demonstrated upon extrusion in the microneedle reservoir [58].

3D printing has also been utilized to fabricate flexible personalized-shaped anti-acne salicylic acid-loaded masks for the treatment of acne. The device was adapted to the nose anatomy of the patient previously scanned. Filaments made of Flex EcoPLA or PCL containing salicylic acid were fabricated by hot-melt extrusion followed by FDM. However, a dramatic drug degradation occurred with both feedstock materials. The same device was printed by SLA using a photopolimerizable resin mixture of PEDGA and PEG avoiding drug degradation. Drug release was sustained for over 3 h, being faster the process for the FDM-printed masks compared to the SLA ones [59].

In summary, 3D printing technologies based on PAM are gaining attention in wound dressing manufacturing due to the ability to construct scaffolds with intricated and complex morphologies as well as only mild conditions are required which avoids degradation of proteins or thermo-sensitive drugs and Maillard reactions in carbohydrate polymers. In contrast, SLA is the most suitable technique due to its greater resolution compared with other printing technologies to fabricate microneedles. SLA also meets the mechanical strength requirements of this type of delivery system. However, achieving a good resolution at the tip of the microneedles is still very challenging even though new printers have reached the market with much greater performance. Most microneedles are coated by inkjet printing after SLA or they are hollowed allowing drug filling after 3D printing. Nevertheless, no reports have been found on 3D-printed polymer-based biodegradable microneedles at the moment.

3.6 NOVEL APPLICATIONS OF 3D PRINTING TECHNOLOGIES FOR PERSONALIZED DRUG DELIVERY: MICROFLUIDIC CHIPS

The microfluidic device is a generic term that refers to a component that has microfluidic channels handling very small fluid volumes. Because the majority of applications requiring fluid handling include biomedical and chemical analyses, microfluidics has been realized as a miniature analytical technology for biomedical and chemical applications [60]. The manufacturing of microfluidic devices has been demonstrated using polymer materials such as polymethylmethacrylate (PMMA), polystyrene (PS), polycarbonate (PC), and polydimethylsiloxane (PDMS). PDMS is one of the most frequently used materials in modern research facilities to create microfluidic devices [61–63].

However, the engineering of microfluidic devices is complex, and there have been some barriers to commercializing these devices that traditional fabrication methods such as injection molding using polydimethylsiloxane (PDMS) have failed to address, e.g., non-standard user interfaces, complex control systems, and high cost. These barriers have been overcome by 3D printing, a more cost-effective technology that has shown a significant improvement in terms of channel resolution [64].

3.6.1 CONTINUOUS MANUFACTURING WITH 3D-PRINTED MICROFLUIDIC CHIPS

The conventional batch manufacturing method still does not have any alternatives in the pharmaceutical industry. However, continuous manufacturing is a process in which raw materials are continually injected into a manufacturing facility, and products are continuously discharged during the operation of the manufacturing processes. Multiple processes are automatically regulated in this manner, which contributes to the overall simplification of the operation and reduces the workload needs of human operators [65]. A handful of medications are already manufactured in continuous production and have already been approved by the FDA [66]. For example, Johnson & Johnson's Janssen drug unit got approved a switchover from batch to continuous manufacturing for the production of HIV drug Prezista. Continuous manufacturing requires a fully automated system and continuous monitoring in real time. Due to its continuous flow, liquid handling, and flexible configuration, multistep flow-based manufacturing technologies are suited for integrating in-line analytics and process control [67]. 3D-printed microfluidic devices enable rapid adaptation of these systems, allowing for continuous flow with real-time screening. Microfluidic devices can easily be adapted to produce personalized medicines, resulting in significant cost savings per unit. 3D printing techniques enable the automated and cost-effective fabrication of microfluidic chips, which reduces the overall cost of the microfluidic device. If necessary, these systems can provide entirely personalized treatment for patients. However, this technology still needs more verification for this customization, but upcoming results seem promising. Microfluidic systems have the potential to be one of the most practical solutions for on-demand production. These devices are easily adaptable to any medical center and can serve as a lifesaver during times of drug shortage.

3.6.2 MANUFACTURING NANOMEDICINES WITH 3D-PRINTED MICROFLUIDIC CHIPS

Manufacturing of nanomedicines is a multistep process with limited scalability, control, high cost, and variability [67]. The feasibility of 3D-printed microfluidic chips for the manufacturing of nanomedicines has recently been demonstrated [68,69]. 3D printing microfluidic devices are capable of high-throughput synthesis of nanomedicines with tunable dimensions resulting in an enormous advantage compared to the conventional batch method [68,70–74].

3.7 FUTURE PERSPECTIVES AND CONCLUSION

Personalized medicine generally consists in tailoring medical treatments to the characteristics, needs, and preferences of every single patient, and it involves purposely run diagnosis, therapy, and follow-up [75]. According to Abraham et al., the personalized medicine is defined as "Providing the right treatment to the right patient, at the right dose, at the right time" [76]. Indeed, the goal of personalized medicine is to drive clinical decision-making by distinguishing in advance those patients who are most likely to benefit from a given treatment from those who would incur costs and side effects without gaining equivalent benefit. The most common uses among the applications of health 3D printing are the personalization of the dose of drugs enabling co-administration of drugs in multi-therapies and avoiding the use of specific excipients and processes involved in intolerances, confirming this way the mantra of "one size does not fit all" (Figure 3.7) [75,77].

Personalization of treatments have multiple benefits. The most remarkable are the reduction of side effects and the maximization of the efficiency and the effectiveness of the use of medicinal resources [78]. It is an undeniable truth that 3D printing techniques have become, increasingly, in really viable methods to obtain versatile drugs for the approach of complex pathologies, increasingly present in the current medical field. The standardization of personalization and differentiation in healthcare management using this kind of technology still is a subject of study [79]. In fact, although conceived to widely ensure the quality of services, the standardization of healthcare does not imply an improvement in effectiveness of individualized service. On the contrary, it generally implies a reduction of the variability, between cases, that patients would appreciate as personalized service [79], which is one of the biggest challenges to overcome in this area.

Some of the questions still to be answered in this field are, what would it mean standardization in healthcare effectiveness on personalized treatments? Can a technology that provides variability be standardized? How healthcare effectiveness, efficiency, and sustainability can be harmonized with personalized treatments? And most of all, what contribution does the 3D printing technology make in harmonizing the effectiveness, efficiency, and sustainability of the personalized treatments?

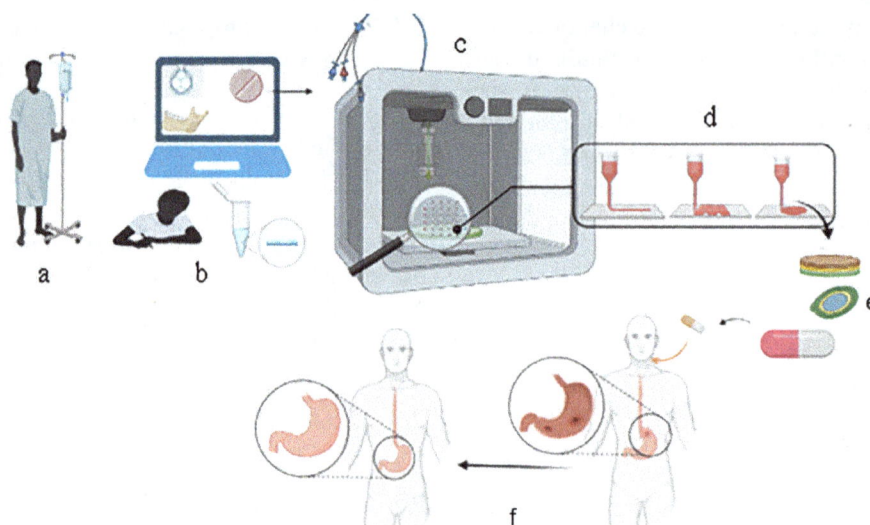

FIGURE 3.7 Implementation of 3D printing personalized medicines in clinical practice: (a) clinical need; (b) medicine design; (c) FDM 3D or (d) PAM 3D printing; (e) 3D-printed personalized medicine; (f) controlled drug delivery system adapted to patient's need.

The continuous advances in all kinds of technologies with the aim of the obstinance of new products, services, or designing new processes are always related to the expectation of eventually inducing an improvement in medical or health applications and outcomes [78,80]; however, they are also likely to introduce different and new challenges in healthcare, especially in pharmaceuticals [79]. 3D-printed medicines may be useful, particularly, for patients who respond to the same drugs in diverse ways. In these cases, a clinician would be able to use patient-related individual information to produce optimal medication doses, rather than relying on a standard set of dosage forms [77,79]; and is then, when understanding how 3D printing technologies works, when fabricating personalized medicines becomes an important matter to attend to. It is hard to deny the intrinsic link that exists between the economic, social, ethical, organizational, and cultural factors of a health system and the 3D-printed drugs, which can hinder the potential progress toward a smarter and more sustainable application of the 3D printed personalized medicines (3DPM) on the healthcare systems [79].

The more evident and possible inconvenient that may present the application of this technique is that (1) pharmaceutical industries worryingly believe that 3DPM may reduce the market size and profits associated with their blockbuster one-size-fits-all drugs; (2) insurers potentially can be afraid of the lack of return on investment of expensive diagnostics and therapeutics and disease prevention; (3) physicians fear that personalized diagnostic and therapeutic strategy is not yet supported in clinical standards or evidence-based medicine; (4) patients who are a potent driving force, but currently are unaware of the benefits correlated to 3DPM and, therefore, the access to 3DPM is not demanded; (5) regulators, health agencies, and authorities, which are not ready to adapt or draft guidances and regulations relative to the new rules required and new safety and security concerns for 3DPM [79].

What does standardization on 3DPM mean? The final product never should be the object of the standardization,

because it has to be different in each circumstance and patient, that is the essence of the concept of personalization. Nevertheless, the opposite happens with the concept of standardization, which focused on giving harmonized responses even under different commands. In the present scenario, a tablet containing more than one drug would be considered a new combination drug formulation, according to the FDA, and would require extensive clinical trials to guarantee patient safety and efficacy, and it gets increasingly complex when factors like live cell or biological substances are involved. And every hospital or community pharmacy that uses a 3D printer for producing and dispensing pharmaceutical products would have to be certified as a "Good Manufacturing Practice" (GMP) facility. Hence, appropriate regulatory guidelines must be developed regarding the manufacture and dispensing of 3DPM [77]. The application of this type of technology seeks to attend to the different variations of patient's needs and their medical conditions cannot leave aside that some processes will eventually need methods to establish continuous fabrication, quality assurance, a common regulatory field, and cost/benefit evaluation methods for each case and patient [81].

The standardization must be in coincidence between the designs and what finally is obtained, and about the final therapeutic results elicited in the patients. For this, it must be considered that the therapeutic and adverse reactions must be contemplated in the original design of the 3D-printed pharmaceutical formulation, the commonly called quality by design (QbD). A high percentage of this coincidence, above 90%, would be a suitable number to start within establishing harmonization in the personalization of 3D-printed therapies.

What type of contribution the 3D printing technology can make in harmonizing the effectiveness, efficiency, and sustainability of personalized treatments? Extemporaneous compounding formulations, orphan illness, kinetics changes, and new versatile formulations are some of the answers to this question [82]. The 3D printing technology has endless potential in the fabrication of patient-specific drug delivery devices (DDD) and dosage forms as technological development is progressing in this field [83]. Control drug release can change the kinetics of absorption and metabolization of a drug, especially important when you are looking for maximizing the effectiveness and minimizing the side effects of a medication or when you are trying to do an extremely specific treatment like in the case of an orphan illness. The integration of nanotechnology-based drugs into 3D printing brings printed personalized nanomedicines within the most innovative perspectives for the next few years [83,84]. Microswimmers, microimplants, and 3D-printed antigen nanoparticles are only some of the novel devices devolved using nano-3D printing. Multiple DDDs have being developed to attend pediatric needs, and some more are under investigation. It is demonstrated that some drugs work well for a group of patients, but the same drug does not elicit any effect in other groups, due to gene expression like the case of Herceptin and Vectibix, which have a special type of regulatory approval, and their use is only allowed in patients who have certain gene overexpression or tumors with a specific type of mutation, respectively [85]. These examples confirm that the same DDD cannot work in the same way on diverse groups of individuals.

To understand how 3D printing technologies work when fabricating personalized medicines is essential to know the possibilities that offers, almost limitless, but also its difficulties. In fact, by fabricating customizable carriers and devices we can deliver drugs with any type of release profiles, drug combination, and any kind of tailoring or personalization, an achievement hard to believe few years ago, what makes us think, how far can we get with more investment and effort in this field?

REFERENCES

1. Konta, A.A., M. Garcia-Pina, and D.R. Serrano, Personalised 3D printed medicines: Which techniques and polymers are more successful? *Bioengineering (Basel)*, 2017. **4**(4): p. 79.
2. Mathew, E., et al., 3D printing of pharmaceuticals and drug delivery devices. *Pharmaceutics*, 2020. **12**(3): p. 266.
3. Fernandez-Garcia, R., et al., Oral fixed-dose combination pharmaceutical products: Industrial manufacturing versus personalized 3D printing. *Pharm Res*, 2020. **37**(7): p. 132.
4. Long, J., et al., Application of fused deposition modelling (FDM) method of 3D printing in drug delivery. *Curr Pharm Des*, 2017. **23**(3): pp. 433–439.
5. Malebari, A.M., et al., Development of advanced 3D-printed solid dosage pediatric formulations for HIV treatment. *Pharmaceuticals (Basel)*, 2022. **15**(4): p. 435.
6. Sanchez-Guirales, S.A., et al., Understanding direct powder extrusion for fabrication of 3D printed personalised medicines: A case study for nifedipine minitablets. *Pharmaceutics*, 2021. **13**(10): p. 1583.
7. Boniatti, J., et al., Direct powder extrusion 3D printing of praziquantel to overcome neglected disease formulation challenges in paediatric populations. *Pharmaceutics*, 2021. **13**(8): p. 1114.
8. El Aita, I., J. Breitkreutz, and J. Quodbach, On-demand manufacturing of immediate release levetiracetam tablets using pressure-assisted microsyringe printing. *Eur J Pharm Biopharm*, 2019. **134**: pp. 29–36.
9. Xu, X., et al., Stereolithography (SLA) 3D printing of a bladder device for intravesical drug delivery. *Mater Sci Eng C Mater Biol Appl*, 2021. **120**: p. 111773.
10. Krkobabic, M., et al., Hydrophilic excipients in digital light processing (DLP) printing of sustained release tablets: Impact on internal structure and drug dissolution rate. *Int J Pharm*, 2019. **572**: p. 118790.
11. Xu, X., et al., Vat photopolymerization 3D printing for advanced drug delivery and medical device applications. *J Control Release*, 2021. **329**: pp. 743–757.
12. Cailleaux, S., et al., Fused deposition modeling (FDM), the new asset for the production of tailored medicines. *J Control Release*, 2021. **330**: pp. 821–841.
13. Fina, F., et al., Selective laser sintering (SLS) 3D printing of medicines. *Int J Pharm*, 2017. **529**(1–2): pp. 285–293.
14. Ayyoubi, S., et al., 3D printed spherical mini-tablets: Geometry versus composition effects in controlling dissolution from personalised solid dosage forms. *Int J Pharm*, 2021. **597**: p. 120336.
15. Khaled, S.A., et al., 3D printing of five-in-one dose combination polypill with defined immediate and sustained release profiles. *J Control Release*, 2015. **217**: pp. 308–314.
16. Herrada-Manchon, H., et al., 3D printed gummies: Personalized drug dosage in a safe and appealing way. *Int J Pharm*, 2020. **587**: p. 119687.
17. Awad, A., et al., 3D printed tablets (printlets) with braille and moon patterns for visually impaired patients. *Pharmaceutics*, 2020. **12**(2): p. 172.
18. Edinger, M., et al., QR encoded smart oral dosage forms by inkjet printing. *Int J Pharm*, 2018. **536**(1): pp. 138–145.
19. Serrano, D.R., J.R. Cerda, R. Fernandez-Garcia, L.F. Perez-Ballesteros, M.P. Ballesteros, A. Lalatsa, Market demands in 3D printing pharmaceuticals products. In: *3D Rpinting Technologies in Nanomedicine*, N. Ahmad, P. Gopinath, and R. Dutta, Editors. 2019: Elsevier, Misouri, Vol. 1, pp. 165–183. Elsevier.
20. Khaled, S.A., et al., 3D printing of tablets containing multiple drugs with defined release profiles. *Int J Pharm*, 2015. **494**(2): pp. 643–650.
21. Haring, A.P., et al., Programming of multicomponent temporal release profiles in 3D printed polypills via core-shell, multilayer, and gradient concentration profiles. *Adv Healthc Mater*, 2018. **7**(16): p. e1800213.
22. Goh, W.J., et al., 3D printing of four-in-one oral polypill with multiple release profiles for personalized delivery of caffeine and vitamin B analogues. *Int J Pharm*, 2021. **598**: p. 120360.
23. Windolf, H., et al., 3D printed mini-floating-polypill for Parkinson's disease: Combination of levodopa, benserazide, and pramipexole in various dosing for personalized therapy. *Pharmaceutics*, 2022. **14**(5): p. 931.
24. Pereira, B.C., et al., 'Temporary Plasticiser': A novel solution to fabricate 3D printed patient-centred cardiovascular 'Polypill' architectures. *Eur J Pharm Biopharm*, 2019. **135**: pp. 94–103.
25. Robles-Martinez, P., et al., 3D printing of a multi-layered polypill containing six drugs using a novel stereolithographic method. *Pharmaceutics*, 2019. **11**(6): p. 274.

26. Ivone, R., Y. Yang, and J. Shen, Recent advances in 3D printing for parenteral applications. *AAPS J*, 2021. **23**(4): p. 87.

27. Shah, V.P., J. DeMuth, and D.G. Hunt, Performance test for parenteral dosage forms. *Dissolution Technol*, 2015. **22**(4): pp. 16–21.

28. Birrer, G.A., et al., Parenteral dosage forms. In *Handbook of Modern Pharmaceutical Analysis*, S. Ahuja and S. Scypinski, Editors. 2001. Elsevier, 566 p.

29. Barbin, T., et al., 3D metal printing in dentistry: An in vitro biomechanical comparative study of two additive manufacturing technologies for full-arch implant-supported prostheses. *J Mech Behav Biomed Mater*, 2020. **108**: p. 103821.

30. Yuste, I., et al., Mimicking bone microenvironment: 2D and 3D in vitro models of human osteoblasts. *Pharmacol Res*, 2021. **169**: p. 105626.

31. Bagshaw, K.R., et al., Pain management via local anesthetics and responsive hydrogels. *Ther Deliv*, 2015. **6**(2): pp. 165–176.

32. Dash, A.K. and G.C. Cudworth, 2nd, Therapeutic applications of implantable drug delivery systems. *J Pharmacol Toxicol Methods*, 1998. **40**(1): pp. 1–12.

33. Gimeno, M., et al., A controlled antibiotic release system to prevent orthopedic-implant associated infections: An in vitro study. *Eur J Pharm Biopharm*, 2015. **96**: pp. 264–271.

34. Stewart, S.A., et al., Implantable polymeric drug delivery devices: Classification, manufacture, materials, and clinical applications. *Polymers (Basel)*, 2018. **10**(12): p. 1379.

35. Fu, J., X. Yu, and Y. Jin, 3D printing of vaginal rings with personalized shapes for controlled release of progesterone. *Int J Pharm*, 2018. **539**(1–2): pp. 75–82.

36. Qiao, X., et al., 3D-printed scaffolds as sustained multi-drug delivery vehicles in breast cancer therapy. *Pharm Res*, 2019. **36**(12): pp. 1–16.

37. Won, J.Y., et al., 3D printing of drug-loaded multi-shell rods for local delivery of bevacizumab and dexamethasone: A synergetic therapy for retinal vascular diseases. *Acta Biomater*, 2020. **116**: pp. 174–185.

38. Ang, H.Y., et al., Mechanical behavior of polymer-based vs. metallic-based bioresorbable stents. *J Thorac Dis*, 2017. **9**(Suppl 9): pp. S923–S934.

39. Guerra, A., 3D-printed bioabsordable polycaprolactone stent: The effect of process parameters on its physical features. *Mater Des*, 2018. **137**: pp. 430–437.

40. Kim, T.H., et al., Development of a 3D-printed drug-eluting stent for treating obstructive salivary gland disease. *ACS Biomater Sci Eng*, 2019. **5**(7): pp. 3572–3581.

41. Lim, S.H., et al., Three-dimensional printing of carbamazepine sustained-release scaffold. *J Pharm Sci*, 2016. **105**(7): pp. 2155–2163.

42. Wang, M., et al., Cold atmospheric plasma (CAP) surface nanomodified 3D printed polylactic acid (PLA) scaffolds for bone regeneration. *Acta Biomater*, 2016. **46**: pp. 256–265.

43. Sommer, A.C. and E.Z. Blumenthal, Implementations of 3D printing in ophthalmology. *Graefes Arch Clin Exp Ophthalmol*, 2019. **257**(9): pp. 1815–1822.

44. Isaacson, A., S. Swioklo, and C.J. Connon, 3D bioprinting of a corneal stroma equivalent. *Exp Eye Res*, 2018. **173**: pp. 188–193.

45. Aguilar-de-Leyva, A., et al., 3D printed drug delivery systems based on natural products. *Pharmaceutics*, 2020. **12**(7): p. 620.

46. Preis, M. and J.M. Rosenholm, Printable nanomedicines: The future of customized drug delivery? *Ther Deliv*, 2017. **8**(9): pp. 721–723.

47. Malda, J., et al., 25th anniversary article: Engineering hydrogels for biofabrication. *Adv Mater*, 2013. **25**(36): p. 5011–5028.

48. van Kogelenberg, S., et al., Three-dimensional printing and cell therapy for wound repair. *Adv Wound Care (New Rochelle)*, 2018. **7**(5): pp. 145–155.

49. Yang, Z., X. Ren, and Y. Liu, Multifunctional 3D printed porous GelMA/xanthan gum based dressing with biofilm control and wound healing activity. *Mater Sci Eng C Mater Biol Appl*, 2021. **131**: p. 112493.

50. Long, J., et al., A 3D printed chitosan-pectin hydrogel wound dressing for lidocaine hydrochloride delivery. *Mater Sci Eng C Mater Biol Appl*, 2019. **104**: p. 109873.

51. Naseri, E., et al., Development of N, O-carboxymethyl chitosan-starch biomaterial inks for 3D printed wound dressing applications. *Macromol Biosci*, 2021. **21**(12): p. e2100368.

52. Muwaffak, Z., et al., Patient-specific 3D scanned and 3D printed antimicrobial polycaprolactone wound dressings. *Int J Pharm*, 2017. **527**(1–2): pp. 161–170.

53. Elahpour, N., et al., 3D printed microneedles for transdermal drug delivery: A brief review of two decades. *Int J Pharm*, 2021. **597**: p. 120301.

54. Luzuriaga, M.A., et al., Biodegradable 3D printed polymer microneedles for transdermal drug delivery. *Lab Chip*, 2018. **18**(8): pp. 1223–1230.

55. Uddin, M.J., et al., 3D printed microneedles for anticancer therapy of skin tumours. *Mater Sci Eng C Mater Biol Appl*, 2020. **107**: p. 110248.

56. Yadav, V., et al., 3D printed hollow microneedles array using stereolithography for efficient transdermal delivery of rifampicin. *Int J Pharm*, 2021. **605**: p. 120815.

57. Pere, C.P.P., et al., 3D printed microneedles for insulin skin delivery. *Int J Pharm*, 2018. **544**(2): pp. 425–432.

58. Farias, C., et al., Three-dimensional (3D) printed microneedles for microencapsulated cell extrusion. *Bioengineering (Basel)*, 2018. **5**(3): p. 59.

59. Goyanes, A., et al., 3D scanning and 3D printing as innovative technologies for fabricating personalized topical drug delivery systems. *J Control Release*, 2016. **234**: pp. 41–48.

60. Song, Y., D. Cheng, and L. Zhao, *Microfluidics: Fundamentals, Devices, and Applications*. 2018: John Wiley & Sons, Hoboken, NJ.

61. Wu, H., et al., Fabrication of complex three-dimensional microchannel systems in PDMS. *J Am Chem Soc*, 2003. **125**(2): pp. 554–559.

62. Xia, Y., et al., Complex optical surfaces formed by replica molding against elastomeric masters. *Science*, 1996. **273**(5273): pp. 347–349.

63. Xia, Y. and G.M. Whitesides, Soft lithography. *Ann Rev Mater Sci*, 1998. **28**(1): pp. 153–184.

64. Gale, B.K., et al., A review of current methods in microfluidic device fabrication and future commercialization prospects. *Inventions*, 2018. **3**: p. 60.

65. Inada, Y. *Continuous Manufacturing Development In Pharmaceutical And Fine Chemicals Industries*. 2020; Available from: https://www.mitsui.com/mgssi/en/report/detail/__icsFiles/afieldfile/2020/04/08/1912m_inada_e_1.pdf.

66. Palmer, E., *FDA Urges Companies to Get on Board with ContinuousManufacturing.*2016,FiercePharma.Availablein: https://www.fiercepharma.com/manufacturing/fda-urges-companies-to-get-on-board-continuous-manufacturing. Accessed date: 12 June 2022

67. Bohr, A., S. Colombo, and H. Jensen, *Future of Microfluidics in Research and in the Market.* 2019, Elsevier. pp. 425–465.

68. Kara, A., et al., Engineering 3D printed microfluidic chips for the fabrication of nanomedicines. *Pharmaceutics*, 2021. **13**(12): p. 2134.

69. Tiboni, M., et al., 3D-printed microfluidic chip for the preparation of glycyrrhetinic acid-loaded ethanolic liposomes. *Int J Pharm*, 2020. **584**: p. 119436.

70. Khadke, S., et al., Formulation and manufacturing of lymphatic targeting liposomes using microfluidics. *J Control Release*, 2019. **307**: pp. 211–220.

71. Shah, V.M., et al., Liposomes produced by microfluidics and extrusion: A comparison for scale-up purposes. *Nanomedicine*, 2019. **18**: pp. 146–156.

72. Bokare, A., et al., Herringbone-patterned 3D-printed devices as alternatives to microfluidics for reproducible production of lipid polymer hybrid nanoparticles. *ACS Omega*, 2019. **4**(3): pp. 4650–4657.

73. Streck, S., et al., Comparison of bulk and microfluidics methods for the formulation of poly-lactic-co-glycolic acid (PLGA) nanoparticles modified with cell-penetrating peptides of different architectures. *Int J Pharm X*, 2019. **1**: p. 100030.

74. Martins, J.P., G. Torrieri, and H.A. Santos, The importance of microfluidics for the preparation of nanoparticles as advanced drug delivery systems. *Expert Opin Drug Deliv*, 2018. **15**(5): pp. 469–479.

75. Zema, L., et al., Three-dimensional printing of medicinal products and the challenge of personalized therapy. *J Pharm Sci*, 2017. **106**(7): pp. 1697–1705.

76. Abrahams, E., Right drug—right patient—right time: Personalized medicine coalition. *Clinical and Translational Science*, 2008. **1**(1): 11–12.

77. Vaz, V.M. and L. Kumar, 3D printing as a promising tool in personalized medicine. *AAPS PharmSciTech*, 2021. **22**(1): p. 49.

78. Beer, N., et al., Scenarios for 3D printing of personalized medicines - A case study. *Explor Res Clin Soc Pharm*, 2021. **4**: p. 100073.

79. Aquino, R.P., et al., Envisioning smart and sustainable healthcare: 3D Printing technologies for personalized medication. *Futures*, 2018. **103**: pp. 35–50.

80. Fong, E.L.S., et al., 3D culture as a clinically relevant model for personalized medicine. *SLAS Technol*, 2017. **22**(3): pp. 245–253.

81. Amekyeh, H., F. Tarlochan, and N. Billa, Practicality of 3D printed personalized medicines in therapeutics. *Front Pharmacol*, 2021. **12**: p. 646836.

82. Horst, D.J., 3D printing of pharmaceutical drug delivery systems. *Arch Org Inorg Chem Sci*, 2018. **1**(2): pp. 1–5.

83. dos Santos, J., et al., 3D printing and nanotechnology: A multiscale alliance in personalized medicine. *Adv Funct Mater*, 2021. **31**(16): p. 2009691.

84. Jain, K., et al., 3D printing in development of nanomedicines. *Nanomaterials (Basel)*, 2021. **11**(2): p. 420.

85. Afsana, et al., 3D printing in personalized drug delivery. *Curr Pharm Des*, 2018. **24**(42): pp. 5062–5071.

4 Bioprinting Biomimetic 3D Constructs for Tissue Modelling and Repair

Luís B. Bebiano
Universidade do Porto
Politécnico do Porto

Flávia Castro
Universidade do Porto

Yi-Chen Ethan Li
Feng Chia University

Bruno Pereira and Rúben F. Pereira
Universidade do Porto

CONTENTS

4.1 INTRODUCTION

Bioprinting is an enabling technology used in the biofabrication field that allows the fabrication of patient-specific grafts for tissue repair and 3D *in vitro* models. The interest in bioprinting is significantly increasing as it allows the spatial arrangement of living and non-living building blocks in 3D with the ultimate goal of recreating key features of native tissues and organs at compositional, structural and functional levels [1]. Building blocks for 3D bioprinting include living cells, biomaterials and bioactive molecules. Depending on the application, building blocks can be bioprinted together or separately, allowing the fabrication of bioengineered tissues for translational applications such as tissue repair and therapeutic drug screening [2].

There are multiple bioprinting technologies available, including extrusion bioprinting, light- and laser-based technologies or inkjet bioprinting [3]. The selection of the

DOI: 10.1201/9781003224464-4

most suited technology depends on multiple considerations such as the resolution of bioprinted constructs, the material properties or the biological question to address. Extrusion bioprinting is the most common technology involving the pressure-driven extrusion of cellular (bioinks) or acellular formulations (biomaterial inks) from a nozzle onto a substrate. Despite extrusion bioprinting allowing the deposition of multiple bioinks with distinct rheological properties, it has lower printing resolution compared to light-based technologies, such as vat photopolymerization and volumetric bioprinting [3]. These emerging technologies use light to trigger the photocrosslinking of a cell-laden hydrogel (bioresin) and are powerful in fabricating complex 3D constructs with high resolution and intricate geometries. Indeed, in recent years, there has been growing interest in developing bioresins that meet both processing and biological requirements, which is essential to create 3D constructs with biological function [4]. Inkjet bioprinting affords the high-throughput deposition of bioink droplets through a nozzle, though it imposes some constraints on the processing of viscous bioinks [3]. Although bioprinting has been primarily explored to create grafts for tissue repair, the lack of human-relevant pre-clinical models for drug discovery and screening has boosted the interest in exploring bioprinting to generate biomimetic 3D *in vitro* models [2]. Such models should ideally recapitulate physiological, pathophysiological or therapeutic responses found in human body, serving as versatile platforms to study fundamental biological questions as well as to evaluate drug safety and therapeutic responses.

Herein, we first introduce bioprinting technologies and strategies used to create grafts with mechanical and biological function, as well as how the integration of 3D (bio)printing and microfluidics has been explored for the fabrication of organ-on-a-chip microfluidic devices. Then, we discuss recent advances in the bioprinting of 3D constructs as grafts for tissue repair and *in vitro* platforms for tissue modelling and drug screening. Focus is given to bioprinted constructs in the context of skin, heart, liver, cancer and involving tissue-derived organoids, which are emerging for regenerative medicine and drug screening applications. Finally, we provide insights into challenges and future opportunities in the field.

4.2 BIOPRINTING: FROM IMPLANTABLE GRAFTS TO DYNAMIC MICROPHYSIOLOGICAL SYSTEMS FOR TISSUE MODELLING AND DRUG SCREENING

3D bioprinting has been primarily used to create anatomically shaped constructs for tissue engineering and regenerative medicine. To accomplish this, in the past decade, a multitude of bioinks and bioprinting strategies have been developed, enabling the fabrication of cell-material constructs that mimic specific features of native tissues and

organs [1,5]. The rational design of printable materials as building blocks for 3D bioprinting represents a major challenge as material properties need to fulfil not only the rheological requirements of a specific bioprinting technology, but also being compatible with cells and yield a biologically functional tissue construct. Several materials such as hydrogels, ceramic powders and thermoplasts are available and can be processed using bioprinting technologies. Hydrogel precursors and pre-crosslinked gels are the first choice for bioink design as they provide the cells with a highly hydrated environment, resembling this feature in native extracellular matrix (ECM). Recent advances in the design of ECM-inspired hydrogel bioinks enable the bioprinting of cell-material constructs in which cell fate and morphogenesis can be instructed via material cues such as the stiffness and presentation of cell adhesion sites [6,7]. Despite these attractive features, bioprinted hydrogel constructs often face limitations regarding the mechanical properties, structural complexity and, in some cases, shape fidelity, making them more suited for soft tissue engineering. To address these issues, emerging bioprinting strategies such as microfluidic, coaxial and embedded bioprinting provide the opportunity to improve the shape fidelity and complexity of 3D constructs [8–10]. Moreover, *in situ* (photo)crosslinking can be explored to improve construct complexity, while dynamic chemistries allow the bioprinting of 3D constructs that capture the dynamic nature and viscoelasticity of native ECMs [4,11]. To create 3D constructs with improved mechanical properties, the bioprinting of biomaterial inks made of thermoplasts is a common approach. Despite the printing conditions not being compatible with cells, mechanically competent constructs can be fabricated for subsequent cell seeding. Depending on the target tissue and bioprinting technology, bioinks and biomaterial inks can be reinforced with ceramics or carbon-based materials and bioprinted using a single technology to improve mechanical and biological outcomes [12,13]. Alternatively, multiple bioprinting technologies can be combined towards the fabrication of multimaterial and multicellular 3D constructs with superior mechanical properties and cellular response [14].

In addition to the fabrication of implantable grafts, 3D (bio)printing technologies have been explored to create static 3D tissue models and dynamic organ-on-a-chip platforms that closely recapitulate human *in vivo*-like conditions. These microphysiological systems offer biomimetic and cost-effective alternatives to two-dimensional (2D) cell cultures and animal models, emerging as human (patho)physiologically relevant platforms for tissue modelling and drug screening. The integration of microfluidic channels in these systems allows for continuous perfusion of tissue constructs, which contributes to improved cell viability and function throughout longer culture periods [8]. Moreover, technological developments allow the fabrication of microfluidic chips capable of dynamically stimulating cultured cells (e.g. fluid flow, mechanical stress, oxygen gradients) similarly to what happens *in vivo*, as well as integrating biosensors for real-time measurement of cellular responses [15]. When compared to conventional techniques for organ-on-a-chip fabrication (e.g. photolithography, soft lithography, replica moulding, etc.), 3D (bio)printing enables the sequential or simultaneous deposition of multiple biomaterials (e.g. hydrogels, thermoplasts), which can be eventually loaded with cells towards the creation of 3D biomimetic tissue constructs [16,17]. Moreover, 3D (bio)printing provides higher precision in the placement of cells and biomaterials and often reduces the complexity and multistep manufacturing approach of conventional techniques. Depending on the application, 3D (bio)printing can be used for different purposes, including the microfluidic chip fabrication (e.g. silicone ink deposition, stereolithography of photosensitive polymers), the printing of sacrificial channels and the bioprinting of cellular formulations inside the chip.

4.3 BIOPRINTED 3D CONSTRUCTS FOR TISSUE MODELLING AND REPAIR

4.3.1 Skin

The skin is a multilayer organ that interfaces with the external environment and, therefore, is prone to a variety of acute and chronic injuries. In Europe, it is estimated that 1.5–2 million people suffer from acute or chronic wounds, while nearly 2.5% of the total population in the United States is affected by chronic wounds [18,19]. Furthermore, according to the American Burn Association, about 486.000 individuals received medical treatments for burn injuries in 2016. From an anatomical point of view, the skin comprises sequential layers of epidermis, dermis and hypodermis, each one containing distinct cell populations and imparting specific properties to the skin. The epidermis is the most superficial skin layer containing keratinocytes, melanocytes, Merkel cells and Langerhans cells. The predominant cells in the dermis are fibroblasts, though cells like mast cells and macrophages are also present. The hypodermis is mainly composed of adipose tissue [20]. As the skin displays a limited self-healing ability upon injury and some wounds do not heal without the application of a wound care product, 3D bioprinting has been employed for the automated fabrication of implantable grafts to assist the healing process and promote *de novo* skin formation, as well as to generate *in vitro* skin models for testing and screening of therapeutics.

4.3.1.1 Bioprinted Grafts for Skin Repair

Bioprinted skin grafts should closely mimic the composition and properties of target skin layer(s), provide a barrier that protects the wound from infection, integrate into the host tissue and stimulate new ECM deposition while minimizing scar tissue formation [20]. Skin constructs are usually fabricated by the sequential deposition of bioinks comprised of a polymer matrix and living cells specific to each skin layer. Depending on the application, constructs can be bioprinted to mimic either a single or multiple skin

layers [21,22]. In the most established strategy, dermal fibroblasts are embedded within either a hydrogel precursor solution or a pre-crosslinked hydrogel and bioprinted into a dermal layer, followed by the bioprinting or seeding of keratinocytes to recreate the epidermis. Keratinocytes are usually bioprinted after the maturation of the dermal region. Following this strategy, laser-assisted bioprinting was used to fabricate bilayered grafts capable of promoting full-thickness cutaneous healing in an animal model [22]. In order to more closely mimic the cellular composition of native skin and enhance the functional properties of regenerated skin, additional skin cells have been included in specific layers of skin constructs, including melanocytes, endothelial cells (ECs), pericytes, follicle dermal papilla cells and adipocytes [23–25].

The extrusion bioprinting of multilayered vascularized skin substitutes was demonstrated through the fabrication of a dermal region composed of human dermal fibroblasts, ECs and placental pericytes embedded within a type I collagen matrix [24]. Human keratinocytes were bioprinted after 4 days of dermis maturation to form the epidermis. Formation of vessel-like structures with open lumens in the dermal region of bioprinted skin was observed *in vitro*, while pre-cultured bioprinted grafts implanted on the dorsum of immunodeficient mice supported the formation of a mature stratified epidermis and the inosculation of human EC-lined structures with mouse microvessels arising from the wound bed 4 weeks post-engraftment. Despite immunodeficient mice not being the most suitable animal model to study skin wound healing, these data provide solid evidence about the importance of cellular composition and vascular network in promoting vascularization and engraftment of bioprinted skin. Using digital light processing (DLP)-based 3D bioprinting, Zhou et al. [26] fabricated a dermal skin graft by the photopolymerization of a gelatin methacrylate (GelMA)/N-(2-aminoethyl)-4-(4-(hydroxymethyl)-2-methoxy-5-nitro-sophenoxy) butanamide (NB)-modified hyaluronic acid (HA-NB) bioink containing human skin fibroblasts and human umbilical vein endothelial cells (HUVECs). Grafts with site-specific microchannels were generated to create an upper dense layer mimicking the epidermis and a lower porous layer resembling the dermis. *In vivo* evaluation of full-thickness skin defects created in the dorsal surface of small (Sprague Dawley rats) and large (pig) animal models showed that bioprinted grafts accelerated wound healing while promoting the formation of a mature epithelial structure and some adnexal structures such as hair follicles, blood vessels and sebaceous glands (Figure 4.1a).

The regeneration of skin appendages remains a major challenge in wound management due to the difficulty in recreating the microenvironmental cues necessary to orchestrate cell fate towards the development of such complex structures. To address this clinical need, extrusion bioprinting was used to create gelatin/alginate hydrogel constructs containing mouse plantar dermis components and epidermal growth factor (EGF) to promote differentiation of epidermal progenitors into sweat gland cells. Upon implantation into mice burned paws, bioprinted constructs restored sweat gland function as indicated by iodine/starch sweat test and histological analysis [27]. Efforts have also been made on the bioprinting of pigmented skin constructs through the incorporation of melanocytes in the epidermal region [28]. Recently, a dermo-epidermal 3D construct was fabricated by the extrusion bioprinting of a dermal region comprised of collagen type I, fibroblasts and ECs, followed by the inkjet bioprinting of keratinocytes and melanocytes at day 14 of culture (Figure 4.1b) [29]. Despite in vivo construct implantation showed the formation of a stratified epidermis and the presence of blood and lymphatic capillaries, additional research is required to demonstrate how to control the pigmentation of newly formed skin.

In addition to the bioprinting of skin grafts in vitro for subsequent in vivo implantation, remarkable advances have been made on in situ bioprinting [23,30,31]. This strategy involves the direct bioprinting of bioinks into the wound bed, surpassing in vitro culture and maturation steps. To inform the characteristics of the graft (e.g. dimension, geometry, layers) and ensure the precise placement of bioinks, a laser scanner is often used to capture wound bed data. Pioneer work using a pressure-driven bioprinting system to deposit a fibrin-collagen gel loaded with bone marrow-derived mesenchymal stem cells (MSCs) and amniotic fluid-derived stem (AFS) cells directly in the injury site showed the ability of cellularized grafts to enhance wound closure and re-epithelialization compared to cell-free hydrogels. Despite results indicated that cells did not permanently integrate into the host tissue, the release of trophic factors by the cells into the wound bed could explain the improved healing outcomes [32]. To enhance the biological function of newly formed skin, recent studies have focused on the development of photocrosslinkable bioinks capable of regulating the sustained release of cell-secreted growth factors [31], the evaluation of functional properties of new skin in a porcine wound model [30], as well as on the bioprinting of three-layer skin grafts consisting of hypodermal (preadipocytes), dermal (dermal fibroblasts and microvascular ECs) and epidermal (keratinocytes and melanocytes) layers [23]. Technological progress has also led to the development of handled bioprinters for the deposition of cell-material formulations into the wound bed. Such compact bioprinting systems do not require the laser scanning of wounds, providing a versatile tool for the rapid delivery of skin grafts onto irregular wounds in vivo. A portable and lightweight handheld extrusion-based bioprinter was developed (Figure 4.1c), and its ability to promote skin repair was demonstrated through the controlled deposition of cell-laden hydrogels into murine and porcine excisional wounds [33].

4.3.1.2 Bioprinted 3D Constructs for Skin Modelling

In recent years, remarkable progress has been made in bioprinting 3D constructs for skin modelling [17,34–36]. Such models have been evaluated for their ability to recapitulate key aspects of healthy and diseased skin, serving as biomimetic platforms for drug screening. The traditional bioprinting strategy involves the deposition of bioinks onto a receiving platform towards the fabrication of skin models that are subsequently cultured under static conditions using standard culture systems. Advanced strategies have been focused on generating dynamic culture systems via the fabrication of perfusable channels and integration with microfluidics. Skin-on-a-chip devices have been mostly established using traditional fabrication techniques [37], but important advances have recently been made by exploring 3D printing to create custom platforms for long-term culture [38] and bioprinting technology to control the location of cells and materials in 3D [17]. One important consideration in bioprinted constructs for skin modelling concerns the evaluation of key functional properties in order to assess their predictive value and biomimicry to the native skin. In this regard, Kim et al. [36] created a perfusable and vascularized multilayer (epidermis, dermis and hypodermis) healthy skin model and evaluated key characteristics such as functional markers in each skin layer and barrier function of the vascular channel. The perfusable model was generated through an integrated biofabrication strategy involving the 3D printing of a customized polycaprolactone (PCL) chamber and the bioprinting of sacrificial channels and bioinks. The model was connected to a peristaltic pump to allow dynamic culture conditions for endothelium formation and perfusion through the endothelium-lined channel to mimic vascular perfusion. Bioprinting has also been used for bioengineering 3D models of diseased skin, including hypertrophic scar [34], diabetic skin [17] and atopic dermatitis [35]. These pathophysiologically relevant models hold potential to assume a pivotal role in the screening of personalized medicine as they can be bioprinted using patients' own cells obtained from tissue biopsies and created in a reasonable time period. In a recent work, a multilayer 3D skin model recapitulating pathophysiological hallmarks of type 2 diabetes (Figure 4.1d) was established by bioprinting bioinks loaded with cells isolated from diabetic donors [17]. It was found that the diabetic skin model was able to reproduce hallmarks of diabetes such as increased insulin resistance, vascular dysfunction, delayed re-epithelialization, adipose hypertrophy and pro-inflammatory responses (Figure 4.1e). Moreover, by administering the drugs metformin and eicosapentaenoic acid through the vascular channel within the hypodermal region, it was possible to demonstrate the responsiveness of the model to these drugs.

4.3.2 HEART

Cardiovascular diseases are a major cause of mortality worldwide with around 17.5 million deaths every year [39]. As cardiomyocytes (CMs) have limited self-renewing potential and are unable to regenerate upon injury, clinical outcomes include scar tissue formation and, ultimately, heart failure. In this regard, induced pluripotent stem cells (iPSCs) hold great promise for cardiac regeneration as they

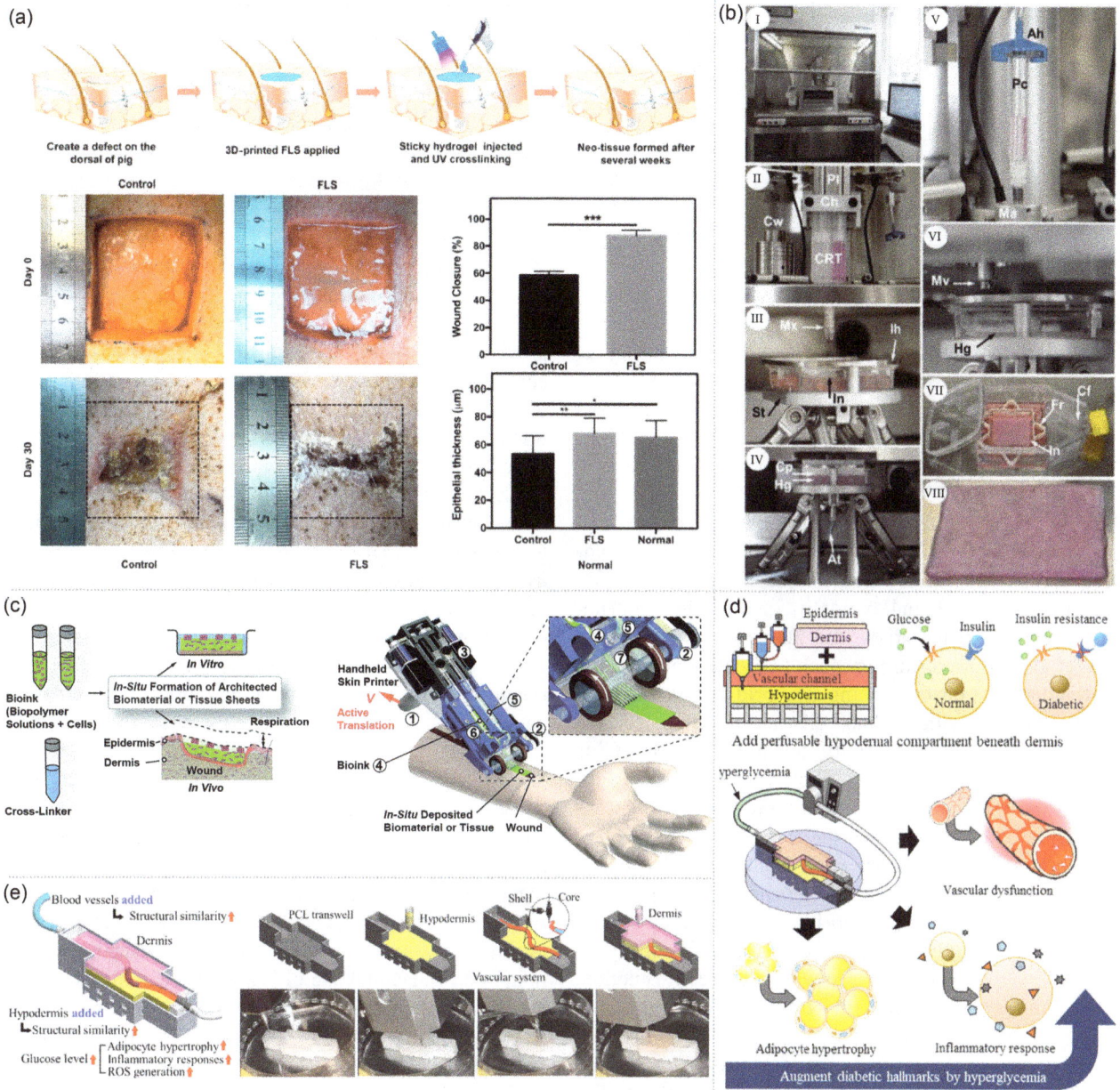

FIGURE 4.1 Bioengineered constructs for skin repair and modelling. (a) Schematic representation of the surgical process; macroscopic images of the skin wound healing; wound closure rate and quantification of the epithelial thickness after 30 days of treatment. (Modified from Ref. [26], with permission.) (b) Illustration of the skin substitute production process. (I) bioprinting equipment; (II) production of the dermal component; (IV) hydrogel plastic compression; (V) inkjet bioprinting of human epidermal cells; (VII) cell culture system; (VIII) pre-vascularized pigmented human dermo-epidermal skin substitute of $6 \times 6 \, cm^2$. (Modified from Ref. [29], with permission.) (c) Handheld extrusion-based bioprinter. A handle (1) enables positioning above the target surface or wound; stepper motor, pulley and drive mechanism (2) define the deposition speed; two on-board syringe pump modules (3) control the dispensing flow rates for bioink (4) and crosslinker solution (5); Syringe holder (6); 3D printed microfluidic cartridge (7) for spatial organization of solutions and sheer formation. (Modified from Ref. [33], with permission.) (d) Bioprinted skin model recreating diabetic hallmarks by the inclusion of perfusable, vascularized hypodermal compartment and (e) augmented diabetic features. (Modified from Ref. [17], with permission.)

are characterized by wide availability and the possibility to differentiate into multiple cell lineages, including CMs [40]. For patients with end-stage heart failure, heart transplantation is a viable treatment. However, donor shortage, potential immune rejection and surgical complications limit its clinical applicability. Therefore, bioprinting has

been employed for the fabrication of cell-laden patches for cardiac regeneration and *in vitro* models of cardiac tissue to test new therapeutic options.

The bioprinting of cardiac tissue is challenging due to the difficulty in creating a 3D construct mimicking the structural complexity and cellular heterogeneity of the native tissue while exhibiting functional and regenerative properties. The heart is divided into three layers, including the pericardium, myocardium and endocardium. The pericardium is the superficial layer that covers the heart. The myocardium is the middle and thickest layer made largely of cardiac muscle cells, a framework of collagenous fibres, blood vessels and nerve fibres. The endocardium is the innermost layer of the heart and is made of simple squamous epithelium called endothelium. Cells in the heart include CMs, cardiac fibroblasts (CFs) and ECs. CMs are contractile cells responsible for myocardium functions, while fibroblasts produce ECM components that support CMs and ECs. CMs are also surrounded by blood vessels and capillaries, increasing the complexity in fabricating cardiac-like tissues. CFs and CMs are the major cells that regulate heart function and respond to pathogenic stimuli [41].

4.3.2.1   Bioprinted Grafts for Cardiac Repair

Biofabrication strategies to create cardiac tissues can be broadly categorized as scaffold-based or scaffold-free approaches [42,43]. Scaffold-based approaches are the most explored and typically involve the bioprinting of cell-laden materials into 3D cardiac-like tissues. Zhu et al. [44] used this approach to create a biomimetic cardiac tissue construct using a hybrid bioink made of GelMA loaded with cardiac cells and gold nanorods to impart electrical conductivity towards inducing CMs maturation and organization. Bioprinted constructs supported the viability, spreading and proliferation of CMs and CFs, while the expression of gap junction protein connexin 43 and synchronized contractile frequency were observed 14 days post-bioprinting. Despite the *in vivo* performance of bioprinted constructs was not assessed, this study demonstrated that electrically conductive gold nanorods improve cardiac cell function. This can be potentially attributed to their inhibition effect on the overproliferation of CFs, which contributes to the maintenance of a balanced CMs to CFs ratio and regulation of cardiac fibrosis. Indeed, previous studies have reported that conductive materials can regulate the behaviour of CFs [45,46]. In another study, a microfluidic bioprinting strategy was used to create multicellular cardiac constructs by dispensing HUVECs and iPSC-CMs embedded in alginate and poly(ethylene glycol) (PEG)-fibrinogen bioink (Figure 4.2a) [9]. The bioprinting strategy allowed the fabrication of myocardial-like constructs supporting the enhanced expression of cardiac proteins (α-sarcomeric actin, cardiac troponin and connexin 43) and inducing preferential orientation of iPSC-derived CMs along the printing direction when compared to manually casted hydrogels. These data suggest superior functional CM organization in bioprinted constructs, which can be attributed to cell geometric confinement and spatial orientation within the fibres. To address the vascularization of cardiac tissue, constructs were bioprinted with controlled spatial distribution of iPSC-derived CMs and HUVECs and implanted *in vivo* subcutaneously in

NOD-SCID mice. After 2 weeks of implantation, "Janus constructs", in which the cells were in close proximity, supported the formation of branched vascular capillaries, integrated into the host tissue and exhibited superior organization and orientation of CMs. These results highlight the role of bioprinting in controlling the interactions and spatial distribution of cells within 3D constructs and the impact of construct heterogeneity in biological function.

The bioprinting of cell-laden cardiac patches for *in vivo* implantation has been evaluated as a strategy to reduce cardiac fibrosis and improve the functional properties of cardiac tissue upon injury. Cellularized cardiac patches can provide a temporary support matrix that provides mechanical support, improves cell retention or engraftment after delivery and encourages stem cell adhesion and proliferation to the damaged site, emerging as an alternative to cell injection [47]. Cardiac patches have been fabricated from a single bioprinting technology [48,49] or by combining multiple bioprinting technologies to improve functional characteristics such as mechanical properties [50,51]. Cui et al. [52] explored beam-scanning stereolithography to create a four-dimensional (4D) cardiac patch for the treatment of myocardial infarction (MI). The printing strategy allowed the fabrication of a GelMA/PEGDA (polyethylene glycol diacrylate) hydrogel patch, in which fibre orientation and mesh pattern were designed to recreate the anisotropic nature of cardiac muscle fibres and adapt to the changes of ventricular curvature during the cardiac cycle (Figure 4.2b). As a decrease in the metabolic activity of iPSC-CMs embedded within the hydrogels was observed in bioprinted contracts, hydrogels were coated with Matrigel for subsequent seeding of iPSC-CMs, ECs and MSCs. To assess long-term functional effects, bioprinted cellular and acellular patches (4 mm Ø, 600 μm thickness) were implanted over the infarcted (ischaemia) site of hearts in a chronic MI mouse model with ischaemia-reperfusion injury. After 4 months, cell-laden patches showed stronger integration within the epicardium, enhanced blood vessel density, higher cell density and smaller infarct area compared to the acellular patch. Despite this study demonstrating that printed patches exerted no adverse effects on the host cardiac function, implanted CMs exhibited immature 3D sarcomeric organization and the patches did not significantly improve the cardiac function. Further evaluation of the impact of patches' mechanical properties on functional properties and implantation in larger animal models are warranted to further evaluate the therapeutic efficacy. Scaffold-free approaches have also been reported for bioprinting cardiac patches by exploring the ability of cells to self-organize and to secret their own ECM without the need for additional biomaterials. This approach was used to create a cardiac patch by using bioprinting for positioning multicellular spheroids (450–550 μm Ø; hiPSC-CMs, CFs and ECs) onto a needle array for subsequent fusion and formation of a mature patch [42]. In this study, bioprinting was used to assure the precise positioning and distance between cardiac spheroids to promote the formation of cardiac patches with uniform shape and thickness. It was found that a close contact between the spheroids is required for spheroid fusion, which is essential to assure proper mechanical integrity and to preclude patch disintegration after decannulation (Figure 4.2c). *In vitro* assays showed that bioprinted cardiac patches were spontaneously beating 3 days after bioprinting and that a higher fibroblast ratio could inhibit the electrical coupling of myocardial cells. Preliminary *in vivo* tests suggested vascularization and engraftment of cardiac patches onto nude rat hearts (Figure 4.2d), though implantation in more relevant animal models and detailed characterization are required. In addition, further evaluation regarding the viability and functionality of cells in the core of cardiac spheroids is also warranted to improve the functional properties of cardiac patches.

4.3.2.2 Bioprinted 3D Constructs for Heart Modelling

A major challenge in the drug development pipeline concerns to the evaluation of cardiotoxicity of drug candidates. Although 2D monolayer cultures and *in vivo* models have been the first choice to assess drug cytotoxicity, the bioprinting of 3D cellularized cardiac models provides the opportunity to generate an appropriate environment capable of recapitulating key aspects of the physiology and biology of the cardiac tissue [53,54]. Moreover, such models offer superior biomimicry and afford easy manipulation of their characteristics when compared to 2D models and animals, respectively. In this regard, 3D bioprinting was used to create endothelialized-myocardium-on-a-chip platform for cardiovascular toxicity evaluation [8]. Using a coaxial printing nozzle, HUVECs were encapsulated within bioprinted microfibrous lattices to induce their migration towards the peripheries of the microfibres during 2 weeks of culture in order to form a layer of confluent endothelium. Then, microfibres were seeded with CMs to induce the formation of myocardium (Figure 4.2e). A microfluidic perfusion bioreactor was designed to create an endothelialized-myocardium-on-a-chip platform to support long-term culture and cell viability. It was found that perfusion at the flow rate of 50 µL/min enhanced the viability of HUVECs and CMs, compared to non-perfused cultures. Endothelialized myocardial constructs showed uniform beating (50–70 bpm) up to at least 2 weeks of culture inside the bioreactor during perfusion culture. When treated with doxorubicin, a common anticancer drug, the endothelialized-myocardium-on-a-chip model showed a dose-dependent response characterized by a decrease in the beating rate of CMs to 70.5% and 1.62% after 6 days of treatment with 10 and 100 µM concentrations, respectively (Figure 4.2f). Similar responses were observed in bioprinted models where hiPSC-CMs were used as the source of CMs, showing the potential of the models to be used in the assessment of drug-induced cardiovascular toxicity and personalized drug screening. In another example, Zhang et al. [55] used a 3D sacrificial bioprinting technique to fabricate an *in vitro* biomimetic human thrombosis-on-a-chip model. In this strategy, sacrificial Pluronic channels were first bioprinted for subsequent casting and photocrosslinking of GelMA solution. Afterwards, Pluronic was removed by immersion in cold phosphate-buffered saline, leaving a hydrogel construct containing microchannels. Then, HUVECs were seeded into the microchannels and cultured until confluency was reached to recreate the endothelium. To induce thrombosis, 0.1 M of $CaCl_2$ in Dulbecco's phosphate-buffered saline was mixed with human whole blood and infused in the microchannels of the bifurcation thrombosis model, leading to red blood cells aggregation and clot formation within ~10 min. To test the model response to tissue plasminogen activator (tPA) therapy, tPA was perfused in the microchannels at 1 mL/h for up to 2 h, resulting in the dissolution of the clots in the microchannels. Furthermore, by culturing fibroblasts embedded within the GelMA hydrogel, authors also demonstrated that their model can be used to study fibrosis of the thrombus through the migration of fibroblasts into the vascular lumen when the integrity of the endothelium is compromised.

4.3.3 Liver

The liver is the largest gland of the human body and plays a pivotal role in the maintenance of good health state of the human body. Specifically, the liver filters blood before entering the bloodstream, detoxifies chemicals and metabolizes drugs, produces bile and synthesizes crucial proteins. The liver is composed not only of a complex and extensive network of blood vessels, but also several types of resident cells, such as hepatocytes, hepatic stellate cells, liver sinusoid ECs, Kupffer cells and biliary epithelial cells, which are spatially arranged in hexagonal blocks named as hepatic lobules [56,57]. Liver diseases represent a major cause of mortality worldwide. According to the Global Burden of Disease project, there were around two million deaths in 2010 due to liver cirrhosis, liver cancer and acute hepatitis [58]. As the self-regeneration ability of the liver is limited, the selection of the most appropriate strategy to enhance liver restoration should take into consideration the degree of the lesion, as well as the pros and cons of therapeutic options. Although the use of autologous grafts or liver transplantation are effective options, they present several drawbacks such as patient morbidity and scarring, limited donor availability or even possibility of patient's rejection [59]. To surpass these limitations and provide biologically functional alternatives for liver tissue engineering, extrusion and DLP-based bioprinting have been used to create 3D constructs to aid liver regeneration and bioengineered liver-on-a-chip platforms [60].

4.3.3.1 Bioprinted Grafts for Liver Repair

In recent years, several works have focused on developing biomaterials for liver regeneration and studying the impact of material cues on cell response and regenerative outcomes both *in vitro* and *in vivo* [61–67]. As an example, Kim et al.

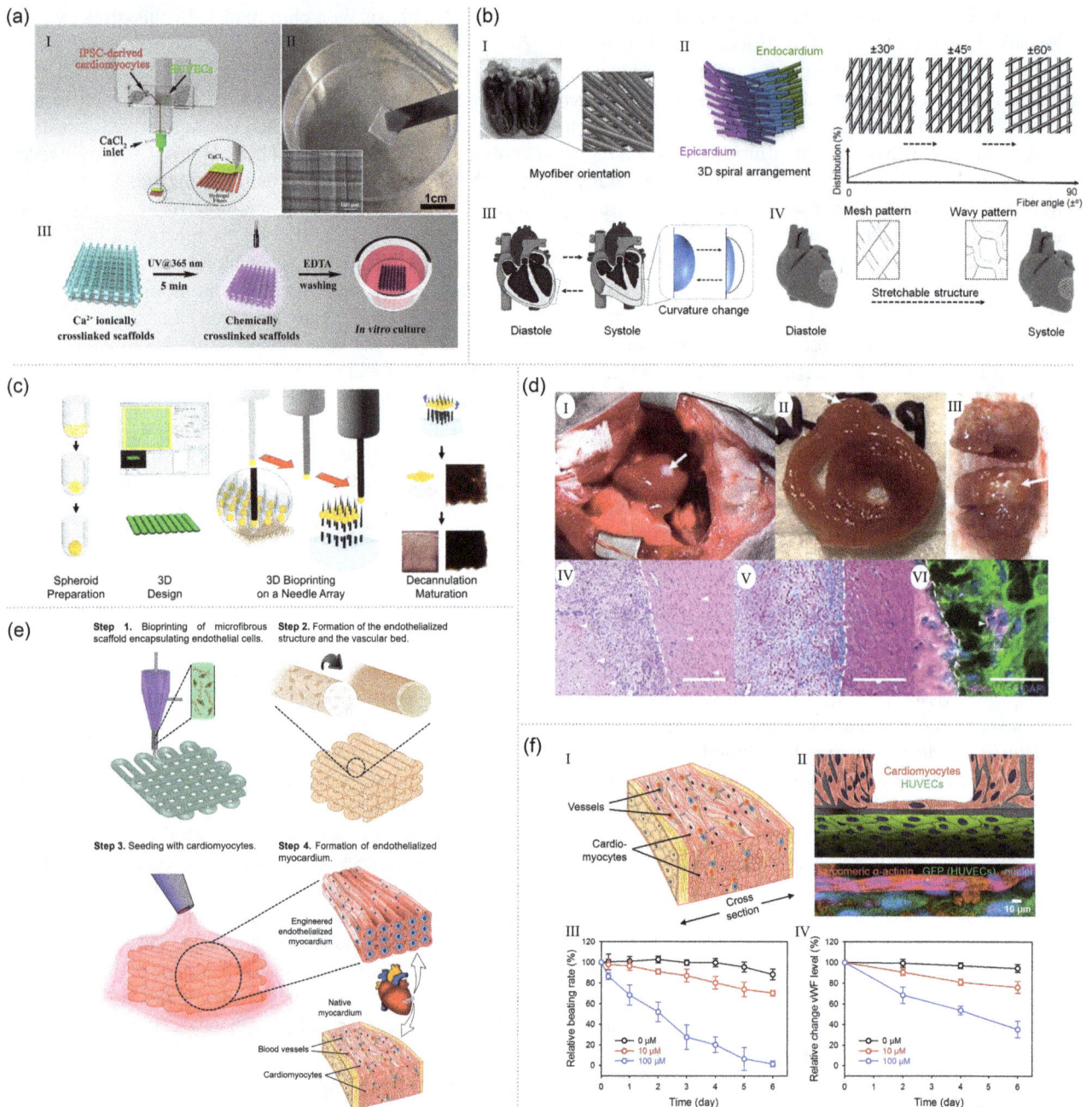

FIGURE 4.2 Bioengineered constructs for heart repair and modelling. (a) Bioprinting process of iPSC-derived CMs and HUVEC cells to create cardiac constructs. (Modified from Ref. [9], with permission.) (b) Bioprinting a 4D cardiac patch; (I) photograph of the anatomical heart and the fibre structure; (II) schematic illustration showing the variation of the fibre angle from the epicardium to the endocardium; (III) curvature change of cardiac muscle at the diastole and systole of the cardiac cycle; (IV) CAD design of the 3D architecture of the heart. (Modified from Ref. [52], with permission.) (c) Representation of the printing process to develop the cardiac patches through the aggregation of cardiac spheroids. (Modified from Ref. [42], with permission.) (d) *In vivo* implantation of bioprinted cardiac patch. (I) implantation of the cardiac patch onto the rat heart; (II, III) cross-section and anterior aspect of the explanted heart, respectively; (IV–VI) H&E, Masson Trichrome and confocal microscopy image of the cardiac patch (left) and the native rat myocardium (right). (Modified from Ref. [42], with permission.) (e) Bioprinting of endothelialized-myocardium-on-a-chip. (f) (I) illustration of the native myocardium; (II) schematic and confocal fluorescence micrograph of the endothelialized myocardial tissue (after 15 days); (III, IV) relative beating and levels of vWF expression upon treatment with different dosages of doxorubicin. (e, f: Modified from Ref. [8], with permission.)

[62] developed a patient-specific hepatic cell sheet from human chemically derived hepatic progenitor cells (hCdHs) on a multiscale fibrous PCL scaffold fabrication through the combination of electrospinning and 3D printing. Most importantly, due to its similarity with the liver tissue-like structure, the hepatic patch effectively repopulated the damaged parenchyma and improved both liver functions and survival rate (> 70%) when implanted in a pre-clinical mouse model. In order to provide cells with an environment that more closely recapitulates the liver tissue, several works have explored liver decellularized extracellular matrix (dECM) for the development of liver-specific bioinks. In this regard, Lee et al. [63] created a liver dECM-based bioink loaded with HepG2 cells and compared its performance to collagen bioinks. The overall results showed higher levels of secreted albumin and urea (up to 7 days) in the liver dECM group, which also enhanced the function of the HepG2 cells and stem cell differentiation potency. These results suggest that the preservation of the liver-specific ECM biochemical cues enhances cell response, highlighting the growing trend for using dECM in bioink design. In a similar approach, DLP-based bioprinting was explored to create a 3D construct made of liver dECM, GelMA and human-induced hepatocytes converted from human fibroblasts [61]. The addition of GelMA in bioink composition precluded dECM modification for photopolymerization, while liver-dECM improved printability and cell viability of GelMA bioinks. In this study, a novel liver microtissue with an inner gear-like structure enabled a larger surface area, improved HepG2 cell function and better stem cell differentiation (Figure 4.3a).

Several studies have been using bioprinting to create vascularized 3D liver constructs in an attempt to restore the native liver vasculature [56]. In a common approach, macroscale channels are printed inside 3D scaffolds using sacrificial materials, and ECs are used to promote vascularization due to their capacity to interconnect and self-assemble through cell-cell interactions, leading to formation of capillary networks. Using this strategy, a hepatic lobule structure with a central vein was created by extrusion bioprinting. The biofabricated hepatic lobules include collagen and alginate solutions loaded with HepG2/C3A and ECs, and a central lumen built using alginate as a sacrificial material. The engineered hepatic lobule construct showed higher albumin and urea secretion, along with higher MRP2 and CD31 protein levels as well as increased cytochrome P450 (CYP)3A4 and CYP1A1 enzyme activity [68].

4.3.3.2 Bioprinted 3D Constructs for Liver Modelling

Over the past years, the mortality risk associated with liver diseases has increased, representing a major healthcare burden worldwide [69]. Thus, it is of utmost importance to develop biomimetic and biologically functional liver models that can be used to test the efficacy and safety of drug candidates to treat liver's diseases and reduce the failure rates of therapeutics in clinical trials [56]. Examples include liver models to study drug-induced liver injury (DILI), tissue-specific infections of different pathogens and hepatocellular carcinoma (HCC), which is the most common type of primary liver cancer.

One of the most common causes of drug development failure and a major problem for the pharmaceutical industry and regulatory authorities is DILI. 3D bioprinting is suitable to create miniaturized constructs that can closely mimic the native liver tissue to study the hepatotoxicity of several compounds in an initial stage of drug development [70]. Nguyen et al. [71] developed a scaffold-free assembly of human hepatocytes, stellate cells and umbilical vein cells towards creating a 3D construct for liver modelling. To evaluate the ability of the model for the assessment of tissue-level DILI, dose responses of hepatotoxic trovafloxacin (inducer of hepatotoxicity) were compared to structurally related non-toxic Levofloxacin. Results showed that after 7 days of dosing, trovafloxacin induced significant decreases in both albumin and ATP, while levofloxacin led to a decrease only in albumin at the top dose tested (100 μM). Despite the model lacking specialized liver cells and the tissue structure needing improvement to recapitulate native liver, it was able to detect drug toxicity. In a similar study, Janani et al. [72] engineered a gelatin/silk fibroin/porcine liver dECM bioink for bioprinting an *in vitro* liver model with a hepatic sinusoid network to evaluate the drug toxicity. The liver model (Figure 4.3b) was fabricated through the deposition of bioinks loaded with human adipose mesenchymal stem cell-derived hepatocyte-like cells (HLCs) and with HUVECs/human hepatic stellate cells (HHSCs) in a 1:1 ratio. Results showed that exposure to non-hepatotoxicants dexamethasone and aspirin had no effect on the viability and metabolism of the bioprinted construct. On the contrary, exposure to the idiosyncratic drug, trovafloxacin mesylate, induced dose-dependent toxicity with a significant decrease in the percentage of cell viability with increased drug concentrations. Overall, the bioprinted liver model was more effective to predict the hepatotoxicity of several drugs than previously reported 3D liver tissues.

Bioprinted liver models have also been established for transduction and infection studies, providing an alternative *in vitro* platform to both animal models and more simplistic multicellular spheroids. To create such a humanized 3D liver model, a bioink comprised of an immortalized hepatic cell line (HepaRG), alginate and gelatin supplemented with human lung-derived ECM was bioprinted into 3D constructs [73]. Despite the ECM was not derived from the liver, it was found that its presence at either 0.5 or 1 mg/mL increased albumin secretion and CYP3A4 activity (Figure 4.3c). Moreover, the model supported viral infection and replication as observed by an increase of infectious adenoviral particles between days 3 and 7 post infection, suggesting that the bioprinted model is suitable to study viral biology in a humanized 3D cell culture.

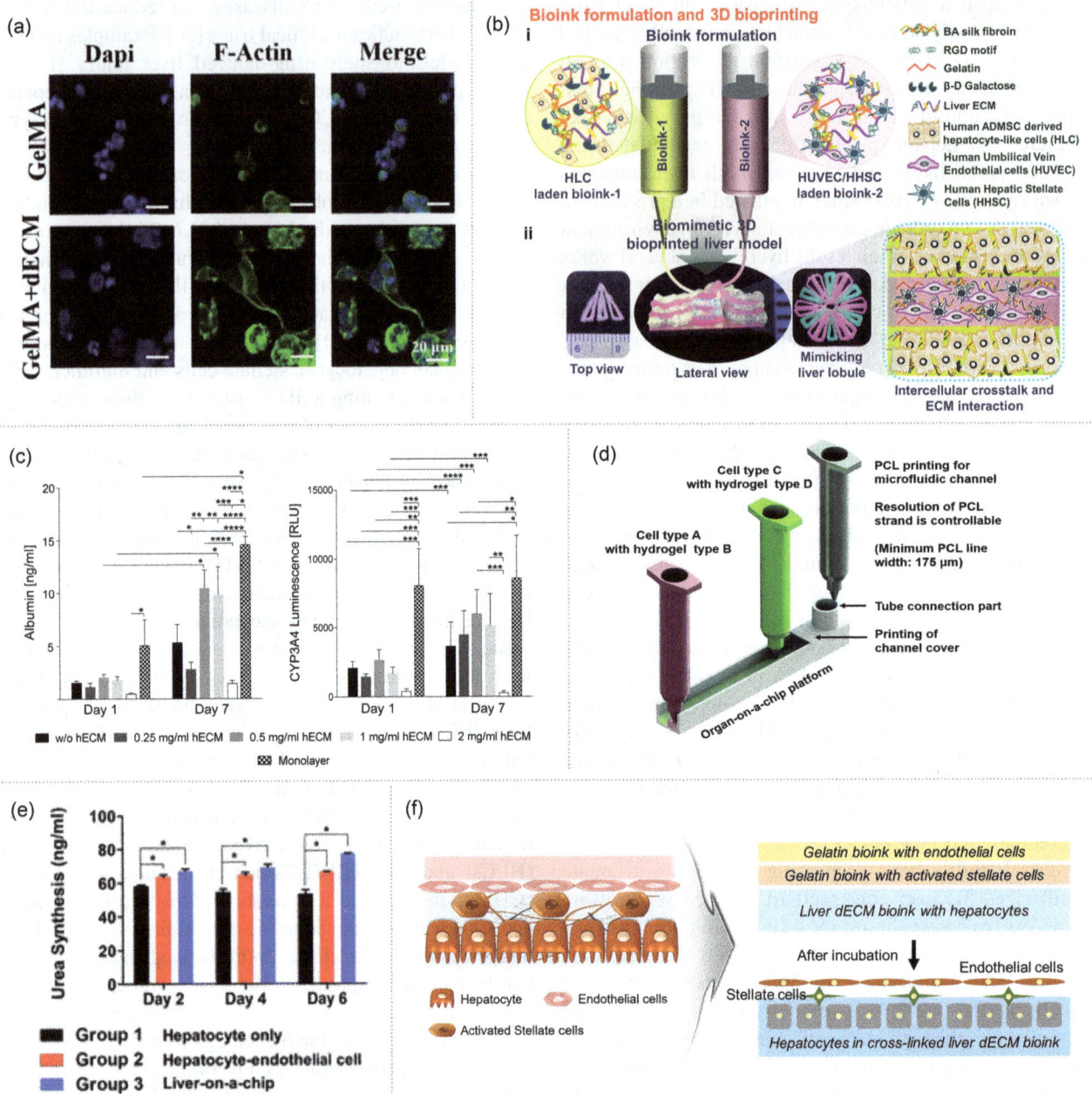

FIGURE 4.3 Bioengineered constructs for liver repair and modelling. (a) Confocal fluorescence microscope image of 3D bioprinted construct (5 days). (Modified from Ref. [61], with permission.) (b) Illustration of the bioink formulation and 3D bioprinting process to create the liver model. (Modified from Ref. [72], with permission.) (c) Albumin secretion and CYP3A4 activity of the bioprinted HepaRG cells. (Modified from Ref. [73], with permission.) (d) Schematic representation of the 3D bioprinting process to develop the liver-on-a-chip and (e) urea synthesis of the liver-on-a-chip. (Modified from Ref. [76], with permission.) (f) Illustration of the spatial arrangement of the liver cells in the liver microenvironment and the 3D bioprinting process to create the liver fibrosis-on-a-chip. (Modified from Ref. [77], with permission.)

Hepatocellular carcinoma is the 6th most common cause of cancer worldwide [69]. In order to generate 3D tissue models capable of providing a spatial cirrhotic architecture with biomimetic native composition and improved predictive value, DLP-based 3D bioprinting was used to create an *in vitro* dECM-based 3D liver platform recapitulating the mechanical properties of cirrhotic liver tissue [74]. Liver dECM was combined with GelMA and HepG2 cell line, and bioprinted in 3D constructs with tunable stiffness.

The stiffness of the models was tailored by changing the photopolymerization time and modulated in a range of 0.5–15 kPa, which is similar to healthy and cirrhotic range matrix environment. Moreover, stiff constructs not only downregulated the expression of the liver-specific markers such as albumin, but significantly induced higher migration and invasion potential of HepG2 liver cancer cells.

The fabrication of liver-on-a-chip has received growing attention due to the possibility of culturing liver cells for long time periods via the continuous perfusion through microfluidic channels [56,75]. As a result, these platforms are promising to understand and monitor liver function and disease progression, as well as to test DILI. In this regard, Lee and Cho [76] established a 3D-bioprinted PCL liver-on-a-chip platform through the encapsulation of HepG2 and HUVECs in collagen type I and gelatin hydrogels, respectively (Figure 4.3d). The liver-on-a-chip exhibited higher secretion values of urea (Figure 4.3e) and albumin in comparison to co-cultured groups, which can be explained by the constant medium perfusion through the printed microfluidic channels, mimicking the native flow environment. In another study, gelatin and liver dECM were used to create a 3D liver fibrosis-on-a-chip through extrusion bioprinting. The model comprises several structured layers of hepatocytes, activated stellate cells and ECs. Specifically, hepatocytes were mixed with liver dECM bioink and bioprinted to provide cells with a 3D liver microenvironment, while gelatin bioinks were used to deliver endothelial or activated stellate cells (Figure 4.3f). The results showed that developed 3D liver fibrosis-on-a-chip exhibited high levels of activated LX2 stellate cells (fibrosis group), reduced gene expression levels of CYP3A4 and recapitulated key hallmarks of liver fibrosis such as collagen deposition and cell apoptosis [77].

4.3.4 BIOPRINTING CANCER MODELS

Cancer remains the second most common cause of death worldwide, with 10 million cancer deaths in 2020 [78]. Despite the advances in diagnosis and new treatments, the therapeutic resistance and the uncertainty of the response to therapy remain huge problems for cancer patients. One of the explanations for therapeutic failure is the lack of experimental models for drug screening that take into consideration the tumour macrostructure and the complex 3D interactions between cancer cells, stromal cells and the ECM. It is well described that gene and protein expression [79,80], protein gradient profiles [81], cell signalling [82], migration [83], proliferation [84] and therapy response [84–86] have been shown to diverge between 2D and 3D cancer models [87]. As consequence, this hinders the development of new therapeutic approaches, highlighting an unmet need for more complex and physiologically relevant 3D cancer models that provide consistent and predictive responses about the *in vivo* efficiency of the drug formulations.

In light of this challenge, the design and development of 3D culture systems have gained a huge relevance once they can gather cancer cells with various cell types in a spatially relevant way, providing cell-cell and cell-ECM interactions that closely mimic the native tumour microenvironment [88,89]. Nevertheless, only a few 3D models have been examined for use in drug discovery and development due to the strict regulatory and validation requirements, and lack of industry validation. One step closer to pre-clinical phase are 3D bioprinted models, a major breakthrough in the field, with key advantages of automation, stability and providing

highly controllable cancer tissue models, with potential to significantly accelerate cancer research and drug discovery and development [90–92] (Figure 4.4a). These models can exhibit a complex 3D architecture of living cells (both cancer and normal cells, in monoculture or co-culture), ECM components and biochemical factors which are bioprinted with microscale resolution, closely reconstituting the native tumour microenvironment [93]. The properties of biomaterials used to recapitulate the ECM are critical to determine not only the bioink printability, but also the fate of embedded cells. For example, in the case of cancer models, the stiffness of these materials should be adjusted to the tissue specificity, as the increased matrix stiffness drives epithelial–mesenchymal transition and tumour metastasis [94] and tumour vasculature phenotype [95] and promotes chemoresistance [96] in different tumour models. Thus, these biomaterials should recreate ECM structural, physicochemical and biological properties to provide characteristic tumour microenvironment within the constructs. Biopolymers such as collagen, hyaluronic acid, alginate and gelatin are widely used in bioinks due to their low cytotoxicity, biocompatibility and high-water content, favouring cell adhesion, proliferation and differentiation. Also, Matrigel [97] and fibrin [98] hybrid matrices have been extensively used for cancer models, providing an ideal microenvironment for cancer cell proliferation and generating a tumour tissue construct with closely similar 3D architecture as well as functional features to *in vivo* tumours.

Many 3D-bioprinted cancer models have been successfully developed, as for breast [99–101] (Figure 4.4b), brain [16,102,103], skin [104], pancreatic [105], ovarian [97], cervical [106] and other cancers [107–109] (Figure 4.4c), and also for metastasis [110,111]. Many of these models produced in low-to-high throughput have been designed to explore their potential as pre-clinical *in vitro* models for evaluating drug efficacy, toxicity, chemotherapy or chemoresistance, which will be summarized below.

4.3.4.1 Breast Cancer

Breast cancer was the most prevalent cancer worldwide in 2020 [78]. Besides its huge heterogeneity, the stromal compartment of breast tumour microenvironment is clearly involved in the tumour progression and also in resistance to therapy, being necessary for more physiologically relevant models to improve the therapeutic strategies design [112,113]. This is likely the type of cancer most frequently reported for 3D-bioprinted cancer models, with several constructs including breast cancer cells, stromal and vascular cells, and some of them being already evaluated as drug screening platforms [114]. Hong and Song [115] recently reported the extrusion bioprinting of a drug-resistant breast cancer spheroid within a gelatin-alginate hydrogel. This model maintained the MCF-7 cells drug resistance profile, expressing CD44^{high}/CD24^{low}/ALDH1^{high} in the gelatin-alginate matrix during 3D culture and exhibited higher expression levels of drug resistance markers, such as GRP78 chaperon and ABCG2 transporter. Further, this model was

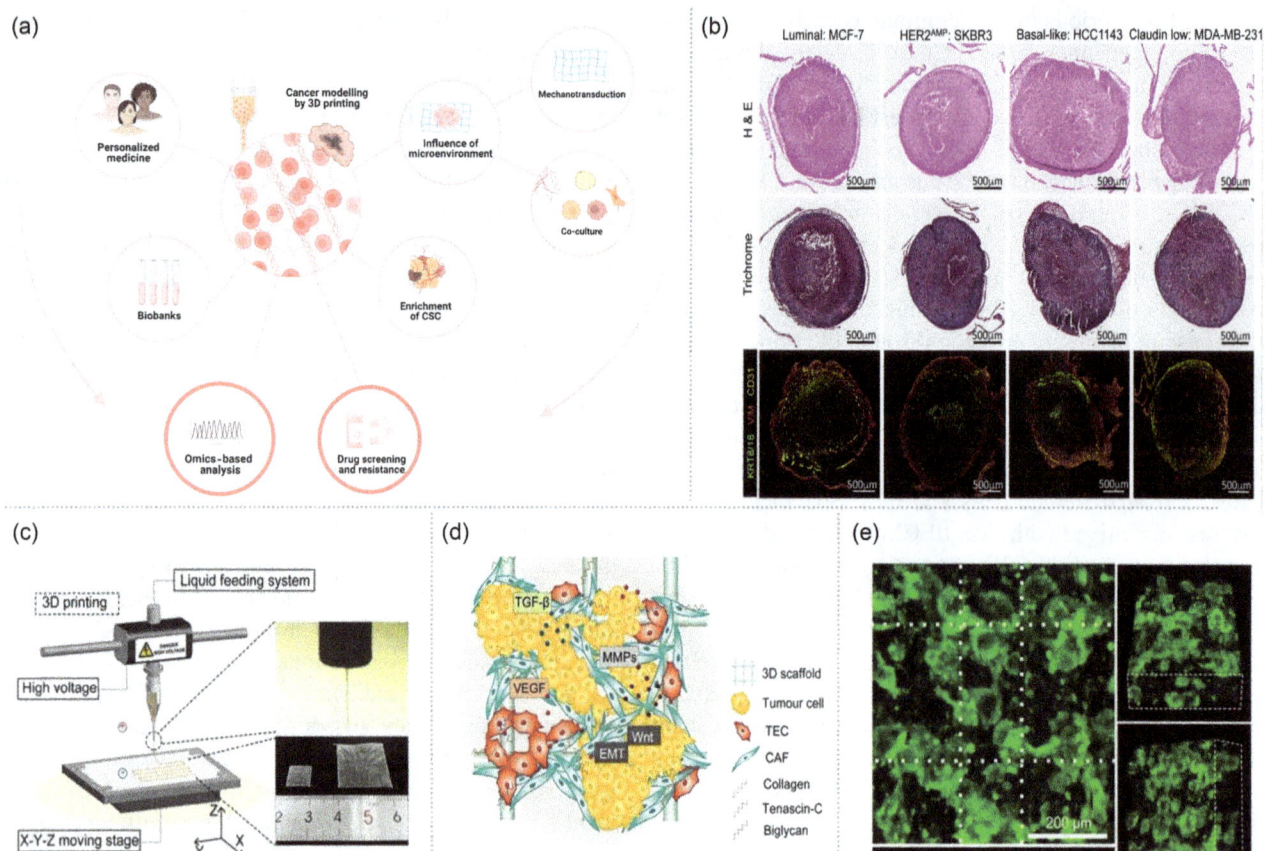

FIGURE 4.4 (a) The potential of 3D-bioprinted cancer models for cancer research. 3D bioprinting offers the ability to create highly complex 3D architecture of living cells (monoculture or co-culture), extracellular matrix (ECM) components and biochemical factors. These models can be used in different studies, ranging from cancer biology to drug screening and resistance approaches, with the advantage of the cancer stem cells (CSCs) enrichment. Further, bioprinted cancer models containing patient-derived cancer and stromal cells are promising for personalized cancer therapy screening. The possibility of testing also a cohort of patients (biobank) is also an attractive tool to test their response and resistance to different therapeutic approaches. (Modified from Ref. [90], with permission.) (b) 3D-bioprinted models from distinct tumour cell subtypes. Representative H&E images of tissues bioprinted with MCF-7, SKBR3, HCC1143 or MDA-MB-231 breast cancer cell lines. Tissues were bioprinted into a stromal mix of human mammary fibroblasts and HUVECs, and fixed on day 10 (scale bars: 500 μm). Trichrome staining of tissues (scale bars: 500 μm). Immunofluorescence on sections from bioprinted tissues, stained for KRT8/18 (green), VIM (red), and CD31 (yellow) (scale bars: 500 μm). (Modified from Ref. [99], with permission.) (c) Illustration of the E-jet 3D bioprinting device. (d) Illustration of the 3D tumour tissue model of colorectal cancer. (e) Fluorescence images of the tumour tissue (green=live cells, scale bar=200 μm). White dotted lines indicate the centre of the scaffold fibres. (c–e: Modified from Ref. [107], with permission.)

used as a screening model for anticancer drugs, and a drug resistance profile was confirmed through higher effective concentration 50 (EC50) values in resistant spheroids than in MCF-7 bulk spheroids. In a more complex model, Wang and colleagues [100] bioprinted constructs with 21PT breast cancer cells and adipose-derived mesenchymal stem/stromal cells (ADMSC), one of the major stromal cells in the breast cancer microenvironment that promote cancer progression. It was observed that the percentage of cleaved Caspase-3 positive cells was significantly lower in the bioprinted constructs with ADMSC and 21PT than in the cancer cell alone constructs, in response to low doxorubicin dose, suggesting that ADMSCs protect cancer cells from doxorubicin antitumor effects. Furthermore, the increased thickness of ADMSCs that mimics the obesity profile was associated with more cells stained negative for cleaved

Caspase-3, indicating less apoptosis, and likely more therapeutic resistance, thus providing a breast cancer model that physiologically is closer to *in vivo* breast tumours. In other study, researchers investigated if a ready-to-use model tissue could be fabricated by bioprinting pre-formed 3D spheroids in alginate-based bioinks [101]. Breast cancer cell lines were bioprinted as individual cells or as pre-formed spheroids, either in monoculture or co-culture with vascular ECs. Bioprinted breast cancer cells only formed spheroids when printed in Matrigel bioink, while pre-formed breast spheroids maintained their viability, architecture and function after bioprinting in alginate-based bioink and also in Matrigel. Furthermore, bioprinted breast spheroids were more resistant to paclitaxel than individually printed breast cells, suggesting that bioprinting has potential to create cancer models that rapidly simulate the tumour microenvironment. Other 3D-bioprinted models combining breast cancer cells with fibroblasts [116] and osteoblasts [117] among other cells have been described, but they were not yet evaluated for drug screening purposes. Notably, recent studies have established bioprinted constructs with patient tissue samples where the cellular proliferation, extracellular matrix deposition and cellular migration were responsive to extrinsic signals or therapies, demonstrating their future potential for personalized therapy [99]. It is still noticeable the lack of models including the immune compartment, being one of the next challenges to be addressed.

4.3.4.2 Pancreatic Cancer

Pancreatic ductal adenocarcinoma has the lowest survival rate among all major cancers and is the 7th leading cause of cancer-related mortality [78]. The disturbing survival statistics and very poor response rates to current treatments urgently claim new therapeutics and pre-clinical models. Hakobyan et al. [105] bioprinted pancreatic exocrine spheroids from rat acinar cells to study early pancreatic tumorigenesis. The authors embedded the spheroids in GelMA hydrogels and showed that bioprinted spheroids, composed of both acinar and ductal cells, replicate the initial stages of PDAC development. This spheroid-based array model can be useful to study the formation of precursor PDAC lesions and cancer progression, and may thus shed light on future PDAC therapy strategies. One step further, pancreatic 3D-bioprinted models have already been established for high-throughput drug screening (HTS) [118]. In a recent study, a cancer-derived spheroid model was produced in flat bottomed well plates with a cell-repellent surface, through the combination of bioprinting technology incorporating magnetic force. The model was validated by evaluating the anticancer activity of 3300 approved drugs on patient-derived pancreatic cancer primary cells, an important milestone to get closer to personalized medicine. The results were compared to 2D assays, being the tested drugs less active in 3D than 2D systems, advancing with homogeneous 3D tumour models compatible with HTS [119].

4.3.4.3 Glioblastoma

Glioblastoma (GBM) is the most common type of brain cancer among adults. GBM are fast-growing and aggressive tumours, and its treatment includes surgery followed by radiotherapy, chemotherapy and other therapeutics. Nevertheless, patients frequently have tumour relapse likely due to cancer drug resistance [120]. The urgency for novel *in vitro* models which mimic the native GBM microenvironment is crucial to overcome tumour drug resistance, with several bioprinted models being reported [121]. In this concern, Dai et al. [102] designed an extrusion 3D-bioprinted glioma stem cell model, using modified porous gelatin/alginate/fibrinogen hydrogel to mimic the tumour ECM. The glioma stem cells presented a high survival and proliferation rate in the bioink as well as the differentiation and vascularization potential. Further, this model was validated for drug studies, with 3D-bioprinted tumour model being more resistant to temozolomide than the 2D monolayer model. In

a more complex model, Yi and colleagues [16] established a "glioblastoma-on-a-chip" using patient-derived glioblastoma cells, co-cultured with vascular endothelial cells on a decellularized ECM from brain tissue. In this model, a compartmentalized cancer–stroma concentric-ring structure was developed, which maintained the radial oxygen gradient and recapitulated the structural, biochemical and biophysical properties of *in vivo* tumour microenvironment. Importantly, this model reproduced the patient-specific resistance to treatment with concurrent chemoradiation and temozolamide, highlighting its potential to be used in the identification of effective treatments for GBM patients, who are resistant to conventional therapies. Further, to study GBM-immune cell interaction and its impact on therapeutics, Heinrich et al. [122] developed 3D-bioprinted minibrains consisting of GBM cells and glioblastoma-associated macrophages (GAMs). In this model, GAMs were actively recruited by GBM cells and polarized into a typical GAM phenotype, which was similar to the transcriptomic analysis of GBM patients. GAMs were able to induce proliferation and invasion of GBM cells, and therapeutics targeting GAMs reduced tumour growth. Others using GBM stem cells alone or with astrocytes and neural precursor cells in a hyaluronic acid-rich hydrogel, with or without macrophages, also showed that constructs incorporating macrophages recapitulate patient transcriptional profiles predictive of patient survival, maintenance of stemness, invasion and drug resistance [123]. Thus, these models can support the study of novel cancer therapeutics taking into consideration the impact of macrophages in the response to therapy. Trying to bring one more variable – blood vessels – to these models and get closer to the tumour microenvironment, Neufeld and colleagues [98] created a fibrin GBM bioink containing patient-derived glioblastoma cells, astrocytes and microglia. Additionally, perfusable blood vessels were created using a bioink coated with brain pericytes and ECs. This model was compared to orthotopic cancer mouse models, and they shared similar growth curves, drug response and genetic signature of GBM cells, in opposite to 2D models. This platform could replace the use of animal models in the future and provide advances with potential to be used for personalized therapy and drug development.

Despite bioprinting cancer models being still in its initial stages, the current advancements show a great pace to have more physiologically relevant models for cancer research and drug development, particularly through the use of patient-derived samples, immune cells and perfusable models. Nevertheless, there are still challenges to overcome, like the choice of a bioink that better mimics the native microenvironment, which is cancer type-dependent, as well as the dimension of bioprinted constructs and the bioprinting time.

4.3.5 Bioprinting of Organoids

Organoids are miniaturized and simplistic stem cell-derived 3D structures that partially replicate tissue-level composition and function. These can now be established from different sources of human tissue, including iPSCs, adult organs and diseased tissue. Undoubtedly, organoid culture development has been a major breakthrough in biomedical research [124]. Still, these advanced culture models present several limitations, namely the presence of typically only one cell component, limited diffusion of nutrients and oxygen, an enclosed lumen that makes access to the apical compartment problematic, as well as intra- and inter-culture heterogeneity regarding organoid shape and size [125]. Moreover, the most frequent choice of ECM for organoid cultures is still Matrigel, a reconstituted basement membrane derived from extracts of Engelbreth-Holm-Swarm mouse sarcoma. Although quite effective in supporting cell growth and differentiation, clinical translation of these models is significantly impaired due to Matrigel's animal origin, poorly defined composition and consequent batch-to-batch variability, offering no control over individual form and function. Recently, 3D bioprinting techniques employing organoids have been put forward trying to solve some of the previous issues by combining physiologically relevant cell sources together with bioinspired engineering strategies to precisely instruct cell architecture and behaviour.

4.3.5.1 Bioprinted Organoids for Regenerative Purposes

Head and neck cancers radiotherapy treatments can lead to the development of xerostomia, a clinical condition whereby saliva output is greatly diminished due to radiation-induced damage in the secretory epithelial and neuronal populations. In this context, bioengineered innervated secretory tissues have significant clinical translation potential. Magnetic 3D bioprinting, employing NanoShuttle-PL magnetized nanoparticles consisting of gold, iron oxide and poly-L-lysine, was used to produce human dental pulp stem cell 3D spheroids of high viability and uniform size [126]. After going through a chemically defined differentiation stage, these spheroids originated salivary-gland-like organoids containing both epithelial and neuronal cell populations. The epithelial compartment was highly polarized and secreted salivary amylase, while the large neuronal network present was functionally responsive to both cholinergic (parasympathetic) and β-adrenergic (sympathetic) neurotransmitters. Notably, transplantation experiments revealed that these salivary gland-like organoids rescued epithelial growth in an irradiated mouse model and stimulated innervation in both irradiated and healthy glands, with neuronal integration arising from both organ and transplanted tissue [126]. Although vascularization was limited, the innervated organoid-like structures mimic to some extent the complex secretory acini architecture of the human salivary gland and hold potential for regenerative purposes aiming to restore saliva flow in disease conditions.

Cardiovascular diseases have a massive impact on human health, and the development of 3D cardiac patches is a promising tool for heart tissue regeneration. A bioink composed of GelMA, collagen methacrylate, fibronectin and laminin-511/111 was optimized to support hiPSCs viability, proliferation and differentiation into CMs [127]. These hiPSCs were bioprinted into 3D human chambered muscle pumps (hChaMPs) organoids with two chambers and a vessel inlet/outlet by free-form reversible embedding of suspended hydrogels and subjected to a CM small molecule-based differentiation protocol. The hChaMPs were maintained for at least 6 weeks, showing a mature CM phenotype and exhibiting contiguous electrical function and pump dynamics [127]. This study adopted a unique approach by performing in situ differentiation, which allowed the coupling between tissue maturation and the formation of stable cell connections, thus achieving robust electromechanical function. Further refinements to the model must focus on increasing the thickness, homogeneity and organization of the cardiac-like muscle wall.

In a seminal work, organoid-forming stem cells' capacity to spontaneously self-organize into a given architecture depending on microenvironmental cues was leveraged through an approach termed bioprinting-assisted tissue emergence (BATE) to create large-scale tissue replicas [128]. As a proof of concept, murine intestinal single-stem cells bioprinted within a viscous Matrigel/collagen precursor solution fused and morphed into fully polarized and lumenized intestinal epithelial tubes, harbouring putative crypt- and villus-like domains [128]. These structures were maintained for long periods of time (over 3 weeks), reached a large diameter (over 400 µm) and width (over 15 mm) and showed physiological responses to chemical stimuli. Moreover, co-culture settings were implemented by simultaneous or sequential bioprinting of intestinal mesenchymal cells, accelerating epithelial morphogenesis. Incidentally, the observed speeding up in intestinal lumen formation and diameter mediated by stromal cells allowed to connect the tissue to a perfusion system for removal of dead cells shed into the lumen during tissue turnover [128]. Through precise control of spatial geometry, cellular density and deposition time, this study introduced a versatile strategy for printing centimetre-scale constructs of diverse nature, such as epithelial, connective and vascular human tissues, as well as to establish organ boundaries in vitro, achieving significant milestones in the field of stem cell biology and regenerative medicine.

Recently, light-driven volumetric bioprinting was used for the establishment of functional human intrahepatic bile duct-derived organoid-laden constructs (Figure 4.5a) [129]. First, optical properties of the bioresin and its effects on bioprinting resolution were optimized with the addition of a biocompatible water-miscible refractive index matching compound termed iodixanol. Next, viability and functional activity of the bioprinted structures under differentiation conditions were compared against extrusion bioprinting and cast-only controls. Volumetric bioprinted organoids displayed superior viability and increased average size. Moreover, this was paired with cell differentiation

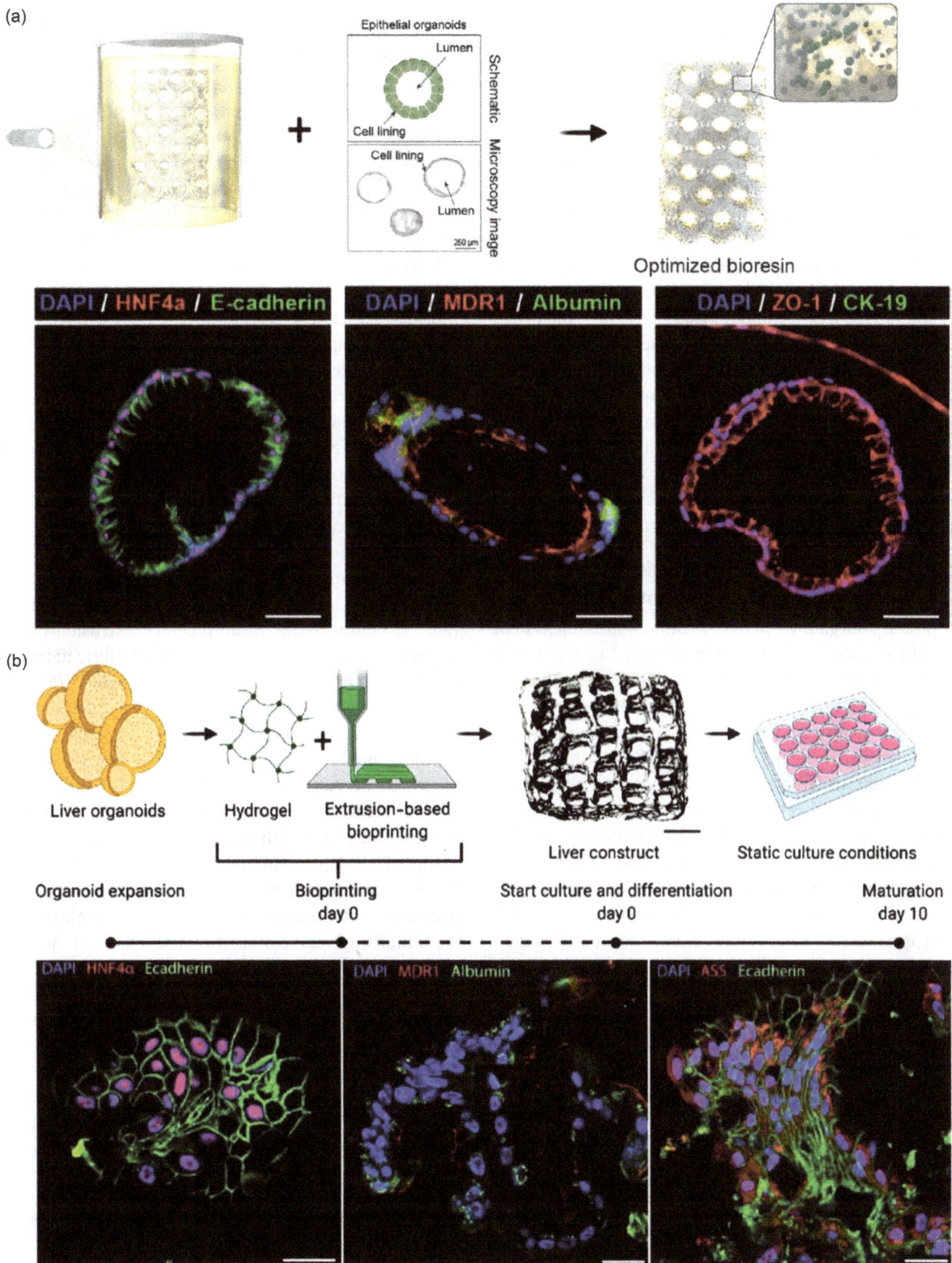

FIGURE 4.5 Bioprinting strategies employing organoids. (a) Graphical representation of light-driven volumetric bioprinting using liver-derived organoids and an optimized GelMA-based bioresin (top panel; scale bar: 250 μm). Representative fluorescence images of the liver-specific markers hepatic nuclear factor 4 alpha (HNF4α) and albumin, as well as of organoid polarization markers E-cadherin, multi-drug resistance protein 1 (MDR1), zonula occludens-1 (ZO-1) and cytokeratin 19 (CK19) after 10 days of hepatic differentiation (bottom panel; scale bars: 50 μm). (Modified from Ref. [129], with permission.) (b) Graphical depiction of the experimental procedure for extrusion-based bioprinting of liver constructs using liver organoid-derived cells and GelMA as a bioink to fabricate porous structures (top; scale bar: 1000 μm). Organoids are guided towards a hepatocyte-like phenotype during culture in differentiation media for 10 days, with expression of the hepatic markers HNF4α, albumin and argininosuccinate synthase (ASS), as well as expression of the polarity markers MDR1 and E-cadherin (bottom; scale bars: 25 μm). (Modified from Ref. [131], with permission.)

properties, such as (1) expression of hepatocyte-specific markers, including hepatocyte nuclear factor 4 alpha (HNF4α) and albumin; (2) liver-specific enzyme activity like aspartate transaminase and glutamate dehydrogenase; (3) glycogen storage; and (4) native-like organoid polarization [129]. Importantly, hepatocyte functionality was combined with dynamic perfusion conditions through the establishment of different lattice structures with interconnected porosity. Volumetric bioprinting allowed fast and high-fidelity bioprinting of these structures, maximizing cell seeding and exchange surface area, which significantly improved ammonia detoxification capacity [129]. As a layerless and nozzle-free technique, volumetric printing offers the ability to print morphologically intact and larger-sized organoids as they are not subjected to potentially damaging shear stresses through a nozzle-based approach (like during extrusion printing), thus potentiating their metabolic activity. On the other hand, a clear limitation seems to be the requirement for high cell densities, which implies the development of bioreactor-like culture systems to support the subsequent volumetric bioprinting process.

4.3.5.2 Bioprinted Organoids for Disease Modelling and Drug Screening

A proprietary Organovo 3D NovoGen Bioprinter system was used to establish a 3D human intestinal tissue mimic with a bilayered structure consisting of primary human ileal epithelial cells and intestinal myofibroblast cells resuspended in a thermo-responsive bioink and printed onto permeable Transwell inserts [130]. This 3D intestinal model could be maintained up to 21 days in culture, presenting specialized epithelial cell types, near-physiological barrier function and permeability, polarized efflux transporters expression, metabolically active cytochrome P450 function and compound-induced cytotoxicity [130]. The main advantages of the model include long-term viability, the printed laminar architecture that assures apical and basal access with bi-directional transport and the possibility of incorporating additional cell types, making it highly suitable for gastrointestinal toxicity testing. Still, the isolated contribution of each cell type used (epithelial and mesenchymal) for morphogenesis and function of the *in vitro* model was not assessed, which is a necessary point to further escalate complexity.

DILI is a significant cause of acute liver failure and the most frequent reason for drug withdrawal after failed clinical trials. In this biological context, human liver-derived epithelial organoids have also been considered a promising cell source to establish more advanced models for *in vitro* pre-clinical testing. Intrahepatic cholangiocyte organoids (ICOs) were resuspended in a GelMA and deposited by extrusion-based bioprinting together with a sacrificial Pluronic F127 hydrogel to create microporous constructs (Figure 4.5b) [131]. These bioprinted structures allowed maintenance of viable and mature ICOs-derived HLCs for at least 10 days, showing cellular polarization, expression of differentiation markers, sustained metabolic activity,

glycogen accumulation and drug-induced toxicity when exposed to the well-known hepatotoxic compound acetaminophen [131]. However, these hepatic cells were not zonally oriented, possibly due to the absence of other cell types, which can be achieved in a co-culture setting.

Pluripotent stem cells are a powerful blank slate to derive different types of organoids and achieve physiological relevance, especially when combined with advanced bioengineering strategies. For instance, hiPSCs were bioprinted within a bioink containing a mixture of alginate, carboxymethyl-chitosan and agarose, which still enabled self-renewal properties after gelation through calcium chloride-mediated crosslinking [132]. The hiPSCs showed proliferative capacity together with contact inhibition, formation of large cell aggregates in the form of spheroids (colony-like structures) and ubiquitous expression of pluripotency cell markers like OCT4 and SOX2. Besides initial expansion, differentiation potential was demonstrated by *in situ* embryoid bodies (EBs) formation, with the expression of specific markers representative of the 3 germ layers – mesoderm, ectoderm and endoderm [132]. It was possible to instruct differentiation of the bioprinted hiPSCs towards a specific lineage, in particular their conversion to functional neural cells by use of alternative media composition to generate 3D neural tissue. More recently, extrusion-based 3D bioprinting was used for the automated fabrication of self-organizing kidney organoids [133]. Single differentiated hiPSCs were deposited as micromasses without any carrier hydrogel on a Transwell filter, showing the spontaneous formation of nephron-like structures after 20 days of culture. The yield and pattern reproducibility of this kidney organoid generation system were advantageous for nephrotoxicity screenings in a high-throughput manner demonstrated through dose-dependent viability responses upon exposure to the chemotherapeutic agent doxorubicin or to broad-spectrum antibiotics of the aminoglycoside class [133]. It was equally observed that the modulation of bioprinting conditions affected organoid conformation, with an impact on patterning, nephron number and maturation. This observation facilitated the biomanufacturing of uniformly patterned kidney-like tissue sheets with functional proximal tubular segments [133]. The establishment of hiPSCs-based printing platforms allowing both self-renewal and differentiation into different cell lineages is a versatile tool to generate physiologically relevant models of disease amenable to personalized therapy.

4.3.5.3 Bioprinting 3D Templates for Organoid-Derived Culture

Several groups have gathered evidence showing that organoid function is intimately linked with its morphology. Thus, efforts have been directed towards establishing more physiologically relevant architectural cues to precisely instruct and control the shape and size of *ex vivo* organ-like structures [134]. Hence, biofabrication techniques are also being harnessed towards the development of specific moulds or templates to be colonized by organoid-derived cells. For instance, sacrificial thioester functionalized PEG elastomers were printed by DLP to create 3D crypt- and villi-like features with high fidelity (<100 μm) that were moulded into soft (<1000 Pa) hydrogel substrates, like Matrigel, supporting the growth of intestinal epithelial layers from mouse single intestinal stem cells [135]. A limitation of the technique is the use of a non-physiological pH for template erosion, which prevents the presence of cells during the process. On the other hand, it is a robust and quick method to fabricate complex 3D features maintaining the integrity of soft tissue culture substrates. In a different approach, using 3D-printing and soft lithography methods, a microwell platform was established to support the development of human embryonic stem cell (ESC)-derived EBs without introducing Matrigel or another ECM [136]. Overall, the high-resolution printed and bottom-curved cell-repulsive moiety methyl ether (m)PEG-silane coated microwells demonstrated the highest efficacy in promoting uniform EB formation and subsequent cerebral organoid maturation [136]. Although cell/tissue dependent, this study opens the possibility of modelling organoid development effectively in an ECM-free environment.

4.4 CONCLUSIONS AND FUTURE PERSPECTIVES

In the last decade, significant progress has been made in the bioprinting of 3D constructs for tissue modelling and repair. This is particularly relevant for skin, cardiac and liver applications with several studies demonstrating not only tissue repair upon the implantation of constructs, but also the ability of bioprinted constructs for modelling key properties of native tissues. Bioprinting is also emerging in cancer and organoid applications, where the unique capabilities of these technologies have been explored to create biomimetic constructs for therapeutic screening and personalized medicine. Despite the bioprinting of bioengineered grafts at clinically relevant dimensions has been demonstrated, the bioprinting of constructs mimicking the biomechanical properties of native tissues and containing a functional vascular network is still a challenge. To address these issues, developments on printable ECM-mimetic biomaterials, isolation and expansion of primary and stem cells, as well as integration of multiple bioprinting technologies, are required. Encouraging results from pre-clinical studies in animal models have demonstrated that anatomically shaped bioprinted grafts can promote tissue repair. However, detailed investigation regarding the functional properties of repaired tissues and rigorous demonstration of the superiority of bioprinted tissues are warranted for clinical translation. In addition, attention should be focused on the selection of cells and animal models for the evaluation of bioprinted constructs, which is essential to enhance the clinical significance of the research findings. Bioprinted constructs for tissue modelling are assuming a central position as potential alternatives to animal models. Although the bioprinting

of biomimetic tissue models and their integration with microfluidics are still in the beginning, it is necessary to demonstrate that such models can accurately replicate key (patho)physiological features of native tissues/organs and their superior performance relative to animal models. Even considering the full replacement of animal models by bioprinted constructs is difficult, such models can be a viable and cost-effective alternative to animals in recreating certain aspects of native tissues/organs and, therefore, can be used for specific applications (e.g. toxicology, drug safety, target identification and validation). To this end, the demonstration of functional properties, biomimicry and validation of predictive value of these models is essential for their acceptance by regulatory agencies, pharmaceutical and biotechnology companies.

REFERENCES

1. Z. Wang, W. Kapadia, C. Li, F. Lin, R.F. Pereira, P.L. Granja, B. Sarmento, W. Cui, Tissue-specific engineering: 3D bioprinting in regenerative medicine, *J Control Release* 329 (2021) 237–256.
2. L. Moroni, J.A. Burdick, C. Highley, S.J. Lee, Y. Morimoto, S. Takeuchi, J.J. Yoo, Biofabrication strategies for 3D in vitro models and regenerative medicine, *Nat Rev Mater* 3 (2018) 21–37.
3. R.F. Pereira, P.J. Bartolo, 3D bioprinting of photocrosslinkable hydrogel constructs, *J Appl Polym Sci* 132(48) (2015) 42458.
4. K.S. Lim, J.H. Galarraga, X. Cui, G.C.J. Lindberg, J.A. Burdick, T.B.F. Woodfield, Fundamentals and applications of photo-cross-linking in bioprinting, *Chem Rev* 120(19) (2020) 10662–10694.
5. A.M. Jorgensen, J.J. Yoo, A. Atala, Solid organ bioprinting: Strategies to achieve organ function, *Chem Rev* 120(19) (2020) 11093–11127.
6. R.F. Pereira, B.N. Lourenco, P.J. Bartolo, P.L. Granja, Bioprinting a multifunctional bioink to engineer clickable 3D cellular niches with tunable matrix microenvironmental cues, *Adv Healthc Mater* 10(2) (2021) e2001176.
7. M. Shin, J.H. Galarraga, M.Y. Kwon, H. Lee, J.A. Burdick, Gallol-derived ECM-mimetic adhesive bioinks exhibiting temporal shear-thinning and stabilization behavior, *Acta Biomater* 95 (2019) 165–175.
8. Y.S. Zhang, A. Arneri, S. Bersini, S.R. Shin, K. Zhu, Z. Goli-Malekabadi, J. Aleman, C. Colosi, F. Busignani, V. Dell'Erba, C. Bishop, T. Shupe, D. Demarchi, M. Moretti, M. Rasponi, M.R. Dokmeci, A. Atala, A. Khademhosseini, Bioprinting 3D microfibrous scaffolds for engineering endothelialized myocardium and heart-on-a-chip, *Biomaterials* 110 (2016) 45–59.
9. F. Maiullari, M. Costantini, M. Milan, V. Pace, M. Chirivì, S. Maiullari, A. Rainer, D. Baci, H.E.-S. Marei, D. Seliktar, C. Gargioli, C. Bearzi, R. Rizzi, A multi-cellular 3D bioprinting approach for vascularized heart tissue engineering based on HUVECs and iPSC-derived cardiomyocytes, *Sci Rep* 8(1) (2018) 13532.
10. S.M. Hull, C.D. Lindsay, L.G. Brunel, D.J. Shiwarski, J.W. Tashman, J.G. Roth, D. Myung, A.W. Feinberg, S.C. Heilshorn, 3D bioprinting using UNIversal Orthogonal Network (UNION) bioinks, *Adv Funct Mater* 31(7) (2021) 2007983.

11. F.L.C. Morgan, L. Moroni, M.B. Baker, Dynamic bioinks to advance bioprinting, *Adv Healthc Mater n/a(n/a)* 9(15) (2020) e1901798.

12. D. Moura, R.F. Pereira, I.C. Gonçalves, Recent advances on bioprinting of hydrogels containing carbon materials, *Mater Today Chem* 23 (2022) 100617.

13. E. Daskalakis, B. Huang, C. Vyas, A.A. Acar, A. Fallah, G. Cooper, A. Weightman, B. Koc, G. Blunn, P. Bartolo, Novel 3D bioglass scaffolds for bone tissue regeneration, *Polymers* 14(3) (2022) 445.

14. H.W. Kang, S.J. Lee, I.K. Ko, C. Kengla, J.J. Yoo, A. Atala, A 3D bioprinting system to produce human-scale tissue constructs with structural integrity, *Nat Biotechnol* 34(3) (2016) 312–319.

15. D.E. Ingber, Human organs-on-chips for disease modelling, drug development and personalized medicine, *Nat Rev Genet* 23 (2022) 1–25. https://doi.org/10.1038/s41576-022-00466-9

16. H.G. Yi, Y.H. Jeong, Y. Kim, Y.J. Choi, H.E. Moon, S.H. Park, K.S. Kang, M. Bae, J. Jang, H. Youn, S.H. Paek, D.W. Cho, A bioprinted human-glioblastoma-on-a-chip for the identification of patient-specific responses to chemoradiotherapy, *Nat Biomed Eng* 3(7) (2019) 509–519.

17. B.S. Kim, M. Ahn, W.-W. Cho, G. Gao, J. Jang, D.-W. Cho, Engineering of diseased human skin equivalent using 3D cell printing for representing pathophysiological hallmarks of type 2 diabetes in vitro, *Biomaterials* 272 (2021) 120776.

18. C.K. Sen, Human wound and its burden: Updated 2020 compendium of estimates, *Adv Wound Care* 10(5) (2021) 281–292.

19. C. Lindholm, R. Searle, Wound management for the 21st century: Combining effectiveness and efficiency, *Int Wound J* 13 Suppl 2 (2016) 5–15.

20. R.F. Pereira, A. Sousa, C.C. Barrias, A. Bayat, P.L. Granja, P.J. Bártolo, Advances in bioprinted cell-laden hydrogels for skin tissue engineering, *Biomanuf Rev* 2(1) (2017) 1.

21. R.F. Pereira, A. Sousa, C.C. Barrias, P.J. Bartolo, P.L. Granja, A single-component hydrogel bioink for bioprinting of bioengineered 3D constructs for dermal tissue engineering, *Mater Horizons* 5(6) (2018) 1100–1111.

22. S. Michael, H. Sorg, C.T. Peck, L. Koch, A. Deiwick, B. Chichkov, P.M. Vogt, K. Reimers, Tissue engineered skin substitutes created by laser-assisted bioprinting form skin-like structures in the dorsal skin fold chamber in mice, *PLoS One* 8(3) (2013) e57741.

23. A.M. Jorgensen, M. Varkey, A. Gorkun, C. Clouse, L. Xu, Z. Chou, S.V. Murphy, J. Molnar, S.J. Lee, J.J. Yoo, S. Soker, A. Atala, Bioprinted skin recapitulates normal collagen remodeling in full-thickness wounds, *Tissue Eng Part A* 26(9–10) (2020) 512–526.

24. T. Baltazar, J. Merola, C. Catarino, C.B. Xie, N.C. Kirkiles-Smith, V. Lee, S. Hotta, G. Dai, X. Xu, F.C. Ferreira, W.M. Saltzman, J.S. Pober, P. Karande, Three dimensional bioprinting of a vascularized and perfusable skin graft using human keratinocytes, fibroblasts, pericytes, and endothelial cells, *Tissue Eng Part A* 26(5–6) (2020) 227–238.

25. B.S. Kim, Y.W. Kwon, J.S. Kong, G.T. Park, G. Gao, W. Han, M.B. Kim, H. Lee, J.H. Kim, D.W. Cho, 3D cell printing of in vitro stabilized skin model and in vivo pre-vascularized skin patch using tissue-specific extracellular matrix bioink: A step towards advanced skin tissue engineering, *Biomaterials* 168 (2018) 38–53.

26. F. Zhou, Y. Hong, R. Liang, X. Zhang, Y. Liao, D. Jiang, J. Zhang, Z. Sheng, C. Xie, Z. Peng, X. Zhuang, V. Bunpetch, Y. Zou, W. Huang, Q. Zhang, E.V. Alakpa, S. Zhang, H. Ouyang, Rapid printing of bio-inspired 3D tissue constructs for skin regeneration, *Biomaterials* 258 (2020) 120287.

27. S. Huang, B. Yao, J. Xie, X. Fu, 3D bioprinted extracellular matrix mimics facilitate directed differentiation of epithelial progenitors for sweat gland regeneration, *Acta Biomater* 32 (2016) 170–177.

28. N. Wei Long, Q. Jovina Tan Zhi, Y. Wai Yee, N. May Win, Proof-of-concept: 3D bioprinting of pigmented human skin constructs, *Biofabrication* 10(2) (2018) 025005.

29. L. Pontiggia, I.A.J. Van Hengel, A. Klar, D. Rütsche, M. Nanni, A. Scheidegger, S. Figi, E. Reichmann, U. Moehrlen, T. Biedermann, Bioprinting and plastic compression of large pigmented and vascularized human dermo-epidermal skin substitutes by means of a new robotic platform, *J Tissue Eng* 13 (2022) 20417314221088513.

30. M. Albanna, K.W. Binder, S.V. Murphy, J. Kim, S.A. Qasem, W. Zhao, J. Tan, I.B. El-Amin, D.D. Dice, J. Marco, J. Green, T. Xu, A. Skardal, J.H. Holmes, J.D. Jackson, A. Atala, J.J. Yoo, In situ bioprinting of autologous skin cells accelerates wound healing of extensive excisional full-thickness wounds, *Sci Rep* 9(1) (2019) 1856.

31. A. Skardal, S.V. Murphy, K. Crowell, D. Mack, A. Atala, S. Soker, A tunable hydrogel system for long-term release of cell-secreted cytokines and bioprinted in situ wound cell delivery, *J Biomed Mater Res B Appl Biomater* 105(7) (2017) 1986–2000.

32. A. Skardal, D. Mack, E. Kapetanovic, A. Atala, J.D. Jackson, J. Yoo, S. Soker, Bioprinted amniotic fluid-derived stem cells accelerate healing of large skin wounds, *Stem Cells Transl Med* 1(11) (2012) 792–802.

33. N. Hakimi, R. Cheng, L. Leng, M. Sotoudehfar, P.Q. Ba, N. Bakhtyar, S. Amini-Nik, M.G. Jeschke, A. Günther, Handheld skin printer: In situ formation of planar biomaterials and tissues, *Lab Chip* 18(10) (2018) 1440–1451.

34. Y. Bin, Z. Dongzhen, C. Xiaoli, E. Jirigala, S. Wei, L. Zhao, H. Tian, Z. Ping, L. Jianjun, W. Yuzhen, Z. Yijie, F. Xiaobing, H. Sha, Modeling human hypertrophic scars with 3D preformed cellular aggregates bioprinting, *Bioact Mater* 10 (2022) 247–254.

35. L. Xue, M. Samuel, B. Kapil, F. Marc, S. Min Jae, A biofabricated vascularized skin model of atopic dermatitis for preclinical studies, *Biofabrication* 12(3) (2020) 035002.

36. B.S. Kim, G. Gao, J.Y. Kim, D.W. Cho, 3D cell printing of perfusable vascularized human skin equivalent composed of epidermis, dermis, and hypodermis for better structural recapitulation of native skin, *Adv Healthc Mater* 8(7) (2019) e1801019.

37. G. Sriram, M. Alberti, Y. Dancik, B. Wu, R. Wu, Z. Feng, S. Ramasamy, P.L. Bigliardi, M. Bigliardi-Qi, Z. Wang, Full-thickness human skin-on-chip with enhanced epidermal morphogenesis and barrier function, *Mater Today* 21(4) (2018) 326–340.

38. N. Mori, Y. Morimoto, S. Takeuchi, Skin integrated with perfusable vascular channels on a chip, *Biomaterials* 116 (2017) 48–56.

39. E.J. Benjamin, P. Muntner, A. Alonso, M.S. Bittencourt, C.W. Callaway, A.P. Carson, A.M. Chamberlain, A.R. Chang, S. Cheng, S.R. Das, F.N. Delling, L. Djousse, M.S.V. Elkind, J.F. Ferguson, M. Fornage, L.C. Jordan, S.S. Khan, B.M. Kissela, K.L. Knutson, T.W. Kwan, D.T. Lackland, T.T. Lewis, J.H. Lichtman, C.T. Longenecker, M.S. Loop,

P.L. Lutsey, S.S. Martin, K. Matsushita, A.E. Moran, M.E. Mussolino, M. O'Flaherty, A. Pandey, A.M. Perak, W.D. Rosamond, G.A. Roth, U.K.A. Sampson, G.M. Satou, E.B. Schroeder, S.H. Shah, N.L. Spartano, A. Stokes, D.L. Tirschwell, C.W. Tsao, M.P. Turakhia, L.B. VanWagner, J.T. Wilkins, S.S. Wong, S.S. Virani, Heart disease and stroke statistics—2019 update: A report From the American Heart Association, *Circulation* 139(10) (2019) e56–e528.

40. R.E. Ahmed, T. Anzai, N. Chanthra, H. Uosaki, A brief review of current maturation methods for human induced pluripotent stem cells-derived cardiomyocytes, *Front Cell Dev Biol* 8 (2020) 178.

41. V. Talman, H. Ruskoaho, Cardiac fibrosis in myocardial infarction—from repair and remodeling to regeneration, *Cell Tissue Res* 365(3) (2016) 563–581.

42. C.S. Ong, T. Fukunishi, H. Zhang, C.Y. Huang, A. Nashed, A. Blazeski, D. DiSilvestre, L. Vricella, J. Conte, L. Tung, G.F. Tomaselli, N. Hibino, Biomaterial-free three-dimensional bioprinting of cardiac tissue using human induced pluripotent stem cell derived cardiomyocytes, *Sci Rep* 7(1) (2017) 4566.

43. E. Yeung, T. Fukunishi, Y. Bai, D. Bedja, I. Pitaktong, G. Mattson, A. Jeyaram, C. Lui, C.S. Ong, T. Inoue, H. Matsushita, S. Abdollahi, S.M. Jay, N. Hibino, Cardiac regeneration using human-induced pluripotent stem cell-derived biomaterial-free 3D-bioprinted cardiac patch in vivo, *J Tissue Eng Regen Med* 13(11) (2019) 2031–2039.

44. K. Zhu, S.R. Shin, T. van Kempen, Y.C. Li, V. Ponraj, A. Nasajpour, S. Mandla, N. Hu, X. Liu, J. Leijten, Y.D. Lin, M.A. Hussain, Y.S. Zhang, A. Tamayol, A. Khademhosseini, Gold nanocomposite bioink for printing 3D cardiac constructs, *Adv Funct Mater* 27(12) (2017) 1605352.

45. K.S. Lee, M.A. El-Sayed, Gold and silver nanoparticles in sensing and imaging: Sensitivity of plasmon response to size, shape, and metal composition, *J Phys Chem B* 110(39) (2006) 19220–192205.

46. G. Chan, D.J. Mooney, New materials for tissue engineering: Towards greater control over the biological response, *Trends Biotechnol* 26(7) (2008) 382–392.

47. R. Gaebel, N. Ma, J. Liu, J. Guan, L. Koch, C. Klopsch, M. Gruene, A. Toelk, W. Wang, P. Mark, F. Wang, B. Chichkov, W. Li, G. Steinhoff, Patterning human stem cells and endothelial cells with laser printing for cardiac regeneration, *Biomaterials* 32(35) (2011) 9218–9230.

48. R. Gaetani, D.A.M. Feyen, V. Verhage, R. Slaats, E. Messina, K.L. Christman, A. Giacomello, P.A.F.M. Doevendans, J.P.G. Sluijter, Epicardial application of cardiac progenitor cells in a 3D-printed gelatin/hyaluronic acid patch preserves cardiac function after myocardial infarction, *Biomaterials* 61 (2015) 339–348.

49. D. Bejleri, B.W. Streeter, A.L.Y. Nachlas, M.E. Brown, R. Gaetani, K.L. Christman, M.E. Davis, A bioprinted cardiac patch composed of cardiac-specific extracellular matrix and progenitor cells for heart repair, *Adv Healthc Mater* 7(23) (2018) 1800672.

50. J. Jang, H.-J. Park, S.-W. Kim, H. Kim, J.Y. Park, S.J. Na, H.J. Kim, M.N. Park, S.H. Choi, S.H. Park, S.W. Kim, S.-M. Kwon, P.-J. Kim, D.-W. Cho, 3D printed complex tissue construct using stem cell-laden decellularized extracellular matrix bioinks for cardiac repair, *Biomaterials* 112 (2017) 264–274.

51. B.-W. Park, S.-H. Jung, S. Das, S.M. Lee, J.-H. Park, H. Kim, J.-W. Hwang, S. Lee, H.-J. Kim, H.-Y. Kim, S. Jung, D.-W. Cho, J. Jang, K. Ban, H.-J. Park, In vivo priming of

human mesenchymal stem cells with hepatocyte growth factor–engineered mesenchymal stem cells promotes therapeutic potential for cardiac repair, *Sci Adv* 6(13) (2020) eaay6994.

52. H. Cui, C. Liu, T. Esworthy, Y. Huang, Z.-X. Yu, X. Zhou, H. San, S.-J. Lee, S.Y. Hann, M. Boehm, M. Mohiuddin, J.P. Fisher, L.G. Zhang, 4D physiologically adaptable cardiac patch: A 4-month in vivo study for the treatment of myocardial infarction, *Sci Adv* 6(26) (2020) eabb5067.

53. M. Abudupataer, N. Chen, S. Yan, F. Alam, Y. Shi, L. Wang, H. Lai, J. Li, K. Zhu, C. Wang, Bioprinting a 3D vascular construct for engineering a vessel-on-a-chip, *Biomed Microdevices* 22(1) (2019) 10.

54. J.U. Lind, T.A. Busbee, A.D. Valentine, F.S. Pasqualini, H. Yuan, M. Yadid, S.-J. Park, A. Kotikian, A.P. Nesmith, P.H. Campbell, J.J. Vlassak, J.A. Lewis, K.K. Parker, Instrumented cardiac microphysiological devices via multimaterial three-dimensional printing, *Nat Mater* 16(3) (2017) 303–308.

55. Y.S. Zhang, F. Davoudi, P. Walch, A. Manbachi, X. Luo, V. Dell'Erba, A.K. Miri, H. Albadawi, A. Arneri, X. Li, X. Wang, M.R. Dokmeci, A. Khademhosseini, R. Oklu, Bioprinted thrombosis-on-a-chip, *Lab Chip* 16(21) (2016) 4097–4105.

56. L. Ma, Y. Wu, Y. Li, A. Aazmi, H. Zhou, B. Zhang, H. Yang, Current advances on 3D-bioprinted liver tissue models, *Adv Healthc Mater* 9(24) (2020) 2001517.

57. Y. Ren, X. Yang, Z. Ma, X. Sun, Y. Zhang, W. Li, H. Yang, L. Qiang, Z. Yang, Y. Liu, C. Deng, L. Zhou, T. Wang, J. Lin, T. Li, T. Wu, J. Wang, Developments and opportunities for 3D bioprinted organoids, *Int J Bioprint* 7(3) (2021) 364–364.

58. P. Byass, The global burden of liver disease: A challenge for methods and for public health, *BMC Med* 12 (2014) 159–159.

59. I. Matai, G. Kaur, A. Seyedsalehi, A. McClinton, C.T. Laurencin, Progress in 3D bioprinting technology for tissue/organ regenerative engineering, *Biomaterials* 226 (2020) 119536.

60. C. Kryou, V. Leva, M. Chatzipetrou, I. Zergioti, Bioprinting for liver transplantation, *Bioengineering (Basel)* 6(4) (2019) 95.

61. Q. Mao, Y. Wang, Y. Li, S. Juengpanich, W. Li, M. Chen, J. Yin, J. Fu, X. Cai, Fabrication of liver microtissue with liver decellularized extracellular matrix (dECM) bioink by digital light processing (DLP) bioprinting, *Mater Sci Eng: C* 109 (2020) 110625.

62. Y. Kim, Y.W. Kim, S.B. Lee, K. Kang, S. Yoon, D. Choi, S.-H. Park, J. Jeong, Hepatic patch by stacking patient-specific liver progenitor cell sheets formed on multiscale electrospun fibers promotes regenerative therapy for liver injury, *Biomaterials* 274 (2021) 120899.

63. H. Lee, W. Han, H. Kim, D.-H. Ha, J. Jang, B.S. Kim, D.-W. Cho, Development of liver decellularized extracellular matrix bioink for three-dimensional cell printing-based liver tissue engineering, *Biomacromolecules* 18(4) (2017) 1229–1237.

64. Y. Wu, A. Wenger, H. Golzar, X. Tang, 3D bioprinting of bicellular liver lobule-mimetic structures via microextrusion of cellulose nanocrystal-incorporated shear-thinning bioink, *Sci Rep* 10(1) (2020) 20648.

65. K. Kang, Y. Kim, H. Jeon, S.B. Lee, J.S. Kim, S.A. Park, W.D. Kim, H.M. Yang, S.J. Kim, J. Jeong, D. Choi, Three-dimensional bioprinting of hepatic structures with directly converted hepatocyte-like cells, *Tissue Eng Part A* 24(7–8) (2017) 576–583.

66. R. Taymour, D. Kilian, T. Ahlfeld, M. Gelinsky, A. Lode, 3D bioprinting of hepatocytes: Core–shell structured co-cultures with fibroblasts for enhanced functionality, *Sci Rep* 11(1) (2021) 5130.

67. C. Zhong, H.-Y. Xie, L. Zhou, X. Xu, S.-S. Zheng, Human hepatocytes loaded in 3D bioprinting generate mini-liver, *Hepatobiliary Pancreat Dis Int* 15(5) (2016) 512–518.

68. D. Kang, G. Hong, S. An, I. Jang, W.-S. Yun, J.-H. Shim, S. Jin, Bioprinting of multiscaled hepatic lobules within a highly vascularized construct, *Small* 16(13) (2020) 1905505.

69. S.K. Asrani, H. Devarbhavi, J. Eaton, P.S. Kamath, Burden of liver diseases in the world, *J Hepatol* 70(1) (2019) 151–171.

70. Y. Zhou, J.X. Shen, V.M. Lauschke, Comprehensive evaluation of organotypic and microphysiological liver models for prediction of drug-induced liver injury, *Front Pharmacol* 10 (2019) 1093.

71. D.G. Nguyen, J. Funk, J.B. Robbins, C. Crogan-Grundy, S.C. Presnell, T. Singer, A.B. Roth, Bioprinted 3D primary liver tissues allow assessment of organ-level response to clinical drug induced toxicity in vitro, *PLoS One* 11(7) (2016) e0158674.

72. G. Janani, S. Priya, S. Dey, B.B. Mandal, Mimicking native liver lobule microarchitecture in vitro with parenchymal and non-parenchymal cells using 3D bioprinting for drug toxicity and drug screening applications, *ACS Appl Mater Interf* 14(8) (2022) 10167–10186.

73. T. Hiller, J. Berg, L. Elomaa, V. Röhrs, I. Ullah, K. Schaar, A.-C. Dietrich, M.A. Al-Zeer, A. Kurtz, A.C. Hocke, S. Hippenstiel, H. Fechner, M. Weinhart, J. Kurreck, Generation of a 3D liver model comprising human extracellular matrix in an alginate/gelatin-based bioink by extrusion bioprinting for infection and transduction studies, *Int J Mol Sci* 19(10) (2018) 3129.

74. X. Ma, C. Yu, P. Wang, W. Xu, X. Wan, C.S.E. Lai, J. Liu, A. Koroleva-Maharajh, S. Chen, Rapid 3D bioprinting of decellularized extracellular matrix with regionally varied mechanical properties and biomimetic microarchitecture, *Biomaterials* 185 (2018) 310–321.

75. J. Deng, W. Wei, Z. Chen, B. Lin, W. Zhao, Y. Luo, X. Zhang, Engineered liver-on-a-chip platform to mimic liver functions and its biomedical applications: A review, *Micromachines (Basel)* 10(10) (2019) 676.

76. H. Lee, D.-W. Cho, One-step fabrication of an organ-on-a-chip with spatial heterogeneity using a 3D bioprinting technology, *Lab Chip* 16(14) (2016) 2618–2625.

77. H. Lee, J. Kim, Y. Choi, D.-W. Cho, Application of gelatin bioinks and cell-printing technology to enhance cell delivery capability for 3D liver fibrosis-on-a-chip development, *ACS Biomater Sci Eng* 6(4) (2020) 2469–2477.

78. H. Sung, J. Ferlay, R.L. Siegel, M. Laversanne, I. Soerjomataram, A. Jemal, F. Bray, Global Cancer Statistics 2020: GLOBOCAN estimates of incidence and mortality worldwide for 36 cancers in 185 countries, *CA Cancer J Clin* 71(3) (2021) 209–249.

79. P.A. Kenny, G.Y. Lee, C.A. Myers, R.M. Neve, J.R. Semeiks, P.T. Spellman, K. Lorenz, E.H. Lee, M.H. Barcellos-Hoff, O.W. Petersen, J.W. Gray, M.J. Bissell, The morphologies of breast cancer cell lines in three-dimensional assays correlate with their profiles of gene expression, *Mol Oncol* 1(1) (2007) 84–96.

80. O. Zschenker, T. Streichert, S. Hehlgans, N. Cordes, Genome-wide gene expression analysis in cancer cells reveals 3D growth to affect ECM and processes associated with cell adhesion but not DNA repair, *PLoS One* 7(4) (2012) e34279.

81. Y.E. Kim, H.J. Jeon, D. Kim, S.Y. Lee, K.Y. Kim, J. Hong, P.J. Maeng, K.R. Kim, D. Kang, Quantitative proteomic analysis of 2D and 3D cultured colorectal cancer cells: Profiling of tankyrase inhibitor XAV939-induced proteome, *Sci Rep* 8(1) (2018) 13255.

82. A. Riedl, M. Schlederer, K. Pudelko, M. Stadler, S. Walter, D. Unterleuthner, C. Unger, N. Kramer, M. Hengstschlager, L. Kenner, D. Pfeiffer, G. Krupitza, H. Dolznig, Comparison of cancer cells in 2D vs 3D culture reveals differences in AKT-mTOR-S6K signaling and drug responses, *J Cell Sci* 130(1) (2017) 203–218.

83. P.H. Wu, A. Giri, S.X. Sun, D. Wirtz, Three-dimensional cell migration does not follow a random walk, *Proc Natl Acad Sci U S A* 111(11) (2014) 3949–3954.

84. K. Stock, M.F. Estrada, S. Vidic, K. Gjerde, A. Rudisch, V.E. Santo, M. Barbier, S. Blom, S.C. Arundkar, I. Selvam, A. Osswald, Y. Stein, S. Gruenewald, C. Brito, W. van Weerden, V. Rotter, E. Boghaert, M. Oren, W. Sommergruber, Y. Chong, R. de Hoogt, R. Graeser, Capturing tumor complexity in vitro: Comparative analysis of 2D and 3D tumor models for drug discovery, *Sci Rep* 6 (2016) 28951.

85. K. Storch, I. Eke, K. Borgmann, M. Krause, C. Richter, K. Becker, E. Schrock, N. Cordes, Three-dimensional cell growth confers radioresistance by chromatin density modification, *Cancer Res* 70(10) (2010) 3925–3934.

86. J.C. Fontoura, C. Viezzer, F.G. Dos Santos, R.A. Ligabue, R. Weinlich, R.D. Puga, D. Antonow, P. Severino, C. Bonorino, Comparison of 2D and 3D cell culture models for cell growth, gene expression and drug resistance, *Mater Sci Eng C Mater Biol Appl* 107 (2020) 110264.

87. M. Kapalczynska, T. Kolenda, W. Przybyla, M. Zajaczkowska, A. Teresiak, V. Filas, M. Ibbs, R. Blizniak, L. Luczewski, K. Lamperska, 2D and 3D cell cultures - a comparison of different types of cancer cell cultures, *Arch Med Sci* 14(4) (2018) 910–919.

88. J. Pape, M. Emberton, U. Cheema, 3D cancer models: The need for a complex stroma, compartmentalization and stiffness, *Front Bioeng Biotechnol* 9 (2021) 660502.

89. S. Pozzi, A. Scomparin, S. Israeli Dangoor, D. Rodriguez Ajamil, P. Ofek, L. Neufeld, A. Krivitsky, D. Vaskovich-Koubi, R. Kleiner, P. Dey, S. Koshrovski-Michael, N. Reisman, R. Satchi-Fainaro, Meet me halfway: Are in vitro 3D cancer models on the way to replace in vivo models for nanomedicine development? *Adv Drug Deliv Rev* 175 (2021) 113760.

90. N. Germain, M. Dhayer, S. Dekiouk, P. Marchetti, Current advances in 3D bioprinting for cancer modeling and personalized medicine, *Int J Mol Sci* 23(7) (2022) 3432.

91. J.L. Albritton, J.S. Miller, 3D bioprinting: Improving in vitro models of metastasis with heterogeneous tumor microenvironments, *Dis Model Mech* 10(1) (2017) 3–14.

92. W. Peng, P. Datta, B. Ayan, V. Ozbolat, D. Sosnoski, I.T. Ozbolat, 3D bioprinting for drug discovery and development in pharmaceutics, *Acta Biomater* 57 (2017) 26–46.

93. P. Datta, M. Dey, Z. Ataie, D. Unutmaz, I.T. Ozbolat, 3D bioprinting for reconstituting the cancer microenvironment, *NPJ Precis Oncol* 4 (2020) 18.

94. S.C. Wei, L. Fattet, J.H. Tsai, Y. Guo, V.H. Pai, H.E. Majeski, A.C. Chen, R.L. Sah, S.S. Taylor, A.J. Engler, J. Yang, Matrix stiffness drives epithelial-mesenchymal transition and tumour metastasis through a TWIST1-G3BP2 mechanotransduction pathway, *Nat Cell Biol* 17(5) (2015) 678–688.

95. F. Bordeleau, B.N. Mason, E.M. Lollis, M. Mazzola, M.R. Zanotelli, S. Somasegar, J.P. Califano, C. Montague, D.J. LaValley, J. Huynh, N. Mencia-Trinchant, Y.L. Negron Abril, D.C. Hassane, L.J. Bonassar, J.T. Butcher, R.S. Weiss, C.A. Reinhart-King, Matrix stiffening promotes a tumor vasculature phenotype, *Proc Natl Acad Sci U S A* 114(3) (2017) 492–497.

96. A.J. Rice, E. Cortes, D. Lachowski, B.C.H. Cheung, S.A. Karim, J.P. Morton, A. Del Rio Hernandez, Matrix stiffness induces epithelial-mesenchymal transition and promotes chemoresistance in pancreatic cancer cells, *Oncogenesis* 6(7) (2017) e352.

97. F. Xu, J. Celli, I. Rizvi, S. Moon, T. Hasan, U. Demirci, A three-dimensional in vitro ovarian cancer coculture model using a high-throughput cell patterning platform, *Biotechnol J* 6(2) (2011) 204–212.

98. L. Neufeld, E. Yeini, N. Reisman, Y. Shtilerman, D. Ben-Shushan, S. Pozzi, A. Madi, G. Tiram, A. Eldar-Boock, S. Ferber, R. Grossman, Z. Ram, R. Satchi-Fainaro, Microengineered perfusable 3D-bioprinted glioblastoma model for in vivo mimicry of tumor microenvironment, *Sci Adv* 7(34) (2021) eabi9119.

99. E.M. Langer, B.L. Allen-Petersen, S.M. King, N.D. Kendsersky, M.A. Turnidge, G.M. Kuziel, R. Riggers, R. Samatham, T.S. Amery, S.L. Jacques, B.C. Sheppard, J.E. Korkola, J.L. Muschler, G. Thibault, Y.H. Chang, J.W. Gray, S.C. Presnell, D.G. Nguyen, R.C. Sears, Modeling tumor phenotypes in vitro with three-dimensional bioprinting, *Cell Rep* 26(3) (2019) 608–623 e6.

100. Y. Wang, W. Shi, M. Kuss, S. Mirza, D. Qi, A. Krasnoslobodtsev, J. Zeng, H. Band, V. Band, B. Duan, 3D bioprinting of breast cancer models for drug resistance study, *ACS Biomater Sci Eng* 4(12) (2018) 4401–4411.

101. S. Swaminathan, Q. Hamid, W. Sun, A.M. Clyne, Bioprinting of 3D breast epithelial spheroids for human cancer models, *Biofabrication* 11(2) (2019) 025003.

102. X. Dai, C. Ma, Q. Lan, T. Xu, 3D bioprinted glioma stem cells for brain tumor model and applications of drug susceptibility, *Biofabrication* 8(4) (2016) 045005.

103. X. Wang, X. Li, X. Dai, X. Zhang, J. Zhang, T. Xu, Q. Lan, Coaxial extrusion bioprinted shell-core hydrogel microfibers mimic glioma microenvironment and enhance the drug resistance of cancer cells, *Colloids Surf B Biointerfaces* 171 (2018) 291–299.

104. J.R. Browning, P. Derr, K. Derr, N. Doudican, S. Michael, S.R. Lish, N.A. Taylor, J.G. Krueger, M. Ferrer, J.A. Carucci, D.S. Gareau, A 3D biofabricated cutaneous squamous cell carcinoma tissue model with multi-channel confocal microscopy imaging biomarkers to quantify anti-tumor effects of chemotherapeutics in tissue, *Oncotarget* 11(27) (2020) 2587–2596.

105. D. Hakobyan, C. Medina, N. Dusserre, M.L. Stachowicz, C. Handschin, J.C. Fricain, J. Guillermet-Guibert, H. Oliveira, Laser-assisted 3D bioprinting of exocrine pancreas spheroid models for cancer initiation study, *Biofabrication* 12(3) (2020) 035001.

106. Y. Zhao, R. Yao, L. Ouyang, H. Ding, T. Zhang, K. Zhang, S. Cheng, W. Sun, Three-dimensional printing of Hela cells for cervical tumor model in vitro, *Biofabrication* 6(3) (2014) 035001.

107. H. Chen, Y. Cheng, X. Wang, J. Wang, X. Shi, X. Li, W. Tan, Z. Tan, 3D printed in vitro tumor tissue model of colorectal cancer, *Theranostics* 10(26) (2020) 12127–12143.

108. Y. Zhang, Z. Wang, Q. Hu, H. Luo, B. Lu, Y. Gao, Z. Qiao, Y. Zhou, Y. Fang, J. Gu, T. Zhang, Z. Xiong, 3D bioprinted GelMA-nanoclay hydrogels induce colorectal cancer stem cells through activating Wnt/beta-catenin signaling, *Small* 18(18) (2022) e2200364.

109. X. Wang, X. Zhang, X. Dai, X. Wang, X. Li, J. Diao, T. Xu, Tumor-like lung cancer model based on 3D bioprinting, *3 Biotech* 8(12) (2018) 501.

110. X. Zhou, W. Zhu, M. Nowicki, S. Miao, H. Cui, B. Holmes, R.I. Glazer, L.G. Zhang, 3D Bioprinting a cell-laden bone matrix for breast cancer metastasis study, *ACS Appl Mater Interfaces* 8(44) (2016) 30017–30026.

111. F. Meng, C.M. Meyer, D. Joung, D.A. Vallera, M.C. McAlpine, A. Panoskaltsis-Mortari, 3D bioprinted in vitro metastatic models via reconstruction of tumor microenvironments, *Adv Mater* 31(10) (2019) e1806899.

112. B.S. Hill, A. Sarnella, G. D'Avino, A. Zannetti, Recruitment of stromal cells into tumour microenvironment promote the metastatic spread of breast cancer, *Semin Cancer Biol* 60 (2020) 202–213.

113. J. Plava, M. Cihova, M. Burikova, M. Matuskova, L. Kucerova, S. Miklikova, Recent advances in understanding tumor stroma-mediated chemoresistance in breast cancer, *Mol Cancer* 18(1) (2019) 67.

114. M. Sharifi, Q. Bai, M.M.N. Babadaei, F. Chowdhury, M. Hassan, A. Taghizadeh, H. Derakhshankhah, S. Khan, A. Hasan, M. Falahati, 3D bioprinting of engineered breast cancer constructs for personalized and targeted cancer therapy, *J Control Release* 333 (2021) 91–106.

115. S. Hong, J.M. Song, 3D bioprinted drug-resistant breast cancer spheroids for quantitative in situ evaluation of drug resistance, *Acta Biomater* 138 (2022) 228–239.

116. T. Jiang, J.G. Munguia-Lopez, S. Flores-Torres, J. Grant, S. Vijayakumar, A. Leon-Rodriguez, J.M. Kinsella, Directing the self-assembly of tumour spheroids by bioprinting cellular heterogeneous models within alginate/gelatin hydrogels, *Sci Rep* 7(1) (2017) 4575.

117. W. Zhu, N.J. Castro, H. Cui, X. Zhou, B. Boualam, R. McGrane, R.I. Glazer, L.G. Zhang, A 3D printed nano bone matrix for characterization of breast cancer cell and osteoblast interactions, *Nanotechnology* 27(31) (2016) 315103.

118. A. Mazzocchi, S. Soker, A. Skardal, 3D bioprinting for high-throughput screening: Drug screening, disease modeling, and precision medicine applications, *Appl Phys Rev* 6(1) (2019) 011302.

119. S. Hou, H. Tiriac, B.P. Sridharan, L. Scampavia, F. Madoux, J. Seldin, G.R. Souza, D. Watson, D. Tuveson, T.P. Spicer, Advanced development of primary pancreatic organoid tumor models for high-throughput phenotypic drug screening, *SLAS Discov* 23(6) (2018) 574–584.

120. M.E. Davis, Glioblastoma: Overview of disease and treatment, *Clin J Oncol Nurs* 20(5 Suppl) (2016) S2–S8.

121. C. Parra-Cantu, W. Li, A. Quinones-Hinojosa, Y.S. Zhang, 3D bioprinting of glioblastoma models, *J 3D Print Med* 4(2) (2020) 113–125.

122. M.A. Heinrich, R. Bansal, T. Lammers, Y.S. Zhang, R. Michel Schiffelers, J. Prakash, 3D-bioprinted mini-brain: A glioblastoma model to study cellular interactions and therapeutics, *Adv Mater* 31(14) (2019) e1806590.

123. M. Tang, Q. Xie, R.C. Gimple, Z. Zhong, T. Tam, J. Tian, R.L. Kidwell, Q. Wu, B.C. Prager, Z. Qiu, A. Yu, Z. Zhu, P. Mesci, H. Jing, J. Schimelman, P. Wang, D. Lee, M.H. Lorenzini, D. Dixit, L. Zhao, S. Bhargava, T.E. Miller, X. Wan, J. Tang, B. Sun, B.F. Cravatt, A.R. Muotri, S.

Chen, J.N. Rich, Three-dimensional bioprinted glioblastoma microenvironments model cellular dependencies and immune interactions, *Cell Res* 30(10) (2020) 833–853.

124. H. Clevers, Modeling development and disease with organoids, *Cell* 165(7) (2016) 1586–1597.

125. J. Pimenta, R. Ribeiro, R. Almeida, P.F. Costa, M.A. da Silva, B. Pereira, Organ-on-chip approaches for intestinal 3D in vitro modeling, *Cell Mol Gastroenterol Hepatol* 13(2) (2022) 351–367.

126. C. Adine, K.K. Ng, S. Rungarunlert, G.R. Souza, J.N. Ferreira, Engineering innervated secretory epithelial organoids by magnetic three-dimensional bioprinting for stimulating epithelial growth in salivary glands, *Biomaterials* 180 (2018) 52–66.

127. M.E. Kupfer, W.-H. Lin, V. Ravikumar, K. Qiu, L. Wang, L. Gao, D.B. Bhuiyan, M. Lenz, J. Ai, R.R. Mahutga, D. Townsend, J. Zhang, M.C. McAlpine, E.G. Tolkacheva, B.M. Ogle, In situ expansion, differentiation, and electromechanical coupling of human cardiac muscle in a 3D bioprinted chambered organoid, *Circ Res* 127(2) (2020) 207–224.

128. J.A. Brassard, M. Nikolaev, T. Hübscher, M. Hofer, M.P. Lutolf, Recapitulating macro-scale tissue self-organization through organoid bioprinting, *Nat Mater* 20(1) (2021) 22–29.

129. P.N. Bernal, M. Bouwmeester, J. Madrid-Wolff, M. Falandt, S. Florczak, N.G. Rodriguez, Y. Li, G. Größbacher, R.-A. Samsom, M. van Wolferen, L.J.W. van der Laan, P. Delrot, D. Loterie, J. Malda, C. Moser, B. Spee, R. Levato, Volumetric bioprinting of organoids and optically tuned hydrogels to build liver-like metabolic biofactories, *Adv Mater* 34(15) (2022) 2110054.

130. L.R. Madden, T.V. Nguyen, S. Garcia-Mojica, V. Shah, A.V. Le, A. Peier, R. Visconti, E.M. Parker, S.C. Presnell, D.G. Nguyen, K.N. Retting, Bioprinted 3D primary human intestinal tissues model aspects of native physiology and ADME/Tox functions, *iScience* 2 (2018) 156–167.

131. M.C. Bouwmeester, P.N. Bernal, L.A. Oosterhoff, M.E. van Wolferen, V. Lehmann, M. Vermaas, M.-B. Buchholz, Q.C. Peiffer, J. Malda, L.J.W. van der Laan, N.I. Kramer, K. Schneeberger, R. Levato, B. Spee, Bioprinting of human liver-derived epithelial organoids for toxicity studies, *Macromol Biosci* 21(12) (2021) 2100327.

132. Q. Gu, E. Tomaskovic-Crook, G.G. Wallace, J.M. Crook, 3D bioprinting human induced pluripotent stem cell constructs for in situ cell proliferation and successive multilineage differentiation, *Ad Healthc Mater* 6(17) (2017) 1700175.

133. K.T. Lawlor, J.M. Vanslambrouck, J.W. Higgins, A. Chambon, K. Bishard, D. Arndt, P.X. Er, S.B. Wilson, S.E. Howden, K.S. Tan, F. Li, L.J. Hale, B. Shepherd, S. Pentoney, S.C. Presnell, A.E. Chen, M.H. Little, Cellular extrusion bioprinting improves kidney organoid reproducibility and conformation, *Nat Mater* 20(2) (2021) 260–271.

134. N. Gjorevski, M. Nikolaev, T.E. Brown, O. Mitrofanova, N. Brandenberg, F.W. DelRio, F.M. Yavitt, P. Liberali, K.S. Anseth, M.P. Lutolf, Tissue geometry drives deterministic organoid patterning, *Science* 375(6576) (2022) eaaw9021.

135. B.J. Carberry, J.E. Hergert, F.M. Yavitt, J.J. Hernandez, K.F. Speckl, C.N. Bowman, R.R. McLeod, K.S. Anseth, 3D printing of sacrificial thioester elastomers using digital light processing for templating 3D organoid structures in soft biomatrices, *Biofabrication* 13(4) (2021). Doi: 10.1088/1758-5090/ac1c98.

136. C. Chen, V. Rengarajan, A. Kjar, Y. Huang, A matrigel-free method to generate matured human cerebral organoids using 3D-Printed microwell arrays, *Bioact Mater* 6(4) (2021) 1130–1139.

5 Advances in Drug Delivery via Electrospun and Electrosprayed Formulations

István Sebe
Semmelweis University
Egis Pharmaceuticals PLC

Edina Szabó
Budapest University of Technology and Economics

Romána Zelkó
Semmelweis University

CONTENTS

5.1 INTRODUCTION

Continuous demand for medicines and unmet medical needs encourage the pharmaceutical industry to develop the portfolio of the pharmaceutical products, which is often achievable via the application of novel technologies. From this point of view, electrohydrodynamic methods, where electrically charged fluids are in focus, have recently gotten a lot of attention. The reason for this is that electrospinning and electrospraying (collectively electrohydrodynamic processes) have several advantages considering both technological aspects and the micro- and nanosize products.

During electrospinning, also called electrostatic fiber formation, fibers are created by electrostatic forces. The process results in a high specific surface area fibrous product from a viscous solution or melts with good electrical conductivity that can be collected in solid-state form. Electrostatic fiber formation has a long history and a great literature. In 1899, J. F. Cooley [1] filed a patent application for the development of the process. It is noteworthy that this was only 2 years after 1897, the year in which the discovery of the electron is known in scientific history. The patent was finally published in 1902. W. J. Morton also describes

DOI: 10.1201/9781003224464-5

a process for electrostatic fiber formation in a patent filed in the same year [2]. The scientific and industrial impetus of the following period led to further advances in the technology. According to the historical review of Nick Tucker et al., Anton Formhals obtained protection for 22 patents on this subject between 1931 and 1934 [3].

Since the late 1990s, electrostatic fiber formation has become an increasingly widespread part of research in other disciplines, including work on drug discovery, tissue regeneration, and other biomedical topics. A positive outcome of this process has been the further differentiation and refinement of the method. Over the decades, other spinning solutions have evolved, such as pneumatic [4] or rotary [5] spinning. In addition, other new processes have emerged from combinations of individual fiber-forming methodologies. In some of the developed techniques, besides the electrostatic force that generates the fibers, other types of forces play a dominant role in fiber formation. For instance, the electrostatic-pneumatic method, also called electroblowing [6], is evolved from the combination of compressed air (pneumatic force) and electrostatic force. Another example is the centrifugal-electrostatic fiber formation such as centrifugal electrospinning, rotary electrospinning, high-speed electrospinning, and electrocentrifugal spinning [7,8], which combines electrostatic and centrifugal forces. Collectively, these can be referred to as electrostatically combined spinning processes.

The formulation and emerging applications of nano- and microfibrous systems are prominent and promising areas of research and development related to novel drug delivery systems and their biomedical utility. Both small molecule and macromolecule active pharmaceutical ingredients (API) can be incorporated in most cases into the polymer matrix. One of the key advantages of the fibrous systems, in addition to the high specific surface area and nanosize range, is the property of fibrous polymer structures that makes it potentially feasible to improve the dissolution and solubility properties of poorly soluble crystalline compounds by keeping them in amorphous form [9]. Amorphous solid dispersions (ASDs) and solid solutions can be formed. It is also possible to use the fibrous web directly or after further processing [10,11].

Examples of industrial applications of electrostatic fiber formation can be found in the textile and packaging industries [12,13], but there are still many challenges and questions to be answered in the field of its application in the pharmaceutical industry. It is because industrial drug development and manufacturing is one of the most strictly regulated areas and patient safety and efficacy are key considerations for licensing. Therefore, thermodynamic stability, minimizing amorphous-solid transformation, ensuring homogeneity, adequate productivity, and economy need to be provided in the case of pharmaceutical electrospun fibers. Addressing these issues is essential to ensure that the process can be effectively translated into industrial practice.

Similar to electrospinning, electrospraying is also an electrohydrodynamic process, during which fine particles can be generated via atomization of liquids/solutions by electrostatic forces. In contrast with electrospinning, less viscous solutions are needed in this case to avoid the formation of fibers. The basic phenomenon relating to electrospraying technology was first described around the 17th century, but its potential for practical application appeared later [14,15]. The first patents about electrospraying, or also called electrohydrodynamic atomization, were published only in the 20th century [16,17]. Since then, the technology has developed a lot, and its application area has increased year by year. Highlighting some examples, electrospraying proved to be a promising method for the food industry [18], tissue engineering [19], electrospray ionization mass spectrometry [20], and in the field of electronics and nanodevices [21]. Furthermore, microscale and nanoscale droplets prepared by electrospraying can also open new routes in innovative drug delivery systems; thus its pharmaceutical applicability is also a widely researched area recently, similar to electrospinning [22].

The nanotechnology approach contributes to the effective formulation of poorly water-soluble drugs since the higher specific surface area achieved by electrospraying can increase the wettability and thus enhance the dissolution rate, similar to the electrospun fibers [23]. Nevertheless, the formulation of ASDs in this way also might further increase the dissolution of APIs with poor water solubility [24]. Besides, the development of nanomedicines results in new solutions in the field of controlled release systems or targeted therapy [25,26]. From the pharmaceutical application point of view, electrosprayed formulation is not only promising in the case of small molecules but is also applicable for biomedicines [27]. All of these examples demonstrate that the capabilities and opportunities of electrospraying (e.g., the preparation of micro-and nanosized particles, the variety of materials that can be used, and the simplicity of the technology) can be exploited by the pharmaceutical industry. Nevertheless, similarly to what was mentioned by the electrospinning, several challenges need to be addressed to make electrospraying suitable for the requirements of the pharmaceutical industry. For instance, the electrostatic chargeability of the small-size electrosprayed particles can cause difficulties during the downstream processing and the preparation of final drug products [28]. On the other hand, ensuring the stability of the prepared formulations can be challenging here as well [29]; therefore, continuous developments of the method and the use of new technological steps during the formulation are essential for its pharmaceutical industry applicability.

In this chapter, besides the basic principles and properties of electrospinning and electrospraying, the recent advances, current trends and future perspectives of these two techniques will be discussed. The authors' main aim is to give a comprehensive view about the electrohydrodynamic processes from the pharmaceutical application point of view.

5.2 ELECTROSPINNING

5.2.1 COMPREHENSIVE DESCRIPTION OF THE PHYSICAL BACKGROUND AND VARIOUS SETUPS

Electrostatic fiber formation is a modern pharmaceutical technology process that offers new perspectives in drug development and can be understood as a technological process for the incorporation of the API. It produces fibers from a viscous solution of a polymer excipient and the API, with a very high specific surface area and a variety of physical, physicochemical, and solid-state properties. In practice, this is achieved by transferring the material from a reservoir containing the solution at a defined volumetric rate into the fiber-forming dispensing head, in the simplest arrangement (Figure 5.1) into a metal needle, to which a high voltage electrical potential is applied. An electrode with a larger surface area, located opposite the needle and at a certain distance from the fiber collection point, is placed at a direct current (DC) potential so that the potential difference between the two points creates the corresponding electrostatic field. The tendency to equalize the potential difference is the driving force for fiber formation, where the material moves from the higher potential point to the DC potential (formerly grounding).

To understand how electrospinning works, it is helpful to break the process down into successive or partially simultaneous sub-processes. A substance flowing into a needle under high voltage appears as a drop at the endpoint of the needle. The repulsive force of the equal charges continuously accumulating on the surface of the liquid droplet results in a conical shape, which is called Taylor cone after the mathematical descriptor of the behavior of the liquid surface [30,31].

The phenomenon is analogous to the peak effect observed in high-frequency electric field forces with a curved surface, where the ion avalanche through the peak produces spectacular corona discharges (Tesla coil). It is interesting to mention this because the corona discharge phenomenon [32] also occurs during electrospinning and is often observed in the characteristic blue light produced by the recombination of excited air. The charged liquid jet is further thinned by the forces and moves toward the collector, initially in a straight line according to the Coulomb forces and then in a whipping motion. The viscoelasticity of the polymer solution or melt helps to keep the thinning jet continuous and tear-free. Thinning results in an increase in specific surface area and a steady decrease in the surface area carrying unit charges. The solvent evaporates more rapidly from the high specific surface area, aided by the charges that are deposited. The solid fibers can be separated from the collector, which are already sufficiently solvent-free or sufficiently cooled. Due to the stochastic nature of some of the processes involved in electrospinning, the resulting fiber web is made up of randomly oriented fibers with interconnected fibers, but it is also possible to produce a product with an oriented fiber arrangement. The Rayleigh cleavage index [33] is used to define the relationships that determine the quality of the fibers (Equation 5.1).

$$X = \frac{q^2}{64\pi^2 \sigma \varepsilon_0 R^3} \leq 1 \qquad (5.1)$$

In Equation 5.1, X is the "fissility," which is a ratio of electrostatic repulsion force to the surface force, q is the charge of the droplet surface, σ is the solution surface tension, and ε_0 is the permittivity of vacuum.

The resulting fibrous structure and the quality of the fiber morphology are determined by the physical and physicochemical parameters specific to the polymer and

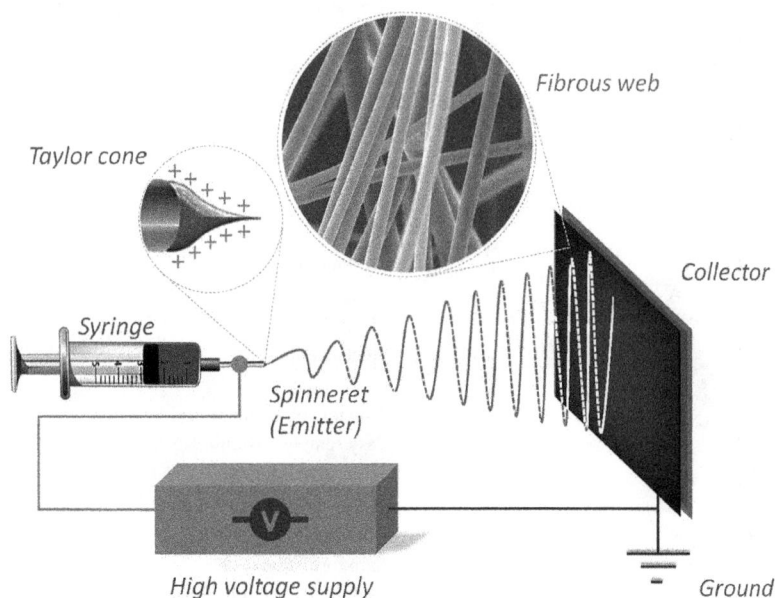

FIGURE 5.1 Schematic design of the basic electrospinning apparatus.

the polymer solution [34]. However, this is also closely related to the technical specifications, construction, and setup of the apparatus used for the production.

Most of the developments that have been made are still aimed at achieving the low yields typical of the technology [35], adding new properties to the fibers, and producing a more uniform and reproducible morphology and structure. The classification should be based on the type and principle of operation of the two main equipment elements: (1) the fiber-forming solution dispenser and (2) the fiber collection unit (Table 5.1). In some of the established techniques, in addition to the electrostatic force that generates the fibers, other types of forces play a dominant role in fiber formation.

5.2.2 Advantages and Application of Electrospinning as Effective Strategies for Drug Delivery

Polymeric structures, drug-polymer matrix systems [58], produced by electrostatic and electrostatically combined fiber formation processes, can offer numerous drug technology and therapeutic advantages, which foresee the translation and proper positioning of the process into actual applications. Considering the physical and physicochemical properties of the nanofibrous system, these advantages include high specific surface area and increased drug dissolution rates, as well as increased apparent solubility of some APIs [59].

TABLE 5.1
Classification of Electrostatic and Electrostatically Combined Spinning Techniques

Method Description	Spinning Force	Featured Property Related to the Final Product Characteristics	Ref.
(a) Operation Principle of the Emitter			
Monoaxial/single-nozzle electrospinning	Electrostatic	Preformulation (lab scale)	[36]
Bubble electrospinning	Electrostatic	Scalability	[37]
Electro-blowing	Electrostatic/pneumatic	Ultrathin fibers	[38]
Roller electrospinning (Nanospider™)	Electrostatic	Scalability	[39]
Spherical ball electrospinning	Electrostatic	There is no highlightable advantage	[40]
Conical wire coil electrospinning	Electrostatic	Higher productivity for lab-scale experiments	[41]
High-speed electrospinning, corona-electrospinning	Electrostatic/ centrifugal	Pharmaceutical scalability	[9,42]
Centrifugal electrospinning	Centrifugal/ electrostatic	Scalability	[43]
Coaxial, core-shell type	Electrostatic	Multiphase fibers	[44]
Triaxial, core-shell type	Electrostatic	Multiphase fibers	[45]
Porous-hollow tube	Electrostatic	Scalability	[35]
Multiple-jet electrospinning	Electrostatic	Scalability	[40]
Stepped pyramid electrospinning	Electrostatic	The decrease of fiber diameter with simultaneous increase of productivity	[46]
Oppositely charged nozzle electrospinning	Electrostatic	Yarn formation	[47]
(b) Collector Type			
Plate collector/grounded static collector	N/A	Preformulation (lab scale)	[48]
Grounded liquid collector	N/A	Preparing complex structure (Limited applicability/not common)	[49]
Parallel electrodes	N/A	Parallel orientation	[50]
Two-pole air gap	N/A	3D fibrous structure formation/highly aligned (tissue regeneration)	[51]
Drum collector			
• Solid cylinder	N/A	Thinning of the fibers	[52–54]
• Wire drum		Easy sample separation	
• Drum with sharp pin inside		Adjustable orientation	
Ring collector	N/A	Yarn formation	[55]
Disk collector	N/A	A more uniform product surface	[56]
Yarn/conical collector	N/A	Yarn formation	[47]
Knife-edge collector	N/A	High-efficiency orientation	[52]
Multiple-field electrospinning	N/A	A more uniform product surface	[57]

The average diameter of electrospun fibers is very small, resulting in a high specific surface area. Ahn et al. in 2006 reported typical diameters of nano- and microfibers in size range <1–50 μm [60]. The larger surface area results in more efficient wettability and higher dissolution rate according to the Noyes–Whitney equation [61].

Nowadays, the pharmaceutical industry is faced with an increase in undesirable physicochemical properties and consequent unfavorable biopharmaceutical behavior of APIs, leading to a progressive decline in the drug approval of small molecules [62–64]. Molecular obesity, namely high molecular weight, high lipophilicity, and poor water solubility, has been recognized as a major problem affecting drug development [65,66]. Approximately 40% of marketed oral drugs fall into the poor water solubility (< 0.1 mg/mL) category [67,68], which poses a significant challenge in drug development. A number of formulation strategies have been used to increase water solubility, such as salt formation [69], cocrystals, particle size reduction [70], or the creation of ASD systems [71,72]. Amorphous materials have higher enthalpies, entropies, and free energies compared to their crystalline forms, resulting in higher apparent solubility. The weaker attractive intermolecular forces of amorphous substances are more easily broken, allowing molecules to move more easily from the surface of the substance into the medium than their crystalline forms. As a result, amorphous materials dissolve better and faster. ASDs are an important strategy to improve the apparent solubility, dissolution rate, and bioavailability of poorly water-soluble drugs [67,73].

In another approach, further advantages can be defined according to the structural design of the API-loaded fibers and its effect on the drug release kinetics [45,74] and the type of application or route of administration. In the simplest process mode, a structure is obtained, which is characterized by a homogeneous and uniform distribution of the fiber-forming polymer and the API. Generations of advances in the technique have led to the creation of core-shell, multi-core-shell [75], and modified surface nanofibers (Figure 5.2), in which the API is located within a discrete domain. The diversity of the structure and the chemical quality of the polymer allow for targeted [76] or controlled drug delivery and protect the drug from premature biodegradation and polymer erosion. For example, Li et al. created a three-layer structure for breast cancer treatment by combining different drugs in different polymers, with which the authors achieved time-programmed release of chemotherapeutic agents [58,77].

Turanli et al. reported in 2019 the electrostatic technique for the sustained release of methylprednisolone [78], which is also known from other studies [79]. It is also advantageous that the direct use of fibrous systems could be used in the future to produce topical, transdermal formulations, medical devices, and conventional dosage forms by further processing of the fibers. In addition to this promising vision (Table 5.2), it is important to note that there is currently no pharmaceutical dosage form approved by the pharmaceutical authorities that can exploit the benefits of electrostatically produced nanofibers. The majority of *in vivo* studies so far have been limited mainly to the preclinical level, and a few human studies have been initiated in the context of medical applications. Vass et al. reported promising results for the future positioning of the technique. In their work, they successfully applied the high-speed electrostatic process as an efficient and scalable alternative to freeze drying for the production of an intravenously administrable reconstitution powder-based drug formulation [80].

5.2.3 IMPACT OF DIFFERENT PARAMETERS ON THE PRODUCT QUALITY

The definition and control of the Critical Process/Production Parameters (CPPs) of this process as a pharmaceutical technology operation is of paramount importance for the manufacture of a good quality product and compliance with stringent regulatory requirements. The CPPs, together with the critical quality attributes (CQAs), have a direct impact on the micro- and macrostructural properties of the fibers and thus on the extent of drug delivery, the kinetics, the bioavailability of the drug, the physicochemical and chemical stability of the drug-polymer system, and the potential for further processing/formulation.

Parameters that are complexly related to each other should be grouped according to the chemical and physicochemical properties of the fiber-forming materials (e.g., conductivity, polymer molecular weight, viscosity), the process (e.g., emitter-collector distance, voltage, flow rate), the technical design of the equipment (e.g., emitter and collector type (Table 5.1), and environmental effects (e.g., temperature, relative humidity) (Table 5.3).

For example, SalehHudin et al. report in their abstract the results of Thompson et al., who, based on a complex study of the effects of 13 different parameters on fiber diameter, highlighted the impact of applied voltage, solution conductivity, emitter-collector distance, and solution viscosity [40,110]. In addition to this, the previously published works highlight that the use of a polymer with the right molecular weight is essential for electrospinning, without which the effect of changing all other parameters is reduced or lost, and a robust and reproducible technology cannot be developed, nor can a product with the desired properties and quality be produced [111].

5.2.4 CHALLENGES RELATING TO ELECTROSPINNING

The examples presented in the previous chapters illustrate the potential benefits of electrospun samples and complex nanofibrous drug-polymer systems, both from a pharmaceutical and therapeutic point of view. Several decades of research explored different aspects (e.g., compliance with pharmaceutical and regulatory quality requirements, chemical and physicochemical stability, chemical compatibility, biocompatibility, drug storage capacity, processability and scalability, compliance with design and development

Multifluid **ELECTROSPINNING / ELECTROSPRAYING**

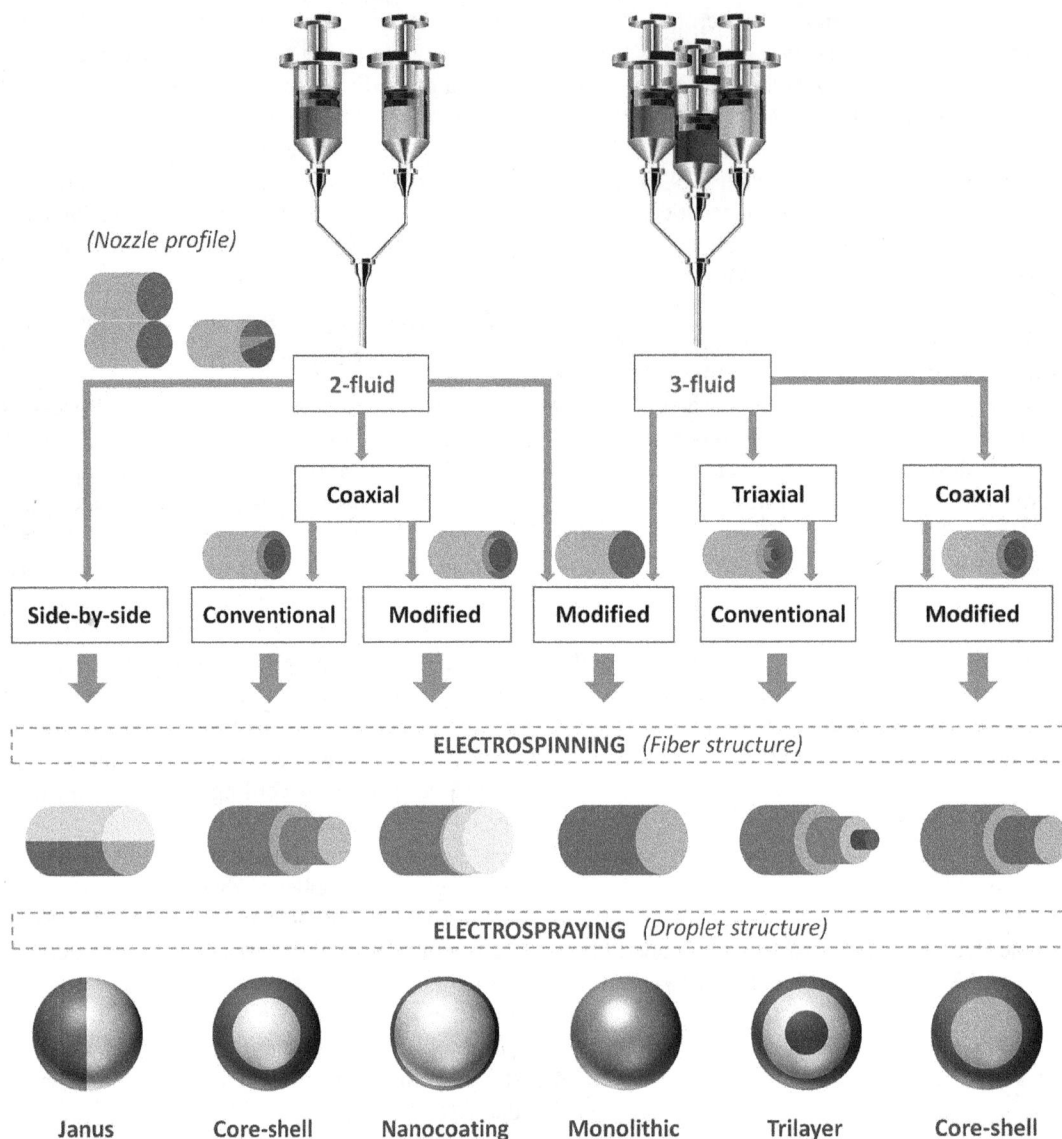

FIGURE 5.2   Overview of the multi-fluid electrospinning and electrospraying methods and the corresponding structures.

requirements) to formulate the issues of transferability of the technology into pharmaceutical industrial practice, its correct positioning, and current limitations and challenges to be addressed.

5.2.4.1 Critical Feasibility Assessment of Pharmaceutical Applicability

5.2.4.1.1 Thermodynamic Stability

In general, the use of nanofibers may be relevant in cases where the fiber structure or the specific physicochemical properties that can be specifically developed (compared to other known processes) through the fiber technology used are crucial for providing the drug technology or therapeutic benefits. With regard to the latter, the electrospinning process from solution has the clear advantage of being an operation that is gentle on the drug-excipient system, without destructive effects (e.g., thermal degradation of the drug, mechanical agitation) that negatively affect the quality parameters. In the majority of the fiber-forming operations described, a favorable change of the poorly soluble APIs can be achieved by bringing the compound to an amorphous-solid state. In these cases, the question of thermodynamic stability and the need for a more thorough and longer storage time investigation of the amorphous-solid transformation justifiably arise [124]. A number of factors can influence the physical stability of the formulation, but the physicochemical and chemical properties of the chosen polymer are particular importance. The results of Tipduangta et al. demonstrate that using hygroscopic polymers leads to poor physical stability during storage due to the sorption of ambient moisture [88]. Their work

TABLE 5.2
Some Examples for Drug-Loaded Electrospun Fibers Showing Their Pharmaceutical Relevance

Targeted Function	Drug Delivery Route	Preparation Method	Therapeutic Indications	APIs (S: Small Molecules, L: Large Molecules)	Polymer/Carrier Matrix	Ref.
BCSI						
• Amorphous form of API • Improving the efficacy of rectal administration	Rectal	Monoaxial electrospinning	• Anaerobe origin infections	METRONIDAZOLE (S)	• Eudragit® S100	[81]
• Achieving pH-controlled release (gastric acid resistant) • Achieving sustained release (~6—22 h) • Enhancing intestinal mucoadhesion	Oral	Coaxial electrospinning (Core-shell)	• Irritable bowel syndrome • Antispasmodic	MEBEVERINE HYDROCHLORIDE (S)	• PEO • Eudragit® S100	[82]
• Improving compatibility of composition • Improving bioavailability (relative: 151.6%) • Achieving fast dissolution (100% within 120s)	Sublingual	Monoaxial electrospinning	• Ischemia • Prevention for angina pectoris	ISOSORBIDE DINITRATE (S)	• PVP K90F • PEG 400	[83]
• Colon-targeted extended drug release • Improving protect the stomach • Improving compatibility of composition	Oral	Triaxial electrospinning (core shell)	• Non-steroidal anti-inflammatory	ACETYLSALICYLIC ACID (S)	• Eudragit® S100	[84]
BCSII						
• Achieving sustained release • Potential alternative administration of poorly soluble API	Subcutaneous implant (*off-label*)	Monoaxial electrospinning	• Adjuvant and first-line treatment of hormone receptor positive invasive early/advance breast cancer (postmenopausal woman)	LETROZOLE (S)	• POL188 • PLLA	[85]
• Increasing *in vitro* dissolution rate (>90% within 10min) • Improving apparent solubility • Scaling up fiber formation (450 g/h) • Amorphous form of API	Not defined	High-speed electrospinning (scaled-up)	• Infections (antifungal agent)	ITRACONAZOLE (S)	• PVP/VA 64	[9]
• Achieving colon-targeted release • Protect the drug from release in acidic conditions • pH-sensitive polymer/lipid nanocomposite • Amorphous form of API and the lipid carrier	Oral	Triaxial electrospinning (Core-shell)	• Non-steroidal anti-inflammatory	DICLOFENAC (S)	• Eudragit® S100 • (Lecithin)	[86]

(Continued)

TABLE 5.2 (Continued)
Some Examples for Drug-Loaded Electrospun Fibers Showing Their Pharmaceutical Relevance

Targeted Function	Drug Delivery Route	Preparation Method	Therapeutic Indications	APIs (S: Small Molecules, L: Large Molecules)	Polymer/Carrier Matrix	Ref.
• Retention of amorphous form of API (15 months stability data are available) • Improving apparent solubility and membrane diffusion • Non-destructive tabletability	Oral (Orodispersible dosage form)	Monoaxial electrospinning	• Osteoarthritis • Rheumatoid arthritis • Ankylosing spondylitis	MELOXICAM (S)	• Eudragit® E • PVP K30 • HP-β-CD	[87]
• Forming ASD • Improving physical stability without compromising dissolution enhancement • Increasing dissolution rate • Decreasing moisture uptake	Oral	Monoaxial electrospinning	• Adjunct to diet and other non-pharmacological treatment (e.g. exercise, weight reduction) for the following: Hypertriglyceridemia, Mixed hyperlipidemia	FENOFIBRATE (S)	• PVP K90F • Eudragit® E • Soluplus® • HPMCAS	[88]
• Potential effective alternative to freeze drying • Scaling up fiber formation (~240 g/h) • Increasing dissolving rate (total within 30 s)	Reconstitution powder for injection	High-speed electrospinning (scaled-up)	• Broad-spectrum antifungal agent	VORICONAZOLE (S)	• SBE-β-CD	[80]
• Amorphous form of API • Improving apparent solubility • Improving homogeneous distribution of the API in the fiber						
• Achieving sustained release • Repositioning of the API	Implantable cardiovascular stent (local, semi-systemic) (off-label)	Monoaxial emulsion electrospinning (core shell)	• Treatment of intracranial aneurysm (off-label) • Promote the in vitro proliferation (off-label)	ATORVASTATIN Ca (S)	• PLA-PCL • PLA/PCL	[89]
• Achieving fast release • Amorphous form of API • Improving apparent solubility	Oral	Monoaxial electrospinning (drum collector)	• Anti-inflammatory • Analgesics	KETOPROFEN (S)	• PVP K90F	[90]
• Achieving fast release • Amorphous form of API • Improving apparent solubility • Increasing dissolving rate	Oral	Monoaxial electrospinning	• Anti-inflammatory • Analgesics • Antipyretic	IBUPROFEN (S)	• HP-β-CD	[91]
• Achieving pH-controlled release • Amorphous form of API • Improving apparent solubility	Oral	Monoaxial electrospinning	• Diuretics	SPIRONOLACTONE (S)	• Eudragit® FS100	[92]

(Continued)

TABLE 5.2 (Continued)
Some Examples for Drug-Loaded Electrospun Fibers Showing Their Pharmaceutical Relevance

Targeted Function	Drug Delivery Route	Preparation Method	Therapeutic Indications	APIs (S: Small Molecules, L: Large Molecules)	Polymer/Carrier Matrix	Ref.
			BCSIII			
• Achieving potential drug delivery system for the treatment of common cancers	Not defined	Monoaxial electrospinning	• Cytotoxic agent (malignancies particularly cancer of the colon and breast) • Insolational and elderly keratosis	5-FLUOROURACIL (S)	• CS • PVP	[93]
• Achieving sustained release (72% within 3 days) • Repositioning of the API	Implantable drug delivery (off-label)	Monoaxial electrospinning	• Postoperative treatment of lung cancer (off-label) • Anti-diabetic/antihyperglycemic agent (type 2 diabetes mellitus)	METFORMIN (S)	• PLG	[94]
• Repositioning of the API	Topical use for wounds (off-label)	Monoaxial electrospinning (Using grounded liquid collector)	• Regeneration of injured peripheral nerves (off-label) • Anti-epileptic (adjunctive therapy) • Treatment of peripheral neuropathic pain	GABAPENTIN (S)	• CA • Gelatin	[95]
• Achieving transbuccal administration • Improving buccal permeability	Oral/transbuccal (off-label)	Monoaxial electrospinning	• Diabetes mellitus	INSULIN (L)	• CS • PEO	[96]
• Achieving nonviral delivery of nucleic acid therapeutics • Achieving axon regeneration	Implantable drug delivery scaffold (off-label)	Monoaxial electrospinning (two-pole air gap electrospinning)	• Spinal cord injury treatment (Gene silencing of NTF3) (off-label)	miRNA-222 (L)	• PCL-PEG • PCL-PPEEA	[97,98]
• Covalently immobilized enzyme • Achieving retention of enzyme activity (~20%–40%) • Achieving antimicrobial effect	Topical / Local (off-label)	Monoaxial electrospinning	• Antimicrobial treatment (off-label)	Glucose oxidase (L)	• CS • PEO	[99]
• Achieving controlled release and nonviral delivery of plasmid DNA • Maintenance of bioactivity • Decreasing the proliferation of breast cancer cell (~40%)	Not defined (site specific) (off-label)	Monoaxial electrospinning	• Treatment of breast cancer/gene therapy (Cdk2 inhibitor) (off-label)	PLASMID DNA (Encoding shRNS) (L)	• PCL	[100]
			BCSIV			
• Achieving sustained release, • Decreasing systemic exposure, • Promote tissue regeneration • Fixed-dose combination • Repositioning of the API	Topical (local) (off-label)	Monoaxial electrospinning	• Antibiotic treatment, • Wound healing (unregistered, new indication)	COLISTIN (S)+A3APO (BCS III) (New chemical entity) (L)	• PVA	[101,102]

(Continued)

TABLE 5.2 (Continued)
Some Examples for Drug-Loaded Electrospun Fibers Showing Their Pharmaceutical Relevance

Targeted Function	Drug Delivery Route	Preparation Method	Therapeutic Indications	APIs (S: Small Molecules, L: Large Molecules)	Polymer/Carrier Matrix	Ref.
• Decreasing dose unit • Sustained release (50% within 24 h) • Repositioning of the API (*drug device*)	Implantable cardiovascular stent (local, semi-systemic) (*off-label*)	Monoaxial electrospinning (using drum collector)	• Minimizing neointima growth after angioplasty (*off-label*) • Reducing restenosis rates (*off-label*)	PACLITAXEL (S)	• PCL	[103]
• Achieving sustained release from tablet dosage form • Immediate release from tablet dosage form	Oral	Monoaxial electrospinning (using drum collector)	• Diuretics	FUROSEMIDE (S)	• Eudragit®	[104]
• Achieving controlled release (~10%–30% within 8 h) • An alternative to the coaxial method for core-shell fiber formation • Repositioning of the API	Topical (local) (*off-label*)	Monoaxial emulsion electrospinning and crosslinking by heat treating (core shell)	• Wound healing (*Off-label*) • Antibiotic treatment	CEPHALEXIN (S)	• PVA • CMC *or* HPC *or* CS *or* CMS • (Corn oil)	[105]
• Achieving sustained release	Not defined	Monoaxial emulsion electrospinning (core shell)	• Antibiotic treatment	CIPROFLOXACIN (S)	• PVP + Dextran • Edible oil	[106]
• Achieving ultrafast dissolution • Improving apparent solubility • Retention of amorphous form of API / stable form of cyclodextrine (3 months stability data available) • Improving membrane permeation • Improving *in vitro–in vivo* correlation	Buccal/mucosal (systemic) (*off-label*)	Monoaxial electrospinning	• Schizophrenia • Bipolar I disorder	ARIPIPRAZOLE (S)	• SBE-β-CD • PEO	[107]
• Potential effective alternative to freeze drying • Scaling up fiber formation (~80 g/h) • Improving homogeneous distribution of the API in the fiber	Intravenous (i.v.) bolus dosage form (*off-label*)	High-speed electrospinning (scaled-up)	• Infections	DOXYCYCLIN (S)	• HP-β-CD	[108]
• Improve stability of API (thermal and irradiation resistance)	Oral	Monoaxial electrospinning	• Folate-deficient megaloblastic anemia • Prophylaxis of drug-induced folate deficiency • Prevention of neural tube defects for woman planning a pregnancy	FOLIC ACID (S)	• Zein	[109]

PLG, poly(DL-lactide-co-glycolide); PVP, polyvinylpyrrolidone; PCL, polycaprolactone; PEO, poly(ethylene oxide); PEG, poly(ethylene glycol); POL, poloxamer; PVP/VA, polyvinylpyrrolidone/vinyl acetate; PVA, polyvinyl alcohol; HPC, hydroxypropyl cellulose; HP-β-CD, hydroxypropyl-β-cyclodextrin; HPMCAS, hypromellose acetate succinate; SBE-β-CD, sulfobuthylether-β-cyclodextrin; PLA-PCL, poly(L-lactide-co-caprolactone; PLLA, poly(L -lactic acid); CS, chitosan; CA, cellulose acetate; CMC, carboxymethyl cellulose; CMS, carboxymethyl starch; PCL-b-PEG, poly(ε-caprolactone)-block-poly(ethylene glycol); PCL-b-PPEEA, poly(ε-caprolactone)-block-poly(2-aminoethylene phosphate); PLA/PCL, L -lactic acid/ε-caprolactone).

TABLE 5.3

Summary of Parameters Influencing the Product Quality of Electrospun Fiber

Parameters	Impact on the Product Quality	Ref.
Composition-Dependent Parameters		
Polymer (*Molecular weight/conductivity*)	The higher molecular weight polymer reduces the formation of beads during electrospinning while increasing the fiber diameter.	[111]
	The type of polymer can have a significant effect on the conductivity of the solution. A higher conductivity increases the magnitude of the fiber-forming force, which tends to decrease the fiber diameter.	[112]
Solution concentration (*Viscosity/surface tension/ relative volatility/conductivity*)	Increasing the concentration, and thus the viscosity and surface tension of the solution, results in fibers with larger diameters. However, too high viscosity leads to bead formation and the formation of fibers with heterogeneous morphology.	[113]
	Increasing the concentration of the solution reduces the relative volatility and thus reduces the formation of a porous microstructure.	[114]
Solvent(s) (*Volatility/ conductivity/dielectric constant*)	The use of more volatile solvents with a higher evaporation rate is preferred for electrospinning. This higher volatility ensures the necessary solvent removal at the needle-to-collector distance. Higher volatility requires higher flow rates to obtain bead-free or less bead-containing fibers. Increasing the conductivity of the solvent by increasing the conductivity of the solution allows thinner fibers to be produced. The dielectric constant of the solution can affect the stability of the fiber-forming beam, resulting in fibers with a more uniform morphology.	[113,115]
Process Parameters		
Flow rate	A higher flow rate results in thicker nanofibers, but too high a flow rate causes the formation of beads.	[116]
Applied voltage	With higher applied voltage, smaller diameter fibers can be created.	[116]
Needle diameter	Some studies have shown that there is no clear correlation between the diameter of the needle changed within the relevant range and the average diameter of the fibers produced.	[117,118]
Spinneret–collector distance	The distance between the needle and the collector influences the morphology of the fibers and its uniformity, the fiber diameter, and the size of the deposition area. For a given operating voltage, the distance directly determines the flight time and thus the time for solvent evaporation. Reducing the needle-to-collector distance increases the diameter of the fibers. Too small distance and too short flight time will lead to bead formation and ellipsoidal cross-sectional fiber formation, so it is important to maintain a critical minimum distance. It should be noted that in some cases (e.g., where high conductivity provides high flight speeds) no or little change in fiber diameter is observed with increasing needle-collector distances.	[119,120]
Type/geometry of collector	The most common types of collectors used in electrospinning are the flat and rotary metal collectors. The different geometric designs of the collectors can affect the macroscopic properties of the nanofibers, the diameter, oriented arrangement, and morphological uniformity of the fibers. These properties can be influenced mainly by the effect of geometry on changes in electric field distribution and local variation of force lines. The rotating drum collector results in thinning of the fibers and reduction of fiber diameter due to mechanical stretching from rotation. An auxiliary electrode coupled to an external potential (e.g., a ring auxiliary electrode) reduces fiber formation instability by focusing the electric flux and helps to form uniform and thinner fibers.	[36]
Environmental Parameters		
Temperature	The higher molecular weight polymer reduces the formation of beads during electrospinning, while increasing the fiber diameter.	[121]
Relative humidity	The type of polymer can have a significant effect on the conductivity of the solution. A higher conductivity increases the magnitude of the fiber-forming force, which tends to decrease the fiber diameter.	[122,123]

highlights the importance of proper packaging and storage conditions. Stability can be improved by using suitable auxiliary polymers or polymer blends, but in such cases, the extent and effect of phase separation should be investigated. In addition to the low hygroscopicity of the polymer, antiplasticizing ability [125] and other molecular interactions [126,127] are important in achieving good stability.

5.2.4.1.2 Chemical Stability, Quality, and Regulatory Compliance

The use of electrospinning in the pharmaceutical industry is subject to strict manufacturing, product quality, and clinical requirements. It must be verified by tests for the uniformity of the API content, the chemical quality of the excipients used, the compatibility of the API with the excipient, and the specification of any degradation products of different origin that may be present in the formulation, the exclusion of genotoxicity, and pharmacokinetic compliance [128]. In another approach, a significant proportion of the experimental work available in the literature involves the use of solvents whose use in pharmaceuticals is either toxicologically unauthorized or not possible or significantly limited due to industrial process considerations [129]. There is also a high proportion of published papers, in which the fiber-forming polymer is not considered to be an accepted pharmaceutical excipient. Based on all these limitations, from the industrial feasibility of manufacturing point of view, preferred drug delivery systems based on aqueous or alcoholic solvents or using pharmaceutical excipients needs to be fulfilled the following main requirements: adequate stability, impurities and degradation products below limits, and suitable *in vitro* release of the API.

5.2.4.1.3 Drug-Loading Capacity and Processability

Fibrous drug delivery systems can be divided into two main groups according to their use. It is possible to use the fibrous web directly or further process it as a phase product to create other classical pharmaceutical dosage forms such as tablets [87,130] or capsules [131]. For both uses, the homogeneous distribution of the API after incorporation into the fiber is important. The percentage of API that can be incorporated into the fibers is low, on average ~5%–20%. If the pharmaceutical form to be created is a tablet, the amount of fibrous phase product for a 200 mg dose formulation would be 1 g, to which the addition of the necessary excipients would result in a tablet of a size that is difficult for the patient to swallow [132]. However, regulatory guidelines on the use of excipients (e.g., FDA IID: FDA Inactive Ingredients Database) may in some cases require lower limits than currently feasible. Based on these factors, the use of fibrous drug delivery systems for tablets is likely to be feasible only for low-dose (~1–10 mg) formulations, which leads back to the issue of homogeneity. Despite the availability of results in the literature on the design of tablets or capsules as the final drug formulation, it is important to continue experiments in the field of fiber milling to achieve the right particle size and adequate tableting and filling properties [133].

5.2.4.1.4 Positioning the Applications of the Technology

On the one hand, the vision of electrospinning is defined by the scientific advances made with the nanofibrous systems produced for use in medical technology, tissue regeneration, and drug delivery systems. On the other hand, the industrial applicability, benefits, and limitations also determine the future of the technology. As stated above, the technology has the potential to deliver low drug content, and the development of drug delivery systems requiring further processing and the addition of excipients positions the technology toward therapeutic areas where low doses of certain highly potent (e.g., letrozole, pomalidomide) [85] agents are required for effective treatment. In the case of direct use, such as topical or buccal formulations (e.g. sheets, inserts) [134], the addition of excipients can be avoided, which helps achieve the required dose. The pharmaceutical technology application of electrospinning is further specialized by the fact that its use can only be justified in cases where the desired beneficial property is strictly the result of this process. Therefore, only in special cases can electrospinning be considered a realistic alternative to classical pharmaceutical technology for incorporating API (e.g. wet granulation for the formulation of low-dose BCS II and BCS IV APIs). Vass and coworkers reported a very promising use of electrospinning in pharmaceutical technology [80]. In their work, they implemented electrospinning as an efficient and preferable alternative to freeze drying for the production of intravenously administrable reconstitution powder-based drug forms. Based on the available literature, it can be concluded that electrospun nanofibrous platforms can represent a relevant added value in the fields of tissue regeneration, wound treatment, and targeted delivery of genetic materials, as supported by a valuable summary published by Puhl et al. [135]. They report on eight different ongoing clinical phase trials in wound treatment and tissue regeneration and summarize the commercially available fiber platform-based products. It also gives an insight into the growing importance of gene therapy-based regenerative medicine, which is supported by the 20 different clinical trials currently ongoing, as mentioned in the summary. All of these are positioning electrospinning toward the manufacture of medical devices and drug device combinations.

5.2.4.2 Scaling-Up Considerations

From a technological point of view, the application of electrospinning in the pharmaceutical industry requires that the fibrous phase or end product can be produced in volume to meet commercial needs, in a reproducible and robust manner, in compliance with the regulatory environment and quality requirements [136], and at an acceptable cost price. The equipment used for the vast majority of work on drug delivery systems based on electrospinning technology is lab-scale and has an average productivity of 0.01–1 g/h

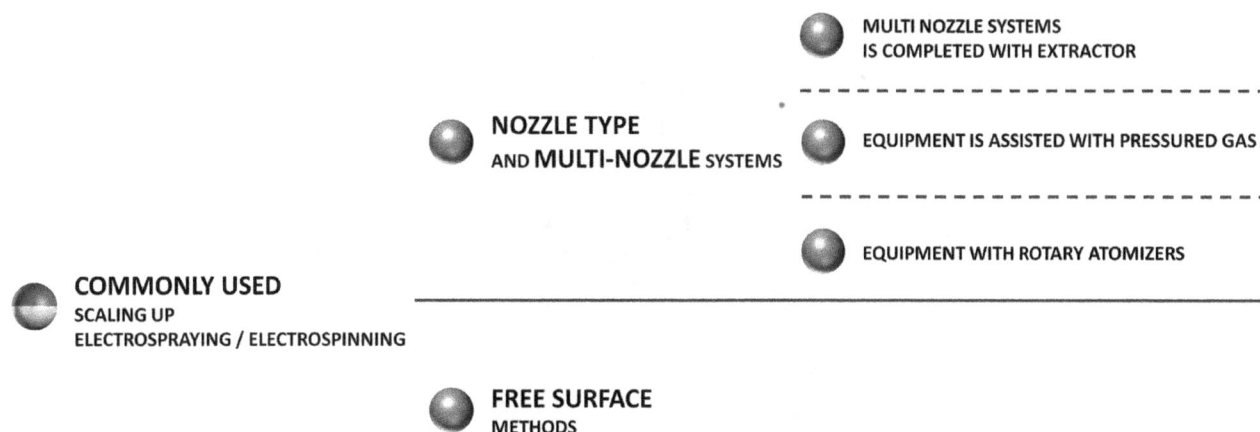

FIGURE 5.3 Overview of scaling-up electrospinning and electrospraying opportunities.

[137]. Therefore, there is a need to scale up the technology and develop a technical solution that meets the requirements. As with any other pharmaceutical technology operation, scale-up is a critical step in electrospinning, as the properties of the formulation developed in the laboratory development phase, the manufacturing conditions, and the manufacturing parameters may change completely during scale-up, and in extreme cases may require a reworking of the development or lead to a halt in development. Significant progress has been made in this area over the last few years; however, the problem of scale-up in the pharmaceutical industry is currently not yet considered solved [13]. Scale-up can be achieved by increasing the number of radii or Taylor cones, which can be achieved by using multiple needle arrangements, free-surface (needle-free, e.g. Nanospider™) designs, or combined fiber-forming forces (e.g. centrifugal or pneumatic forces) [132] (Figure 5.3). Nagy et al. have demonstrated the applicability of high-speed electrospinning to the pharmaceutical industry. They achieved a feed rate of 1500 mL/h and a productivity of 450 g/h using a free-surface fiber-forming device [9]. Farkas and coworkers achieved two orders of magnitude higher productivity compared to the classical single-needle process using corona alternating current electrospinning (C-ACES) using a single-needle process [138]. Nozzle electrospinning devices are limited solution surface systems where the solution is directly introduced into needles or multi-hole emitters. In contrast, free-surface electrospinning processes involve the formation of significantly more Taylor cones and thus significantly more liquid jets leaving the open surface of the solution surface [137]. An additional difficulty with multi-nozzle electrospinning processes is that repulsion and interference between adjacent nozzles can lead to fiber unevenness and loss of fiber quality. These effects can be partially compensated by proper geometry of the nozzles and the use of auxiliary electrodes [40]. There are several studies in the literature that provide an overview of companies that have aimed to increase the size of electrospinning [13,132]. In summary, the scaling-up of the fiber-forming process for pharmaceutical and, more dominantly, medical

device applications is currently at the milestone of pilot-scale manufacturing, with continued work on process control and process validation needed to achieve the next major milestone in the pharmaceutical implementation field.

5.2.4.3 Formulation of the Final Dosage Forms

Electrospinning can be used to create excipient structures with extremely high specific surface areas and diverse micro- and macrostructures. The process is gentle in terms of undesirable chemical and physical effects, and the rapid and efficient solvent evaporation during the process is also beneficial for the stability of the final product. However, taking all these factors into account, the processability of the resulting fibrous web and the advantages (e.g., physico-chemical, morphological) that the system can provide are the factors that will determine which type of administration route the technology is most suitable for (Figure 5.4).

One possible way to produce an electrospinning-based pharmaceutical formulation is to use fibers collected and processed (e.g., milling) in the primary step, and then additional excipients are added to form the final oral pharmaceutical form, such as a tablet or capsule. The use of fibers for tablets or capsules may be justified in cases where the main purpose of the electrospinning technique from the point of the active substance is to achieve a stable amorphous solid dispersion (ADS) system or to preserve the biological activity of the active substance [87]. Examples of direct compression of fibers can also be found in the literature; however, it should be noted that for industrial solid drug development, robust manufacturability of a fiber-based formulation to meet the requirements, and to achieve the appropriate pharmacokinetic behavior, the use of excipients with different functions is inevitable in the vast majority of cases [130,139].

Electrospun fibers can be also used as a reconstitution dosage form for intravenous injection, which can be prepared after processing the fibers by milling or without the use of additional excipients [80]. In this case, in addition to the design of the ASD system or the preservation

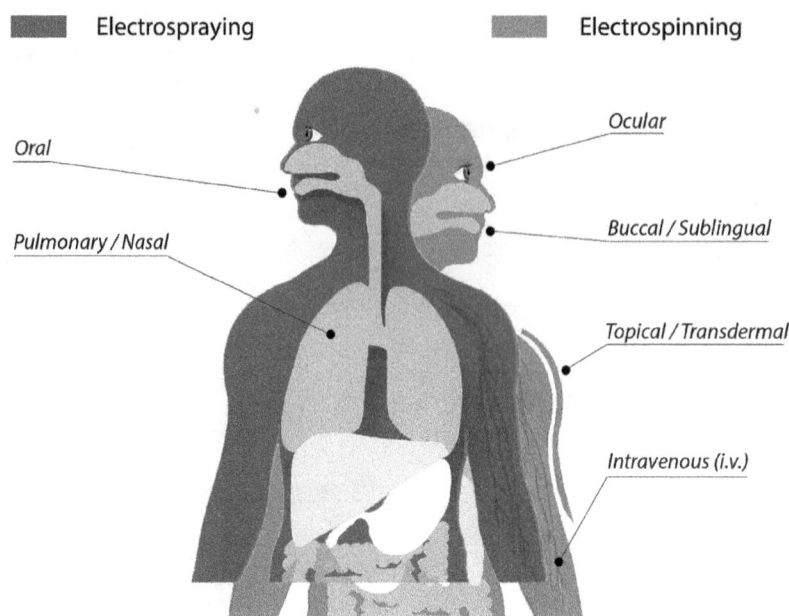

FIGURE 5.4 Highlighting the most common administration routes with respect to electrospinning and electrospraying.

of biological activity, the high specific surface area of the fibers is of particular importance to ensure a suitable dissolution rate.

Another use is direct use, where the fibrous product can be used to produce the desired pharmaceutical form or medical device without further processing. This category also includes those solutions where additional steps are necessary to create a formulation with the right shape and dimensions, but these do not affect the behavior and performance of the drug delivery system or device as primarily designed. In general, for direct application, the fiber shape, morphological characteristics, specific surface area, and mechanical properties play a major role in the efficacy of the drug product and the performance or usability of the medical device.

Such direct use can be applied to produce ophthalmic inserts and therapeutic contact lenses [140], for which several advances have been made in recent years [141]. The advantages of fibrous ophthalmic inserts include efficient in situ gelation and up to several hours of retention time due to the high specific surface area [142]. By using suitable polymer or polymer composites, complex release kinetics can be achieved, where an initial rapid release providing a saturating dose is followed by a prolonged extended release. Fibrous ophthalmic inserts can be promising and non-invasive alternatives for the intravitreal treatment of certain eye diseases.

The direct use of fibers can be for the formulation of buccal sheets and transdermal patches [143]. In a pharmaceutical collaboration, Edmans et al. have developed a fiber-based mucoadhesive formulation for the treatment of chronic oral inflammation under the brand name Rivelin®, which successfully met the primary endpoint in a phase 2b clinical trial in 2020. In addition to the local treatment of oral cavity diseases, the additional advantage of

the electrospun buccal dosage form is that in some cases it allows systemic treatment without significant metabolic transformation of the drug [134].

A growing number of literature demonstrate that the morphological properties of nano- and microfibers have a highly beneficial effect on the regeneration of certain tissue types [144]. This fiber characteristic has led to the success of several developments, e.g., medical devices for skin or bone tissue regeneration or medical devices for drug delivery. Nowadays, there are several electrospun medical devices and drug device combinations for wound management (e.g., Phoenix Wound Matrix®, NeoDura®), but also fiber-based vascular graft (e.g., AVflo™) coronary stents (PK Papyrus®) [135]. Of particular note is the development of Nanomedic Technologies Ltd. under the SpinCare™ brand. The company has developed a portable electrospinning device for the efficient and rapid extemporaneous treatment of minor skin tissue injuries and burns [145].

5.3 ELECTROSPRAYING

5.3.1 COMPREHENSIVE DESCRIPTION OF THE PHYSICAL BACKGROUND AND VARIOUS SETUPS

Electrospraying is a gentle way for preparing fine particles, but the process parameters can highly influence the quality of the electrosprayed products. Therefore, the knowledge of the theoretical background and the operation of the technology is essential to achieve a conscious design of electrosprayed formulations. To better understand the physical background of the electrospraying, the key elements of a basic equipment will be introduced at first. The simplest electrospraying setup, similarly to the electrospinning, consists of a feeding system (usually a syringe pump), a syringe, a metal nozzle (often a metal needle) connecting

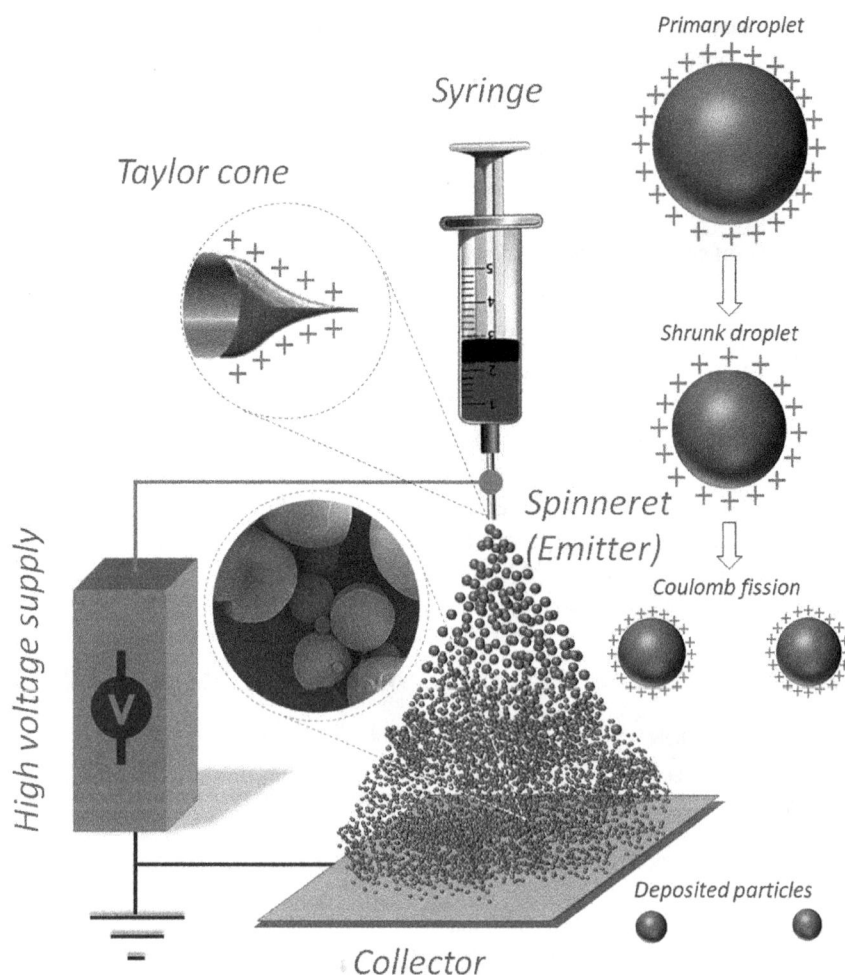

FIGURE 5.5 Schematic design of the basic electrospraying apparatus.

to a high voltage power supply (sometimes a tube for linking the syringe and the spinneret), and a grounded collector opposite to the nozzle (Figure 5.5).

During the process, the solution is pumped toward the charged metal needle. Similar to electrospinning, the outgoing solution drop mostly takes a conical shape due to the potential difference between the nozzle and the grounded collector. In contrast with electrospinning, the liquid jet breaks up into droplets during electrospraying due to the lower solution concentration. The size of these so-called primary droplets is reducing while the surface charge density is increasing continuously due to the quick solvent evaporation resulted by the electrostatic field. Finally, the fission of the primary droplets is realized when the electrostatic forces overcome the surface tension forces, and the complete evaporation of solvents leads to the formation of the smaller secondary droplets [146]. The limit in the surface density is also called the Rayleigh limit, and this part of the electrospraying is usually named Rayleigh disintegration or Coulomb fission.

The presented droplet formation is usually mentioned as Taylor cone-jet mode regarding electrospraying since this method is also based on the formation of the cone shape in the presence of the electrostatic field, described previously in Section 5.2.1 [31]. Although the applied voltage can influence the spraying mode [147], the Taylor cone-jet mode is the most preferred in the field of micro- and nanosized particle preparation [148]. This can be explained by the fact that the stable production of monodispersed droplets is feasible in this way. To find the suitable applied voltage, the calculation of the critical voltage according to Equation 5.2 can give a good starting point if the radius of the needle (R) is much less than the distance between the nozzle and the grounded collector (l) [149].

$$V_{critical} \approx \sqrt{\frac{2\gamma R \cos\theta}{\varepsilon_0}} \ln\frac{4l}{R} \qquad (5.2)$$

In the Equation, γ is the solution surface tension, ε_0 is the permittivity of vacuum, and θ is the semi-cone angle, which is influenced by several factors such as surface tension stress, electric stress, space charge, flow rate, liquid loss through the cone, and ambient pressure.

Adjusting the appropriate process parameters can be even harder if multi-fluid electrospraying is applied to prepare particles with special properties [150]. To better understand the differences between the single-fluid and multi-fluid methods, the basic operation of the latter one needs to be overviewed at first. Similar to electrospinning, the most common multi-fluid electrospraying techniques are the coaxial (usually called dual-capillary) and triaxial (commonly mentioned as tri-capillary) methods, which consist of two and three capillaries, respectively (Figure 5.2). Besides, Janus spinneret also appeared in the field of electrospraying as an alternative bi-fluid electrospraying method.

One of the key challenges relating to all multi-fluid techniques is that the selected solutions must be compatible with each other to avoid precipitations during the electrospraying process [151]. On the other hand, the feeding rates of the solutions need to be synchronized to achieve the Taylor cone formation and thus a stable condition for fine particle production. If the above-mentioned criteria are met, the multi-fluid setups make feasible the production of multi-component systems in the form of hollow microspheres or particles with core-shell structure, which can be beneficial from the drug delivery point of view as well.

5.3.2 ADVANTAGES AND APPLICATION OF ELECTROSPRAYING AS EFFECTIVE STRATEGIES FOR DRUG DELIVERY

The micro- and nanosized particles prepared by electrospraying have great potential in the field of drug delivery. Particles with small diameter can be characterized by good wettability while their dissolution and permeability are also fast and high. Therefore, electrospraying is a promising technology regarding nanoformulations, which was proved in many publications. For instance, Thakkar and Misra used electrospraying to prepare a more effective formulation of the antitumor docetaxel [152]. The prepared nanocrystals showed better properties both *in vitro* and *in vivo* than the pure drug and physical mixture. Their research also highlighted that the applied triblock copolymer contributed to the higher wettability since its hydrophilic part was located outward.

Besides the advantages of small-scale particles, preparation of ASDs via electrospraying is also feasible; thus, similarly to electrospinning, it is usually presented as a promising technology for the formulation of poorly water-soluble APIs. Zhang et al. took advantage of this and successfully improved the dissolution of griseofulvin [153]. During their work, coaxial electrospraying was applied, which enabled the preparation of particles with a diameter of around 1 μm and the formation of core-shell structure. According to their results, the small particle size, the improved dispersity, and the amorphization might result in not only good dissolution but increased *in vivo* oral absorption as well. Furthermore, multi-fluid equipment such as coaxial or triaxial electrospraying make it possible to produce complex formulations, which can be used for targeted therapy or in controlled release systems. These advantages were demonstrated by Bai and Liu, who prepared

drug-loaded electrosprayed particles and bioconjugated them with a given type of monoclonal antibody to reach targeted delivery to ovarian cancer cells [154].

It is also noteworthy that electrospraying, similarly to electrospinning, works without any special environmental conditions (no high temperature and pressure are needed). It can be especially advantageous in the case of sensitive biomolecules where the damage of the macromolecule must be avoided. Zamani et al. revealed that core-shell structure is also recommended for the encapsulation of proteins [155]. Bovine serum albumin was used as a model protein during their work, and the influence of process parameters was investigated. Their results highlighted that the production conditions have an impact on the material properties since the flow rate of the core solution and the concentration of the protein in the core influenced the encapsulation efficiency.

All of the above-mentioned advantages resulted in that electrosprayed formulations were tested in the case of several drug delivery routes such as oral, nasal, or intravenous delivery. In addition, a wide range of therapeutic effects can be achieved with the small particle size electrosprayed products. Table 5.4 summarizes some examples to present the variability of the electrospraying technology from the pharmaceutical application point of view.

5.3.3 IMPACT OF DIFFERENT PARAMETERS ON THE PRODUCT QUALITY

Considering the strict requirements of the pharmaceutical industry, the preparation of high-quality, reproducible product has to be achieved in the case of electrospraying as well. However, different factors can affect the various quality attributes of the electrosprayed products, including the particle size, the particle shape, and structure, residual solvent content, or physical stability. In drug-loaded systems, optimizing these properties is crucial since the bioavailability, efficiency, and applicability of the electrospraying-based formulation may depend on it. The influencing parameters can be divided according to the Critical Material Attributes (CMAs), the CPPs, and the ambient factors, similar to the factors described at electrospinning (Table 5.5).

Although the influencing factors are similar as they were presented previously in the case of electrospinning, their interpretation differs from some perspectives. The main difference lies in that the solution concentration determines whether fibers or small-size particles are formed. Jahangiri et al. applied an electrohydrodynamic method for preparing triamcinolone acetonide-PLGA nanoformulations [171]. They found that bead formation can be performed at lower concentrations while higher solution concentration results in fiber formation. Their experiments highlighted how the surface tension and viscoelastic forces affect the disruption of liquid jets.

If the fiber formation is avoided, mostly spherical or similar particles are forming based on electrospraying principles. Therefore, during the electrospraying, the various

TABLE 5.4
Some Examples for Drug-Loaded Electrosprayed Particles Showing Their Pharmaceutical Relevance

Targeted Function	Drug Delivery Route	Preparation Method	Therapeutic Indications	APIs (S: Small Molecules, L: Large Molecules)	Polymer/Carrier Matrix	Ref.
• Achieving controlled release	Oral	Monoaxial electrospraying with water bath used as collector	• Antibiotic treatment	DOXYCYCLINE HYCLATE (S)	• PLGA	[156]
• Achieving sustained release • Improved drug loading and efficiency	Porous microparticles for gastroretentive systems or for pulmonary drug delivery	Monoaxial electrospraying with aluminum foil collector	• Widely used in therapies for Helicobacter pylori infection • Antibiotic treatment	METRONIDAZOLE (S)	• PLGA	[157,158]
			BCSI			
• Dissolution improvement through preparation of ASD	Oral	Monoaxial electrospraying with aluminum foil collector	• Non-steroidal anti-inflammatory drug with analgesic and antipyretic effects	KETOPROFEN (S)	• PVP K30	[159]
			BCSII			
• Bioavailability enhancement via the production of nanocrystals • Excipient-free formulation	Oral (capsules)	Monoaxial electrospraying with aluminum foil collector	• For treatment of malaria	ATOVAQUONE (S)	N/A	[160]
• Bioavailability enhancement via the production of amorphous particles • Suitable product without excipient for dry powder inhalers • Excipient-free formulation	Pulmonary delivery (inhalation therapy) (amorphous)	Monoaxial electrospraying with a stationary collector	• Antibiotic treatment	AZITHROMYCIN (S)	N/A	[161]
• Increasing oral bioavailability • Achieving controlled release	Oral	Coaxial electrospraying with aluminum foil collector	• Antihypertensive effect	FELODIPINE (inner layer) (S)	• PVP K30 (inner layer), HPMC K4M (outer layer)	[162]
• Controlled release • Targeted delivery seemed to be achievable • Amorphous form of the API	Oral	Coaxial electrospraying with a metal collector plate	• Treatment of inflammatory bowel disease • Treatment of a number of inflammatory and autoimmune conditions	PREDNISOLON (S)	• Eudragit L100-55	[163]
• Colon targeting delivery system	N/A	Monoaxial electrospraying with aluminum foil collector	• Non-steroidal anti-inflammatory drug	INDOMETHACIN (S)	• Inulin acetate	[164]
• Colon targeting delivery system through the preparation of core-shell structure	oral	Coaxial electrospraying	• Prevention and treatment of a chronic, recurrent, and debilitating inflammatory disease	CURCUMIN (S)	• Zein protein (inner layer) • Shellac (outer layer)	[165]

(Continued)

TABLE 5.4 (Continued)
Some Examples for Drug-Loaded Electrosprayed Particles Showing Their Pharmaceutical Relevance

Targeted Function	Drug Delivery Route	Preparation Method	Therapeutic Indications	APIs (S: Small Molecules, L: Large Molecules)	Polymer/Carrier Matrix	Ref.
• Multidrug delivery • Enhancing oral absorption	oral	Coaxial electrospraying	• Both applied APIs are protease inhibitors	• LOPINAVIR (BCSII) (S) • RITONAVIR (BCSII) (S)	• PVP (inner with the drugs) • Eudragit L100 (outer)	[166]
• Multidrug delivery • Controlled release	N/A	Coaxial tri-capillary electrospray system with a ground ring immersed in olive oil bath used as collector	• Budesonide: treatment of asthma and non-infectious rhinitis, prevention of lung cancer tumors (*off-label*) • EGCG: antitumor and antimutagenic activities (*off-label*)	• BUDESONIDE (BCSII) (S) • EPIGALLOCATECHIN GALLATE - EGCG (BCSIII) (S)	• PLGA • PCADK	[167]
• Multidrug delivery • Controlled release	N/A	Single-nozzle emulsion electrospraying	• Naproxen: non-steroidal anti-inflammatory drug with analgesic and antipyretic effects **BCSIII**	• NAPROXEN (BCSII) (S) • RHODAMINE B (S)	• PVP K30 • Chitosan	[168]
• Achieving controlled release • Preparation of protein-type drug	N/A	Coaxial electrospraying with aluminum foil collector	N/A	• BOVINE SERUM ALBUMIN (L)	• PLGA	[155]
• Targeted delivery • Controlled release	Intravenous application	Monoaxial electrospraying	• Cancer treatment	• DOXORUBICIN (S)	• PLGA	[169]
• Multidrug delivery • Controlled release	Local delivery	Single-nozzle emulsion electrospraying	• Treatment of osteosarcoma **BCSIV**	• DOXORUBICIN (core) (S) • METHOTREXATE (outer shell) (BCSIV) (S)	• hydroxyapatite (core) • PCL+chitosan (outer shell)	[170]
• Achieving controlled release	N/A	Monoaxial electrospraying	• Steroidal anti-inflammatory drug	• TRIAMCINOLONE ACETONIDE (S)	• PLGA	[171]
• Targeted delivery	Local drug delivery (*off-label*)	Single-needle electrospraying coupled with a ring on which lower high voltage was applied	• Anti-cancer chemotherapeutic	• PACLITAXEL (S)	• PLGA • PDLA • PLLA • PCL	[172]

PLGA, poly(lactic-co-glycolic acid); PVP, polyvinylpyrrolidone; HPMC, hydroxypropyl methylcellulose; PCADK, poly(cyclohexane-1,4-diyl acetone dimethylene ketal); PCL, polycaprolactone; PDLA, poly(DL-lactide); PLLA, poly(L-lactide).

TABLE 5.5
Summary of Parameters Influencing the Product Quality of Electrosprayed Particles

Parameters	Impact on the Product Quality	Ref.
	Composition-Dependent Parameters	
Polymer	Polymers with lower molecular weights result in smaller particle size products.	[173]
	The type of the polymer can influence the conductivity of the solution and thus the particle size. The absorbed moisture in the case of more hygroscopic polymers can increase the electrical conductivity, and higher conductivity may lead to a decrease of the particle size.	[28,174]
Solution concentration	Increasing solution concentration results in higher viscosity and higher surface tensions, which facilitate the preparation of larger particles. Too high solution concentration might lead to fiber formation.	[171,175]
Solvent(s)	The boiling point, the conductivity, the viscosity, and the surface tension of the applied solvent(s) all can affect particle formation during electrospraying. More volatile solvents with less viscosity and surface tension, and higher conductivity might contribute to the formation of smaller size particles.	[176,177]
	Process Parameters	
Flow rate	Smaller particle size can be achieved via decreasing the flow rate.	[178,179]
Applied voltage	Increasing voltage lead to the formation of smaller particles.	[178,180]
Needle diameter	Nozzles with smaller inner diameter may facilitate the formation of spherical shape particles with lower size, while agglomerations can occur if a needle with a larger diameter is applied.	[180,181]
Spinneret-collector distance	At a lower working distance, the electric field is stronger, which results in smaller particles. However, if the spinneret is too close to the collector, electric discharge prevents particle formation.	[180]
Collector type	In most cases, aluminum foil is used for collecting the electrosprayed materials, which has no effect on the particle size and shape. However, there are some examples, where the electrosprayed particles are collected in liquid media, where the concentration of the media can influence the particle characteristic. Increasing surface tension of the collection liquid might result in increased particle diameter and deformation in the particle shape.	[182,183]
	Environmental Parameters	
Temperature	Products with higher particle sizes can be prepared at higher temperatures.	[184]
Relative humidity	Humidity has no significant impact on the diameter of the particles. However, it can influence the structure of the electrosprayed samples. Pores can be formatted on the surface of the particles at higher relative humidity.	[185]

material, process, and environmental parameters influence mainly the diameter and the structure of these particles. At this point, it is important to note that the formation of the final electrospray products highly depends on the solvent evaporation and the diffusion rate of the solid material. The correlation between these two properties can be determined with a modified Peclet number (Pe) (Equation 5.3) [186].

$$Pe = \frac{\frac{\partial r_d}{\partial t} \cdot r_d}{D_{AB}} \qquad (5.3)$$

In Equation 5.3, $\frac{\partial r_d}{\partial t}$ indicates the rate of change of the droplet radius caused by solvent evaporation, r_d means the radius of the droplet, and D_{AB} is the diffusion coefficient of the polymeric solute inside the droplet. Yao et al. used this Peclet number to predict qualitatively the morphology of PLGA particles prepared by electrospraying [187]. They found that hollow particles can be formed if the Pe value is around 100, while decreasing Pe results in porous particles. As the Pe reduced to about a tenth smooth spherical particles were observed. These results demonstrated well that a slow diffusion rate (low D_{AB} and thus high Pe) leads to the concentration of the polymer on the droplet surface, which causes the formation of the hollow morphology. With an

increasing diffusion rate, the polymers are able to diffuse to the center of the droplets and generate a porous or smooth spherical structure. In the context of drug delivery systems, finding the suitable morphology can be crucial since it can influence the release of the APIs. Almería et al. showed the significance of this approach in the pharmaceutical field [188,189]. In one of their works, PLGA microparticles were prepared with different morphology to select the good products for drug delivery [189]. Then they developed an electrosprayed formulation containing PLGA, and doxorubicin as model API [188]. During the preparation of drug-loaded samples, great emphasis was placed on the effect of particle size on the release. Similar examinations were performed by Bohr et al., who highlighted that selecting the solvents is also very important [190]. According to their results, application of anti-solvent can lead to more compact PLGA particles, from which quick release can be achieved.

It is clearly visible from Table 5.5 that besides the solution-related parameters, the process parameters and the environmental factors also affect the morphology of the prepared electrosprayed particles [187]. In addition, adjusting these parameters can determine the presence or absence of residual solvents, which is important to check in the case of drug delivery systems due to the strict pharmaceutical requirements [160]. Too high feeding rate and too small working distance can cause wet products; thus determination of the optimal process parameters is essential during the early development. On the other hand, if amorphous form is planned to achieve, exploration of the factors influencing the appearance of crystalline traces can be crucial since the physical stability can affect the drug release [29]. Consequently, it is worth designing the preparation processes taking into account the influencing parameters summarized in Table 5.5.

5.3.4 CHALLENGES RELATING TO ELECTROSPRAYING

The previous sections presented the potential of electrospraying in the pharmaceutical formulations. However, some challenges need to be addressed if the industrial application wants to be achieved. From the product quality point of view, conscious design and ensuring reproducibility of the CMAs such as particle morphology, particle size, composition, or physical state (amorphous or crystalline) are essential since these factors determine the efficacy and safety, and thus the authorization of a potential final dosage form. Besides, technological aspects also need to keep in mind, such as rational selection of the process parameters, scaling up, and development of the downstream processing steps to prepare marketable medicine products in satisfying amounts. The next subchapters will overview the main difficulties in this field to highlight the barriers connecting to the electrospraying and present the current solutions for answering the critical points relating to the technology. As the critical feasibility assessments are similar to what was detailed in Section 5.2.4.1, here only the considerations connecting strongly to electrospraying will be discussed more.

5.3.4.1 Designing Suitable Electrosprayed Products

During the development of electrospraying-based formulations, the first challenge is to prepare effective electrosprayed particles, which can reach the planned therapeutic goal while staying physically and chemically stable during production, storage, and usage. Therefore, the influencing factors mentioned in Section 5.3.3 need to be considered during the design of suitable pharmaceutical products containing electrosprayed particles. If the selection of the process parameters is appropriate, it is feasible to create stable submicron-sized crystals [191]. The addition of right polymers can further increase the thermal stability of the APIs; thus the application of complex compositions is a widely used strategy in this field [192]. All the more so because proper excipients might facilitate the formation of matrix systems, in which the polymer can stabilize the amorphous nature of the API [159,193–195].

The role of polymers is even higher in the case of multilayered systems since their properties affect the rate and the location of the API release. Yu and his coworkers pointed out that immediate release could be achieved if PVP is used next to the helicid (applied as a model drug) in the shell of the electrosprayed particles [196]. Their results showed that from this polymer-drug composition, the preparation of

solid product was not feasible via the application of simple single-fluid electrospraying. Therefore, they demonstrated a possible multi-fluid solution, during which a shellac core provided the formulation of the solid particles. In this way, the preparation of an easy-to-handle product and the immediate release of the API are realized at the same time. Although nanocoating with the drug-loaded solution can be useful in some special cases, the production of controlled released systems is more common during the application of multi-fluid methods. The main advantage of this is that core-shell structures can be formed, where the composition of the two layers determines the release kinetic of the applied API(s). For instance, Wang et al. accomplished a dual-stage release of ketoprofen with PVP and ethyl cellulose nanoparticles [197]. During their work, the shell contained one part of the drug and PVP; thus a fast release was achieved at first. To reach a sustained release after that, the core contained the other part of the API mixed with ethyl cellulose. The release rate can also decrease through the production of lipid-based microparticles, which can be a suitable formulation for intraperitoneal chemotherapy [198]. It is worth mentioning here that electrosprayed samples with core-shell structure are very promising from the oncology application point of view since not only reaching sustained release is feasible but targeted delivery [199] and combinational delivery [170] can also be achieved. Furthermore, the preparation of simple spherical electrosprayed particles with appropriate excipients is a widely researched area in the context of different cancer treatments [169,200,201].

The above-described examples demonstrate well that the application of different compositions and process parameters enable the production of diverse electrosprayed samples for application in drug delivery systems [202]. Rational selection of suitable materials and settings depends on the desired therapeutic effect and final dosage form. Consequently, the Quality by Design (QbD) approach does not avoid this area of drug development either. Although the designing strategy connecting to electrospraying is only a small component of the secondary manufacturing, the basic QbD-based thinking is transferable from the related ICH guideline or similar technological processes [203–206]. Therefore, determining the Quality Target Product Profile (QTPP) can be the first step even if an electrospraying-based formulation is planned to achieve. Then the CQAs need to be identified, which are influenced by the CPPs and the CMAs. It means that the parameters influencing electrospraying and the properties of the relevant raw materials must be explored during the early developments to find CPPs and CMAs, which have the greatest impact on the CQAs. It can be stated that the QbD approach can facilitate the design of drug delivery systems containing electrosprayed particles. To make its steps clearly visible and easy to interpret, the QbD-based thinking converted to electrospraying (and also to electrospinning) is summarized in Figure 5.6.

Overall, formulations prepared by electrospraying can be applied to fulfill several medical needs, as shown in

○ Type of the applied excipients, polymers

○ Drug-excipient ratio

○ Type of solvents

○ Concentration of the solution

CMA

QTPP ──────────── **CQA** ──────────────────────

CPP

○ Therapeutic target / indication   ○ Physical and chemical stability

○ Routes of drug administration   ○ Drug release

○ Final dosage form and dose   ○ Powder properties

○ Preparation method   *(e.g. flow properties of the fibers*
(type of electrospinning / electrospraying) *or particles)*

○ Product quality and stability   ○ Residual solvents

○ Release profile

○ Possible stress factors (*e.g. presure, temperature, moisture, etc.*) during the production

○ Applied voltage

○ Applied flow rate

○ Applied nozzle / spinneret

○ Applied collector, nozzle-collector distance

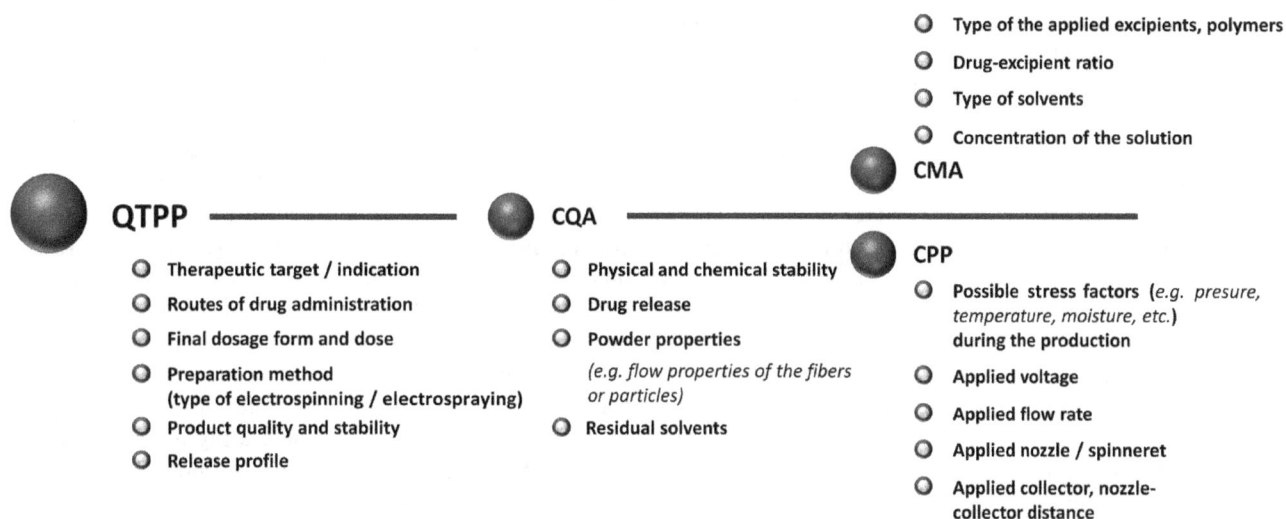

FIGURE 5.6   Flowchart of the QbD-based thinking with regard to electrospinning and electrospraying.

Table 5.4. However, selecting the proper excipients for a given purpose and finding the most fitting electrospraying methods and process parameters are the key elements of electrospraying-based formulation developments.

5.3.4.2 Scaling-Up Considerations

The productivity of the single-nozzle electrospraying methods (including both single-fluid and multi-fluid systems) is low, similar to the basic electrospinning apparatus. Several developments were performed to reach higher throughput [207], which are analogous to the techniques described above by electrospinning.

The simplest way of scaling up is here also to increase the number of Taylor cones with the usage of multi-nozzle systems (Figure 5.3), which was already achieved in 1993 by Rulison and Flagan via the application of a linear array of more capillary electrodes [208]. Since then, Almekinders and Jones demonstrated that electrospraying can be accomplished with an increased number of nozzles, exactly with forming 24 Taylor cones [209]. Besides, Parhizkar et al. highlighted in their publication the role of the arrangements of the nozzles [210]. During their research, similar productivity was achieved with both circular and rectangular plate configurations (the latter corresponded to the linear array). However, a more uniform size of the particles was prepared with the circular arrangement while it needed lower applied voltage. These results assume that the nozzle configuration selection is important from a technological and thus industrial applicability point of view. In addition, the research of Parhizkar et al. presents a critical property of the multi-nozzle electrospraying that the interaction of the nozzles and electrically charged liquid droplets might influence the technological efficiency of the process and the quality of the products. For instance, decreasing the distance between the nozzles and increasing the applied voltage can lead to the mixing of the droplets leaving the neighboring needles, resulting in larger particles [211,212]. To handle the space charging issues, nozzles with holes can be applied

for multi-needle electrospraying [188,213–216]. The equipment is usually supplemented in this case with a so-called "extractor" functionalizing as an intermediate electrode. The extractor contributes to the operation of the given electrospraying apparatus in cone-jet mode while preventing the electrostatic interferences and ensuring the electrostatic shielding of the Taylor cones. Instead of applying an extractor, the operation of a multi-nozzle electrospraying equipment coupled with a pressurized gas system also proved to be a promising way to increase the productivity of the technology [217–220]. Similar to electro-blowing, the application of pressurized gas facilitates the solvent evaporation and thus the nebulization from the solution; therefore, the interactions between the charged droplets can be avoided. Another potential solution for effective electrospray particle preparation can be the application of a rotary atomizer with orifices or nozzles. The operation principle of the spinneret is similar to that was mentioned by the electrospinning [221,222]. It means that besides the electrostatic forces centrifugal forces also speed up the drying and thus influence the particle formation.

Furthermore, free-surface methods such as the previously mentioned Nanospider™ also can be applied in the case of electrospraying if the clogging of the nozzle wants to be avoided [223,224]. In addition, setups with various shape, free surface, and charged spinneret have been published over the years, including for example a spiral tower, a rotating cylinder, or a special form of metallic wire [225–227]. Although the free-surface techniques have similar advantages to those presented by the electrospinning, only a few pharmaceutical examples can be found in the literature [228]. One of the main limitations can be that besides the scaling up the strict quality and safety requirements are also needed to keep in mind. During pharmaceutical application of the technology, volatile solvents are usually used, which quick evaporation rate can cause changes in the solution concentration. In addition, it means high risk due to the flammable nature of the volatile organic solvents.

Overall, however, it can be said that both nozzle-type and free-surface methods seem to be applicable for preparing small-size electrosprayed particles with higher productivity. Despite this, the majority of the published scaling-up techniques deal with single-fluid systems and less with multi-fluid processes. Since the potential in coated and core-shell type particles is quite significant in the field of drug delivery, it would be useful to reach higher throughputs for industrial application. According to the results of Yan et al., it is possible to produce microparticles with core-shell structure using multi-nozzle equipment [229]. However, similarly to the single-fluid systems, it is worth applying higher nozzle-to-nozzle distance or "dummy" nozzles (empty nozzles, without fluid flowing) to avoid the interactions of the neighboring nozzles' electric field. Nevertheless, increasing the number of the coaxial electrospray emitters proved to be also feasible based on the results of Olvera-Trejo and Velásquez-García, which can open new ways in the field of the electrospraying for drug delivery systems [230].

5.3.4.3 Formulation of the Final Dosage Forms

Electrospraying results in smaller particles due to the quick solvent evaporation as the widely used spray drying. However, the electrostatic chargeability of the electrosprayed products can make harder the handling of the prepared solid particles during the formulation of final dosage forms. As the route of administration largely determines what kind of delivery system is worth to be produced, the downstream processing steps after the electrospraying can be varied according to this. Figure 5.4 highlights the most common drug delivery routes with regard to the electrospraying. Similar to the electrospun fibers, the simplest way of electrospraying-based formulation could be if the particles are utilized for drug delivery right after the preparation process (it is usually called direct method) [231]. It might be an obvious solution if the final dosage forms are aerosols and appropriate pulmonary delivery devices are applied. Several publications can be found in the literature about electrosprayed particles designed for pulmonary delivery routes [232,233]. Furthermore, some patented inhalers and suitable equipment also appeared in this field [234,235]. Although the examples demonstrated the applicability of electrospraying-based aerosol dosage forms, commercially available products have not come to the market so far. Consequently, developing a patient compliant, effective, and precise devices working on the principles of electrohydrodynamic atomization is one of the challenges in the context of formulation of final dosage forms containing electrosprayed particles.

During the development of other dosage forms, the first processing step is usually the collection, which can be different depending on the desired quality of the particles and the therapeutic aim of the medicine. In general, metal collectors, mainly aluminum foil, are used to host the solid particles coming from the charged solution. However, the particles seek to discharge as soon as possible; thus, not all of them reach the collector. A significant part of the solid material sticks to the other elements of the setup. To increase collection efficiency, Kim et al. combined the electrospraying process with an electrostatic precipitator [236]. From an industrial point of view, even more promising is the collection method of Grafahrend et al., who presented a device assisted by airflow for the continuous collection of electrosprayed samples [237]. They proved that the equipment is able to operate for 24 h continuously while a 79.2% yield was achieved. Although the collection efficiency was more than twofold higher compared to a basic electrospraying setup, only circa 300 mg product was prepared during 24 h. Therefore, increasing the productivity and accomplishing the scaling-up of both electrospraying and the collection are essential for reaching the industrially sufficient amount of the electrosprayed particles. To achieve the material need of the pharmaceutical industry, the pilot-plant device of the company called Bioinicia can be a suitable solution [238]. A case study with carvedilol demonstrated well that a much higher feeding rate was achieved as in the case of the previously mentioned method. While Grafahrend et al. have applied only 0.08 mL/h feeding rate, 240 mL/h solution flowrate was adjusted during the application of the Bioinicia's equipment. In addition, the continuous collection of carvedilol-loaded electrosprayed particles was accomplished with a cyclonic collector, which also allowed industrially effective collection of the product. The applicability of the presented electrospraying and cyclonic collection system was further proved with another drug-polymer system, which also demonstrated the equipment developed by Bioinicia could be suitable for the pharmaceutical industry [219].

The next challenge relating to the formulation of the final dosage form is the dosing of the poorly flowable small particles. It can be especially a crucial point if parenteral administration via injection dosage forms is planned to be applied [239,240]. To handle the poor flowability, different excipients can be applied, which can even help the preparation of tablets containing electrosprayed particles [241,242]. Besides the scalability, it is also important that the samples prepared by electrospraying keep their advantageous properties even after the compression. Therefore, the production of stable and high amount of final dosage forms containing electrosprayed samples can be challenging during each processing step. Probably, these are the main reasons that the majority of the publications only focus on the preparation of the small-scale particles via electrospraying while the further industrially relevant formulation steps are not discussed in general.

5.3.5 Industrial Approach of Electrospraying

The process of electrospraying can be found in various application fields since its basic operation unit is quite simple while the atomization and preparation of small particles are effective [243]. One of its most common usages relating to the mass spectrometry, where electrospray ionization is usually coupled to the system [244]. Besides several commercialized products for different purposes, electrospraying has become a widely researched method from drug delivery point of view as well. The examples discussed in Section 5.3.2 demonstrated the pharmaceutical applicability of the technique, and its medical possibilities were also examined. For instance, electrospraying proved to be a promising tool for coating stents to reach controlled release of the API [245], but it is also well-seeming for gene therapy application [246].

As the main difference between electrospinning and electrospraying is related to the solution parameters but the operation principles are the same, the equipment used for electrospinning can also be suitable for electrospraying. In terms of the commercially available devices, it can be stated that more than 20 companies distribute electrospinning apparatus, about half of which offer industrial-scale equipment too [132]. This high number of manufacturers assumes that the industrial approach of electrospraying is also feasible using marketed electrospinning devices besides appropriate solution concentration and process parameters. It is confirmed by the fact that some suppliers suggest their electrospinning equipment for electrospraying experiments as well. What is more, the above-mentioned Bioinicia company is also developing pharmaceutical and biomedical devices using electrosprayed particles [247]. According to their pipelines, there are products under early-stage developments for immunotherapy, hypertension, antibiotic injection, oncology, nicotine addiction, and pain. In addition, particles prepared for handling hypertensions are in late development status.

It is important to note that besides the high productivity of a technology, its financial implications need to be also considered from industrial perspective. In turn of the electrospraying, usually similar cost might be occurred as during electrospinning. Application of a higher amount of solvents to form spherical particles, or ensuring pressurized gas in the case of some equipment to facilitate scaling-up production, might mean an extra cost during manufacturing small-scale particles via electrospraying. Despite this, the large number of marketed electrospinning/electrospraying equipment suggest that both technologies can be operated economically and thus can be promising for pharmaceutical application. Although there are some GMP and ISO-validated devices that fit the pharmaceutical industry's strict safety and quality requirements, no marketed pharmaceutical product containing electrosprayed particles has appeared so far. In addition, compared to electrospinning, fewer examples can be found for electrospraying-based medical devices. However, taking into account the increasing number of publications in this field, the opportunities relating to multi-fluid electrospraying systems, and the developments of the equipment manufacturers, commercially available products for drug delivery are expected to appear in the near future.

5.4 COMPREHENSIVE COMPARISON OF ELECTROSPINNING AND ELECTROSPRAYING

All of the presented examples demonstrate well that both electrospinning and electrospraying can be applied effectively to prepare drug delivery systems with advanced properties. The high specific surface area of the micro- and nanoscale fibers and particles contribute to the increased wettability and enhanced dissolution of the poorly water-soluble APIs, while the multi-axial techniques open new ways into the direction of multidrug delivery systems and controlled release formulations.

From the product point of view, the morphology and the dimension of the fibers and particles are the most vivid difference. While electrospun products can be characterized by length and diameter, electrosprayed particles only have the diameter dimension [248]. As it was described before, several administration route and formulation were tested in both cases, but it is worth considering which physical structure is preferred and suitable for the production of the planned final dosage form. For instance, the advantages of the fibrous structure can be utilized excellently in wound healing treatments or drug-loaded patches [249]. In contrast, small-size particles prepared by electrospraying are especially beneficial for pulmonary delivery routes, where spraying the API-loaded powders is a common and widely used strategy [248].

If both electrohydrodynamic processes seem promising for a given purpose, different technological considerations need to be kept in mind during the formulation development (Table 5.6). As electrospinning and electrospraying are solvent-based techniques, rational selection of the solvents' type and determination of the solvent needs are crucial factors due to the strict quality requirements and limits relating to residual solvents. In addition, the solvents can influence the quality of the products since the drying kinetics can differ according to the boiling point of the applied solvents or solvent mixtures. Consequently, the morphology of the fibers or particles can be varied, or in the case of ASD systems, phase separation can occur [250,251]. Besides, the planned final dosage forms require the conscious design of each further formulation step, where the difficult handling of the nano- or microscale fibers and particles, the commercially available devices, and the energetic and economic needs must be considered.

Based on the increasing literature on electrohydrodynamic processes, it can be stated that both electrospinning and electrospraying have great potential to result in marketed pharmaceutical products. Moreover, the equipment manufacturers also support such efforts in order to make feasible the reproducible and high-quality production of fibers and particles on the micro- and nanosize scale. Although the introduction of new technology in the pharmaceutical industry is always challenging, some publications proved that electrospinning and electrospraying are competitive technologies relative to the currently used solvent-based

TABLE 5.6

Different Considerable Technological Aspects during the Electrohydrodynamic-Based Formulation Developments

Method	Electrospinning	Electrospraying
Solvent need	**Less solvent** is needed compared to the electrospraying since fiber formation starts only from solutions with higher viscosity.	**More solvent** is needed compared to electrospinning to avoid the fiber formation.
Limit of composition	**The solubility of the applied component (especially of the drugs) can limit the process** because fibers can be produced only from viscous solutions, which usually require the use of the maximum possible API amount (mainly in the case of high drug-loaded systems).	**Not limited by the solubility** properties because electrosprayed particles can be produced from dilute solutions.
Need for post-drying	Less solvent needs to evaporate during the process thus **post-drying can be skipped** if optimal process parameters are adjusted during the electrospinning.	Although the electrostatic field facilitates the drying, the **higher amount of applied solvents can require the integration of a post-drying step** after the electrospraying.
Energy consumption	The electrical energy need is determined only by the feeder, the high voltage supply, and the collector system, which is **similar** in the case of electrospraying.	The energy consumption of the electrospraying process is **similar** to the energy needs of electrospinning. **Post-drying can increase the energy needs.**
Commercially available equipment, scaling-up opportunities	Several lab-scale devices are **commercially available** and some manufacturers offer **scale-up equipment**.	The commercially available **electrospinning equipment can apply for electrospraying** as well.
Handling of the products/ challenges during the formulation processes to reach final dosage forms	The fibers' collection **is feasible** both by batch and continuous techniques. • If fibers need to dose for packaging or reaching tablet or capsule forms, **grinding** of the fibrous products **is usually needed.** • The low bulk densities and poor flow properties can make more complex the formulation processes (e.g., blending and tableting); thus **application of excipient with good flowability is often needed.**	• Several collection methods can be applied but collecting the small-size particles can be challenging due to the high electrostatic chargeability of the samples. • **Dosing toward further formulation steps or dosing for packaging can mean a huge challenge** due to the poor flowability, high electrostatic chargeability, and high hygroscopicity caused by the large specific surface area.

formulation methods. According to the results of Browne et al., electrospraying can be as effective as spray drying in the context of dissolution enhancement while smaller particles and more uniform particle size were produced via electrospraying [28]. Besides, electrospinning is also applicable for similar purposes than spray drying. Moreover, the preparation of stable ASDs without phase separation might be easier, thanks to the quick solvent evaporation facilitated by the electrostatic field [221,252]. The advantage of electrostatic forces can also be exploited during the formulation of sensitive biologics since electrohydrodynamic methods could be potential alternatives to freeze drying or spray drying [80,253,254].

5.5 CONCLUSION AND FUTURE PERSPECTIVES

The significant growth of the literature in the field of electrohydrodynamic techniques foresees a promising future of electrospinning- and electrospraying-based innovative drug delivery systems. Both technologies are suitable for incorporating APIs into a suitable carrier matrix in a simple and one-step way, leading to nano- or microsize fibers or

particles with advantageous properties. Despite their great popularity, there is no approved pharmaceutical product in the market up to now, which contains electrospun fibers or electrosprayed particles. The lack of these formulations can be explained by the above-mentioned design and technological challenges. The technologists need to first explore and understand the physical and physicochemical behaviors of the electrospun and electrosprayed systems to develop a stable, reproducible, and effective pharmaceutical product. In addition, the realization of the electrohydrodynamic technologies in the industrial environment also has to solve to reach the appearance of a marketed product containing electrospun fibers or electrosprayed particles. Finally, it is important to mention that new technologies are mostly introduced in the pharmaceutical industry only if they are able to provide a solution to problems that traditional methods cannot answer. Consequently, applications, where the nano- and micro sizes are especially preferred, can take the advantages of electrospinning or electrospraying.

The current state of the art shows that electrospun products are well applicable for wound healing and also promising in the case of medical devices. Therefore, combining these types of products with drug delivery systems has a great role in healthcare. Besides, electrospraying can have great significance in the administration of small-scale particles through the pulmonary system. Thus, electrospraying-based medical devices, which target similar areas, might also have industrial relevance.

Although the currently marketed products, which are somewhat related to drug delivery systems, are mainly developed for medical applications and clinical usage, electrohydrodynamic methods are expected to extend to the production of medicines in the future. This assumption is based on the fact that these electrostatic methods allow quick, gentle, and energy-efficient drying while stable, solid products are generated; thus, sensitive APIs such as biologics or thermolabile small molecules can be formulated effectively. Given these advantages, it may be realistic to consider these methods as alternatives to the current techniques (e.g., freeze drying or spray drying). Furthermore, electrospun and electrosprayed formulations can be promising in controlled release systems or targeted drug delivery; therefore, dosage forms containing fibers or electrosprayed particles can be applicable for special treatments, personalized medicines, or innovative products with higher patient compliance.

The publications summarized in this chapter highlighted that both electrospinning and electrospraying have numerous advantages from the pharmaceutical application point of view while the devices are developing year by year in the interest of satisfying the needs of the industry. Consequently, the appearance of electrospun and electrosprayed products in medicines is already imminent. Even if it is not always necessary to replace proven methods with these new technologies, they can open up new avenues in product development in several critical cases.

REFERENCES

1. Cooley JF. Apparatus for electrically dispersing fluids patent US692631A. 1902.
2. Morton WJ. Method of dispersing fluids patent US705691A. 1902.
3. Tucker N, Stanger JJ, Staiger MP, et al. The history of the science and technology of electrospinning from 1600 to 1995. *Journal of Engineered Fibers and Fabrics*. 2012;7(-2):63–73. https://doi.org/10.1177/155892501200702S10.
4. Behrens AM, Casey BJ, Sikorski MJ, et al. In situ deposition of PLGA nanofibers via solution blow spinning. *ACS Macro Letters*. 2014;18(3):249–254. https://doi.org/10.1021/mz500049x.
5. Badrossamay MR, McIlwee HA, Goss JA, et al. Nanofiber assembly by rotary jet-spinning. Nano Letters. 2010;10(-6):2257–2261. https://doi.org/10.1021/nl101355x.
6. Medeiros ES, Glenn GM, Klamczynski AP, et al. Solution blow spinning: A new method to produce micro- and nanofibers from polymer solutions. *Journal of Applied Polymer Science*. 2009;113(4):2322–2330. https://doi.org/10.1002/app.30275.
7. Khamforoush M, Asgari T. A modified electro-centrifugal spinning method to enhance the production rate of highly

8. Dabirian F, Hosseini Ravandi SA, Pishevar AR, et al. A comparative study of jet formation and nanofiber alignment in electrospinning and electrocentrifugal spinning systems. *Journal of Electrostatics*. 2011;69(6):540–546. https://doi.org/10.1016/j.elstat.2011.07.006.
9. Nagy ZK, Balogh A, Démuth B, et al. High speed electrospinning for scaled-up production of amorphous solid dispersion of itraconazole. *International Journal of Pharmaceutics*. 2015;480(1):137–142. https://doi.org/10.1016/j.ijpharm.2015.01.025.
10. Sebe I, Bodai Z, Eke Z, et al. Comparison of directly compressed vitamin B12 tablets prepared from micronized rotary-spun microfibers and cast films. *Drug Development and Industrial Pharmacy*. 2015;41(9):1438–1442. https://doi.org/10.3109/03639045.2014.956112.
11. Vass P, Hirsch E, Kóczián R, et al. Scaled-up production and tableting of grindable electrospun fibers containing a protein-type drug. *Pharmaceutics*. 2019;11(7). https://doi.org/10.3390/pharmaceutics11070329.
12. Torres-Giner S, Busolo M, Cherpinski A, et al. Chapter 10: Electrospinning in the packaging industry. In: Kny E, Ghosal K, Thomas S, editors. *Electrospinning: From Basic Research to Commercialization*. London: The Royal Society of Chemistry; 2018. pp. 238–260.
13. Kannan B, Cha H, Hosie IC. Electrospinning—Commercial applications, challenges and opportunities. In: Fakirov S, editor. *Nano-size Polymers: Preparation, Properties, Applications*. Cham: Springer International Publishing; 2016. pp. 309–342.
14. Colchester WGo. De Magnete. Londini1600.
15. Grimm RL. Fundamental studies of the mechanisms and applications of field-induced droplet ionization mass spectrometry and electrospray mass spectrometry. California Institute of Technology; 2006.
16. Imperial Chemical Industries Ltd, assignee. Atomisation of liquids patent GB1569707A. 1976.
17. Kelly AJ, inventor; ExxonMobil Research and Engineering Co, assignee. Electrostatic atomizing device patent US4581675A. 1983.
18. Bhushani JA, Anandharamakrishnan C. Electrospinning and electrospraying techniques: Potential food based applications. *Trends in Food Science & Technology*. 2014;38(-1):21–33. https://doi.org/10.1016/j.tifs.2014.03.004.
19. Maurmann N, Sperling L-E, Pranke P. Electrospun and electrosprayed scaffolds for tissue engineering. *Cutting-Edge Enabling Technologies for Regenerative Medicine*. 2018:79–100. https://doi.org/10.1007/978-981-13-0950-2_5.
20. Fenn J. Electrospray ionization mass spectrometry: How it all began. *Journal of Biomolecular Techniques: JBT*. 2002;13(3):101.
21. Jaworek A, Sobczyk AT. Electrospraying route to nanotechnology: An overview. *Journal of Electrostatics*. 2008;66(-3–4):197–219. https://doi.org/10.1016/j.elstat.2007.10.001.
22. Nguyen DN, Clasen C, Van den Mooter G. Pharmaceutical applications of electrospraying. *Journal of Pharmaceutical Sciences*. 2016;105(9):2601–2620. https://doi.org/10.1016/j.xphs.2016.04.024.
23. Rostamabadi H, Falsafi SR, Rostamabadi MM, et al. Electrospraying as a novel process for the synthesis of particles/nanoparticles loaded with poorly water-soluble bioactive molecules. Advances *in Colloid and Interface Science*. 2021;290:102384. https://doi.org/10.1016/j.cis.2021.102384.

24. Wu Y-H, Yu D-G, Li J-J, et al. Medicated multiple-component polymeric nanocomposites fabricated using electrospraying. *Polymers and Polymer Composites*. 2017;25(1):57–62. https://doi.org/10.1177/096739111702500109.

25. Malik SA, Ng WH, Bowen J, et al. Electrospray synthesis and properties of hierarchically structured PLGA TIPS microspheres for use as controlled release technologies. *Journal of Colloid and Interface Science*. 2016;467:220–229. https://doi.org/10.1016/j.jcis.2016.01.021.

26. Khan M, Hasan M, Barnett A, et al. Co-axial electrospraying of injectable multi-cancer drugs nanocapsules with polymer shells for targeting aggressive breast cancers. *Cancer Nanotechnology*. 2022;13(1):1–18. https://doi.org/10.1186/s12645-022-00114-1.

27. Wang J, Jansen JA, Yang F. Electrospraying: Possibilities and challenges of engineering carriers for biomedical applications—a mini review. *Frontiers in Chemistry*. 2019;7:258. https://doi.org/10.3389/fchem.2019.00258.

28. Browne E, Charifou R, Worku ZA, et al. Amorphous solid dispersions of ketoprofen and poly-vinyl polymers prepared via electrospraying and spray drying: A comparison of particle characteristics and performance. *International Journal of Pharmaceutics*. 2019;566:173–184. https://doi.org/10.1016/j.ijpharm.2019.05.062.

29. Kawakami K, Zhang S, Chauhan RS, et al. Preparation of fenofibrate solid dispersion using electrospray deposition and improvement in oral absorption by instantaneous post-heating of the formulation. *International Journal of Pharmaceutics*. 2013;450(1–2):123–128. https://doi.org/10.1016/j.ijpharm.2013.04.006.

30. Taylor GI, Van Dyke MD. Electrically driven jets. *Proceedings of the Royal Society of London A Mathematical and Physical Sciences*. 1969;313(1515):453–475. https://doi.org/10.1098/rspa.1969.0205.

31. Taylor GI. Disintegration of water drops in an electric field. *Proceedings of the Royal Society of London Series A Mathematical and Physical Sciences*. 1964;280(1382):383–397. https://doi.org/10.1098/rspa.1964.0151.

32. Uematsu I, Uchida K, Nakagawa Y, et al. Direct observation and quantitative analysis of the fiber formation process during electrospinning by a high-speed camera. *Industrial & Engineering Chemistry Research*. 2018;57(36):12122–12126. https://doi.org/10.1021/acs.iecr.8b02352.

33. Salata VO. Tools of nanotechnology: Electrospray. *Current Nanoscience*. 2005;1(1):25–33. http://dx.doi.org/10.2174/1573413052953192.

34. Huang C, Soenen SJ, Rejman J, et al. Stimuli-responsive electrospun fibers and their applications. *Chemical Society Reviews*. 2011;40(5):2417–2434. https://doi.org/10.1039/C0CS00181C.

35. Varabhas JS, Chase GG, Reneker DH. Electrospun nanofibers from a porous hollow tube. *Polymer*. 2008;49(19):4226–4229. https://doi.org/10.1016/j.polymer.2008.07.043.

36. Gupta A, Ayithapu P, Singhal R. Study of the electric field distribution of various electrospinning geometries and its effect on the resultant nanofibers using finite element simulation. *Chemical Engineering Science*. 2021;235:116463. https://doi.org/10.1016/j.ces.2021.116463.

37. Patel S, Patel G. A review and analysis on recent advancements in bubble electrospinning technology for nanofiber production. *Recent Patents on Nanotechnology*. 2019;13(2):80–91. https://doi.org/10.2174/1872210513666190306154923.

38. Alghoraibi I, Alomari S. Different methods for nanofiber design and fabrication. In: Barhoum A, Bechelany M, Makhlouf A, editors. *Handbook of Nanofibers*. Cham: Springer International Publishing; 2018. pp. 1–46.

39. Zhang Y, Zhang L, Cheng L, et al. Efficient preparation of polymer nanofibers by needle roller electrospinning with low threshold voltage. *Polymer Engineering & Science*. 2019;59(4):745–751. https://doi.org/10.1002/pen.24993.

40. Saleh Hudin HS, Mohamad EN, Mahadi WNL, et al. Multiple-jet electrospinning methods for nanofiber processing: A review. *Materials and Manufacturing Processes*. 2018;33(-5):479–498. https://doi.org/10.1080/10426914.2017.1388523.

41. Wang X, Xu W. Effect of experimental parameters on needleless electrospinning from a conical wire coil. *Journal of Applied Polymer Science*. 2012;123(6):3703–3709. https://doi.org/10.1002/app.35044.

42. Molnar K, Nagy ZK. Corona-electrospinning: Needleless method for high-throughput continuous nanofiber production. *European Polymer Journal*. 2016;74:279–286. https://doi.org/10.1016/j.eurpolymj.2015.11.028.

43. Chen J, Yu Z, Li C, et al. Review of the principles, devices, parameters, and applications for centrifugal electrospinning. *Macromolecular Materials and Engineering*. 2022:2200057. https://doi.org/10.1002/mame.202200057.

44. Pant B, Park M, Park S-J. Drug delivery applications of core-sheath nanofibers prepared by coaxial electrospinning: A review. *Pharmaceutics*. 2019;11(7):305. https://doi.org/10.3390/pharmaceutics11070305.

45. Ghosal K, Augustine R, Zaszczynska A, et al. Novel drug delivery systems based on triaxial electrospinning based nanofibers. *Reactive and Functional Polymers*. 2021;163:104895. https://doi.org/10.1016/j.reactfunctpolym.2021.104895.

46. Jiang G, Zhang S, Qin X. Effect of processing parameters on free surface electrospinning from a stepped pyramid stage. *Journal of Industrial Textiles*. 2016;45(4):483–494. https://doi.org/10.1177/1528083714537101.

47. Buzol Mülayim B, Göktepe F. Analysis of polyacrylonitrile nanofiber yarn formation in electrospinning by using a conical collector and two oppositely charged nozzles. *The Journal of the Textile Institute*. 2021;112(3):494–504. https://doi.org/10.1080/00405000.2020.1768772.

48. Nangare S, Jadhav N, Ghagare P, et al. Pharmaceutical applications of electrospinning. *Annales Pharmaceutiques Françaises*. 2020;78(1):1–11. https://doi.org/10.1016/j.pharma.2019.07.002.

49. Park SM, Eom S, Kim W, et al. Role of grounded liquid collectors in precise patterning of electrospun nanofiber mats. *Langmuir*. 2018;34(1):284–290. https://doi.org/10.1021/acs.langmuir.7b03547.

50. İçoğlu Hİ, Ceylan Ş, Yıldırım B, et al. Production of aligned electrospun polyvinyl alcohol nanofibers via parallel electrode method. *The Journal of the Textile Institute*. 2021;112(6):936–945. https://doi.org/10.1080/00405000.2020.1789274.

51. Jha BS, Colello RJ, Bowman JR, et al. Two pole air gap electrospinning: Fabrication of highly aligned, three-dimensional scaffolds for nerve reconstruction. *Acta Biomaterialia*. 2011;7(1):203–215. https://doi.org/10.1016/j.actbio.2010.08.004.

52. Sahay R, Thavasi V, Ramakrishna S. Design modifications in electrospinning setup for advanced applications. *Journal of Nanomaterials*. 2011:317673. https://doi.org/10.1155/2011/317673.

53. Li Y, Zhu J, Cheng H, et al. Developments of advanced electrospinning techniques: A critical review. *Advanced*

Materials Technologies. 2021;6(11):2100410. https://doi.org/10.1002/admt.202100410.

54. Angammana CJ, Jayaram SH. Fundamentals of electrospinning and processing technologies. *Particulate Science and Technology.* 2016;34(1):72–82. https://doi.org/10.1080/02726351.2015.1043678.

55. Shuakat MN, Lin T. Direct electrospinning of nanofibre yarns using a rotating ring collector. *The Journal of the Textile Institute.* 2016;107(6):791–799. https://doi.org/10.1080/00405000.2015.1061785.

56. Park S, Park K, Yoon H, et al. Apparatus for preparing electrospun nanofibers: Designing an electrospinning process for nanofiber fabrication. *Polymer International.* 2007;56(-11):1361–1366. https://doi.org/10.1002/pi.2345.

57. Deitzel JM, Kleinmeyer JD, Hirvonen JK, et al. Controlled deposition of electrospun poly(ethylene oxide) fibers. *Polymer.* 2001;42(19):8163–8170. https://doi.org/10.1016/S0032-3861(01)00336-6.

58. Luraghi A, Peri F, Moroni L. Electrospinning for drug delivery applications: A review. *Journal of Controlled Release.* 2021;334:463–484. https://doi.org/10.1016/j.jconrel.2021.03.033.

59. Aidana Y, Wang Y, Li J, et al. Fast dissolution electrospun medicated nanofibers for effective delivery of poorly water-soluble drug. *Current Drug Delivery.* 2022;19(4):422–435. https://doi.org/10.2174/1567201818666210215110359.

60. Ahn YC, Park SK, Kim GT, et al. Development of high efficiency nanofilters made of nanofibers. *Current Applied Physics.* 2006;6(6):1030–1035. https://doi.org/10.1016/j.cap.2005.07.013.

61. Noyes A, Whitney W. The rate of solution of solid substances in their own solutions. *J Am Chem Soc.* 1897;19(-12):930–934. https://doi.org/10.1021/ja02086a003.

62. Allison M. Reinventing clinical trials. *Nature Biotechnology.* 2012;30:562. https://doi.org/10.1038/nbt0612-562a.

63. Amidon G, Lennernäs H, Shah V, et al. A theoretical basis for a biopharmaceutic drug classification: The correlation of in vitro drug product dissolution and in vivo bioavailability. *Pharmaceutical Research.* 1995;12(3):413–420. https://doi.org/10.1023/A:1016212804288.

64. Hann M, Keseru G. Finding the sweet spot: The role of nature and nurture in medicinal chemistry. *Nature Reviews Drug discovery.* 2012;11(5):355–365. https://doi.org/10.1038/nrd3701.

65. Hann M. Molecular obesity, potency and other addictions in drug discovery. In: Scapin G, Patel D, Arnold E, editors. *Multifaceted Roles of Crystallography in Modern Drug Discovery.* Berlin: Springer; 2015. pp. 183–196.

66. Lipinski C, Lombardo F, Dominy B, et al. Experimental and computational approaches to estimate solubility and permeability in drug discovery and development settings. *Advanced Drug Delivery Reviews.* 2012;64:4–17. https://doi.org/10.1016/j.addr.2012.09.019.

67. Modica de Mohac L, Keating AV, De Fátima Pina M, et al. Engineering of nanofibrous amorphous and crystalline solid dispersions for oral drug delivery. *Pharmaceutics.* 2019;11(1). https://doi.org/10.3390/pharmaceutics11010007.

68. Vo CL-N, Park C, Lee B-J. Current trends and future perspectives of solid dispersions containing poorly water-soluble drugs. *European Journal of Pharmaceutics and Biopharmaceutics.* 2013;85(3):799–813. https://doi.org/10.1016/j.ejpb.2013.09.007.

69. Serajuddin ATM. Solid dispersion of poorly water-soluble drugs: Early promises, subsequent problems, and recent breakthroughs. *Journal of Pharmaceutical Sciences.* 1999;88(10):1058–1066. https://doi.org/10.1021/js980403l.

70. Loh ZH, Samanta AK, Sia Heng PW. Overview of milling techniques for improving the solubility of poorly water-soluble drugs. *Asian Journal of Pharmaceutical Sciences,* 10(4):255–274. https://doi.org/10.1016/j.ajps.2014.12.006.

71. Paudel A, Worku ZA, Meeus J, et al. Manufacturing of solid dispersions of poorly water soluble drugs by spray drying: Formulation and process considerations. *International Journal of Pharmaceutics.* 2013;453(1):253–284. https://doi.org/10.1016/j.ijpharm.2012.07.015.

72. Corrigan OI. Retardation of polymeric carrier dissolution by dispersed drugs: Factors influencing the dissolution of solid dispersions containing polyethlene glycols. *Drug Development and Industrial Pharmacy.* 1986;12(11–13):1777–1793. https://doi.org/10.3109/03639048609042609.

73. Van den Mooter G. The use of amorphous solid dispersions: A formulation strategy to overcome poor solubility and dissolution rate. *Drug Discovery Today: Technologies.* 2012;9(-2):e79–e85. https://doi.org/10.1016/j.ddtec.2011.10.002.

74. Puppi D, Chiellini F. Chapter 12: Drug release kinetics of electrospun fibrous systems. In: Focarete ML, Tampieri A, editors. *Core-Shell Nanostructures for Drug Delivery and Theranostics.* Sawstone: Woodhead Publishing; 2018. pp. 349–374.

75. Moydeen AM, Ali Padusha MS, Aboelfetoh EF, et al. Fabrication of electrospun poly(vinyl alcohol)/dextran nanofibers via emulsion process as drug delivery system: Kinetics and in vitro release study. *International Journal of Biological Macromolecules.* 2018;116:1250–1259. https://doi.org/10.1016/j.ijbiomac.2018.05.130.

76. Khodadadi M, Alijani S, Montazeri M, et al. Recent advances in electrospun nanofiber-mediated drug delivery strategies for localized cancer chemotherapy. *Journal of Biomedical Materials Research Part A.* 2020;108(7):1444–1458. https://doi.org/10.1002/jbm.a.36912.

77. Li X, He Y, Hou J, et al. A time-programmed release of dual drugs from an implantable trilayer structured fiber device for synergistic treatment of breast cancer. *Small.* 2019;16(9):1902262.

78. Yaqoubi S, Adibkia K, Nokhodchi A, et al. Co-electrospraying technology as a novel approach for dry powder inhalation formulation of montelukast and budesonide for pulmonary co-delivery. *International Journal of Pharmaceutics.* 2020;591:119970. https://doi.org/10.1016/j.ijpharm.2020.119970.

79. Turanlı Y, Tort S, Acartürk F. Development and characterization of methylprednisolone loaded delayed release nanofibers. *Journal of Drug Delivery Science and Technology.* 2019;49:58–65. https://doi.org/10.1016/j.jddst.2018.10.031.

80. Padmakumar S, Paul-Prasanth B, Pavithran K, et al. Long-term drug delivery using implantable electrospun woven polymeric nanotextiles. *Nanomedicine: Nanotechnology, Biology and Medicine.* 2019;15(1):274–284. https://doi.org/10.1016/j.nano.2018.10.002.

81. Vass P, Démuth B, Farkas A, et al. Continuous alternative to freeze drying: Manufacturing of cyclodextrin-based reconstitution powder from aqueous solution using scaled-up electrospinning. *Journal of Controlled Release.* 2019;298:120–127. https://doi.org/10.1016/j.jconrel.2019.02.019.

82. Rade PP, Giram PS, Shitole AA, et al. Physicochemical and in vitro antibacterial evaluation of metronidazole loaded Eudragit S-100 nanofibrous mats for the intestinal

drug delivery. *Advanced Fiber Materials*. 2022;4(1):76–88. https://doi.org/10.1007/s42765-021-00090-y.

83. Jia D, Gao Y, Williams GR. Core/shell poly(ethylene oxide)/Eudragit fibers for site-specific release. *International Journal of Pharmaceutics*. 2017;523(1):376–385. https://doi.org/10.1016/j.ijpharm.2017.03.038.

84. Chen J, Wang X, Zhang W, et al. A novel application of electrospinning technique in sublingual membrane: Characterization, permeation and in vivo study. *Drug Development and Industrial Pharmacy*. 2016;42(8):1365–1374. https://doi.org/10.3109/03639045.2015.1135939.

85. Ding Y, Dou C, Chang S, et al. Core–shell Eudragit S100 nanofibers prepared via triaxial electrospinning to provide a colon-targeted extended drug release. *Polymers*. 2020;12(9):2034. https://doi.org/10.3390/polym12092034.

86. Wang H, Huang J, Liu Y, et al. Poloxamer188 composite electrospun poly L-lactic acid fibrous nonwoven: Sustained in vitro and in vivo release letrozole as a subcutaneous implant. *Journal of Industrial Textiles*. 2022:15280837211062055. https://doi.org/10.1177/15280837211062055.

87. Yang C, Yu D-G, Pan D, et al. Electrospun pH-sensitive core–shell polymer nanocomposites fabricated using a triaxial process. *Acta Biomaterialia*. 2016;35:77–86. https://doi.org/10.1016/j.actbio.2016.02.029.

88. Casian T, Borbás E, Ilyés K, et al. Electrospun amorphous solid dispersions of meloxicam: Influence of polymer type and downstream processing to orodispersible dosage forms. *International Journal of Pharmaceutics*. 2019;569:118593. https://doi.org/10.1016/j.ijpharm.2019.118593.

89. Tipduangta P, Belton P, McAuley WJ, et al. The use of polymer blends to improve stability and performance of electrospun solid dispersions: The role of miscibility and phase separation. *International Journal of Pharmaceutics*. 2021;602:120637. https://doi.org/10.1016/j.ijpharm.2021.120637.

90. Chu J, Chen L, Mo Z, et al. An atorvastatin calcium and poly(L-lactide-co-caprolactone) core-shell nanofiber-covered stent to treat aneurysms and promote reendothelialization. *Acta Biomaterialia*. 2020;111:102–117. https://doi.org/10.1016/j.actbio.2020.04.044.

91. Geng Y, Zhou F, Williams GR. Developing and scaling up fast-dissolving electrospun formulations based on poly(vinylpyrrolidone) and ketoprofen. *Journal of Drug Delivery Science and Technology*. 2021;61:102138. https://doi.org/10.1016/j.jddst.2020.102138.

92. Celebioglu A, Uyar T. Fast dissolving oral drug delivery system based on electrospun nanofibrous webs of cyclodextrin/ibuprofen inclusion complex nanofibers. *Molecular Pharmaceutics*. 2019;16(10):4387–4398. https://doi.org/10.1021/acs.molpharmaceut.9b00798.

93. Balogh A, Farkas B, Domokos A, et al. Controlled-release solid dispersions of Eudragit® FS 100 and poorly soluble spironolactone prepared by electrospinning and melt extrusion. *European Polymer Journal*. 2017;95:406–417. https://doi.org/10.1016/j.eurpolymj.2017.08.032.

94. Grant JJ, Pillai SC, Perova TS, et al. Electrospun fibres of chitosan/PVP for the effective chemotherapeutic drug delivery of 5-fluorouracil. *Chemosensors*. 2021;9(4):70. https://doi.org/10.3390/chemosensors9040070.

95. Samadzadeh S, Mousazadeh H, Ghareghomi S, et al. In vitro anticancer efficacy of Metformin-loaded PLGA nanofibers towards the post-surgical therapy of lung cancer. *Journal of Drug Delivery Science and Technology*. 2021;61:102318. https://doi.org/10.1016/j.jddst.2020.102318.

96. Farzamfar S, Naseri-Nosar M, Vaez A, et al. Neural tissue regeneration by a gabapentin-loaded cellulose acetate/gelatin wet-electrospun scaffold. *Cellulose*. 2018;25(2):1229–1238. https://doi.org/10.1007/s10570-017-1632-z.

97. Lancina Iii MG, Shankar RK, Yang H. Chitosan nanofibers for transbuccal insulin delivery. *Journal of Biomedical Materials Research Part A*. 2017;105(5):1252–1259. https://doi.org/10.1002/jbm.a.35984.

98. Stojanov S, Berlec A. Electrospun nanofibers as carriers of microorganisms, stem cells, proteins, and nucleic acids in therapeutic and other applications. *Frontiers in Bioengineering and Biotechnology*. 2020;8:130. https://doi.org/10.3389/fbioe.2020.00130.

99. Nguyen LH, Gao M, Lin J, et al. Three-dimensional aligned nanofibers-hydrogel scaffold for controlled non-viral drug/gene delivery to direct axon regeneration in spinal cord injury treatment. *Scientific Reports*. 2017;7(1):42212. https://doi.org/10.1038/srep42212.

100. Bösiger P, Tegl G, Richard IMT, et al. Enzyme functionalized electrospun chitosan mats for antimicrobial treatment. *Carbohydrate Polymers*. 2018;181:551–559. https://doi.org/10.1016/j.carbpol.2017.12.002.

101. Achille C, Sundaresh S, Chu B, et al. Cdk2 silencing via a DNA/PCL electrospun scaffold suppresses proliferation and increases death of breast cancer cells. *PLoS One*. 2012;7(12):e52356. https://doi.org/10.1371/journal.pone.0052356.

102. Sebe I, Ostorhazi E, Fekete A, et al. Polyvinyl alcohol nanofiber formulation of the designer antimicrobial peptide APO sterilizes Acinetobacter baumannii-infected skin wounds in mice. *Amino Acids*. 2016;48(1):203–211. https://doi.org/10.1007/s00726-015-2080-4.

103. Sebe I, Ostorházi E, Bodai Z, et al. In vitro and in silico characterization of fibrous scaffolds comprising alternate colistin sulfate-loaded and heat-treated polyvinyl alcohol nanofibrous sheets. *International Journal of Pharmaceutics*. 2017;523(1):151–158. https://doi.org/10.1016/j.ijpharm.2017.03.044.

104. Nazarkina ZK, Chelobanov BP, Kuznetsov KA, et al. Influence of elongation of paclitaxel-eluting electrospun-produced stent coating on paclitaxel release and transport through the arterial wall after stenting. *Polymers*. 2021;13(7):1165. https://doi.org/10.3390/polym13071165.

105. Vlachou M, Kikionis S, Siamidi A, et al. Development and characterization of eudragit®-based electrospun nanofibrous mats and their formulation into nanofiber tablets for the modified release of furosemide. *Pharmaceutics*. 2019;11(9):480. https://doi.org/10.3390/pharmaceutics11090480.

106. Abdul Hameed MM, Mohamed Khan SAP, Thamer BM, et al. Core-shell nanofibers from poly(vinyl alcohol) based biopolymers using emulsion electrospinning as drug delivery system for cephalexin drug. *Journal of Macromolecular Science, Part A*. 2021;58(2):130–144. https://doi.org/10.1080/10601325.2020.1832517.

107. Moydeen AM, Padusha MSA, Thamer BM, et al. Single-nozzle core-shell electrospun nanofibers of PVP/dextran as drug delivery system. *Fibers and Polymers*. 2019;20(10):2078–2089. https://doi.org/10.1007/s12221-019-9187-2.

108. Borbás E, Balogh A, Bocz K, et al. In vitro dissolution–permeation evaluation of an electrospun cyclodextrin-based formulation of aripiprazole using µFlux™. International Journal of Pharmaceutics. 2015;491(1):180–189. https://doi.org/10.1016/j.ijpharm.2015.06.019.

109. Kiss K, Vass P, Farkas A, et al. A solid doxycycline HP-β-CD formulation for reconstitution (i.v. bolus) prepared

by scaled-up electrospinning. *International Journal of Pharmaceutics*. 2020;586:119539. https://doi.org/10.1016/j.ijpharm.2020.119539.

110. do Evangelho JA, Crizel RL, Chaves FC, et al. Thermal and irradiation resistance of folic acid encapsulated in zein ultrafine fibers or nanocapsules produced by electrospinning and electrospraying. *Food Research International*. 2019;124:137–146. https://doi.org/10.1016/j.foodres.2018.08.019.

111. Thompson CJ, Chase GG, Yarin AL, et al. Effects of parameters on nanofiber diameter determined from electrospinning model. *Polymer*. 2007;48(23):6913–6922. https://doi.org/10.1016/j.polymer.2007.09.017.

112. Ewaldz E, Patel R, Banerjee M, et al. Material selection in electrospinning microparticles. *Polymer*. 2018;153:529–537. https://doi.org/10.1016/j.polymer.2018.08.015.

113. Angammana CJ, Jayaram SH. Analysis of the effects of solution conductivity on electrospinning process and fiber morphology. *IEEE Transactions on Industry Applications*. 2011;47(3):1109–1117. https://doi.org/10.1109/TIA.2011.2127431.

114. Ahmadian A, Shafiee A, Aliahmad N, et al. Overview of nano-fiber mats fabrication via electrospinning and morphology analysis. *Textiles*. 2021;1(2):206–226. https://doi.org/10.3390/textiles1020010.

115. Haider A, Haider S, Kang I-K. A comprehensive review summarizing the effect of electrospinning parameters and potential applications of nanofibers in biomedical and biotechnology. *Arabian Journal of Chemistry*. 2018;11(8):1165–1188. https://doi.org/10.1016/j.arabjc.2015.11.015.

116. Lasprilla-Botero J, Álvarez-Láinez M, Lagaron JM. The influence of electrospinning parameters and solvent selection on the morphology and diameter of polyimide nanofibers. *Materials Today Communications*. 2018;14:1–9. https://doi.org/10.1016/j.mtcomm.2017.12.003.

117. Bakar SSS, Fong KC, Eleyas A, et al. Effect of voltage and flow rate electrospinning parameters on polyacrylonitrile electrospun fibers. *IOP Conference Series: Materials Science and Engineering*. 2018;318:012076. https://doi.org/10.1088/1757-899x/318/1/012076.

118. Macossay J, Marruffo A, Rincon R, et al. Effect of needle diameter on nanofiber diameter and thermal properties of electrospun poly(methyl methacrylate). *Polymers for Advanced Technologies*. 2007;18(3):180–183. https://doi.org/10.1002/pat.844.

119. Liu S, Reneker DH. Droplet-jet shape parameters predict electrospun polymer nanofiber diameter. *Polymer*. 2019;168:155–158. https://doi.org/10.1016/j.polymer.2019.01.082.

120. Hekmati AH, Rashidi A, Ghazisaeidi R, et al. Effect of needle length, electrospinning distance, and solution concentration on morphological properties of polyamide-6 electrospun nanowebs. *Textile Research Journal*. 2013;83(14):1452–1466. https://doi.org/10.1177/0040517512471746.

121. Zhang C, Yuan X, Wu L, et al. Study on morphology of electrospun poly(vinyl alcohol) mats. *European Polymer Journal*. 2005;41(3):423–432. https://doi.org/10.1016/j.eurpolymj.2004.10.027.

122. Rodoplu D, Mutlu M. Effects of electrospinning setup and process parameters on nanofiber morphology intended for the modification of quartz crystal microbalance surfaces. *Journal of Engineered Fibers and Fabrics*. 2012;7(2):155892501200700217. https://doi.org/10.1177/155892501200700217.

123. De Vrieze S, Van Camp T, Nelvig A, et al. The effect of temperature and humidity on electrospinning. *Journal of Materials Science*. 2009;44(5):1357–1362. https://doi.org/10.1007/s10853-008-3010-6.

124. Topuz F, Satilmis B, Uyar T. Electrospinning of uniform nanofibers of Polymers of Intrinsic Microporosity (PIM-1): The influence of solution conductivity and relative humidity. *Polymer*. 2019;178:121610. https://doi.org/10.1016/j.polymer.2019.121610.

125. Jermain SV, Brough C, Williams RO. Amorphous solid dispersions and nanocrystal technologies for poorly water-soluble drug delivery – An update. *International Journal of Pharmaceutics*. 2018;535(1):379–392. https://doi.org/10.1016/j.ijpharm.2017.10.051.

126. Huang J, Wigent RJ, Schwartz JB. Drug–polymer interaction and its significance on the physical stability of nifedipine amorphous dispersion in microparticles of an ammonio methacrylate copolymer and ethylcellulose binary blend. *Journal of Pharmaceutical Sciences*. 2008;97(1):251–262. https://doi.org/10.1002/jps.21072.

127. Chokshi RJ, Shah NH, Sandhu HK, et al. Stabilization of low glass transition temperature indomethacin formulations: Impact of polymer-type and its concentration. *Journal of Pharmaceutical Sciences*. 2008;97(6):2286–2298. https://doi.org/10.1002/jps.21174.

128. Choi J-Y, Yoo JY, Kwak H-S, et al. Role of polymeric stabilizers for drug nanocrystal dispersions. *Current Applied Physics*. 2005;5(5):472–474. https://doi.org/10.1016/j.cap.2005.01.012.

129. Regulska K, Michalak M, Murias M, et al. Genotoxic impurities in pharmaceutical products – regulatory, toxicological and pharmaceutical considerations. *Journal of Medical Science*. 2021;90(1):e502. https://doi.org/10.20883/medical.e502.

130. Démuth B, Farkas A, Pataki H, et al. Detailed stability investigation of amorphous solid dispersions prepared by single-needle and high speed electrospinning. *International Journal of Pharmaceutics*. 498(1):234–244. https://doi.org/10.1016/j.ijpharm.2015.12.029.

131. Pisani S, Friuli V, Conti B, et al. Tableted hydrophilic electrospun nanofibers to promote meloxicam dissolution rate. *Journal of Drug Delivery Science and Technology*. 2021;66:102878. https://doi.org/10.1016/j.jddst.2021.102878.

132. Radacsi N, Giapis KP, Ovari G, et al. Electrospun nanofiber-based niflumic acid capsules with superior physicochemical properties. *Journal of Pharmaceutical and Biomedical Analysis*. 2019;166:371–378. https://doi.org/10.1016/j.jpba.2019.01.037.

133. Omer S, Forgách L, Zelkó R, et al. Scale-up of electrospinning: Market overview of products and devices for pharmaceutical and biomedical purposes. *Pharmaceutics*. 2021;13(2):286. https://doi.org/10.3390/pharmaceutics13020286.

134. Szabó E, Záhonyi P, Galata DL, et al. Powder filling of electrospun material in vials: A proof-of-concept study. *International Journal of Pharmaceutics*. 2022;613:121413. https://doi.org/10.1016/j.ijpharm.2021.121413.

135. Edmans JG, Clitherow KH, Murdoch C, et al. Mucoadhesive electrospun fibre-based technologies for oral medicine. *Pharmaceutics*. 2020;12(6):504. https://doi.org/10.3390/pharmaceutics12060504.

136. Puhl DL, Mohanraj D, Nelson DW, et al. Designing electrospun fiber platforms for efficient delivery of genetic material

and genome editing tools. *Advanced Drug Delivery Reviews.* 2022;183:114161. https://doi.org/10.1016/j.addr.2022.114161.

137. Burke LD, Blackwood KA, Zomer Volpato F. Reproducibility and robustness in electrospinning with a view to medical device manufacturing. In: Almodovar J, editor. *Electrospun Biomaterials and Related Technologies.* Cham: Springer International Publishing; 2017. pp. 1–19.

138. Vass P, Szabó E, Domokos A, et al. Scale-up of electrospinning technology: Applications in the pharmaceutical industry. *WIREs Nanomedicine and Nanobiotechnology.* 2020;12(4):e1611. https://doi.org/10.1002/wnan.1611.

139. Farkas B, Balogh A, Cselkó R, et al. Corona alternating current electrospinning: A combined approach for increasing the productivity of electrospinning. *International Journal of Pharmaceutics.* 2019;561:219–227. https://doi.org/10.1016/j.ijpharm.2019.03.005.

140. Vass P, Nagy ZK, Kóczián R, et al. Continuous drying of a protein-type drug using scaled-up fiber formation with HP-β-CD matrix resulting in a directly compressible powder for tableting. *European Journal of Pharmaceutical Sciences.* 2020;141:105089. https://doi.org/10.1016/j.ejps.2019.105089.

141. Hosseinian H, Hosseini S, Martinez-Chapa SO, et al. A meta-analysis of wearable contact lenses for medical applications: Role of electrospun fiber for drug delivery. *Polymers.* 2022;14(1):185. https://doi.org/10.3390/polym14010185.

142. Omer S, Zelkó R. A systematic review of drug-loaded electrospun nanofiber-based ophthalmic inserts. *Pharmaceutics.* 2021;13(10):1637. https://doi.org.10.3390/pharmaceutics 13101637.

143. Andreadis II, Karavasili C, Thomas A, et al. In situ gelling electrospun ocular films sustain the intraocular pressure-lowering effect of timolol maleate: In vitro, ex vivo, and pharmacodynamic assessment. *Molecular Pharmaceutics.* 2022;19(1):274–286. https://doi.org.10.1021/acs.molpharmaceut.1c00766.

144. Mofidfar M, Prausnitz MR. Electrospun transdermal patch for contraceptive hormone delivery. *Current Drug Delivery.* 2019;16(6):577–583. https://doi.org.10.217 4/1567201816666190308112010.

145. Rahmati M, Mills DK, Urbanska AM, et al. Electrospinning for tissue engineering applications. *Progress in Materials Science.* 2021;117:100721. https://doi.org/10.1016/j.pmatsci.2020.100721.

146. Schulz A, Fuchs PC, Heitzmann W, et al. Our initial experience in the customized treatment of donor site and burn wounds with a new nanofibrous temporary epidermal layer. *Ann Burns Fire Disasters.* 2021;34(1):58–66.

147. Rayleigh L. On the equilibrium of liquid conducting masses charged with electricity. *The London, Edinburgh, and Dublin Philosophical Magazine and Journal of Science.* 1882;14(-87):184–186. https://doi.org/10.1080/14786448208628425.

148. Cloupeau M, Prunet-Foch B. Electrostatic spraying of liquids: Main functioning modes. *Journal of Electrostatics.* 1990;25(-2):165–184. https://doi.org/10.1016/0304-3886(90)90025-Q.

149. He T, Jokerst J V. Structured micro/nanomaterials synthesized via electrospray: A review. *Biomaterials Science.* 2020;8(-20):5555–5573. https://doi.org/10.1039/D0BM01313G.

150. Smith DP. The electrohydrodynamic atomization of liquids. *IEEE Transactions on Industry Applications.* 1986;3:527–535. https://doi.org/10.1109/TIA.1986.4504754.

151. Yu D-G, Zhou F, Parker GJ, et al. Innovations and advances in electrospraying technology. In: Kasoju N, Ye H, editors. *Biomedical Applications of Electrospinning and Electrospraying.* Sawstone: Elsevier; 2021. pp. 207–228.

152. Mei F, Chen D-R. Operational modes of dual-capillary electrospraying and the formation of the stable compound cone-jet mod. *Aerosol and Air Quality Research.* 2008;8(-2):218–232. https://doi.org/10.4209/aaqr.2008.01.0003.

153. Thakkar S, Misra M. Electrospray drying of docetaxel nanosuspension: A study on particle formation and evaluation of nanocrystals thereof. *Journal of Drug Delivery Science and Technology.* 2020;60:102009. https://doi.org/10.1016/j.jddst.2020.102009.

154. Zhang S, Kawakami K, Yamamoto M, et al. Coaxial electrospray formulations for improving oral absorption of a poorly water-soluble drug. *Molecular pharmaceutics.* 2011;8(3):807–813. https://doi.org/10.1021/mp100401d.

155. Bai M-Y, Liu S-Z. A simple and general method for preparing antibody-PEG-PLGA sub-micron particles using electrospray technique: An in vitro study of targeted delivery of cisplatin to ovarian cancer cells. *Colloids and Surfaces B: Biointerfaces.* 2014;117:346–353. https://doi.org/10.1016/j.colsurfb.2014.02.051.

156. Zamani M, Prabhakaran MP, San Thian E, et al. Protein encapsulated core–shell structured particles prepared by coaxial electrospraying: Investigation on material and processing variables. *International Journal of Pharmaceutics.* 2014;473(-1–2):134–143. https://doi.org/10.1016/j.ijpharm.2014.07.006.

157. Wang J, Helder L, Shao J, et al. Encapsulation and release of doxycycline from electrospray-generated PLGA microspheres: Effect of polymer end groups. *International Journal of Pharmaceutics.* 2019;564:1–9. https://doi.org/10.1016/j.ijpharm.2019.04.023.

158. Hao S, Wang Y, Wang B, et al. Formulation of porous poly (lactic-co-glycolic acid) microparticles by electrospray deposition method for controlled drug release. *Materials Science and Engineering: C.* 2014;39:113–119. https://doi.org/10.1016/j.msec.2014.02.014.

159. Prabhakaran MP, Zamani M, Felice B, et al. Electrospraying technique for the fabrication of metronidazole contained PLGA particles and their release profile. *Materials Science and Engineering: C.* 2015;56:66–73. https://doi.org/10.1016/j.msec.2015.06.018.

160. Yu D-G, Williams GR, Wang X, et al. Polymer-based nanoparticulate solid dispersions prepared by a modified electrospraying process. *Journal of Biomedical Science and Engineering.* 2011;4(12):741. https://doi.org/10.4236/jbise.2011.412091.

161. Darade A, Pathak S, Sharma S, et al. Atovaquone oral bioavailability enhancement using electrospraying technology. *European Journal of Pharmaceutical Sciences.* 2018;111:195–204. https://doi.org/10.1016/j.ejps.2017.09.051.

162. Arauzo B, Lopez-Mendez TB, Lobera MP, et al. Excipient-free inhalable microparticles of azithromycin produced by electrospray: A novel approach to direct pulmonary delivery of antibiotics. *Pharmaceutics.* 2021;13(12):1988. https://doi.org/10.3390/pharmaceutics13121988.

163. Yin X, Pan H, Liu H. A novel micron-size particulate formulation of felodipine with improved release and enhanced oral bioavailability fabricated by coaxial electrospray. *AAPS PharmSciTech.* 2019;20(7):1–8. https://doi.org/10.1208/s12249-019-1495-8.

164. Shams T, Illangakoon U, Parhizkar M, et al. Electrosprayed microparticles for intestinal delivery of prednisolone. *Journal of the Royal Society Interface.* 2018;15(-145):20180491. https://doi.org/10.1098/rsif.2018.0491.

165. Jain AK, Sood V, Bora M, et al. Electrosprayed inulin microparticles for microbiota triggered targeting of colon.

Carbohydrate Polymers. 2014;112:225–234. https://doi.org/10.1016/j.carbpol.2014.05.087.

166. Zhang C, Chen Z, He Y, et al. Oral colon-targeting core–shell microparticles loading curcumin for enhanced ulcerative colitis alleviating efficacy. *Chinese Medicine.* 2021;16(1):1–14. https://doi.org/10.1186/s13020-021-00449-8.

167. Sakuma S, Matsumoto S, Ishizuka N, et al. Enhanced boosting of oral absorption of lopinavir through electrospray coencapsulation with ritonavir. *Journal of Pharmaceutical Sciences.* 2015;104(9):2977–2985. https://doi.org/10.1002/jps.24492.

168. Lee Y-H, Bai M-Y, Chen D-R. Multidrug encapsulation by coaxial tri-capillary electrospray. *Colloids and Surfaces B: Biointerfaces.* 2011;82(1):104–110. https://doi.org/10.1016/j.colsurfb.2010.08.022.

169. Wang Y, Zhang Y, Wang B, et al. Fabrication of core–shell micro/nanoparticles for programmable dual drug release by emulsion electrospraying. *Journal of Nanoparticle Research.* 2013;15(6):1–12. https://doi.org/10.1007/s11051-013-1726-y.

170. Hsu M-Y, Huang Y-T, Weng C-J, et al. Preparation and in vitro/in vivo evaluation of doxorubicin-loaded poly [lactic-co-glycol acid] microspheres using electrospray method for sustained drug delivery and potential intratumoral injection. *Colloids and Surfaces B: Biointerfaces.* 2020;190:110937. https://doi.org/10.1016/j.colsurfb.2020.110937.

171. Prasad SR, Jayakrishnan A, Kumar TS. Combinational delivery of anticancer drugs for osteosarcoma treatment using electrosprayed core shell nanocarriers. *Journal of Materials Science: Materials in Medicine.* 2020;31(5):1–11. https://doi.org/10.1007/s10856-020-06379-5.

172. Jahangiri A, Davaran S, Fayyazi B, et al. Application of electrospraying as a one-step method for the fabrication of triamcinolone acetonide-PLGA nanofibers and nanobeads. *Colloids and Surfaces B: Biointerfaces.* 2014;123:219–224. https://doi.org/10.1016/j.colsurfb.2014.09.019.

173. Xie J, Marijnissen JC, Wang C-H. Microparticles developed by electrohydrodynamic atomization for the local delivery of anticancer drug to treat C6 glioma in vitro. *Biomaterials.* 2006;27(17):3321–3332. https://doi.org/10.1016/j.biomaterials.2006.01.034.

174. Abyadeh M, Aghajani M, Gohari Mahmoudabad A, et al. Preparation and optimization of chitosan/pDNA nanoparticles using electrospray. *Proceedings of the National Academy of Sciences, India Section B: Biological Sciences.* 2019;89(3):931–937. https://doi.org/10.1007/s40011-018-1009-6.

175. Smeets A, Clasen C, Van den Mooter G. Electrospraying of polymer solutions: Study of formulation and process parameters. *European Journal of Pharmaceutics and Biopharmaceutics.* 2017;119:114–124. https://doi.org/10.1016/j.ejpb.2017.06.010.

176. Xu Y, Skotak M, Hanna M. Electrospray encapsulation of water-soluble protein with polylactide. I. Effects of formulations and process on morphology and particle size. *Journal of Microencapsulation.* 2006;23(1):69–78. https://doi.org/10.1080/02652040500435048.

177. Bohr A, Yang M, Baldursdóttir S, et al. Particle formation and characteristics of Celecoxib-loaded poly (lactic-co-glycolic acid) microparticles prepared in different solvents using electrospraying. *Polymer.* 2012;53(15):3220–3229. https://doi.org/10.1016/j.polymer.2012.05.002.

178. Trotta M, Cavalli R, Trotta C, et al. Electrospray technique for solid lipid-based particle production. *Drug Development and Industrial Pharmacy.* 2010;36(4):431–438. https://doi.org/10.3109/03639040903241817.

179. Abyadeh M, Zarchi AAK, Faramarzi MA, et al. Evaluation of factors affecting size and size distribution of chitosan-electrosprayed nanoparticles. *Avicenna Journal of Medical Biotechnology.* 2017;9(3):126.

180. Hao S, Wang B, Wang Y, et al. Enteric-coated sustained-release nanoparticles by coaxial electrospray: Preparation, characterization, and in vitro evaluation. *Journal of Nanoparticle Research.* 2014;16(2):1–11. https://doi.org/10.1007/s11051-013-2204-2.

181. Songsurang K, Praphairaksit N, Siraleartmukul K, et al. Electrospray fabrication of doxorubicin-chitosan-tripolyphosphate nanoparticles for delivery of doxorubicin. *Archives of Pharmacal Research.* 2011;34(4):583–592. https://doi.org/10.1007/s12272-011-0408-5.

182. Jafari-Nodoushan M, Barzin J, Mobedi H. Size and morphology controlling of PLGA microparticles produced by electro hydrodynamic atomization. *Polymers for Advanced Technologies.* 2015;26(5):502–513. https://doi.org/10.1002/pat.3480.

183. Gao Y, Zhao D, Chang M-W, et al. Morphology control of electrosprayed core–shell particles via collection media variation. *Materials Letters.* 2015;146:59–64. https://doi.org/10.1016/j.matlet.2015.02.013.

184. Yao Z-C, Jin L-J, Ahmad Z, et al. Ganoderma lucidum polysaccharide loaded sodium alginate micro-particles prepared via electrospraying in controlled deposition environments. *International Journal of Pharmaceutics.* 2017;524(1–2):148–158. https://doi.org/10.1016/j.ijpharm.2017.03.064.

185. Park CH, Chung N-O, Lee J. Monodisperse red blood cell-like particles via consolidation of charged droplets. *Journal of Colloid and Interface Science.* 2011;361(2):423–428. https://doi.org/10.1016/j.jcis.2011.06.003.

186. Huang X, Gao J, Zheng N, et al. Influence of humidity and polymer additives on the morphology of hierarchically porous microspheres prepared from non-solvent assisted electrospraying. *Colloids and Surfaces A: Physicochemical and Engineering Aspects.* 2017;517:17–24. https://doi.org/10.1016/j.colsurfa.2017.01.003.

187. Fiegel J, Garcai-Contreras L, Elbert K, et al., editors. Dry powder aerosols for multi-drug resistant tuberculosis (Mdr-Tb) treatment. *AIChE Annual Meeting 2005*; 2005.

188. Yao J, Lim LK, Xie J, et al. Characterization of electrospraying process for polymeric particle fabrication. *Journal of Aerosol Science.* 2008;39(11):987–1002. https://doi.org/10.1016/j.jaerosci.2008.07.003.

189. Almería B, Fahmy TM, Gomez A. A multiplexed electrospray process for single-step synthesis of stabilized polymer particles for drug delivery. *Journal of Controlled Release.* 2011;154(2):203–210. https://doi.org/10.1016/j.jconrel.2011.05.018.

190. Almería B, Deng W, Fahmy TM, et al. Controlling the morphology of electrospray-generated PLGA microparticles for drug delivery. *Journal of Colloid and Interface Science.* 2010;343(1):125–133. https://doi.org/10.1016/j.jcis.2009.10.002.

191. Bohr A, Wan F, Kristensen J, et al. Pharmaceutical microparticle engineering with electrospraying: The role of mixed solvent systems in particle formation and characteristics. *Journal of Materials Science: Materials in Medicine.* 2015;26(2):1–13. https://doi.org/10.1007/s10856-015-5379-5.

192. Ambrus R, Radacsi N, Szunyogh T, et al. Analysis of submicron-sized niflumic acid crystals prepared by

electrospray crystallization. *Journal of Pharmaceutical and Biomedical Analysis*. 2013;76:1–7. https://doi.org/ 10.1016/j.jpba.2012.12.001.

193. Martín Giménez VM, Russo MG, Narda GE, et al. Synthesis, physicochemical characterisation and biological activity of anandamide/ε-polycaprolactone nanoparticles obtained by electrospraying. *IET Nanobiotechnology*. 2020;14(1):86–93. https://doi.org/10.1049/iet-nbt.2019.0108.

194. Cavalli R, Bisazza A, Bussano R, et al. Poly (amidoamine)-cholesterol conjugate nanoparticles obtained by electrospraying as novel tamoxifen delivery system. *Journal of Drug Delivery*. 2011;2011. https://doi.org/10.1155/2011/587604.

195. Bohr A, Kristensen J, Dyas M, et al. Release profile and characteristics of electrosprayed particles for oral delivery of a practically insoluble drug. *Journal of the Royal Society Interface*. 2012;9(75):2437–2449. https://doi.org/10.1098/rsif.2012.0166.

196. Liu Z-P, Zhang L-L, Yang Y-Y, et al. Preparing composite nanoparticles for immediate drug release by modifying electrohydrodynamic interfaces during electrospraying. *Powder Technology*. 2018;327:179–187. https://doi.org/10.1016/j.powtec.2017.12.066.

197. Yu D-G, Zheng X-L, Yang Y, et al. Immediate release of helicid from nanoparticles produced by modified coaxial electrospraying. *Applied Surface Science*. 2019;473:148–155. https://doi.org/10.1016/j.apsusc.2018.12.147.

198. Wang P, Wang M, Wan X, et al. Dual-stage release of ketoprofen from electrosprayed core-shell hybrid polyvinyl pyrrolidone/ethyl cellolose nanoparticles. *Mater Highlights*. 2020;1(2). https://dx.doi.org/10.2991/mathi.k.200825.001.

199. Han S, Dwivedi P, Mangrio FA, et al. Sustained release paclitaxel-loaded core-shell-structured solid lipid microparticles for intraperitoneal chemotherapy of ovarian cancer. *Artificial Cells, Nanomedicine, and Biotechnology*. 2019;47(1):957–967. https://doi.org/10.1080/21691401.2019.1576705.

200. Xu S, Xu Q, Zhou J, et al. Preparation and characterization of folate-chitosan-gemcitabine core–shell nanoparticles for potential tumor-targeted drug delivery. *Journal of Nanoscience and Nanotechnology*. 2013;13(1):129–138. https://doi.org/10.1166/jnn.2013.6794.

201. Lee SY, Lee J-J, Park J-H, et al. Electrosprayed nanocomposites based on hyaluronic acid derivative and Soluplus for tumor-targeted drug delivery. *Colloids and Surfaces B: Biointerfaces*. 2016;145:267–274. https://doi.org/10.1016/j.colsurfb.2016.05.009.

202. Varshosaz J, Ghassami E, Noorbakhsh A, et al. Poly (butylene adipate-co-butylene terephthalate) nanoparticles prepared by electrospraying technique for docetaxel delivery in ovarian cancer induced mice. *Drug Development and Industrial Pharmacy*. 2018;44(6):1012–1022. https://doi.org/10.1080/03639045.2018.1430819.

203. Ali A, Zaman A, Sayed E, et al. Electrohydrodynamic atomisation driven design and engineering of opportunistic particulate systems for applications in drug delivery, therapeutics and pharmaceutics. *Advanced Drug Delivery Reviews*. 2021;176:113788. https://doi.org/10.1016/j.addr.2021.04.026.

204. Colombo S, Beck-Broichsitter M, Bøtker JP, et al. Transforming nanomedicine manufacturing toward quality by design and microfluidics. *Advanced Drug Delivery Reviews*. 2018;128:115–131. https://doi.org/10.1016/j.addr.2018.04.004.

205. Zhang C, Yang L, Wan F, et al. Quality by design thinking in the development of long-acting injectable PLGA/PLA-based

microspheres for peptide and protein drug delivery. *International Journal of Pharmaceutics*. 2020;585:119441. https://doi.org/10.1016/j.ijpharm.2020.119441.

206. Rantanen J, Khinast J. The future of pharmaceutical manufacturing sciences. *Journal of Pharmaceutical Sciences*. 2015;104(11):3612–3638. https://doi.org/10.1002/jps.24594

207. Group IEW, editor ICH Harmonized tripartite guideline. Pharmaceutical development Q8 (R2). *International Conference on Harmonization of Technical Requirements for Registration of Pharmaceuticals for Human Use*; 2009.

208. Xie J, Jiang J, Davoodi P, et al. Electrohydrodynamic atomization: A two-decade effort to produce and process micro-/nanoparticulate materials. *Chemical Engineering Science*. 2015;125:32–57. https://doi.org/10.1016/j.ces.2014.08.061.

209. Rulison AJ, Flagan RC. Scale-up of electrospray atomization using linear arrays of Taylor cones. *Review of Scientific Instruments*. 1993;64(3):683–686. https://doi.org/10.1063/1.1144197.

210. Almekinders J, Jones C. Multiple jet electrohydrodynamic spraying and applications. *Journal of Aerosol Science*. 1999;30(7):969–971. https://doi.org/10.1016/S0021-8502(98)00755-1.

211. Parhizkar M, Reardon P, Knowles J, et al. Performance of novel high throughput multi electrospray systems for forming of polymeric micro/nanoparticles. *Materials & Design*. 2017;126:73–84. https://doi.org/10.1016/j.matdes.2017.04.029.

212. Snarski S, Dunn P. Experiments characterizing the interaction between two sprays of electrically charged liquid droplets. *Experiments in Fluids*. 1991;11(4):268–278. https://doi.org/10.1007/BF00192755.

213. Regele J, Papac M, Rickard M, et al. Effects of capillary spacing on EHD spraying from an array of cone jets. *Journal of Aerosol Science*. 2002;33(11):1471–1479. https://doi.org/10.1016/S0021-8502(02)00093-9.

214. Bocanegra R, Galán D, Márquez M, et al. Multiple electrosprays emitted from an array of holes. *Journal of Aerosol Science*. 2005;36(12):1387–1399. https://doi.org/10.1016/j.jaerosci.2005.04.003.

215. Deng W, Klemic JF, Li X, et al. Increase of electrospray throughput using multiplexed microfabricated sources for the scalable generation of monodisperse droplets. *Journal of Aerosol Science*. 2006;37(6):696–714. https://doi.org/10.1016/j.jaerosci.2005.05.011.

216. Deng W, Gomez A. Influence of space charge on the scale-up of multiplexed electrosprays. *Journal of Aerosol Science*. 2007;38(10):1062–1078. https://doi.org/10.1016/j.jaerosci.2007.08.005.

217. Zhang C, Chang M-W, Ahmad Z, et al. Stable single device multi-pore electrospraying of polymeric microparticles via controlled electrostatic interactions. *RSC Advances*. 2015;5(-107):87919–87923. https://doi.org/10.1039/C5RA18482G.

218. Sochorakis N, Grifoll J, Rosell-Llompart J. Scaling up of extractor-free electrosprays in linear arrays. *Chemical Engineering Science*. 2019;195:281–298. https://doi.org/10.1016/j.ces.2018.09.006.

219. Busolo M, Torres-Giner S, Prieto C, et al. Electrospraying assisted by pressurized gas as an innovative high-throughput process for the microencapsulation and stabilization of docosahexaenoic acid-enriched fish oil in zein prolamine. *Innovative Food Science & Emerging Technologies*. 2019;51:12–19. https://doi.org/10.1016/j.ifset.2018.04.007.

220. Prieto C, Evtoski Z, Pardo-Figuerez M, et al. Nanostructured valsartan microparticles with enhanced bioavailability

produced by high-throughput electrohydrodynamic room-temperature atomization. *Molecular Pharmaceutics*. 2021;18(8):2947–2958. https://doi.org/10.1021/acs.molpharmaceut.1c00098.

221. José María Lagaron Cabello SCR, Valle JM, Galan Nevado D, inventor; Installation and procedure of industrial encapsulation of termolabels substances patent ES2674808A1. 2016.

222. Szabó E, Záhonyi P, Brecska D, et al. Comparison of amorphous solid dispersions of spironolactone prepared by spray drying and electrospinning: The influence of the preparation method on the dissolution properties. *Molecular Pharmaceutics*. 2020;18(1):317–327. https://doi.org/10.1021/acs.molpharmaceut.0c00965.

223. Sato M. Formation of uniformly sized liquid droplets using spinning disk under applied electrostatic field. *IEEE Transactions on Industry Applications*. 1991;27(2):316–322. https://doi.org/10.1109/28.73619.

224. Böttjer R, Grothe T, Wehlage D, et al. Electrospraying poloxamer/(bio-) polymer blends using a needleless electrospinning machine. *Journal of Textiles and Fibrous Materials*. 2018;1:2515221117743079. https://doi.org/10.1177/2515221117743079.

225. Peng Q, Yang K, Venkataraman M, et al. Preparation of electrosprayed composite coated microporous filter for particulate matter capture. *Nano Select*. 2021. https://doi.org/10.1002/nano.202100186.

226. Wang H, Chen Z, Liu B, et al. Needleless electrospray of magnetic film from magnetization-induced cone array. *Materials and Manufacturing Processes*. 2018;33(10):1115–1120. https://doi.org/10.1080/10426914.2017.1415446.

227. Shaid A, Wang L, Padhye R, et al. Needleless electrospinning and electrospraying of mixture of polymer and aerogel particles on textile. *Advances in Materials Science and Engineering*. 2018;2018. https://doi.org/10.1155/2018/1781930.

228. Xiong X, Venkataraman M, Yang T, et al. Preparation and characterization of electrosprayed aerogel/polytetrafluoroethylene microporous materials. *Polymers*. 2021;14(1):48. https://doi.org/10.3390/polym14010048.

229. Loepfe M, Duss A, Zafeiropoulou K-A, et al. Electrospray-based microencapsulation of epigallocatechin 3-gallate for local delivery into the intervertebral disc. *Pharmaceutics*. 2019;11(9):435. https://doi.org/10.3390/pharmaceutics11090435.

230. Yan WC, Tong YW, Wang CH. Coaxial electrohydrodynamic atomization toward large scale production of core-shell structured microparticles. *AIChE Journal*. 2017;63(12):5303–5319. https://doi.org/10.1002/aic.15821.

231. Olvera-Trejo D, Velásquez-García L. Additively manufactured MEMS multiplexed coaxial electrospray sources for high-throughput, uniform generation of core–shell microparticles. *Lab on a Chip*. 2016;16(21):4121–4132. https://doi.org/10.1039/C6LC00729E.

232. Kavadiya S, Biswas P. Electrospray deposition of biomolecules: Applications, challenges, and recommendations. *Journal of Aerosol Science*. 2018;125:182–207. https://doi.org/10.1016/j.jaerosci.2018.04.009.

233. Ijsebaert JC, Geerse KB, Marijnissen JC, et al. Electrohydrodynamic atomization of drug solutions for inhalation purposes. *Journal of Applied Physiology*. 2001;91(6):2735–2741. https://doi.org/10.1152/jappl.2001.91.6.2735.

234. Ronald Alan Coffee ABP, Davies DN, inventor; Battelle Memorial Institute Inc, assignee. Inhaler patent US6684879B1. 1998.

235. Stenzler A, inventor; Pulmonary drug delivery device patent US20040065321A1. 2000.

236. Kim J-H, Lee H-S, Kim H-H, et al. Electrospray with electrostatic precipitator enhances fine particles collection efficiency. *Journal of Electrostatics*. 2010;68(4):305–310. https://doi.org/10.1016/j.elstat.2010.03.002.

237. Grafahrend D, Jungbecker P, Seide G, et al. Development and optimization of an electrospraying device for the continuous collection of nano-and microparticles. *The Open Chemical and Biomedical Methods Journal*. 2010;3(1). https://doi.org/10.2174/1875038901003010001.

238. Prieto C, Evtoski Z, Pardo-Figuerez M, et al. Bioavailability enhancement of nanostructured microparticles of carvedilol. *Journal of Drug Delivery Science and Technology*. 2021;66:102780. https://doi.org/10.1016/j.jddst.2021.102780.

239. Aldossary AM, Ekweremadu CS, Offe IM, et al. A guide to oral vaccination: Highlighting electrospraying as a promising manufacturing technique toward a successful oral vaccine development. *Saudi Pharmaceutical Journal*. 2022. https://doi.org/10.1016/j.jsps.2022.03.010.

240. Steipel RT, Gallovic MD, Batty CJ, et al. Electrospray for generation of drug delivery and vaccine particles applied in vitro and in vivo. *Materials Science and Engineering: C*. 2019;105:110070. https://doi.org/10.1016/j.msec.2019.110070.

241. Anup N, Thakkar S, Misra M. Formulation of olanzapine nanosuspension based orally disintegrating tablets (ODT); comparative evaluation of lyophilization and electrospraying process as solidification techniques. *Advanced Powder Technology*. 2018;29(8):1913–1924. https://doi.org/10.1016/j.apt.2018.05.003.

242. Moreno MA, Gómez-Mascaraque LG, Arias M, et al. Electrosprayed chitosan microcapsules as delivery vehicles for vaginal phytoformulations. *Carbohydrate Polymers*. 2018;201:425–437. https://doi.org/10.1016/j.carbpol.2018.08.084.

243. Chen D-R, Pui DY. Electrospray and its medical applications. In: Marijnissen JC, Gradon L, editor. *Nanoparticles in Medicine and Environment*. Berlin: Springer; 2010. pp. 59–75.

244. Kebarle P. A brief overview of the present status of the mechanisms involved in electrospray mass spectrometry. *Journal of Mass Spectrometry*. 2000;35(7):804–817. https://doi.org/10.1002/1096-9888(200007)35:7<804::AID-JMS22>3.0.CO;2-Q.

245. Guo Q, Knight PT, Mather PT. Tailored drug release from biodegradable stent coatings based on hybrid polyurethanes. *Journal of Controlled Release*. 2009;137(3):224–233. https://doi.org/10.1016/j.jconrel.2009.04.016.

246. Chen D-R, Wendt CH, Pui DY. A novel approach for introducing bio-materials into cells. *Journal of Nanoparticle Research*. 2000;2(2):133–139. https://doi.org/10.1023/A:1010084032006.

247. Bioinicia. [cited 2022]. Available from: https://bioinicia.com/pharma/.

248. Nikolaou M, Krasia-Christoforou T. Electrohydrodynamic methods for the development of pulmonary drug delivery systems. *European Journal of Pharmaceutical Sciences*. 2018;113:29–40. https://doi.org/10.1016/j.ejps.2017.08.032.

249. Bombin ADJ, Dunne NJ, McCarthy HO. Electrospinning of natural polymers for the production of nanofibres for wound healing applications. *Materials Science and Engineering: C*. 2020;114:110994. https://doi.org/10.1016/j.msec.2020.110994.

250. Chen S, Huang X, Cai X, et al. The influence of fiber diameter of electrospun poly (lactic acid) on drug delivery. *Fibers and Polymers.* 2012;13(9):1120–1125. https://doi.org/10.1007/s12221-012-1120-x.

251. Bohr A, Wang Y, Beck-Broichsitter M, et al. Influence of solvent mixtures on HPMCAS-celecoxib microparticles prepared by electrospraying. *Asian Journal of Pharmaceutical Sciences.* 2018;13(6):584–591. https://doi.org/10.1016/j.ajps.2018.01.007.

252. Sóti PL, Nagy ZK, Serneels G, et al. Preparation and comparison of spray dried and electrospun bioresorbable drug delivery systems. *European Polymer Journal.* 2015;68:671–679. https://doi.org/10.1016/j.eurpolymj.2015.03.035.

253. Moayyedi M, Eskandari MH, Rad AHE, et al. Effect of drying methods (electrospraying, freeze drying and spray drying) on survival and viability of microencapsulated Lactobacillus rhamnosus ATCC 7469. *Journal of Functional Foods.* 2018;40:391–399. https://doi.org/10.1016/j.jff.2017.11.016.

254. Domján J, Vass P, Hirsch E, et al. Monoclonal antibody formulation manufactured by high-speed electrospinning. *International Journal of Pharmaceutics.* 2020;591:120042. https://doi.org/10.1016/j.ijpharm.2020.120042.

6 Microfluidic Manufacture of Polymeric Nanoparticles

Enrica Chiesa, Ida Genta and Rossella Dorati
University of Pavia

Silvia Pisani
IRCCS Foundation Policlinico S.Matteo,

Bice Conti
University of Pavia

CONTENTS

6.1   INTRODUCTION: BACKGROUND AND DRIVING FORCES

Microfluidics refers to the science of dealing with fluids in microchannel; the miniaturization confers additional characteristics that can be exploited in order to perform processes not otherwise possible in macroscale, offering primarily new skills in controlling the molecular concentration in space and time.

The microfluidics advantages derive from the microscopic amount of fluid a microfluidic device can handle. Regardless of the size of the surrounding instrumentation and the material of which the microfluidic device is made, only the space where the fluid is processed has to be miniaturized. The miniaturization of the whole system is not a requirement of a microfluidic system. The microscopic quantity of fluid is the key issue in microfluidics. Therefore, the term "microfluidics" here refers not to link the fluid mechanics to any length scale, such as micron, but rather to refer in general to situations in which a small-size scale causes changes in fluid behaviour. Microfluidic devices enable the manipulation of microquantities of liquid solutions in the order of a few milliseconds, where mixing and demixing processes can be controlled to improve the efficacy of the mass transfer to achieve efficient chemical

reactions. These can be successfully and widely used for applications in pharmacy, biotechnology and chemical industries. With time, microfluidic applications in pharmaceutical area spread to drug discovery, design and delivery (i.e. separation and sample analysis, synthesis, organ-on-chip creation and nanoparticles (NPs) manufacturing). The last application involves the synthesis of nanoparticulate drug delivery systems, such as drug nanocrystals, lipid-based systems (e.g. liposomes and solid lipid nanoparticles), polymeric-based systems (e.g. nanospheres, nanocapsules, polymer-drug conjugates and dendrimers) and inorganic nanoparticles (e.g. gold, porous silica, porous silicon and fullerenes).

As long as analytical purposes are concerned (Figure 6.1a), focusing on biomedical areas for rapid and efficient diagnosis is an important way to improve human life expectancy. In this area, microfluidics can improve both extraction of biomarkers from complex biological matrices (or body) and their analysis, achieving fast and effective sample preparation techniques and rapid analysis, thanks to the small size of sample required together with separation high resolution (Nge, Rogers, and Woolley 2013). For this reason, biomarker analysis has been one of the most actively pursued applications of the miniaturization of chemical analysers (Nge, Rogers, and Woolley 2013;

DOI: 10.1201/9781003224464-6

Sonker, Sahore, and Woolley 2017; Wu et al. 2016). On-chip sample preparation can be used to selectively extract, pre-concentrate and label target analytes in an automated way such as with antibodies or aptamers fixed on a solid support in order to purify complex matrices, such as blood (Cui et al. 2015). Moreover, analyte labelling is a fundamental sample preparation step that can be performed on a chip. On-chip labelling requires loading, reacting and purifying and typically uses a solid support inside the microchannels; fluorescent labelling is the most common approach being explored. The application of microfluidic to capillary electrophoresis for protein separation and analysis seems to be a topic of great interest, and it has been recently experimented and reviewed (Vitorino et al. 2021; Rodriguez-Ruiz et al. 2018).

Organs-on-chips are microscale devices, combining conventional cell culture methods with microfabrication and microfluidics technologies and modelling the functional units of human organs (Figure 6.1b) (Low et al. 2021; Huh et al. 2013; Balijepalli and Sivaramakrishan 2017). Such chips resemble more closely complex interactions between different tissues and organs, in contrast with mimicking interactions between cells in the same type of 3D model. The term "organs-on-chips" is used interchangeably to describe a single or multiple organ(s) on a chip. Microfluidics enables the control of gas exchange (dissolved oxygen sensors can be added), precise regulation of flow by drastically reducing fluid turbulence and enabling the laminar flow of cell culture media that act as blood substitutes (mimicking the circulatory system). The laminar flow achieved generates chemical gradients, which have an important bearing on cell migration and differentiation. Organ-on-chips have been developed to study the behaviour of various cell types in different pathologic situations such as breast cancer and neurodegeneration (Song et al. 2009; Park et al. 2006). They have been also proposed for testing drug safety and efficacy, and to preliminarily screen new drugs, thus reducing animal model tests (Huh et al. 2013; Jang and Suh 2010). Microengineered cell culture systems that mimic complex organ physiology have the potential to be used for the development of human-relevant disease models, which are more predictive of drug efficacy and toxicity in patients. They can also provide greater insight into the drug mechanism of action and the accurate determination of drug pharmacokinetic, accelerating drug-development process, meanwhile developing safer and more effective drugs at lower costs (Huh, Hamilton, and Ingber 2011).

Due to the relatively new technique, organ-on-chip has a lack of regulation for their official use as drug evaluation systems. Currently, and to date, pharmacodynamic studies can exploit organ-on-chip potential because no models exist for these studies. The same does not occur for pharmacokinetic studies where specific animal models are required for the tests. From a regulatory standpoint, organs-on-chips can be considered analytical instruments for the purpose of drug testing. Therefore, they should comply with Good Laboratory Practices regulations in terms of calibration, validation and qualification. However, not all regulations referring to analytical instruments can be applied to organs-on-chips, and regulation is still needed.

With respect to the microfluidic application in nanomedicine, a variety of micro/nanoscale drug delivery systems can be produced by using the microfluidics technology (Bjornmalm, Yan, and Caruso 2014). The main advantage of using the microfluidic technique is the opportunity to control the reaction environment in a very refined way.

The control of the reaction environment leads to improving the quality of nanoparticulate drug delivery systems (NPs, liposomes, lipid nanoparticles (LNP), better modulating their size and drug loading and eventually improving the process yield of nanoparticles. Microfluidic reactors enable rapid mixing of reagents, temperature control and the precise spatio-temporal manipulation of reactions to obtain an efficient and homogeneous mass transfer to achieve monodisperse nanostructure formulations.

All these factors are difficult, if not impossible, to be controlled in conventional, bulk synthesis methods where mixing is heterogeneous and typically occurs on a time scale

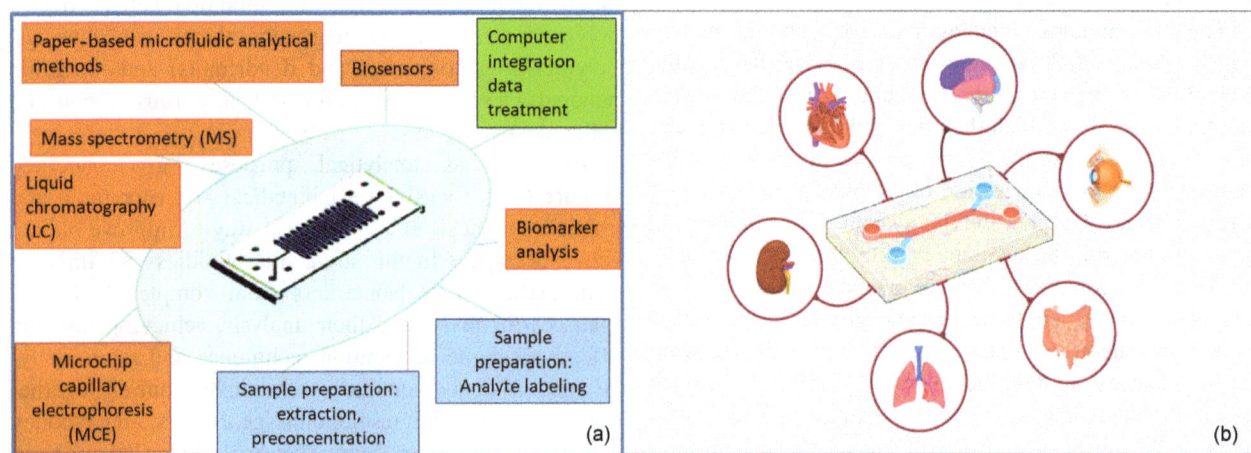

FIGURE 6.1 Microfluidic technique applications in pharmaceutical and biomedical field: (a) analytical chemistry, sample preparation separation and quantification; (b) organ on chip.

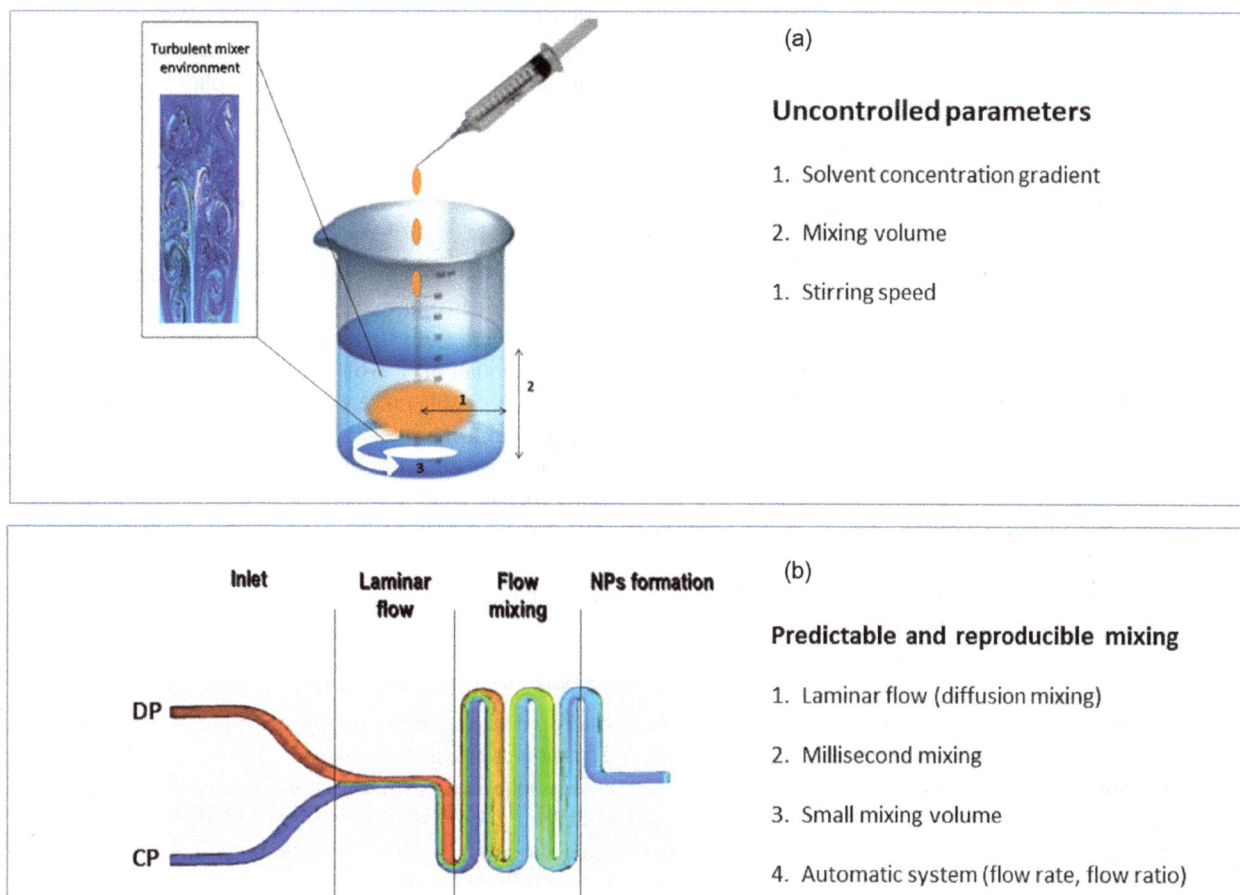

FIGURE 6.2 Schematic representation of (a) traditional nanoprecipitation method performed in a bench scale, (b) passive mixing in a microchannel.

longer than the characteristic time scale for self-assembly, achieving large and polydisperse nanosystems (Marre and Jensen 2010; Zhao et al. 2011). Therefore, microfluidic technology has been revealed as one of the most promising approaches for reproducible and reliable design and development of nanoparticulate systems due to the refined control on the morphology, size and polydispersity by setting a few process parameters. The physicochemical properties of nanosystems, determining their in vivo fate, can be precisely tuned in a reproducible manner (Lim et al. 2014; Lim and Karnik 2014). As long as NPs are concerned, the traditional batch methods lack precise control over the mixing and supersaturation level, leading to uncontrolled nucleation and growth processes, resulting in poor control over final particle characteristics (Figure 6.2). Otherwise in microfluidic devices, the small dimension enables homogeneous, fast and reproducible mixing conditions, heat and mass transfer that can dramatically improve NPs process yield, drug loading and size distribution while reducing the formation of undesirable by-products. These characteristics are fundamental because they, in turn, determine the physio-chemical properties of the produced nanomaterials (Capretto et al. 2013; Valencia et al. 2012).

A differentiation can be found in the literature between "bottom-up" and "top-down" microfluidic methods. Bottom-up methods aim at creating NPs and LNPs by combining nanoscale components to form larger structures in the nanometric range. This approach aims to drive particle growth from individual subunits. A number of microfluidic strategies are included in bottom-up methods (Nie and Takeuchi 2018; Desai et al. 2021): depending on the geometry of microfluidic channels, the mixing pattern changes, and in this chapter, the type of microfluidic mixing devices to prepare nanoparticles will be critically summarized (see Section 6.3.2.)

Otherwise, top-down approaches typically involve the breakdown of initial macroscopic material to the desired size through sequential processing steps (Aryal et al. 2019; Desai et al. 2021). Most nanoparticles fabricated using conventional methods use a top-down approach and require further processing downstream to obtain a uniform and monodisperse population.

6.2 THEORETICAL BASIS OF MICROFLUIDIC TECHNIQUE

The basic principles of the microfluidic technique reside in the mechanisms and laws governing mixing in a narrow environment, as investigated and reported by several authors (Ward and Fan 2015; Capretto et al. 2013; Suh and Kang 2010; Tice et al. 2003; Pilkington, Seddon, and Elani 2021). The main feature of this technique is the ability to translate bulk mixing techniques into a microchannel with a width in the rank between 100 and 800 nm, where mixing takes place through physical characteristics and diffusion-based mass transfer is different from macroscale. Therefore, microfluidic mixing is governed by laws different from those applied at the macroscale, and microfluidic devices are not the miniaturization of macroscale ones. The main feature of microfluidic technique is laminar flow where the streamlines are well defined and mixing across the streamlines can only be achieved by molecular diffusion, and it is obtained by the predominant role of viscous forces with respect to inertial ones. Briefly, fluid flow can be described by two regimens: turbulent and laminar. The formation of a flow vortex occurs in the turbulent regimen; meanwhile, fluid runs in parallel layers in the laminar flow condition, and there is no current perpendicular to the flow direction (Figure 6.2b). The main difference between the fluid flow in a microfluidic device and its macroscale counterpart is the greater surface-to-volume ratio that affects the forces acting on the fluid. Inertial forces which play a dominant role in macroscopic fluid flow process became negligible in microfluidic device. Therefore, macroscale techniques applied to NPs production involve generation of turbulent flow, and as shown in Figure 6.2a, in this regimen there is low control on solvent concentration gradient and diffusion rate, over time. These two differential regimens can be revealed by Reynolds number (Re) as defined by equation 6.1, which is a dimensionless number representing the ratio of inertial force to viscous force:

$$Re = \text{inertial forces} / \text{viscous forces} = \rho u D / \eta = u D / v \quad (6.1)$$

where ρ and η are the fluid density and dynamic viscosity, respectively; v is the fluid viscosity; u is the mean fluid velocity; and D is the hydraulic diameter of the channel. D depends on the cross-sectional geometry of the channel and is given by equation 6.2:

$$D = 4A / P \quad (6.2)$$

where A is the channel cross-sectional area and P is the channel wet perimeter.

Reynolds number greater than 100 corresponds to turbulent flow, which is dominated by random motion, where mass transport phenomena happen in all spatial directions. Reynolds number lower than 100 is associated to laminar flow, where viscous forces overcome inertial ones, and mixing between streams is regulated by molecular diffusion. A transitional flow regimen usually occurs between laminar and turbulent regimens. The exact transitional Reynolds number is a function of many parameters, including channel geometry and channel surface roughness. In a microfluidic device, mainly due to the small hydraulic diameters of channels, Re values can be lower than 100 and flow mixing

is dominated by passive molecular diffusion. The phenomenon resides in molecular transport from higher to lower concentration areas, and it can be explained by Fick's law (6.3):

$$J = -D * d\varphi / dx \quad (6.3)$$

where φ is the concentration of a chemical substance, x is position, the dimension of which is length, and D is the molecular diffusion coefficient, whose dimension is area per unit time.

In the equation, J, which is the diffusion flux, measures the amount of substance that will flow through a unit area during a unit time interval.

When two different but miscible liquids are in contact in microchannel in laminar flow regimes, the required time (t) for species diffusion is in quadratic correlation with the distance covered (x). This concept is described by the following equation 6.4:

$$t_{\text{mix}} \alpha x^2 / D \quad (6.4)$$

where x and D are the diffusion length and the molecular diffusion coefficient (m^2/s).

For spherical particles (embryonic particle or liposome) undergoing Brownian motion in liquids, D can be calculated as self-diffusion coefficient in liquids, by Stokes–Einstein Equation (6.4):

$$D = kT / 6\pi\eta r \quad (6.5)$$

where k is the Boltzmann constant, T is absolute temperature, η is the fluid flow dynamic viscosity, and r is the particle radius.

On a microscale, the diffusion distance can be extremely small, especially if fluid streams are hydro-dynamically focused. Because time depends on the square power of x, a decrease in the diffusion distance dramatically reduces the time required for complete mixing to be achieved.

Summarizing, the theoretical basis of microfluidic technique resides in the ability of controlling solvent diffusion rate through laminar flow in a fast solvent diffusion rate promoted by the micromixing environment. This leads to very fast nanoparticle formation in a reproducible way.

6.3 MICROFLUIDIC PLATFORMS MATERIALS AND GEOMETRIES

Micromachining process is one of the manufacture methods for microfluidic devices, and it can be applied to silicon, glass, plastics, and stainless as substrates. The micromachining process starts with a design of microchannel created by a computer-aided design (CAD) software, and then the microchannel was patterned onto the substrate directly. Microfluidic device made of silicon and glass are fabricated by chemical etching that allow to definite and smooth microchannel structures. In this process the microchannel design is patterned by photolithography before to start the chemical etching process (Duffy et al. 1998)

Essential properties that must be considered when choosing the material to make a microfluidic device are durability, ease of fabrication, transparency, biocompatibility, chemical compatibility with the implied reagents, meeting the temperature and pressure conditions needed for the reaction and the potential of the surface functionalization. The topic was recently nicely reviewed by Adelina-Gabriela Niculescu and colleagues (Niculescu et al. 2021).

A lot of materials have been studied and are used to manufacture microfluidic devices, such as glass borosilicate, metals, monocrystalline silicon, and polydimethylsiloxane (PDMS). The material choice depends on which product is intended to be manufactured and which are the process conditions the microfluidic chip will undergo. Here below some frequent examples are discussed and tabulated (Table 6.1).

Glass, namely soda lime, and borosilicate glasses are suitable materials for making microfluidic devices because glass is chemically inert, thermostable, electrically insulating, rigid and biologically compatible; it can be easily functionalized at its surface; and it withstands harsh conditions in terms of solvents, pHs, temperature and pressure (Niculescu et al. 2021). Moreover, glass is transparent and cheap, and it can be integrated to components of other materials (i.e. valves and pumps). Microcapillary glass reactors achieve higher resolution at the micrometer scale than that the one achieved with other materials. For these reasons, glass devices can be used for the better-controlled synthesis of emulsions and polymeric nanoparticles. Drawbacks of microcapillary glass reactors are microfabrication difficulties, time-consuming manufacturing process and their brittleness (Hwang et al. 2019; Shakeri et al. 2019; Niculescu et al. 2021).

Microfluidic devices made of metals have been demonstrated useful in nanomaterial synthesis. The most common metals used to make microfluidic devices are aluminium, copper and iron, namely their alloys with other metals, which can improve their chemical resistance. These materials have several advantages: they are cheap, widely accessible and easy to be processed. Moreover, metals can withstand high temperature and pressure, and harsh solvents (except strong acids); their resistance to robust

TABLE 6.1
Materials Commonly Used for Microfluidic Device Chips

Microfluidic Device Materials	Advantages	Disadvantages	References
Glass (soda lime or borosilicate)	Biocompatible, inert, transparent, resistant to harsh conditions, cheap. It can be integrated to other materials	Microfluidic device preparation process is time-consuming, hence expensive	Huang et al. (2016); Shakeri et al. (2019); Hwang et al. (2019); Niculescu et al. (2021)
Metals (aluminium, copper, and iron and their alloys)	Cheap, widely accessible, easy to be processed. Resistant high temperature, high pressure, harsh solvents (except strong acids), robust handling	Lack of transparency	Singh et al. (2010); James et al. (2020); Niculescu et al. (2021)
Monocrystalline silicon (mono-Si or c-Si)	Semiconducting material, ease of fabrication, design flexibility, surface modifications	Opaque, fragile, expensive material, sensitive to most common organic solvents	Chiesa et al. (2018)
Polydimethylsiloxane (PDMS)	Biocompatible material, low permeable to water, easily mouldable, transparent	Sensitive to most common organic solvents	Greener et al. (2010); Shakeri, Khan, and Didar (2021); Kim et al. (2009); Deguchi et al. (2021); Niculescu et al. (2021)
Acrylic polymers obtained by photopolymerization; Polymethylmethacrylae (PMMA)	UV-transparent; low auto-fluorescence background; biocompatible; low water absorption; excellent water resistance	Low resistance to various organic solvents such as ethanol, isopropyl alcohol, acetone	Greener et al. (2010); van Midwoud et al. (2012); Ballacchino et al. (2021)
Thermoplastic polyurethane (TPU material)	UV-transparent, biocompatible polymer	Sagging of polymer layers	Ballacchino et al. (2021); Mehta and Rath (2021)
Cyclo olefin polymer (COP)	UV-Vis transparent; low viscosity at elevated temperatures; low water absorption; low fluorescence background; surface activation by O_2 plasma treatment	Possible cytotoxicity due to solvent residual from fabrication process	Greener et al. (2010); Liu et al. (2018); Wen et al. (2021); van Midwoud et al. (2012)
Polycarbonate (PC)	Biocompatible; quite hydrophilic, favours microchannel filling	Low UV-transparency, poor resistance to chlorinated organic solvents	Greener et al. (2010); van Midwoud et al. (2012)

handling is convenient for cleaning operations (Singh et al. 2010; James et al. 2020).

In the latest years, polymers gained interest becoming the most common materials making microfluidic apparatuses.

Silicone (polysiloxane), in particular monocrystalline silicone and poly(dimethylsiloxane) (PDMS), are among the most used polymers for microfluidic systems fabrication due to their ready availability, chemical compatibility, and thermostability (Singh et al. 2010; Martins, Torrieri, and Santos 2018).

The main advantages of monocrystalline silicone are its semiconducting properties, ease of fabrication and design flexibility including surface modifications; meanwhile, opacity is its most evident limitation. When optical detection is required, along nanoparticles synthesis process, a portion of the device must be non-silicon (Nielsen et al. 2020; Martins, Torrieri, and Santos 2018; Ren, Zhou, and Wu 2013; Campbell et al. 2021). Other limitations are silicone mechanical properties and its rather high costs. Besides, being a fragile material with high elastic modulus, it is difficult to connect silicone devices to valves and pumps. Nonetheless, silicon microfluidic platforms find use in biological applications, such as point-of-care medical diagnostics and organ-on-chip devices for drug toxicity screening (Nielsen et al. 2020).

Poly(dimethylsiloxane) (PDMS) has become the favourite material for the fabrication of microfluidic devices. It can be easily moulded and patterned into channels, it can reproduce micrometre-sized features with high fidelity, it is optically transparent, and it has low permeability to water. Furthermore, PDMS is a biocompatible polymer suitable for biological and cellular applications; however, the low compatibility of PDMS with organic solvents such as amines, strong acids and hydrocarbons restricts its use to aqueous solutions, and it's not suitable for oxygen-sensitive applications. These limits make this material suitable for lab-on-chip applications, or whenever a biologic environment is needed (Kim et al. 2009; Greener et al. 2010; Shakeri, Khan, and Didar 2021; Deguchi et al. 2021). Soft lithography is the most widely used microfabrication method for PDMS devices leading to microstructures with 10–200 μm. PDMS prepolymer was poured on a mould created by photolithography and then heated to recover a PDMS replica. The deeper microstructure can be obtained by using a master mould made by thicker photoresist layer. Other materials reported in the literature for microreactor fabrication are cycloolefin polymer (COP), thermoplastic polyurethane (TPU), UV-transparent acrylic polymer and polycarbonate (PC). All these are thermoplastic materials with T_g in the range ≥113°C ≤149°C.

Greener and colleagues and Liu and colleagues highlighted COP has advantages owing to the combination of high transmission in the UV-Vis, relatively low viscosity at elevated temperatures, low water absorption, low fluorescence background and its ability to activate its surface by O_2 plasma treatment. Moreover, Liu and colleagues propose laser ablation as a cheap and accurate method to fabricate the microchannels (Greener et al. 2010; Liu et al. 2018). Injection moulding and hot embossing methods are the most suitable microfabrication methods of disposable microfluidic devices. In the last years, 3D printing technique has been studied to fabricate low-cost microfluidic devices that can also be tailored according to specific needs (Tiwari, Bhat, and Mahato 2020). In these aims, TPU and UV-transparent acrylic resin were successfully applied by Ballacchino and colleagues to manufacture microfluidic devices by 3D printing. The polymers were chosen because of their elastomeric behaviour and processability through fuse deposition modelling (FDM) and liquid crystal display 3D printing techniques (Ballacchino et al. 2021; Pitaru et al. 2020). A preliminary comparison between the microfluidic devices produced by two different 3D printing techniques showed LCD was the more promising one irrespective to the polymer tested (Ballacchino et al. 2021).

Regarding the microfluidic device structure, micromixers are classified into active and passive mixers. Active mixers are provided with external sources that physically agitate the liquid in a microchannel. They can be acoustic or ultrasonic waves with acoustic actuators, magnetic disturbances with magnetic particles and an external magnetic field, pressure disturbance with the repeated stopping and flowing of the fluids, etc. (Kwak et al. 2016; Bau, Zhong, and Yi 2001).

As explained above in Section 6.2, mixing induced by laminar flow is one of the most important characteristics improving the performances of microscale devices.

A main difference is between active and passive micromixer devices, where active micromixers are those requiring external energy sources, such as pressure fields, electrical fields, sound fields, magnetic fields and thermal fields, while passive micromixers don't require external energy input except the energy for driving the fluids, and mixing is promoted by microchannel geometries. In an active micromixer, the key parameter is the external energy source that enables fluid mixing, diffusion and eventually nanoparticle formation.

Active mixers can be activated on demand by a user, and controllable mixing may be carried out using pressure gradients, electrical voltages across the fluid or integrated mixing elements like stirring bars. The use of external energy makes mixing rather independent from fluid properties such as viscosity. For example, pressure field–driven micromixers and acoustofluidic micromixers can perform rapid and homogeneous mixing of highly viscous fluids (Orbay et al. 2017; Cai et al. 2017). In general, active micromixers are shown to get very high mixing efficiency of about 93% or even higher, in very short times. An interesting example is a three-dimensional electroosmotic micromixer based on the Koch fractal principle described and fabricated by Siyue Xiong and colleagues (Xiong, Chen, and Wang 2021). Although they can be useful in enhancing mixing, particularly for chamber mixing, active mixers are harder to be fabricated than passive mixers due to movable parts and often require an external power source. For these reason, active micromixers are more used for research purposes and less applied to nanoparticle manufacturing systems. Some recent review papers can be found in the literature, which detail active micromixer types and mechanisms (Ward and Fan 2015; Cai et al. 2017; Bayareh, Ashani, and Usefian 2020).

Passive micromixers use complex channel geometries to enhance the diffusion or chaotic advection. Passive mixing is achieved by altering the structure or configuration of fluid channels, and the system is minimally externally controlled by the users. The theoretical basis is explained in Section 6.2, considering that the required time for species diffusion is in quadratic correlation with the distance covered (see equation 6.4), and on a microscale, the diffusion distance can be extremely small especially if fluid streams are hydro-dynamically focused. Passive mixers exploit the hydrodynamic manipulation of the fluids: chaotic advection, enhanced molecular diffusion, utilization of surface tension, fluid lamination, sequential splitting and the combining of fluids inside the microchannel are typical of passive mixers. The key benefit of passive mixing is that there are no moving parts within the mixer, resulting in easier fabrication and operation.

In the different types of microfluidic devices, the developed channel geometry results in different droplet generation mechanisms. At least five structures can be found in microchannel-based devices and are worth to be discussed (Figure 6.3): terrace-like microchannel, T- and Y-junction microchannel, hydrodynamic flow focusing devices (HFF) and staggered herringbone micromixer (SHM).

In the terrace-like microchannel devices, several microchannels deliver the dispersed phase at the top and from both sides of the main channel, where the continuous phase flows (Figure 6.3a). The compressed dispersed phase passes onto to a flat surface (terrace), and the droplet forms as the dispersed phase thread falls from the flat surface in a deeper well.

In T-junction microchannel devices (Figure 6.3b), the to-be-dispersed phase is delivered through a microchannel perpendicular to a main channel, in which the continuous phase flows. Thread breaks up and droplet formation takes place at the junction of the two microchannels, or further downstream. Droplets form perpendicular to the main channel and are dragged out by parallel continuous phase flow. The parameters such as injection angle, density ratio and viscosity were investigated from mathematical and experimental standpoint showing that injection angle close to perpendicular and parallel conditions worked to increase the diameter of droplets. In these conditions, the reduction of the diameter of droplets could be obtained by increasing flow rate and density ratio (Jamalabadi et al. 2017).

FFDs (Figure 6.3c) are based on the principle of hydrodynamic focusing. The disperse phase flows in a central microchannel while the continuous phase is delivered through two side channels. Mixing of the continuous phase with the disperse phase happens at the restriction point, and at the same time, the droplet is generated. In this microfluidic device, mixing takes place in a further reduced size environment, speeding up droplets formation rate, with controlled laminar flow conditions. The HFF developed as two-dimensional (2-D) HFF and three-dimensional (3-D) HFF is the most used fluid mixing approach in microfluidic: 2-D HFF is a planar device where the central stream of fluid

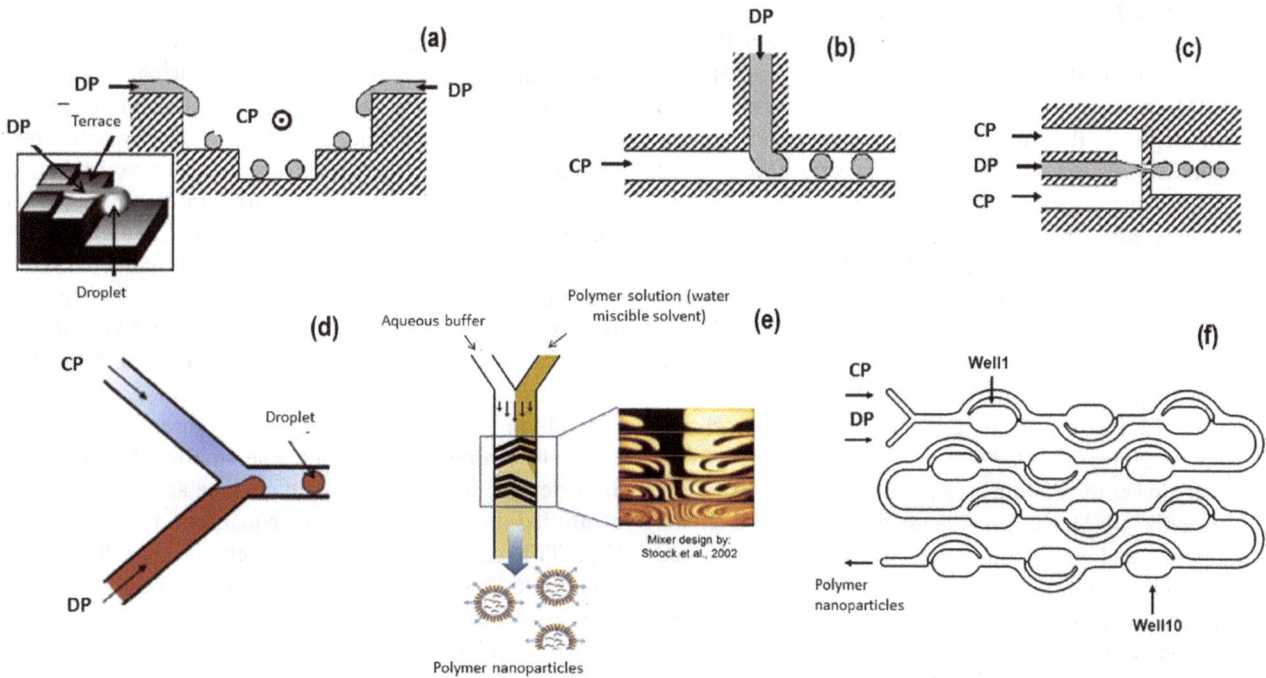

FIGURE 6.3 Examples of passive microfluidic devices geometries: (a) terrace-like device; (b) T-junction device; (c) flow-focusing microchannel device (FFD); (d) Y-junction device; (e) staggered herringbone micromixer (SHM); and (f) small geometry device (SGD). CP and DP are the continuous and dispersed phases, respectively.

is constrained by lateral flow, whereas in 3-D HFF the central stream is both horizontally and vertically focused into a small fluid filament.

Y-junction microchannel devices represent a variation of the T-junction microchannel device.

The disperse phase is delivered through a microchannel perpendicular to another channel, in which the continuous phase flows (Figure 6.3d). These two channels merge to form the main channel. At first, the contact angle between the disperse phase and continuous phase is 90°, and then the disperse phase undergoes about 120° incline, merging into the main channel. According to mathematical models shown by Jamalabadi and colleagues, these conditions should favour the formation of small-sized droplets. The mixing primarily occurs in the contact surface between the two inlet fluids. It is the diffusion rate at the inlet fluid interface that determines the rate of mixing (mixing time).

Mixers based on passive chaotic advection have been widely studied and are very common because of their simple design and fabrication, in addition to their higher mixing efficiency. Chaotic micromixers consist of geometries embedded into the microchannel to disturb the laminar flow. One of the most studied chaotic micromixers is the staggered herringbone mixer (SHM) (Figure 6.3e), originally developed by Stroock (Stroock et al. 2002) and commercialized by Precision Nanosystem®.

The SHM consists of Y-shaped device with a rectangular cross-section area with embossed grooves on the bottom which have an asymmetrical V-shape (the peak of the V is not in the middle of the geometry's width) and are periodically reversed after a group of some grooves. The length at which the two groups of the asymmetrical grooves are spread is a mixing cycle or a unit cell of the SHM. An SHM is composed of many unit cells in a series. Studies on SHM geometry outlined that mixing is more affected by the depth of the groove than the angle, and numerical simulations verified the highest mixing efficiency to be in the range $0 < Re < 100$ (Stroock et al. 2002; Kee and Gavriilidis 2008; Kwak et al. 2016). Moreover, the number of chaotic mixer structures affects the fluid dynamics. Maeki et al. (2017) studied the mechanism of LNPs formation into a microfluidic channel by varying the number of chaotic mixer structures (depth kept constant at 31 μm) and the relationship between the number of chaotic mixer structures and LNP size demonstrating that at least 10 cycles of mixed structures were needed to obtain LNPs with mean diameter lower than 50 nm.

Recently, SHM has been studied from computational standpoint, and new designs of the SHM are proposed which increase the mixing efficiency. In particular, adding symmetric grooves and making the herringbone sculpture at the bottom wall of the SHM slippery were found to improve the performance of the SHM and reduce the length required for mixing from ~9% up to ~50% compared to the design of Kee and Gavriilidis (Hadjigeorgiou, Boudouvis, and Kokkoris 2021).

Another example of microfluidic device geometry is the small geometry device (SGD) recently introduced by Desai and colleagues (Desai et al. 2021). The mixers have two inlets connected to independent fluid streams that merge in a mixing "channel" composed of fluidic traps (wells). Mixing is promoted by vortices that generate inside the well and further enhanced by the continuous bifurcation of fluid, as a portion of the fluid stream enters the well while the rest bypasses the trap to recombine downstream. Moreover, the fluidic traps are alternated and inverted, and the main channel turns every three wells.

6.4 NANOPARTICLES FOR DRUG DELIVERY MANUFACTURING BY MICROFLUIDICS

Microfluidics does not introduce new nanoparticulate drug delivery systems preparation methods, but applying already known bulk techniques in a microfluidic environment leads to improve nanoparticle rate and yield of production, together with drug encapsulation efficiency and process reproducibility.

The nanoparticle manufacturing processes starting from already-formed nanoparticles precursors are based on:

a. self-assembling of lipids into liposomes or lipid nanoparticles (LNP) promoted by lipid solvent diffusion;
b. polymer nanoprecipitation through solvent diffusion leading in a microenvironment with spontaneous self-assembling of nanoconstructs; and
c. polymers interaction leading to form a nanostructured polyelectrolyte complex.

When they apply to a microfluidic device for nanoparticulate drug delivery systems manufacturing, the size of NPs can be controlled through the optimization of process parameters as well as the formulation and physicochemical parameters. The main process parameters ascribed and affecting nanoparticles formation are total flow rate (TFR) and flow rate ratio (FRR). TFR represents the flow rate of mixing solvents and solutions in the mixing channel (mixing in Figure 6.2b); FRR is the ratio between the flow rates of dispersed phase (DP) and continuous phase (CP) in the two inlet channels, respectively (inlet in Figure 6.2b). Typically, the minimum and maximum values of these parameters depend on microfluidic apparatus features, namely the channels' hydraulic diameters. The smaller the hydraulic diameter, the higher the flow rate can be achieved keeping controlled the laminar flow (see equation 6.1, Section 6.2). High TFR can be achieved with high viscosity fluids (see equation 6.1, Section 6.2.) keeping controlled the laminar flow. As long as FFR is concerned, ratios different from 1:1 are useful when mixing fluids with different dynamic viscosities, in such a way to balance mixing forces.

Among the formulation and physicochemical properties parameters, the DP and CP chemical nature, the respective solubility of the materials in both phases, their concentration and their action on the surface tension should be taken into consideration.

Both in (a) self-assembling of liposomes and (b) polymer nanoprecipitation through solvent diffusion, nanoparticle formation is controlled by solvent diffusion (see equations 6.3 and 6.5, Section 6.2), mainly related to the miscibility between solvent and nonsolvent that rules the speed of mass transfer (Ismagilov et al. 2000). Table 6.2 reports some recent examples of lipids, polymer solvents and dispersed phases combination used to manufacture liposomes, LNP and polymer nanoparticles by microfluidics. Here below the a), b), c) above indicated cases are introduced and discussed.

6.4.1 Liposomes, Lipid Nanoparticles (LNP) through Self-Assembling

Liposome and LNP manufacturing by microfluidics have been thoroughly investigated, and several recent papers can be found in the literature (Anderluzzi et al. 2021; Anderluzzi et al. 2020; Ballacchino et al. 2021; Roces, Christensen, and Perrie 2020; Roces et al. 2020). As known, liposomes and LNP are vesicular structures with at least one lipid bilayer. In a microfluidic manufacturing process, liposomes are

quite homogeneous with size not greater than 500 nm. They spontaneously form by mixing the lipid alcoholic solution with an aqueous phase.

The most used lipid solvents are ethanol, isopropyl alcohol and methanol; the aqueous phase can be phosphate buffer pH 7.4 or citrate buffer pH 6, depending on the type of lipids.

The lipid phase is traditionally made from phosphatidylcholine (e.g. 1,2-distearoyl-sn-glycero-3-phosphocholine (DSPC), 2-dimyristoyl-sn-glycero-3-phosphocholine (DMPC), 1,2-dioleoyl-sn-glycero-3-phosphocholine (DOPC)) and cholesterol (Chol), as structural components. Recently, cationic and pegylated lipids gained a lot of interest, also because of COVID-19 vaccine formulations. Y. Perrie's research group did a lot of research on microfluidic manufacturing of cationic and ionizable LNP. In a recent work (Roces et al. 2020), they prepared LNP by a microfluidic device based on SHM. The lipid components namely DSPC, cholesterol, DMG-PEG2000 and a cationic lipid (either DDAB, DOTAP or MC3) were dissolved in ethanol at different molar percentages, and the aqueous phase was either citrate buffer 100 mM or Tris buffer pH 7.4. The authors evaluated the effect of formulation parameters, such as lipid and aqueous phase composition, and

TABLE 6.2
Lipids Polymer Solvents and Dispersed Phases Combination Used to Manufacture Liposomes, LNP and NPs by Microfluidics

Continuous Phase (CP)		Dispersed Phase (DP)		
Nanoparticle Polymer/Lipids	Solvent	Drug or Polymer	Solvent	References
DOPE, DOTAP or DDA	Methanol 1:1 w/w	mRNA	Tris buffer pH 7.4 10 mM	Anderluzzi et al. (2021); Anderluzzi et al. (2020)
DOTAP) or DDA (1:1 w/w), 2 Mol % DMG-PEG2000	Ethanol (70°C)	mRNA	Tris buffer pH 7.4 10 mM	Anderluzzi et al. (2021); Anderluzzi et al. (2020)
DOTAP	Ethanol (10 mg/mL)	–	25 mM acetate buffer (pH 4.0)	Haseda et al. (2020)
DMPC:Chol 2:1 ratio Curcumin	Ethanol	–	Phosphate-buffered saline (PBS, pH 7.4)	Ballacchino et al. (2021)
DSPC:Chol:DOTAP: DMGPEG2000	Ethanol	PolyA or ssDNA, or mRNA	Citrate buffer pH 6 100 mM	Roces et al. (2020)
DSPC:Chol:MC3:DMG-PEG2000	Ethanol	PolyA, or ssDNA, or mRNA	Citrate buffer pH 6 100 mM	
PLGA (50:50), mol/w 30,000–60,000, DOTAP or DDA 1:1w/w	DMSO	mRNA	Acetate buffer 100 mM, pH 6	Anderluzzi et al. (2020)
PLGA 85:15 (Mw: 50,000–75,000) or 75:25 (Mw: 66,000–107,000) or 50:50 (Mw: 30,000–60,000)	Acetonitrile 1% w/v	–	Tris buffer	
PLGA 85:15 (Mw: 50,000–75,000) or 75:25 (Mw: 66,000–107,000) or 50:50 (Mw: 30,000–60,000)	Acetonitrile 1% w/v	OVA or BSA	Water for injection	Roces et al. (2020)
PLGA (L:G 50:50, Mw. 25–35 kDa, PLGA-PEG (L:G 50:50, Mw. 30 kDa and 5 kDa 1'-Dioctadecyl-3,3,3',3'-Tetramethylindocarbocyanine perchlorate, Nile Red and Doxorubicin	Acetonitrile 1% w/v, ethanol, DMSO	–	Purified water	Mares et al. (2021); Desai et al. (2021)
HA and tripolyphosphate	Water for injection	CS	Water for injection	Chiesa et al. (2020)
HA and tripolyphosphate	Water for injection	CS, Myo	Water for injection	Chiesa et al. (2021)

process conditions, such as TFR and FRR, on LNP size, zeta potential and PDI. They concluded that cationic (DDAB and DOTAP) and ionizable (MC3) LNPs tested have high sensitivity to the microfluidic operating parameter: high (TFR ≥ 10 mL/min) and high FRR gave stable results in terms of low LNP size (< 100 nm). Moreover, formulation parameters buffer compostion and molar lipid content are important considerations for LNP manufacture, as these influenced LNP size and stability.

6.4.2 POLYMER NANOPARTICLES (NPS) THROUGH NANOPRECIPITATION

When the polymeric precursors undergo a change in solvent quality, these polymers can self-assemble into NPs. The self-assembly of block copolymer takes place in three stages: (1) nucleation, (2) nanoparticles growth through aggregation and (3) locked nanoparticles after a characteristic aggregation time scale (t_{agg}) (Karnik et al. 2008). Indeed, the NPs formation triggered by the change of solvent quality is strictly correlated to the mixing of polymer and aqueous solutions. The mixing time (equation 6.4) has been highlighted as the main factor affecting the average particles size and the NPs size distribution (Chiesa et al. 2018; Capretto et al. 2013): briefly, when mixing occurs faster than the duration of nucleation ($t_{mix} < t_{agg}$), the nucleation of block copolymers as well as the growth of nuclei occurs in a homogeneous solvent environment. Otherwise, if the mixing rate is slower ($t_{mix} > t_{agg}$), the NPs nucleation occurs when the mixing is incomplete leading to larger NPs size and wide size distribution. NPs manufacturing processes by microfluidics starting from already-formed polymers and exploiting nanoprecipitation have been performed with polyesters such as polylactide-co-glycolide (PLGA) (Roces, Christensen, and Perrie 2020; Mares et al. 2021). This polymer is already approved by regulatory agencies for human use in injectable products; it is biodegradable and biocompatible; and pharmaceutical products made by PLGA injectable microspheres loaded with either small molecules or protein drugs are already on the market (Operti et al. 2021). It has been evaluated in NPs manufacturing by microfluidics either alone or in various PLGA molecular weight blends, or blended with a cationic lipid. A further advantage of using PLGA in microfluidics is its hydro/lipophilicity that can be modulated depending on its composition, i.e. lactide:glycolide ratio, and it is soluble in acetone and acetonitrile (ACN) which are generally recognized as safe (GRAS) solvents that are soluble in water.

Typically, the dispersed phase (DP, see Figure 6.2) is made by polymer solution, where the solvent needs to be miscible with the CP in order to diffuse in it reducing polymer solubility and triggering their nanoprecipitation. A lot of literature studies report PLGA NPs produced by nanoprecipitation; however, the microfluidic chip design can change and affect the final NPs quality attributes.

As an example, a recently published study of Roces and colleagues (Roces, Christensen, and Perrie 2020) shows the fabrication of protein-loaded poly(lactic-co-glycolic acid) nanoparticles from bench to scale-independent production using microfluidics exploiting an SHM microfluidic device. PLGA was dissolved in acetonitrile at a concentration of 10 mg/mL (1% w/v), and the aqueous phase containing ovalbumin or bovine serum albumin or tuberculosis vaccine candidate "Hybrid56" (H56) at 0.2, 0.5 or 1 mg/mL was the DPs. The authors tested different production parameters, total flow rates (TFRs) 5, 10 and 15 mL/min and flow rate ratios (FRR, aqueous-to-organic ratio) 1:1, 3:1 and 5:1 were selected. The results showed that the very fast NPs process formation by microfluidics allows to avoid the addition of stabilizing agents such as polyvinyl alcohol (PVA) that is commonly used in bulk preparation methods for PLGA NPs. Increasing TFR from 5 to 15 mL/min led to decrease in NPs size from 150 to 94 nm, respectively. Moreover, the authors were able to load in a single step proteins/antigens or peptides into polymeric nanoparticles, and their payloads were modulated in a range of 8%–50% depending on PLGA composition, with best FFR 3:1.

Moreover, microfluidic devices based on HFF are widely used for the production of PLGA NPs starting from the pioneer study carried out by Karnik et al. (2008) who reported the controlled generation of NPs made of PLGA loaded with docetaxel demonstrating that microfluidic channel offers high mixing rate, thus avoiding large polymer aggregation. More recently, Rhee et al. (2011) demonstrated the advantages of using 3-D configuration of HFF devices for the PLGA NPs production, suggesting that 3D technology significantly improves the mixing polymer/aqueous phases allowing to smaller NPs size.

6.4.3 POLYMER NANOPARTICLES (NPS) THROUGH POLYELECTROLYTE COMPLEXATION

Polyelectrolyte complex formation by ionotropic gelation into a microfluidic channel is an example of a reaction taking place between two already-formed polymers and leading to solubility change and nanoparticle precipitation as polyelectrolyte complex. This process has been thoroughly investigated by Chiesa and colleagues (Chiesa et al. 2021; Chiesa et al. 2020) who have studied the microfluidic-assisted manufacturing process of hydrophilic polymer NPs made from chitosan (CS, M_w 110 kDa) and hyaluronic acid (HA, M_w 710 kDa) nanoparticles by using the SHM microfluidic device. The authors set up the conditions process in order to achieve HA-coated CS NPs through a design of experiment approach and found out that HA concentration affects NPs sizes; in particular, its increase triggers NPs enlargement, meanwhile HA:CS ratio higher than 1.5 is liable of NPs negative surface charge, suggesting a predominant HA deposition on NPs surface that is crucial for targeting CD44 cell transmembrane receptor. Best results in terms of NPs size below 200 nm, PDI 0.306 and zeta potential about – 20 mV were obtained when HA concentration in feeding solution was 0.150 mg/mL and HA:Cs ratio 1:3 (Chiesa et al. 2020). More recently, the same authors investigated the role of HA weight average molecular weight (Mw) ranging from 280 to 820 kDa (280, 540, 710 and 820 kDa) in NPs formation inside a microfluidic device. Based on the relationship between polymer Mw and solution viscosity, they proposed a methodological approach to ensure critical quality attributes (size of 200 nm, PDI ≤ 0.3). The feasibility of the protein encapsulation was demonstrated by using myoglobin, as a model neutral protein, with an encapsulation efficiency always higher than 50%. Lastly, all NPs samples were successfully internalized by CD44-expressing cells (Chiesa et al. 2021).

6.5 MISCELLANEOUS

A side application of microfluidics involves separation of extracellular vesicles (EVs).

EVs are cell-derived lipid bilayer structures released by cells that are gaining interest for their critical role in the regulation of various biological processes and pathologies, and they have significant utility as potential diagnostic biomarkers and drug delivery systems. EVs are classified by size and origin into two primary groups: exosomes and microvesicles. Exosome size is between 50 and 150 nm, and they originate from intraluminal vesicles generated within endosomal-derived multivesicular bodies; microvesicles size ranges from 50 to 500 nm, and they are shed from the plasma membrane.

Microfluidic-based techniques have been recently exploited for EV separation by "size-based" or "immunoaffinity-based" mechanisms. Moreover, microfluidics can be used to synthesize EV mimetics as drug delivery systems (DDSs) by bottom-up and top-down technologies. Top-down approaches typically use nanometre-sized structures to break apart cells into EV mimetics. For example, Jo and coworkers used an array of hydrophilic microfluidic channels to extrude cells and produce EVs mimetics with an average diameter of 100 nm for mRNA delivery (Jo et al. 2014).

A recent review by Y. Meng and colleagues (Meng et al. 2021) introduces and thoroughly discusses this interesting topic.

6.6 CONCLUSIONS AND OUTLOOK

An overview of the main topics related to microfluidics for pharmaceutical and biomedical applications has been provided highlighting the technological ability to miniaturize many large-scale complex processes. With respect to pharmaceutical application, microfluidics systems ensure fast processing and low sample quantity achieving high-quality and high-throughput data that are very important to speed up the development of nanoparticles formulations and their clinical translation.

The microfluidics devices are prospective apparatus for drug delivery system production because the precise size-controlled NPs production reached by microfluidic device may accelerate the development of next-generation nanomedicine also considering that classical methods, primarily rely on bulk mixing, suffer from poor reproducibility, batch-to-batch variability and poor scale-up feasibility.

Despite the recent advances in fabrication of nanoparticulate DDSs using microfluidic, more efforts should be channelled for the translation of this technology to industrial application increasing the production rate up to kg/day. The close collaborations among several researchers from different areas such as biologists, chemists and engineers will accelerate the process.

ABBREVIATIONS

DOPE:	1,2-dioleoyl-sn-glycero-3-phosphoethanolamine
DOTAP:	1,2-dioleoyl-3-trimethylammonium-propane (chloride salt)
DDA:	dimethyldioctadecylammonium bromide
DMG-PEG2000:	2-dimyristoylrac-glycero-3-methoxypolyethylene glycol-2000
DMPC:	1,2-dimyristoyl-sn-glycero-3-phosphocholine
Chol:	cholesterol
DSPC:	1,2-distearoyl-sn-glycero-3-phosphocholine
MC3:	heptatriaconta-6, 9, 28, 31-tetraen-19-yl 4-(dimethylamino) butanoate
PLGA:	Poly(D,L-lactide-co-glycolide) lactide: glycolide
HA:	Hyaluronic acid
PolyA:	Polyadenylic acid
OVA:	Ovalbumin
BSA:	Bovine serum albumin
CS:	Chitosan
Myo:	Myoglobin

REFERENCES

Anderluzzi, G, G Lou, S Gallorini, M Brazzoli, R Johnson, DT O'Hagan, BC Baudner, and Y Perrie. 2020. "Investigating the impact of delivery system design on the efficacy of self-amplifying RNA vaccines." *Vaccines* 8 (2). https://doi.org/10.3390/vaccines8020212.

Anderluzzi, G, ST Schmidt, R Cunliffe, S Woods, CW Roberts, D Veggi, I Ferlenghi, DT O'Hagan, BC Baudner, and Y Perrie. 2021. "Rational design of adjuvants for subunit vaccines: The format of cationic adjuvants affects the induction of antigen-specific antibody responses." *Journal of Controlled Release* 330: 933–944. https://doi.org/10.1016/j.jconrel.2020.10.066.

Aryal, S, H Park, JF Leary, and J Key. 2019. "Top-down fabrication-based nano/microparticles for molecular imaging and drug delivery." *International Journal of Nanomedicine* 14: 6631–6644. https://doi.org/10.2147/IJN.S212037.

Balijepalli, A, and V Sivaramakrishan. 2017. "Organs-on-chips: Research and commercial perspectives." *Drug Discovery Today* 22 (2): 397–403. https://doi.org/10.1016/j.drudis.2016.11.009. https://www.ncbi.nlm.nih.gov/pubmed/27866008.

Ballacchino, G, E Weaver, E Mathew, R Dorati, I Genta, B Conti, and DA Lamprou. 2021. "Manufacturing of 3D-printed microfluidic devices for the synthesis of drug-loaded liposomal formulations." *International Journal of Molecular Sciences* 22 (15). https://doi.org/10.3390/ijms22158064.

Bau, HH, JH Zhong, and MQ Yi. 2001. "A minute magneto hydro dynamic (MHD) mixer." *Sensors and Actuators B-Chemical* 79 (2–3): 207–215. https://doi.org/10.1016/s0925-4005(01)00851-6.

Bayareh, M, MN Ashani, and A Usefian. 2020. "Active and passive micromixers: A comprehensive review." *Chemical Engineering and Processing-Process Intensification* 147. https://doi.org/10.1016/j.cep.2016.107771.

Bjornmalm, M, Y Yan, and F Caruso. 2014. "Engineering and evaluating drug delivery particles in microfluidic devices." *Journal of Controlled Release* 190: 139–146. https://doi.org/10.1016/j.jconrel.2014.04.030.

Cai, GZ, L Xue, HL Zhang, and JH Lin. 2017. "A review on micromixers." *Micromachines* 8 (9). https://doi.org/10.3390/mi8090274.

Campbell, SB, QH Wu, J Yazbeck, CA Liu, S Okhovatian, and M Radisic. 2021. "Beyond polydimethylsiloxane: Alternative materials for fabrication of organ-on-a-chip devices and microphysiological systems." *ACS Biomaterials Science & Engineering* 7 (7): 2880–2896. https://doi.org/10.1021/acsbiomaterials.0c00640.

Capretto, L, D Carugo, S Mazzitelli, C Nastruzzi, and X Zhang. 2013. "Microfluidic and lab-on-a-chip preparation routes for organic nanoparticles and vesicular systems for nanomedicine applications." *Advanced Drug Delivery Reviews* 65 (11–12): 1496–1532. https://doi.org/10.1016/j.addr.2013.08.002. https://www.ncbi.nlm.nih.gov/pubmed/23933616.

Chiesa, E, R Dorati, T Modena, B Conti, and I Genta. 2018. "Multivariate analysis for the optimization of microfluidics-assisted nanoprecipitation method intended for the loading of small hydrophilic drugs into PLGA nanoparticles." *International Journal of Pharmaceutics* 536 (1): 165–177. https://doi.org/10.1016/j.ijpharm.2017.11.044.

Chiesa, E, A Greco, F Riva, R Dorati, B Conti, T Modena, and I Genta. 2021. "Hyaluronic acid-based nanoparticles for protein delivery: Systematic examination of microfluidic production conditions." *Pharmaceutics* 13 (10). https://doi.org/10.3390/pharmaceutics13101565. https://www.scopus.com/inward/record.uri?eid=2-s2.0-85116048485&doi=10.3390/pharmaceutics13101565&partnerID=40&md5=0f9a3b03a0211856d67c397f66c04796.

Chiesa, E, F Riva, R Dorati, A Greco, S Ricci, S Pisani, M Patrini, T Modena, B Conti, and I Genta. 2020. "On-chip synthesis of hyaluronic acid-based nanoparticles for selective inhibition of CD44+ human mesenchymal stem cell proliferation." *Pharmaceutics* 12 (3). https://doi.org/10.3390/pharmaceutics12030260. https://www.ncbi.nlm.nih.gov/pubmed/32183027.

Cui, F, M Rhee, A Singh, and A Tripathi. 2015. "Microfluidic sample preparation for medical diagnostics." *Annual Review of Biomedical Engineering* 17: 267–286. https://doi.org/10.1146/annurev-bioeng-071114-040538. https://www.ncbi.nlm.nih.gov/pubmed/26290952.

Deguchi, S, M Tsuda, K Kosugi, A Sakamoto, N Mimura, R Negoro, E Sano, T Nobe, K Maeda, H Kusuhara, H Mizuguchi, F Yamashita, Y Torisawa, and K Takayama. 2021. "Usability of polydimethylsiloxane-based microfluidic devices in pharmaceutical research using human hepatocytes." *ACS Biomaterials Science & Engineering* 7 (8): 3648–3657. https://doi.org/10.1021/acsbiomaterials.1c00642.

Desai, D, YA Guerrero, V Balachandran, A Morton, L Lyon, B Larkin, and DE Solomon. 2021. "Towards a microfluidics platform for the continuous manufacture of organic and inorganic nanoparticles." *Nanomedicine-Nanotechnology Biology and Medicine* 35. https://doi.org/10.1016/j.nano.2021.102402.

Duffy, DC, JC McDonald, OJA Schueller, and GM Whitesides. 1998. "Rapid prototyping of microfluidic systems in poly(dimethylsiloxane)." *Analytical Chemistry* 70 (23): 4974–4984. https://doi.org/10.1021/ac980656z.

Greener, J, W Li, J Ren, D Voicu, V Pakharenko, T Tang, and E Kumacheva. 2010. "Rapid, cost-efficient fabrication of microfluidic reactors in thermoplastic polymers by combining photolithography and hot embossing." *Lab on a Chip* 10 (4): 522–4. https://doi.org/10.1039/b918834g. https://www.ncbi.nlm.nih.gov/pubmed/20126695.

Hadjigeorgiou, AG, AG Boudouvis, and G Kokkoris. 2021. "Thorough computational analysis of the staggered herringbone micromixer reveals transport mechanisms and enables mixing efficiency-based improved design." *Chemical Engineering Journal* 414. https://doi.org/10.1016/j.cej.2021.128775.

Haseda, Y, L Munakata, J Meng, R Suzuki, and T Aoshi. 2020. "Microfluidic-prepared DOTAP nanoparticles induce strong T-cell responses in mice." *PLoS One* 15 (1). https://doi.org/10.1371/journal.pone.0227891.

Huang, JR, YT Zhu, LN Xu, JW Chen, W Jiang, and XA Nie. 2016. "Massive enhancement in the thermal conductivity of polymer composites by trapping graphene at the interface of a polymer blend." *Composites Science and Technology* 129: 160–165. https://doi.org/10.1016/j.compscitech.2016.04.029.

Huh, D, GA Hamilton, and DE Ingber. 2011. "From 3D cell culture to organs-on-chips." *Trends in Cell Biology* 21 (12): 745–754. https://doi.org/10.1016/j.tcb.2011.06.005.

Huh, D, HJ Kim, JP Fraser, DE Shea, M Khan, A Bahinski, GA Hamilton, and DE Ingber. 2013. "Microfabrication of human organs-on-chips." *Nature Protocols* 8 (11): 2135–2157. https://doi.org/10.1038/nprot.2013.137. https://www.ncbi.nlm.nih.gov/pubmed/24113786.

Hwang, J, YH Cho, MS Park, and BH Kim. 2019. "Microchannel fabrication on glass materials for microfluidic devices." *International Journal of Precision Engineering and Manufacturing* 20 (3): 479–495. https://doi.org/10.1007/s12541-019-00103-2.

Ismagilov, RF, AD Stroock, PJA Kenis, G Whitesides, and HA Stone. 2000. "Experimental and theoretical scaling laws for transverse diffusive broadening in two-phase laminar flows in microchannels." *Applied Physics Letters* 76 (17): 2376–2378. https://doi.org/10.1063/1.126351.

Jamalabadi, MYA, M DaqiqShirazi, A Kosar, and MS Shadloo. 2017. "Effect of injection angle, density ratio, and viscosity on droplet formation in a microfluidic T-junction." *Theoretical and Applied Mechanics Letters* 7 (4): 243–251. https://doi.org/10.1016/j.taml.2017.06.002.

James, M, RA Revia, Z Stephen, and M Zhang. 2020. "Microfluidic synthesis of iron oxide nanoparticles." *Nanomaterials*

(Basel) 10 (11). https://doi.org/10.3390/nano10112113. https://www.ncbi.nlm.nih.gov/pubmed/33114204.

Jang, KJ, and KY Suh. 2010. "A multi-layer microfluidic device for efficient culture and analysis of renal tubular cells." *Lab on a Chip* 10 (1): 36–42. https://doi.org/10.1039/b907515a. https://www.ncbi.nlm.nih.gov/pubmed/20024048.

Jo, W, D Jeong, J Kim, S Cho, SC Jang, C Han, JY Kang, YS Gho, and J Park. 2014. "Microfluidic fabrication of cell-derived nanovesicles as endogenous RNA carriers." *Lab on a Chip* 14 (7): 1261–1266. https://doi.org/10.1039/c3lc50993a.

Karnik, R., F Gu, P Basto, C Cannizzaro, L Dean, W Kyei-Manu, R Langer, and OC Farokhzad. 2008. "Microfluidic platform for controlled synthesis of polymeric nanoparticles." *Nano Letters* 8 (9): 2906–2912. https://doi.org/10.1021/nl801736q.

Kee, SP, and A Gavriilidis. 2008. "Design and characterisation of the staggered herringbone mixer." *Chemical Engineering Journal* 142 (1): 109–121. https://doi.org/10.1016/j.cej.2008.02.001.

Kim, BY, LY Hong, YM Chung, DP Kim, and CS Lee. 2009. "Solvent-resistant PDMS microfluidic devices with hybrid inorganic/organic polymer coatings." *Advanced Functional Materials* 19 (23): 3796–3803. https://doi.org/10.1002/adfm.200901024.

Kwak, TJ, YG Nam, MA Najera, SW Lee, JR Strickler, and WJ Chang. 2016. "Convex grooves in staggered herringbone mixer improve mixing efficiency of laminar flow in microchannel." *PLoS One* 11 (11): e0166068. https://doi.org/10.1371/journal.pone.0166068. https://www.ncbi.nlm.nih.gov/pubmed/27814386.

Lim, JM., N Bertrand, PM Valencia, M Rhee, R Langer, S Jon, OC Farokhzad, and R Karnik. 2014. "Parallel microfluidic synthesis of size-tunable polymeric nanoparticles using 3D flow focusing towards in vivo study." *Nanomedicine-Nanotechnology Biology and Medicine* 10 (2): 401–406. https://doi.org/10.1016/j.nano.2013.08.003.

Lim, JM, and R Karnik. 2014. "Optimizing the discovery and clinical translation of nanoparticles: Could microfluidics hold the key?" *Nanomedicine (Lond)* 9 (8): 1113–1116. https://doi.org/10.2217/nnm.14.73. https://www.ncbi.nlm.nih.gov/pubmed/25118702.

Liu, SC, YQ Fan, KX Gao, and YJ Zhang. 2018. "Fabrication of Cyclo-olefin polymer-based microfluidic devices using CO_2 laser ablation." *Materials Research Express* 5 (9). https://doi.org/10.1088/2053-1591/aad72e.

Low, LA, C Mummery, BR Berridge, CP Austin, and DA Tagle. 2021. "Organs-on-chips: Into the next decade." *Nature Reviews Drug Discovery* 20 (5): 345–361. https://doi.org/10.1038/s41573-020-0079-3.

Maeki, M, Y Fujishima, Y Sato, T Yasui, N Kaji, A Ishida, H Tani, Y Baba, H Harashima, and M Tokeshi. 2017. "Understanding the formation mechanism of lipid nanoparticles in microfluidic devices with chaotic micromixers." *PLoS One* 12 (11). https://doi.org/10.1371/journal.pone.0187962.

Mares, AG, G Pacassoni, JS Marti, S Pujals, and L Albertazzi. 2021. "Formulation of tunable size PLGA-PEG nanoparticles for drug delivery using microfluidic technology." *PLoS One* 16 (6). https://doi.org/10.1371/journal.pone.0251821.

Marre, S, and KF Jensen. 2010. "Synthesis of micro and nano-structures in microfluidic systems." *Chemical Society Reviews* 39 (3): 1183–1202. https://doi.org/10.1039/b821324k.

Martins, JP, G Torrieri, and HA Santos. 2018. "The importance of microfluidics for the preparation of nanoparticles as advanced drug delivery systems." *Expert Opinion on Drug Delivery* 15 (5): 469–476. https://doi.org/10.1080/17425247.2018.1446936.

Mehta, V, and SN Rath. 2021. "3D printed microfluidic devices: A review focused on four fundamental manufacturing approaches and implications on the field of healthcare." *Bio-Design and Manufacturing* 4 (2): 311–343. https://doi.org/10.1007/s42242-020-00112-5.

Meng, YC, M Asghari, MK Aslan, A Yilmaz, B Mateescu, S Stavrakis, and AJ deMello. 2021. "Micro fluidics for extra-cellular vesicle separation and mimetic synthesis: Recent advances and future perspectives." *Chemical Engineering Journal* 404. https://doi.org/10.1016/j.cej.2020.126110.

Nge, PN, CI Rogers, and AT Woolley. 2013. "Advances in microfluidic materials, functions, integration, and applications." *Chemical Reviews* 113 (4): 2550–2583. https://doi.org/10.1021/cr300337x.

Niculescu, AG, C Chircov, AC Birca, and AM Grumezescu. 2021. "Fabrication and applications of microfluidic devices: A review." *International Journal of Molecular Sciences* 22 (4). https://doi.org/10.3390/ijms22042011.

Nie, MH, and S Takeuchi. 2018. "Bottom-up biofabrication using microfluidic techniques." *Biofabrication* 10 (4). https://doi.org/10.1088/1758-5090/aadef9.

Nielsen, JB, RL Hanson, HM Almughamsi, C Pang, TR Fish, and AT Woolley. 2020. "Microfluidics: Innovations in materials and their fabrication and functionalization." *Analytical Chemistry* 92 (1): 150–168. https://doi.org/10.1021/acs.analchem.9b04986.

Operti, MC, A Bernhardt, S Grimm, A Engel, CG Figdor, and O Tagit. 2021. "PLGA-based nanomedicines manufacturing: Technologies overview and challenges in industrial scale-up." *International Journal of Pharmaceutics* 605. https://doi.org/10.1016/j.ijpharm.2021.120807.

Orbay, S, A Ozcelik, J Lata, M Kaynak, MX Wu, and TJ Huang. 2017. "Mixing high-viscosity fluids via acoustically driven bubbles." *Journal of Micromechanics and Microengineering* 27 (1). https://doi.org/10.1088/0960-1317/27/1/015008.

Park, JW, B Vahidi, AM Taylor, SW Rhee, and NL Jeon. 2006. "Microfluidic culture platform for neuroscience research." *Nature Protocols* 1 (4): 2128–2136. https://doi.org/10.1038/nprot.2006.316.

Pilkington, CP, JM Seddon, and Y Elani. 2021. "Microfluidic technologies for the synthesis and manipulation of biomimetic membranous nano-assemblies." *Physical Chemistry Chemical Physics* 23 (6): 3693–3706. https://doi.org/10.1039/d0cp06226j.

Pitaru, AA, JG Lacombe, ME Cooke, L Beckman, T Steffen, MH Weber, PA Martineau, and DH Rosenzweig. 2020. "Investigating commercial filaments for 3D printing of stiff and elastic constructs with ligament-like mechanics." *Micromachines* 11 (9). https://doi.org/10.3390/mi11090846.

Ren, KN, JH Zhou, and HK Wu. 2013. "Materials for microfluidic chip fabrication." *Accounts of Chemical Research* 46 (11): 2396–2406. https://doi.org/10.1021/ar300314s.

Rhee, M, PM Valencia, MI Rodriguez, R Langer, OC Farokhzad, and R Karnik. 2011. "Synthesis of size-tunable polymeric nanoparticles enabled by 3D hydrodynamic flow focusing in single-layer microchannels." *Advanced Materials* 23 (12): H79–H83. https://doi.org/10.1002/adma.201004333.

Roces, CB, D Christensen, and Y Perrie. 2020. "Translating the fabrication of protein-loaded poly(lactic-co-glycolic acid) nanoparticles from bench to scale-independent production using microfluidics." *Drug Delivery and Translational Research* 10 (3): 582–593. https://doi.org/10.1007/s13346-019-00699-y.

Roces, CB, GS Lou, N Jain, S Abraham, A Thomas, GW Halbert, and Y Perrie. 2020. "Manufacturing considerations for the development of lipid nanoparticles using microfluidics." *Pharmaceutics* 12 (11). https://doi.org/10.3390/pharmaceutics12111095.

Rodriguez-Ruiz, I, V Babenko, S Martinez-Rodriguez, and JA Gavira. 2018. "Protein separation under a microfluidic regime." *Analyst* 143 (3): 606–616. https://doi.org/10.1039/c7an01568b.

Shakeri, A, N Abu Jarad, A Leung, L Soleymani, and TF Didar. 2019. "Biofunctionalization of glass- and paper-based microfluidic devices: A review." *Advanced Materials Interfaces* 6 (19). https://doi.org/10.1002/admi.201900940.

Shakeri, A, S Khan, and TF Didar. 2021. "Conventional and emerging strategies for the fabrication and functionalization of PDMS-based microfluidic devices." *Lab on a Chip* 21 (16): 3053–3075. https://doi.org/10.1039/d1lc00288k.

Singh, A, CK Malek, and SK Kulkarni. 2010. "Development in microreactor technology for nanoparticle synthesis." *International Journal of Nanoscience* 9: 93–112.

Song, JW, SP Cavnar, AC Walker, KE Luker, M Gupta, YC Tung, GD Luker, and S Takayama. 2009. "Microfluidic endothelium for studying the intravascular adhesion of metastatic breast cancer cells." *PLoS One* 4 (6). https://doi.org/10.1371/journal.pone.0005756.

Sonker, M, V Sahore, and AT Woolley. 2017. "Recent advances in microfluidic sample preparation and separation techniques for molecular biomarker analysis: A critical review." *Analytica Chimica Acta* 986: 1–11. https://doi.org/10.1016/j.aca.2017.07.043.

Stroock, AD, SKW Dertinger, A Ajdari, I Mezic, HA Stone, and GM Whitesides. 2002. "Chaotic mixer for microchannels." *Science* 295 (5555): 647–651. https://doi.org/10.1126/science.1066238.

Suh, YK, and S Kang. 2010. "A review on mixing in microfluidics." *Micromachines* 1 (3): 82–111. https://doi.org/10.3390/mi1030082.

Tice, JD, H Song, AD Lyon, and RF Ismagilov. 2003. "Formation of droplets and mixing in multiphase microfluidics at low values of the Reynolds and the capillary numbers." *Langmuir* 19 (22): 9127–9133. https://doi.org/10.1021/la030090w.

Tiwari, SK, S Bhat, and KK Mahato. 2020. "Design and fabrication of low-cost microfluidic channel for biomedical application." *Scientific Reports* 10 (1). https://doi.org/10.1038/s41598-020-65995-x.

Valencia, PM, OC Farokhzad, R Karnik, and R Langer. 2012. "Microfluidic technologies for accelerating the clinical translation of nanoparticles." *Nature Nanotechnology* 7 (10): 623–626. https://doi.org/10.1038/nnano.2012.168.

van Midwoud, PM, A Janse, MT Merema, GMM Groothuis, and E Verpoorte. 2012. "Comparison of biocompatibility and adsorption properties of different plastics for advanced microfluidic cell and tissue culture models." *Analytical Chemistry* 84 (9): 3938–3944. https://doi.org/10.1021/ac300771z.

Vitorino, R, S Guedes, JP da Costa, and V Kasicka. 2021. "Microfluidics for peptidomics, proteomics, and cell analysis." *Nanomaterials* 11 (5). https://doi.org/10.3390/nano11051118.

Ward, K, and ZH Fan. 2015. "Mixing in microfluidic devices and enhancement methods." *Journal of Micromechanics and Microengineering* 25 (9). https://doi.org/10.1088/0960-1317/25/9/094001.

Wen, XP, S Takahashi, K Hatakeyama, and KI Kamei. 2021. "Evaluation of the effects of solvents used in the fabrication of microfluidic devices on cell cultures." *Micromachines* 12 (5). https://doi.org/10.3390/mi12050550.

Wu, J, Z He, Q Chen, and J-M Lin. 2016. Biochemical analysis on microfluidic chips. *TrAC Trends in Analytical Chemistry* 80: 213–231.

Xiong, SY, XY Chen, and JY Wang. 2021. "A novel three-dimensional electroosmotic micromixer based on the Koch fractal principle." *RSC Advances* 11 (21): 12860–12865. https://doi.org/10.1039/d1ra00218j.

Zhao, CX, L. Z. He, S. Z. Qiao, and A. P. J. Middelberg. 2011. "Nanoparticle synthesis in microreactors." *Chemical Engineering Science* 66 (7): 1463–1476. https://doi.org/10.1016/j.ces.2010.08.039.

7 Microfluidics for Drug Discovery and Development

Maryam Parhizkar
University College London

Dimitrios Tsaoulidis
University of Surrey

CONTENTS

7.1 INTRODUCTION

Drug development process from discovery to bedside is often very long and on average takes 10–15 years, costing up to $2.5bn [1]. The current workflow for developing a new drug prior to reaching the patient follows a number of established stages that are mainly defined as early drug discovery, preclinical research and clinical trials followed by regulatory review and approval, manufacturing and post-marketing safety surveillance. At the early discovery stage, a biological target is identified, and a library of molecules are screened against the target to select the lead candidates with more potency and selectivity and optimised pharmacokinetic properties. This is followed by the preclinical testing using both in-vitro and in-vivo models to evaluate the lead candidates' efficacy and safety and biological behaviour. Ultimately, the new drug candidate is evaluated in three phases during the clinical trials, prior to approval by the regulatory authority. In 2021, FDA approved 50 new molecular entities (NMEs) despite the impact of the COVID-19 pandemic [2,3]. This is a significant increase in the number of approvals compared to decades ago with only 24 FDA approvals per year. Whilst this shows a great promise in the drug development pipeline, the process is still hindered by a variety of challenges.

These include the gap in analysing drug candidates in a more rapid and accurate manner, the inefficiency of the conventional methods of synthesising drug compounds and delivery systems and in-vitro and in-vivo models that are not representative of the human body. Therefore, this poses a great opportunity to develop and foster novel tools to address these challenges and aid the drug discovery and development process.

In light of this, microfluidics has become a fast-growing technology that is being employed through the entire drug discovery and development workflow [4–7]. Microfluidics is the science and technology of systems that facilitate the manipulation of fluids in channels with dimensions of tens to hundreds of micrometres [8]. Microfluidics exploits intrinsic features such as laminar flows in small microchannels and hence offers multiple capabilities in many applications [9–12]. The earliest application of microfluidics was in analysis [13], where very small sample quantities and reagents were used to perform high-resolution separation and detection, leading to the development of low-cost analytical devices that deliver short-time analysis with small footprints. Microfluidics has also emerged as a transformative technology in the drug discovery and development process benefiting from features such as precise control of fluid

DOI: 10.1201/9781003224464-7

chemical and physical properties, fast mixing and actuation and low reagent consumption [14].

Drug discovery and development process, as the name suggests, consists of two main phases of discovery and development. Traditional tools used in the discovery phase are standardised systems that are often costly and require high processing times to meet the throughput demands. High-throughput screening (HTS) method is an example of an established process for screening new chemical entities using multiple-well plates with large volumes of samples [15]. Microfluidics can potentially replace these systems in the future to enable HTS studies in a miniaturised manner, reducing the need for large samples [16]. On the other hand, drug development processes can equally benefit from microfluidics technology from multiple angles. Drug development involves the synthesis of drug and therapeutic compounds followed by the development of drug delivery systems for better biological uptake and enhanced therapeutic effect. At a final stage, the drug delivery systems are evaluated for their safety, toxicity, pharmacokinetic and physiochemical properties.

Conventional methods for drug development use various batch chemical reactors for drug synthesis, whilst complex animal and cell models are used for efficacy and safety evaluations [17,18]. Microfluidics devices can be exploited in many of the processes involving drug synthesis and evaluation. This includes the use of microfluidics for the chemical synthesis of drug compounds, where sample preparation, separation and detection occur in a highly controllable manner in small microchannels [19,20]. Microfluidics is also capable of manufacturing drug carriers with tuneable morphology, size and surface functionality [21]. Microfluidic devices can directly act as drug delivery systems for targeted and site-specific drug release [22]. Microfluidics can also contribute to the evaluation process of drugs in form of small molecules, macromolecules and drug delivery systems. Moreover, traditional in-vitro approaches using 2D cell culture often fail to represent complex physiological systems, while animal models are costly and extrapolating these models to human disease conditions where the physiological and pathogenic mechanisms are quite different. Recently, microfluidic organ-on-a-chip (organ chip) technology is being exploited in a number of biomedical research studies such as the modelling of mechanics in metastasis cancer and human lung inflammation [23–25].

It is therefore timely to review the state-of-the-art microfluidic technologies that have emerged in the drug discovery and development field in recent years. Thus, the following sections will evaluate the impact of microfluidics throughout the entire drug discovery and development process. Each section will discuss the applications of microfluidics, from early drug discovery stage to drug development process and subsequently preclinical studies. Finally, a perspective on the challenges in their translation to industrial settings is presented with an outlook on the future directions of microfluidics in drug discovery and development pipeline.

7.2 MICROFLUIDICS TECHNOLOGY

The first application of microfluidics dates back to the 1970s, when miniaturised gas analysis systems were introduced by Terry et al. [26] and more prominently on 'micro total analysis systems' (µTAS) by Manz et al. in 1990 [13]. In general, microfluidics technology offers a number of advantages such as reduced reagent consumption, rapid mass and heat transfer due to high surface-to-volume ratio, faster reaction times, miniaturisation, automation and better control of fluid flow parameters at small scales [27]. More relevant advantages to the field of drug discovery and development are the reduction in the cost per reaction or assay point due to the use of significantly less reagents and greater control on experimental conditions (e.g. single cell manipulation) [28,29]. Moreover, with the integration of components such as fluid pumps and detection instrument in to the microfluidic device, a macroscale-sized laboratory can be miniaturised, where the process can be replicated on a 'lab-on-a-chip' [30].

Transport of mass in fluids is governed by viscous and inertial effects, where the latter gives rise to several instabilities and commonly turbulence. In microfluidics, due to flow of fluids in miniaturised channels, the inertial effects are negligible and therefore the flow is mainly laminar. Although different types of microfluidic technologies have been developed over the years, microfluidic devices are generally categorised into two types of continuous-flow and segmented-flow (droplet-based) devices, depending on the characteristics of the fluid flow [19,31].

7.2.1 CONTINUOUS-FLOW MICROFLUIDICS

Continuous-flow microfluidic devices generally process laminar, continuous streams of single or multiple fluids that pass through the microchannels, and as a result, mixing is achieved through molecular diffusion. The flow is usually controlled by external pressure sources such as mechanical pumps, electrokinetic actuation and surface tension–driven flow, among other actuation methods. Due to its simplicity, continuous flow microfluidics is the most common approach that has been employed at different stages of drug discovery and development process, such as drug compound synthesis and drug evaluation and nanoparticle synthesis [32,33]. Although continuous-flow microfluidics has been extensively used in these applications, such systems have a number of limitations and are less effective for tasks requiring a high degree of mixing (due to their laminar flow nature where only diffusion is possible), flexibility and fluid manipulation. In addition, there have been events of occasional channel clogging, whilst scale-up and integration of continuous-flow microfluidics are also hindered by their application-specific usage.

7.2.2 DROPLET-BASED MICROFLUIDICS

In segmented flow (often referred to as droplet-based) microfluidic devices at least two immiscible fluids are processed. Droplet-based microfluidics are capable of droplet generation and manipulation. Droplet-based microfluidic devices can produce highly monodisperse fluid droplets when minute volumes (μL to fL range) of the dispersed phase are enclosed in an immiscible, continuous liquid-phase carrier fluid by using passive or active techniques. In passive technique, hydrodynamic forces (pressure-driven flow) and intricate geometrical design of the microchannel define the local fluid behaviour, deformation and droplet break-up. Different channel geometries are often classified into co-flow, cross-flow and flow-focusing methods. Droplet formation by active techniques relies on the application of external force, and depending on the type of energy source, they are categorised into electrical, magnetic, thermal and mechanical forces [31]. Monodisperse droplets in microfluidics are capable of acting as mini-reactors for chemical reactions or compartments for encapsulation of various entities such as cells, drugs and nucleic acids. Droplet microfluidics is also popular in pharmaceutical industry and bears extensive value in applications such as drug delivery and drug encapsulation, single-cell study, nanoparticle synthesis and diagnostics [34]. Droplet-based microfluidic technique offers greater advantages over continuous-flow microfluidics. Improved mixing rates, lower reagent consumption, reduced channel clogging, high throughput and scalability are among the advantages of droplet-based microfluidics.

7.3 MICROFLUIDICS FOR DRUG SYNTHESIS

Drug synthesis plays an integral part in the drug development process. Drug synthesis involves the performance of a series of advanced chemical reactions to synthesise the drug compounds. The most important attribute when performing these reactions is the ability to accurately control and change the essential parameters (e.g. concentration of reagents, residence time, temperature, structure, mixing degree, etc.). Current practices for drug synthesis include macroscale systems (e.g. batch reactors) that suffer from, among others, large inventories and residence times, batch-to-batch variability, uncontrollability and inefficiency [5,35].

Microfluidics can precisely manipulate small volume of fluids and achieve high control of mixing and reaction dynamics; consequently, this gives the ability to control the quality of the drugs at every stage of the manufacturing process, where the margin for error is extremely low. Although microfluidic technologies have primarily been exploited for their analytical and screening potential, features such as design versatility, control accuracy and high heat and mass transfer rates due to the thin fluidic films make them ideal candidates for preparative applications. Coupled with artificial intelligence (AI), microfluidics can

be a powerful tool to deliver automated and self-optimising technologies that enable excellent process controllability and quality assurance [36]. Microfluidic platforms can be designed to accommodate a variety of procedures, from simple to very complex synthesis, thus providing a powerful tool for controlled and automated synthesis of drugs. In general, the synthesis performed in microfluidic devices will yield to higher conversion rates and improved selectivity compared to conventional methods.

Reaction time, mixing time and residence time distribution are important parameters that should be considered when designing the reactor for synthesis applications. Shortening the reaction and mixing time and obtaining a narrow residence time distribution can significantly improve drug synthesis in microfluidic reactors which would have not been achievable in conventional systems. Numerous examples of fast reactions that suffer from mixing limitations have been reported in the literature [37–39] and demonstrated the benefits of small-scale reactors by significantly reducing the time from hours to minutes and seconds.

Microfluidic technologies, both continuous-flow and droplet-based, have demonstrated many advantages over conventional batch systems. The advances in digitisation enable the coupling with online analysis systems (e.g. IR and Raman spectroscopy), which results in real-time reaction optimisation. Platforms can be designed in such a way for the various reactants to be mixed with adjustable degree of mixing (e.g. in the cases where shear-sensitive substances are involved). The concentration of the reagents involved can be controlled both with respect to position and time, resulting in the increase in performance efficiency.

7.3.1 MICROFLUIDICS FOR SINGLE-STEP DRUG SYNTHESIS

To characterise the structure of the drug molecules, protein separation and crystallisation are carried out, with crystallisation being often the rate-limiting step. The ability of microfluidic devices to integrate multiple functions in one platform and incorporate, for example, the use of electric field makes them attractive for protein separation to overcome issues of low throughput and sensitivity, and limited amounts of samples. In addition, high heat dissipation can be achieved due to the high surface area-to-volume ratio, which gives the ability, by using higher electric fields, to achieve higher separation resolution and selectivity. Integrated approaches have been used to separate peptide mixtures, by using micellar electrokinetic chromatography and capillary zone electrophoresis [40], or two-dimensional gel electrophoresis [41], and have shown sufficiently better overall performance over conventional methods.

The application of photochemical processes for the synthesis of lead compounds or the synthesis of functionalised proteins have also been widely used; however, there are still some limitations with the existing methods due to the limited reactive irradiation volume brought by the increased path length [42]. Microfluidic devices have been implemented successfully to increase the efficiency of pho- tochemical reactions (e.g. for the photochemical synthesis of allylic trifluoromethanes from enhanced light penetra- tion due to short lengths of microchannels) [43,44] or for the preparation of radiolabelled protein conjugates (e.g. 89Zr- radiolabelled proteins) [45].

For more complicated enzymatic reactions, , microfluidic devices can enable different immobilisation approaches, i.e., direct coating of the microfluidc walls with a thin layer of the enzyme or immobilisation of enzymes on carriers followed by the integration of the carriers in the microflu- idic devices. These approaches have been used for glucose precursor (3-PGA) production [46,47] and showed that enzyme consumption and contamination were prevented, and the separation and recovery of enzyme could happen along with product generation. Enzymatic catalysis could successfully be implemented in microchannels to overcome bottlenecks such as the fast deactivation of enzymes.

7.3.2 Microfluidics for Multi-step Drug Synthesis

Combinatorial chemistry coupled with microfluidic tech- nology can provide a powerful tool for the discovery of new drugs. In general, solution-based combinatorial synthesis requires large space and quantities [48]. A combinatorial formation of amides from two different amines and two dif- ferent acid chlorides (2×2 parallel organic synthesis) has been performed in an integrated microreactor alleviating problems such as equal flow distribution among the chan- nels, continuous flow and consistent product quality [49]. A similar design of microreactors could have the ability to perform combinatorial synthesis for larger libraries by designing three-dimensional networks and achieve high- throughput analysis by scaling up.

In the case of multi-step reactions, generally, cross- contamination and carry-over of species from the previous steps are an undesirable phenomenon and could affect the yield of conversion. Microfluidic reactors have demon- strated great potential in synthesising many compounds (e.g. for the multi-step formation of dipeptides and tripep- tides) [50–52]. They are also capable of obtaining accurate control of every step and providing new information about the kinetic mechanisms of the intermediate steps (e.g. for the synthesis of multicomponent reaction products) [53]. Microfluidics has been employed for the production of ibu- profen in a three-step chemical reaction without the need for complicated purification and isolation steps. A micro- fluidic platform consisting of three individual reactors was designed that enabled the isolation of the three steps and prevented the carry-over of species [54]. Advanced inte- grated process systems have been used for the continu- ous multi-step synthesis of structural complex APIs. An electro-flow reactor containing patterned electrodeposited Ni/Pt metal over a copper electrode has been developed to perform a multi-step daclatasvir (DCV) synthesis and showed the capabilities of examining multi-step reaction sequences [55]. Although the complexity of the reactor

design increases with the number of compounds, microflu- idic technologies can decrease footprint, increase resource efficiency and allow for operating flexibilities.

7.4 MICROFLUIDICS FOR DRUG DELIVERY

Drug delivery systems are imperative in providing thera- pies to patients that can be administered in an efficient and effective manner. One important aspect of the drug devel- opment process is the formulation of drug molecules into a number of delivery systems, for example, oral preparations, creams and emulsions for topical and transdermal delivery, controlled-release systems through various administration routes and polymeric and lipid-based systems for injections (for administration via the intravenous (IV), intramuscular (IM), intraductal (ID)) [56]. Whilst these delivery systems are effective in many instances, there are still several chal- lenges associated with the delivery of drugs such as in-vivo instability, poor bioavailability and solubility, off-target delivery, poor absorption in the body and possible adverse effects of drugs [57].

In order to overcome these limitations, it is essential to develop novel drug delivery systems that enable a sustained and controlled drug delivery while maintaining efficiency and simultaneously reducing side effects. Advances in drug delivery technologies can pave the way for improved pharmacological effects, including enhanced efficacy and bioavailability of compounds, leading to the discovery and development of more effective therapies for better patient outcomes and quality of life. Microfluidics technology facilitates the advancements in drug delivery through a number of ways. Microfluidics enable the fabrication of drug delivery carriers with highly tunable properties. In addition, microfluidics devices can be employed for site- specific delivery and selective release of drugs.

7.4.1 Microfluidics for Production of Drug Carriers

Provoked by the compelling advancements in the fields of nanotechnology and biomaterials, nanoscale platforms and vesicular structures are amongst the novel strate- gies enabling targeted delivery and triggered release [58]. Nanoparticulate drug delivery systems have shown to improve the efficacy of treatments and reduce the side effects associated with drugs that are already on the mar- ket, allowing further investigation into new methods of delivery of drugs [59]. This is due to their high volume- to-surface ratio, tunable chemical surface properties and great potential to load active pharmaceutical compounds. For the synthesis of nanoparticulate drug delivery sys- tems, conventional small-scale batch reactors are utilised. Conventional synthetic methods are poorly controllable and not well-characterised systems and suffer from a variety of drawbacks including insufficient mixing, the generation of non-uniform structures that impact the drug release profile and the need to use large volumes of valuable drugs or chemicals [60].

Recent advances in microfluidics technologies have enabled the next generation of fabrication techniques for highly optimised drug carriers. By leveraging their capacity to manipulate small sample quantities and their high level of control on the emulsification process, microfluidics allows for the versatile generation of monodispersed droplets in the microchannels, leading to the production of nano and microparticles with tunable properties (size, shape and surface composition) and high reproducibility for the delivery of compounds (e.g. cytotoxic drugs, proteins and peptides) and their controlled release in-vitro [61].

Microfluidic devices enable the creation of both organic and inorganic particles as drug carriers. Amongst the most common types of carriers that have been developed through microfluidics in recent years are liposomes, polymeric and lipid nanoparticles and emulsions [62,63]. The overwhelming success of lipid-nanoparticle-based mRNA vaccines at an extraordinary speed in addressing the COVID-19 pandemic highlights the need for novel tools to produce these drug carriers more efficiently, whilst reducing the associated costs. Microfluidic hydrodynamic focusing was first employed to prepare nanoscale lipid vesicles in microfluidic channels [64]. This was followed by the application of micromixers and droplet-based microfluidics to enhance the homogeneity and tunability of liposomes and lipid nanoparticles [65]. Polymeric nanoparticles are also widely synthesised by microfluidic platforms. Droplet-based microfluidics (e.g. hydrodynamic flow focusing) and micromixers have been broadly employed to generate drug-loaded monodisperse size-controlled polymeric nanoparticles. A wide range of synthetic (e.g. poly(lactide-co-glycolide) (PLGA), polycaprolactone (PCL)) and natural (e.g. chitosan, alginate) polymers are processed in microfluidic devices to prepare drug-loaded nanoparticles [66]. PLGA-based nanoparticles prepared by microfluidics are amongst the most studied in the drug delivery field. It has been established that the biological activity of the nanoparticles (e.g. biodistribution, cellular update and clearance) is highly dependent on the size and shape of nanoparticles. Therefore, preparation methods that have high control over these parameters are fundamental to design drug delivery systems with high performance and minimised adverse effects. Microfluids has the capability to modulate the size and shape of nanoparticles by controlling the mixing process through appropriate adjustment of flow rates. Microfluidic techniques have demonstrated superiority over batch method processes in preparing highly reproducible nano-sized drug carriers with the potential for their scale-up production.

7.4.2 Microfluidic Devices as Drug Delivery Systems

Microfluidic devices can be utilised to directly deliver the drug molecules. They have the capability to efficiently transport drugs to a target site, allowing for increased localisation and reduction of side effects caused by the interaction of drugs in other organs or tissues. Ever since the success of microelectromechanical systems (MEMS) in the development of sensors and actuators into microchips without the need for electrical devices, microfluidics has been increasingly utilised for controlled drug delivery [67]. Microfluidic drug delivery platforms are often comprised of a drug reservoir, a pump or an actuator, a valve and a membrane that controls the release rate of the drug. Depending on the mechanism of actuation, drug delivery can either occur through a stable and prolonged release through passive means such as diffusion and convection or drug release can be precisely controlled through active means from a pre-pressurised reservoir by either mechanical, electrochemical, thermal or magnetic actuation [68]. MEMS micropumps enable active control of drug release where the release rate and infusion volumes can be precisely tuned to supply the drugs continuously through a reservoir.

Microfluidics integrated with MEMS technology allows for the development of smart release systems with wireless transmission and accurate control of drug doses over multiple occasions. Over the last decade, there have been a number of efforts in developing implantable, oral and minimally invasive drug delivery devices to the brain, eye, skin and GI tract [69]. In addition, microneedle technology for transdermal delivery is another successful application of microfluidics for the delivery of RNA and vaccines and other therapeutic compounds [70].

Although microfluidic platforms hold a great promise in localised and direct delivery to the disease site, there are still a few drawbacks in their full translation. The use of degradable material that could be absorbed without the need for follow-up surgeries would significantly improve patients' compliance. Immune response and inflammation and the interaction of the tissues with the device are also a few other challenges that can lead to device rejection. These can be prevented by coating the device with anti-inflammatory agents that help with extending the reliability of the system over the course of treatment. Finally, designing a reliable actuation and precision flow regulating systems would enable prolonged and on-demand drug administration.

7.5 MICROFLUIDICS FOR DRUG SCREENING AND EVALUATION

Once drug candidates are synthesised, drug screening procedures are adopted to identify the therapeutic effects of the drugs. Amongst different approaches, target-based biochemical screening and phenotypic cell-based screening and animal models are the various models used in drug screening and evaluation. HTS of a large number of sample reagents is often supported by utilising robotics, data-processing software and AI, where a library of compounds can be tested in parallel. An essential aspect of the drug development process is the determination of the cytotoxicity, potency and efficacy of new molecules. Although a plethora of diverse new drug molecules and drug delivery systems have been developed around the world and whilst many of them proved effective during the development process, only a handful have made it successfully past the preclinical studies. This is mainly caused by substantial drawbacks in conventional in-vivo animal models and in-vitro cell culture platforms.

There are still significant discrepancies between preclinical models and clinical data. This necessitates the need for models that recapitulate the human physiology for a more realistic analysis of drug compounds. For example, traditional 96 or 384-well plates for drug screening are incapable of analysing small volumes of patient biopsy samples [71]. Due to its merits of low sample input, microfluidics technology is the best alternative to replace these conventional screening models. There is also a great initiative in decreasing the number of animal models for preclinical testing due to rising ethical concerns. On the other hand, 2D and static 3D cell cultures do not accurately mimic the cellular microenvironment. Therefore, new technologies such as microfluidic organ-on-chip platforms that allow testing in a 3D structure by continuous perfusion have become increasingly popular in recent years.

7.5.1 Microfluidics for Drug Screening

Drug screening is a key step in the drug discovery process, which mainly aims at identifying the drug composition and concentration on a molecular scale. HTS methods offer a great help in understanding the role of certain drug candidates and their selective binding to a target molecule. However, due to their high complexity, costs and consumption of reagents and time, these techniques can be replaced with new solutions for effective drug screening. Microfluidics has emerged as a new platform to overcome these shortcomings. In general, two types of continuous flow and droplet-based microfluidics are applied for drug screening. Continuous-flow microfluidics has been used as drug dilution generators for dose-response screening and programmable cell culture arrays to evaluate the efficiency of the drugs [72]. However, the throughput in continuous-flow microfluidics is still very low and can only be improved by parallelisation of the microchannels. Droplet-based microfluidics is also used for applications of single-cell drug screening and concentration-gradient generators for dose-response studies [73]. Droplet-based microfluidics is mostly favoured due to their capability in parallel processing and experimentation and ease of scalability and increased high throughput without the need to add complexity to the design of microchannels.

7.5.2 Microfluidic for Drug Evaluation in Preclinical Studies

Preclinical evaluations for drug candidates combine in-vitro cell culture protocols and in-vivo animal testing. Animal models are essential in the drug evaluation process. However, there are still many challenges associated with using complex animal models. In-vitro 2D cell culture has been employed for drug screening as a bridge between in-vivo testing and biochemical assays. Alternative 3D cell culture methods have been more powerful in matching the results seen in the clinic. The progress in tissue engineering techniques has enabled the generation of biologically relevant data by creating 3D cultures in microfluidic devices. In recent years, 3D cell-on-chip and organ-on-chip devices have severed as functional tissue and organ analogues. Cell growth in in-vitro models can be accomplished through the synergy between various complex internal and external environmental parameters. Organ-on-chip technology has the capability to control these parameters through various means such as generating fluid shear force, mechanical strain and biochemical concentration gradient. These characteristics enable the cells to self-assemble in response to stimuli and exhibit more realistic physiological functions. In addition, fluid flow can be manipulated to produce a shear force that can mimic the fluid flow in the body. Microfluidics can also provide dynamic mechanical strain (e.g. blood, lung or bone pressure) in order to maintain body's physiological functions such as tissue formation, cell differentiation and even tumour growth. A range of microfluidic organ-on-a-chip models have since been developed, mimicking diverse biological functions by culturing cells from blood vessels, muscles, bones, airways, liver, brain, gut and kidneys [74]. These microfluidic platforms provide a microenvironment similar to the environment that cells undergo in real organs or tissues. It is, therefore, expected that organ-on-chip devices become an integral part of the drug development process, once their regulatory pathways are identified.

7.6 CONCLUSION AND FUTURE PERSPECTIVES

As evidenced by a diverse range of applications, microfluidic technology provides powerful strategies and tools to enhance the entire drug development process from drug synthesis to delivery and evaluation. The last decade has seen a burst of technological innovation in the field of microfluidics in drug carrier development, organ-on-chip devices and high-throughput screening. In order for these technologies to have a more significant impact on the drug discovery and development process, they need to be widely adopted by the pharmaceutical industry. This process has been slow, despite the advances in the field, due to lack of regulatory pathways and difficulty in their translation to commercial settings. There still remains much work to be done to address many challenges that stand in the way of translating these technologies to the pharmaceutical industry. More specifically, their ease of fabrication and use and increase in assay analysis efficiencies can be improved. Microfluidic devices are not yet scalable or robust enough to replace well-based HTS assays. Through continuous flow synthesis and parallelisation, microfluidics has shown superiority in scale-up and time-efficient production compared with batch synthesis methods. However, replacing batch reactors with microfluidic technologies would require a large investment. One area that microfluidics can realise its full potential is the screening of drugs in preclinical studies prior to the use of expensive animals and clinical trials through organ-on-chip systems. Continued research efforts and collaborations between academia and industry would enable the development of more practical microfluidic platforms for drug development and screening and provide a clear technology validation pathway for them. Microfluidic devices could have a significant impact on the future drug discovery and development process, and they continue to raise interest in the field in the years to come.

REFERENCES

1. Wouters, O.J., M. McKee, and J. Luyten, Estimated research and development investment needed to bring a new medicine to market, 2009–2018. *JAMA*, 2020. **323**(9): pp. 844–853.
2. Mullard, A., 2021 FDA approvals. *Nature Reviews Drug Discovery*, 2022. **21**: pp. 83–88.
3. FDA, *Advancing Health through Innovation: New Drug Approvals 2021*. 2022.
4. Li, W., et al., Microfluidic fabrication of microparticles for biomedical applications. *Chemical Society Reviews*, 2018. **47**(15): pp. 5646–5683.
5. Dittrich, P.S. and A. Manz, Lab-on-a-chip: Microfluidics in drug discovery. *Nature Reviews Drug Discovery*, 2006. **5**(3): pp. 210–218.
6. Liu, D., et al., Current developments and applications of microfluidic technology toward clinical translation of nanomedicines. *Advanced Drug Delivery Reviews*, 2018. **128**: pp. 54–83.
7. Zheng, W. and X. Jiang, Synthesizing living tissues with microfluidics. *Accounts of Chemical Research*, 2018. **51**(12): pp. 3166–3173.
8. Whitesides, G.M., The origins and the future of microfluidics. *Nature*, 2006. **442**(7101): pp. 368–373.
9. Ahmadi, S., et al., Chapter 8: Microfluidic devices for gene delivery systems, in *Biomedical Applications of Microfluidic Devices*, M.R. Hamblin and M. Karimi, Editors. 2021, Academic Press: Cambridge, pp. 187–208.
10. Neethirajan, S., et al., Microfluidics for food, agriculture and biosystems industries. *Lab on a Chip*, 2011. **11**(9): pp. 1574–1586.
11. Burklund, A., et al., Advances in diagnostic microfluidics. *Advances in Clinical Chemistry*, 2020. **95**: pp. 1–72.
12. Song, Y., J. Hormes, and C.S. Kumar, Microfluidic synthesis of nanomaterials. *Small*, 2008. **4**(6): pp. 698–711.
13. Manz, A., N. Graber, and H.M. Widmer, Miniaturized total chemical analysis systems: A novel concept for chemical sensing. *Sensors and Actuators B: Chemical*, 1990. **1**(1): pp. 244–248.
14. Liu, Y. and X. Jiang, Why microfluidics? Merits and trends in chemical synthesis. *Lab on a Chip*, 2017. **17**(23): pp. 3960–3978.
15. Macarron, R., Critical review of the role of HTS in drug discovery. *Drug Discovery Today*, 2006. **11**(7–8): pp. 277–279.
16. Yang, Y., et al., Microfluidics for biomedical analysis. *Small Methods*, 2020. **4**(4): p. 1900451.
17. Ciociola, A.A., et al., How drugs are developed and approved by the FDA: Current process and future directions. *Official journal of the American College of Gastroenterology | ACG*, 2014. **109**(5): pp. 620–623.
18. Taylor, D., The pharmaceutical industry and the future of drug development, in *Pharmaceuticals in the Environment*, R.E. Hester and R.M. Harrison, Editors. 2016, The Royal Society of Chemistry: London, pp. 1–33.
19. Malet-Sanz, L. and F. Susanne, Continuous flow synthesis. A pharma perspective. *Journal of Medicinal Chemistry*, 2012. **55**(9): pp. 4062–4098.
20. Hughes, D.L., Applications of flow chemistry in drug development: Highlights of recent patent literature. *Organic Process Research & Development*, 2018. **22**(1): pp. 13–20.
21. Zhang, L., et al., Microfluidic methods for fabrication and engineering of nanoparticle drug delivery systems. *ACS Applied Bio Materials*, 2019. **3**(1): pp. 107–120.
22. Cobo, A., R. Sheybani, and E. Meng, MEMS: Enabled drug delivery systems. *Advanced Healthcare Materials*, 2015. **4**(7): pp. 969–982.
23. Yesil-Celiktas, O., et al., Mimicking human pathophysiology in organ-on-chip devices. *Advanced Biosystems*, 2018. **2**(10): p. 1800109.
24. Chen, M.B., et al., On-chip human microvasculature assay for visualization and quantification of tumor cell extravasation dynamics. *Nature Protocols*, 2017. **12**(5): pp. 865–880.
25. Benam, K.H., et al., Small airway-on-a-chip enables analysis of human lung inflammation and drug responses in-vitro. *Nature Methods*, 2016. **13**(2): pp. 151–157.
26. Terry, S.C., J.H. Jerman, and J.B. Angell, A gas chromatographic air analyzer fabricated on a silicon wafer. *IEEE Transactions on Electron Devices*, 1979. **26**(12): pp. 1880–1886.
27. Dietzel, A., A brief introduction to microfluidics, in *Microsystems for Pharmatechnology: Manipulation of Fluids, Particles, Droplets, and Cells*, A. Dietzel, Editor. 2016, Springer International Publishing: Cham. pp. 1–21.
28. Poulsen, C.E., et al., A Microfluidic platform for the rapid determination of distribution coefficients by gravity-assisted droplet-based liquid–liquid extraction. *Analytical Chemistry*, 2015. **87**(12): pp. 6265–6270.
29. Tavakoli, H., et al., Recent advances in microfluidic platforms for single-cell analysis in cancer biology, diagnosis and therapy. *TrAC Trends in Analytical Chemistry*, 2019. **117**: pp. 13–26.
30. Francesko, A., V.F. Cardoso, and S. Lanceros-Méndez, Chapter 1: Lab-on-a-chip technology and microfluidics, in *Microfluidics for Pharmaceutical Applications*, H.A. Santos, D. Liu, and H. Zhang, Editors. 2019, William Andrew Publishing: New York, pp. 3–36.
31. Shang, L., Y. Cheng, and Y. Zhao, Emerging droplet microfluidics. *Chemical Reviews*, 2017. **117**(12): pp. 7964–8040.
32. Antfolk, M. and T. Laurell, Continuous flow microfluidic separation and processing of rare cells and bioparticles found in blood - A review. *Analytica Chimica Acta*, 2017. **965**: pp. 9–35.
33. Balbino, T.A., et al., Continuous flow production of cationic liposomes at high lipid concentration in microfluidic devices for gene delivery applications. *Chemical Engineering Journal*, 2013. **226**: pp. 423–433.
34. Chou, W.-L., et al., Recent advances in applications of droplet microfluidics. *Micromachines*, 2015. **6**(9): pp. 1249–1271.

35. Liu, Y., et al., Microfluidics for drug development: From synthesis to evaluation. *Chemical Reviews*, 2021. **121**(13): pp. 7468–7529.

36. Kang, L., et al., Microfluidics for drug discovery and development: From target selection to product lifecycle management. *Drug Discovery Today*, 2008. **13**(1–2): pp. 1–13.

37. Kappe, C.O. and D. Dallinger, The impact of microwave synthesis on drug discovery. *Nature Reviews Drug Discovery*, 2006. **5**(1): pp. 51–63.

38. Larhed, M. and A. Hallberg, Microwave-assisted high-speed chemistry: A new technique in drug discovery. *Drug Discovery Today*, 2001. **6**(8): pp. 406–416.

39. Yoshida, J.-I., *Flash Chemistry: Fast Organic Synthesis in Microsystems*. 2008: John Wiley & Sons, Hoboken, Canada.

40. Cooper, J.W., Y. Wang, and C.S. Lee, Recent advances in capillary separations for proteomics. *Electrophoresis*, 2004. **25**(23–24): pp. 3913–3926.

41. Chen, H. and Z.H. Fan, Two-dimensional protein separation in microfluidic devices. *Electrophoresis*, 2009. **30**(5): pp. 758–765.

42. Sun, A.C., et al., A droplet microfluidic platform for high-throughput photochemical reaction discovery. *Nature Communications*, 2020. **11**(1): p. 6202.

43. Kreis, L.M., et al., Photocatalytic synthesis of allylic trifluoromethyl substituted styrene derivatives in batch and flow. *Organic Letters*, 2013. **15**(7): pp. 1634–1637.

44. Wu, M., et al., Photon upconversion for the enhancement of microfluidic photochemical synthesis. *RSC Advances*, 2019. **9**(45): pp. 26172–26175.

45. Earley, D.F., et al., Microfluidic preparation of 89Zr-radiolabelled proteins by flow photochemistry. *Molecules*, 2021. **26**(3): p. 764.

46. Gong, A., et al., Moving and unsinkable graphene sheets immobilized enzyme for microfluidic biocatalysis. *Scientific Reports*, 2017. **7**(1): pp. 1–15.

47. Zhu, Y., et al., Continuous artificial synthesis of glucose precursor using enzyme-immobilized microfluidic reactors. *Nature Communications*, 2019. **10**(1): pp. 1–9.

48. Neils, C., et al., Combinatorial mixing of microfluidic streams. *Lab on a Chip*, 2004. **4**(4): pp. 342–350.

49. Kikutani, Y., et al., Glass microchip with three-dimensional microchannel network for 2× 2 parallel synthesis. *Lab on a Chip*, 2002. **2**(4): pp. 188–192.

50. Watts, P., et al., Investigation of racemisation in peptide synthesis within a micro reactor. *Lab on a Chip*, 2002. **2**(3): pp. 141–144.

51. Watts, P., et al., The synthesis of peptides using micro reactors, in *Microreaction Technology*, M. Matlosz, W. Ehrfeld, and J. P. Baselt, Editors. 2001, Springer. pp. 508–517.

52. Watts, P., et al., Solution phase synthesis of β-peptides using micro reactors. *Tetrahedron*, 2002. **58**(27): pp. 5427–5439.

53. Mitchell, M.C., V. Spikmans, and A.J. de Mello, Microchip-based synthesis and analysis: Control of multicomponent reaction products and intermediates. *Analyst*, 2001. **126**(1): pp. 24–27.

54. McQuade, D.T., A. Bogdan, and S.L. Poe, *Method and apparatus for continuous flow synthesis of ibuprofen*. 2013, Google Patents.

55. Mahajan, B., et al., Micro-electro-flow reactor (μ-EFR) system for ultra-fast arene synthesis and manufacture of daclatasvir. *Chemical Communications*, 2019. **55**(79): pp. 11852–11855.

56. Langer, R., New methods of drug delivery. *Science*, 1990. **249**(4976): pp. 1527–1533.

57. Allen, T.M. and P.R. Cullis, Drug delivery systems: Entering the mainstream. *Science*, 2004. **303**(5665): pp. 1818–1822.

58. Laffleur, F. and V. Keckeis, Advances in drug delivery systems: Work in progress still needed? *International Journal of Pharmaceutics*, 2020. **590**: p. 119912.

59. Anselmo, A.C. and S. Mitragotri, Nanoparticles in the clinic. *Bioengineering & Translational Medicine*, 2016. **1**(1): pp. 10–29.

60. Kim, Y. and R. Langer, Microfluidics in nanomedicine. *Translational Medicine: Cancer, 2 Volumes*, 2016: p. 409.

61. Liu, D., et al., Microfluidic-assisted fabrication of carriers for controlled drug delivery. *Lab on a Chip*, 2017. **17**(11): pp. 1856–1883.

62. Carugo, D., et al., Liposome production by microfluidics: Potential and limiting factors. *Scientific Reports*, 2016. **6**(1): p. 25876.

63. Karnik, R., et al., Microfluidic platform for controlled synthesis of polymeric nanoparticles. *Nano Letters*, 2008. **8**(9): pp. 2906–2912.

64. Jahn, A., et al., Controlled vesicle self-assembly in microfluidic channels with hydrodynamic focusing. *Journal of the American Chemical Society*, 2004. **126**(9): pp. 2674–2675.

65. Zhang, G. and J. Sun, Lipid in chips: A brief review of liposomes formation by microfluidics. *International Journal of Nanomedicine*, 2021. **16**: p. 7391.

66. Fabozzi, A., et al., Polymer based nanoparticles for biomedical applications by microfluidic techniques: From design to biological evaluation. *Polymer Chemistry*, 2021. **12**(46): pp. 6667–6687.

67. Nguyen, N.-T., X. Huang, and T.K. Chuan, MEMS-micropumps: A review. *Journal of Fluids Engineering*, 2002. **124**(2): pp. 384–392.

68. Riahi, R., et al., Microfluidics for advanced drug delivery systems. *Current Opinion in Chemical Engineering*, 2015. **7**: pp. 101–112.

69. Lee, H.J., et al., MEMS devices for drug delivery. *Advanced Drug Delivery Reviews*, 2018. **128**: pp. 132–147.

70. Mansoor, I., et al., Microneedle-based vaccine delivery: Review of an emerging technology. *AAPS PharmSciTech*, 2022. **23**(4): pp. 1–12.

71. Sun, J., A.R. Warden, and X. Ding, Recent advances in microfluidics for drug screening. *Biomicrofluidics*, 2019. **13**(6): p. 061503.

72. Zhai, J., et al., Cell-based drug screening on microfluidics. *TrAC Trends in Analytical Chemistry*, 2019. **117**: pp. 231–241.

73. Tsui, J.H., et al., Microfluidics-assisted in-vitro drug screening and carrier production. *Advanced Drug Delivery Reviews*, 2013. **65**(11): pp. 1575–1588.

74. Zhang, B., et al., Advances in organ-on-a-chip engineering. *Nature Reviews Materials*, 2018. 3(8): pp. 257–278.

8 Biosensors for Diagnosis

Victor C. Diculescu, Madalina M. Barsan, and Teodor A. Enache
National Institute of Materials Physics

CONTENTS

8.1 INTRODUCTION

The recent advances in biosensor technologies enabled the development of point-of-care (POC) testing devices for health management, capable of monitoring target analytes for rapid and early disease diagnosis, as well as for the treatment/therapy efficiency of individual patients. The developments in biosensor technologies and data analytics will soon ensure a better disease management and clinical treatment.

DEFINITION. According to *IUPAC Gold Book* a biosensor is "a device that uses specific biochemical reactions mediated by isolated enzymes, immunosystems, tissues, organelles or whole cells to detect chemical compounds usually by electrical, thermal or optical signals" (Nagel et al., 1992). Generally, a biosensor contains two connected components: a (bio)chemical recognition system (receptor) and a physicochemical transducer. The biosensor transforms the chemical information from the interaction between the analyte and the bioreceptor into a useful analytical, measurable signal transferred by the transducer to a measuring device. Biosensors can be classified depending on their biorecognition element and the target as well as considering their detection principle, as depicted in Figure 8.1.

This chapter summarizes current developments, challenges and future directions for the use of biosensors in the diagnosis and monitoring of various diseases through the detection/identification of their related biomarkers. Biomarkers are molecules, which are intermediates and/or products of metabolism, and include fragments associated with energy storage and utilization, precursors to proteins and carbohydrates, regulators of gene expression and signalling molecules, most representing the downstream outputs of global cellular processes. Their detection can be selectively carried out through bioaffinity interaction with specific recognition elements such as nucleic acids, proteins or cells at the biosensor surface. This chapter focuses on the recent developments in the detection methodology and underlying applicability of DNA-, protein- and cell-based biosensors for disease monitoring and progression.

8.2 DNA-BASED BIOSENSORS

Apart from its biological function, DNA has also been identified as a nanomaterial for biosensors development due to some properties among which biocompatibility, high stability in mild conditions, specificity given not only by the strict base-pairing interaction but mostly because of the special recognition of some target (bio)molecules due to its programmability that allows the construction of different structures by artificial control (Wang et al., 2022). DNA biosensors have been used as diagnostic tools for the detection and identification of biomarkers such as small molecules, other nucleic acids or proteins, and some of the most recent advances are presented below.

8.2.1 SMALL MOLECULES

Metabolomics describes the "quantitative measurement of time-related multiparametric metabolic responses of multicellular systems to pathophysiological stimuli or genetic modification" (Davis et al., 2011; Nicholson et al., 2008). Contrary to *Genomics* and *Proteomics*, which focus on upstream gene and protein products, metabolomics involves the characterization of patterns of metabolites (since changes in metabolism result in alterations of the

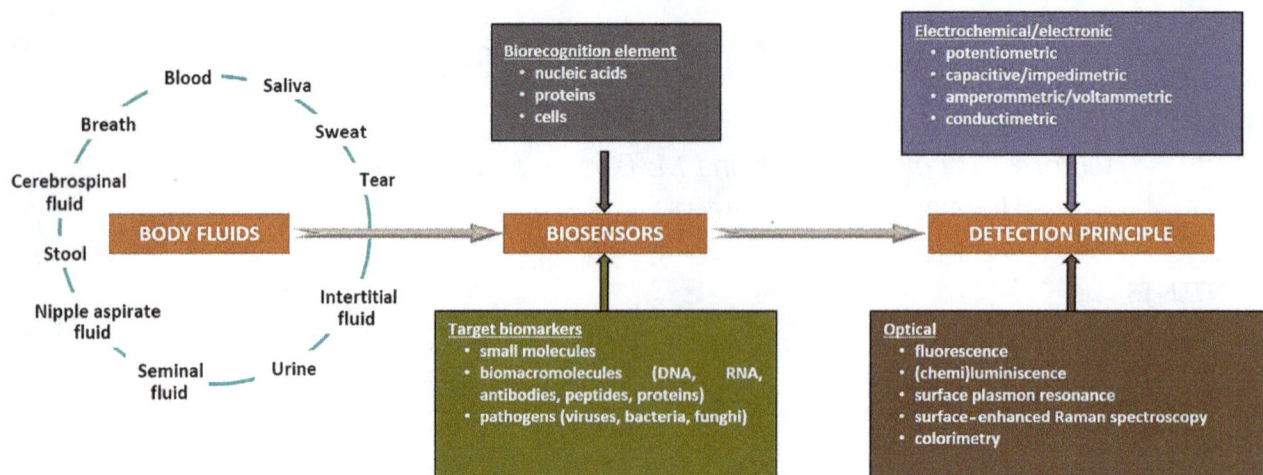

FIGURE 8.1   Schematic representation of the use of biosensors for the detection of various targets extracted from body fluids and their classification considering the detection principle.

abundance of groups of metabolites) representing the true cellular phenotype.

The necessity of identifying metabolites in tissue extracts and bodily fluids as a reflection of overall health status had led to the development of different analytical assays and procedures based on separation approaches (liquid or gas chromatography, capillary electrophoresis), magnetic resonance technologies (NMR and magnetic resonance spectroscopy) and immunoassays (Whiley and Legido-Quigley, 2011), among others.

DNA-based biosensors were largely applied to the detection of small molecules, especially after the invention of SELEX (systematic evolution of ligands by exponential enrichment), which entails the selection of DNA or RNA oligonucleotides, denominated as aptamers that have the ability to rearrange into a three-dimensional unique structure for highly specific binding to their particular targets ranging from small molecules to even a whole cell (Alkhamis et al., 2019). When compared to other biorecognition elements such as antibodies, there are some advantages related first to higher stability, reduced fabrication cost and ease of chemical synthesis but also to the amenability to chemical modification for sensing purposes, or tunable affinity and specificity (Dunn et al., 2017). Most of the sensing strategies of small molecules with aptamers are based on optical (Hu et al., 2018; Kim and Yoo, 2021) and electrochemical methods (Amor-Gutiérrez et al., 2020; Hashem et al., 2021; Jin et al., 2018a), although field-effect transistors (FET) have been also applied for this purpose (Hwang et al., 2021). The majority of the proposed architectures rely on aptamers labelled with chemical probes such as a fluorophore-quencher pair (Zhao et al., 2021) or redox-active molecule (Amor-Gutiérrez et al., 2020). The binding of the target small molecule to the aptamer results in a conformational change that is then transduced into a measurable signal due to changes in the environment of the chemical probe. Another strategy involves label-free sensing which employs a fully folded aptamer in combination with a signalling

probe such as redox-active (Jin et al., 2018b) and dye molecules (Deore et al., 2019), enzyme (Canoura et al., 2018; Lin et al., 2016) or the target itself (Zhang et al., 2019). Some of the target molecules are nucleic acids bases, their nucleosides and nucleotides, neurotransmitters and saccharides, among others.

Beyond its remarkable genome editing ability, the CRISPR/Cas9 effector has also been utilized in biosensing applications. Several CRISPR/Cas systems have been established for detecting various targets, including small molecules (Liang et al., 2019; Mahas et al., 2022). In general, such biosensing systems make use of bacterial allosteric transcription factors (aTF) in combination with CRISPR-Cas12a. In this architecture, a functional dsDNA strand bound to the specific aTF domain is released in the presence of the target molecule. The liberated dsDNA then binds to the Cas12a–crRNA complex, activating the nonspecific ssDNA substrate cleavage activity of Cas12a, which cleaves a fluorophore quencher (FQ), thus providing the analytical signal of the device, Figure 8.2.

For electrochemical detection, redox-reactive molecules conjugated to an ssDNA or ssRNA substrate tethered to an electrode surface served as detection reporters (Bruch et al., 2019; Dai et al., 2019). On another hand, the high affinity of aptamers for the target molecules was also used in synergy with the CRISPR/Cas systems. In this configuration, the aptamer is immobilized at the surface of the electrode and hybridized with the CRISPR nucleic acid substrate, which is released upon target binding by the aptamer and then cleaved by the Cas system (Niu et al., 2021). Different kinds of small molecules involved in the diagnosis of metabolic diseases were tested, among which uric and p-hydroxybenzoic acids (Liang et al., 2019), adenosine-5′-triphosphate (Niu et al., 2021) or Na^+ (Xiong et al., 2020).

Nonetheless, the ability of CRISPR/Cas systems was also used for the detection of other biomarkers including proteins and miRNA.

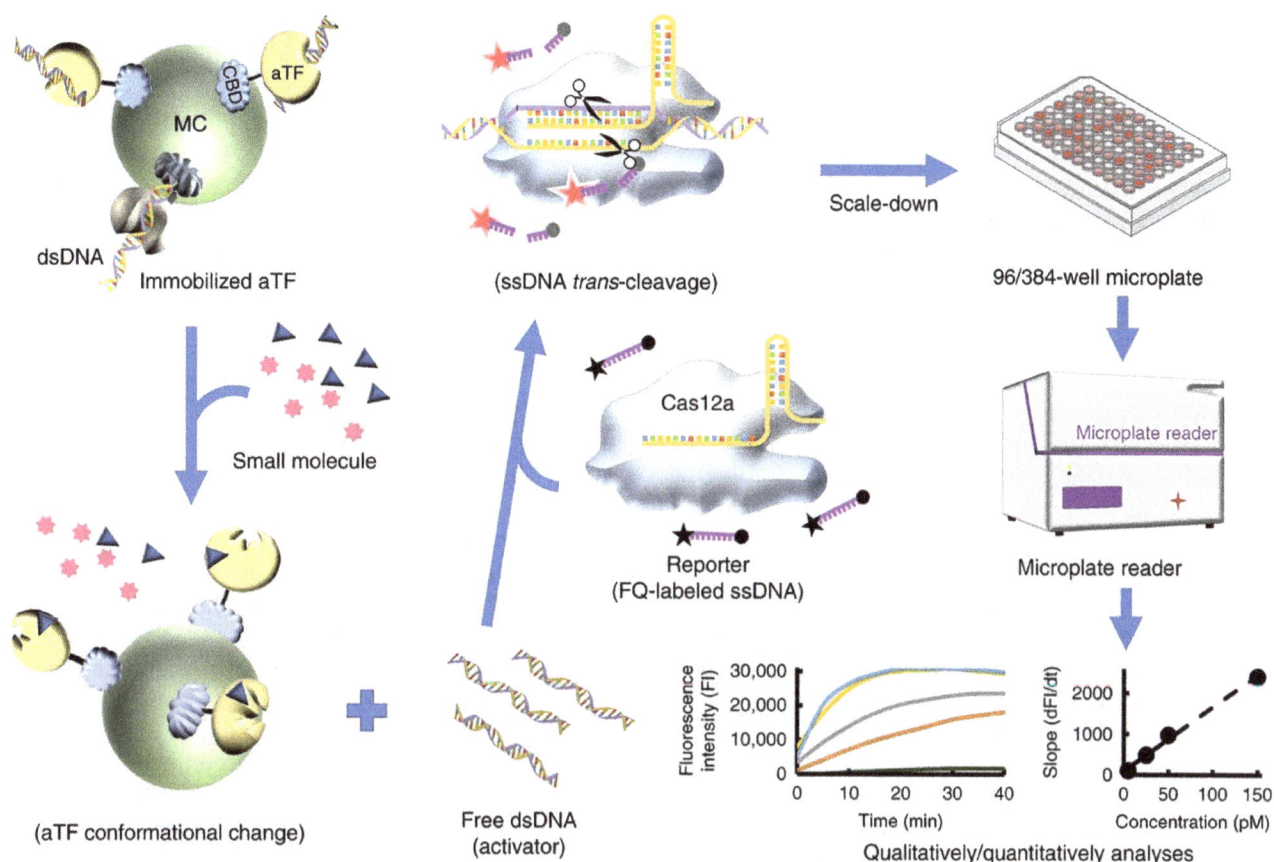

FIGURE 8.2 Schematic diagram of a biosensing platform based on CRISPR/Cas systems for the highly sensitive detection of diverse small molecules. MC microcrystalline cellulose, aTF allosteric transcription factor, CBD cellulose-binding domain, dsDNA double-stranded DNA, FQ-labelled ssDNA fluorophore-quencher-labelled single-stranded DNA. (Reprint with permission from Liang et al., 2019. Copyright American Chemical Society.)

8.2.2 NUCLEIC ACIDS

Genomics is the research domain that describes the structure, function, evolution, mapping and editing of genomes, which is the complete set of an organism's DNA. The DNA is constantly under attack of endo- and exogenous factors, which frequently leads to DNA damages such as single and double strand breaks, base alkylation or base oxidation, among others (Chatterjee and Walker, 2017). From this point of view, certain sequences of DNA are considered common biomarkers for various diseases such as cancer, neurodegenerative and infectious diseases (Gilboa et al., 2020).

DNA biomarkers include insertions, deletions, translocations, single nucleotide polymorphisms (SNPs) and short tandem repeats, among other types (Sui et al., 2020). A DNA biomarker can be measured directly in cells or as cell-free DNA after its release into the bloodstream upon apoptosis (Rahat et al., 2020). DNA detection has an outstanding potential in molecular diagnostics of the above-mentioned medical disorders. Among the most useful modern technologies are polymerase chain reaction (PCR) and its derivatives, gel and capillary electrophoresis (CE), DNA microarrays, fluorescent in situ hybridization (FISH) and Southern blot (Kolpashchikov and Gerasimova, 2013).

Nonetheless, continuing efforts have been made to seek ideal tools for fast, sensitive, low-cost and easy-to-use detection of nucleic acids.

DNA-based biosensors for the recognition of nucleic acids represent a rapid alternative to those technologies above described (Hua et al., 2022). A DNA biosensor for the detection of nucleic acids consists of a probe which is generally a short synthetic oligonucleotide of known sequence immobilized at the surface of an electrode, specially designed to hybridize with the target nucleic acid in cell or upon its release into the bloodstream.

The possibility to synthesize DNA strands with programmed sequences allowed also various particular modifications for improved transduction of the hybridization processes. Two different strategies for designing the DNA biosensors capable to detect nucleic acid sequences can be identified.

i. The first one is the *amplification-free strategies* (Wang et al., 2022) in which the transduction event is enhanced with the aid of nanostructured materials or their hybrid structures with various types of labels and/or markers. For example, a FET biosensor based on low-dimensional chemical vapour deposition (CVD)-grown monolayer of MoS$_2$ films decorated with gold nanoparticles (NPs) for probe DNA immobilization were applied for the non-invasive prenatal testing for trisomy 21 (Down) syndrome, Figure 8.3A (Liu et al., 2019). The fabricated DNA-FET biosensor reliably detected target DNA fragments (chromosome 21 or 13) with a detection limit below 100 aM, or 300 DNA molecules. Graphene was also utilized for the development of FET DNA biosensing devices (Figure 8.3B). The modification of graphene channel with a poly-L-lysine layer allowed for viral RNA detection to a limit of 1 fM and an entire detection time within 20 min using 2 µL of human serum and throat swab samples (Gao et al., 2022).

A different strategy involved an electrochemical genosensor based on gold nanoparticles coated with probe DNA strands functionalized with copper-porphyrins, in which an increased density of redox labels on the electrode surface compared to sensors without nanoparticles allowed the detection of only 23 DNA molecules, approaching single molecule detection desiderate (Kaur et al., 2018). Also, hybrid nanostructured systems of carbon nanotubes, gold nanoparticles and polymers were employed for DNA discrimination in the fM range (Han et al., 2020). Layer-by-layer films of chitosan and carbon nanotubes were used for the immobilization of DNA probes corresponding to human papillomavirus 16 (HPV16). This architecture was investigated for the detection of HPV16 through electrochemical impedance spectroscopy, with LOD = 18.5 pM (Soares et al., 2020).

Circulating tumour DNA (ctDNA) that includes DNA mutations, epigenetic alterations and other forms of tumour-specific abnormalities is a promising biomarker for noninvasive cancer assessment. Peptide nucleic acids, in which the deoxyribose phosphate backbone is replaced by a pseudo-peptide polymer to which the nucleobases are linked, represent a powerful tool for biosensing applications (Pellestor and Paulasova, 2004). A dual biomarker detection platform based on peptide nucleic acids probe-gold nanoparticles hybrid material in synergy with lead phosphate apoferritin (LPA)-based was developed to quantify ctDNA by detection of tumour-specific mutations and methylation of PIK3CA gene (Cai et al., 2018). The proposed DNA biosensor yielded a detection limit of 10 fM and was successfully applied for the detection of ctDNA collected from cancer patient serum.

In its turn, optical biosensing was also explored for signal amplification strategies that convert a single molecular recognition event into a response equivalent to hundreds of fluorescent dyes (Krishnan et al., 2019). Forster resonance energy transfer (FRET) was applied with 40 nm dye-loaded poly(methyl methacrylate-co-methacrylic acid) (PMMA-MA) NPs functionalized with oligonucleotides specific to survivin cancer marker and allowed a detection limit of 0.25 pM, and ~23 hybridization events at the surface of the single NP-probe (Melnychuk and Klymchenko, 2018). Another technique that allows similar detection limit values is based on localized surface plasmon resonance (LSPR) sensing transduction of hybridization events of probe oligonucleotides immobilized on gold nanoislands with RNA target of SARS-CoV2 virus (Qiu et al., 2020).

ii. The second strategy relies on the *amplification* of the probe nucleic acid sequences. One alternative is the well-known PCR and its derivatives such as real-time quantitative PCR (RT-qPCR). A PCR-assisted impedimetric biosensor was developed for the detection of the clbN gene responsible for the production of colibactin, a harmful genotoxin that has been associated with colorectal cancer, in *E. coli* samples. The protocol involved the immobilization of forward primers on Au electrodes followed by the PCR cycling and target DNA identification by monitoring changes in the system's charge transfer resistance values with a LOD = 17 ng/mL (Solis-Marcano et al., 2021). Also, a paper-based lateral flow nucleic acid (LFNA) test platform was developed for visual detection of the Epstein-Barr virus (EBV) nucleic acids. The system is based on capture probe (CP)/gold nanoparticles (AuNPS) and silicon dioxide (SiO2) (AuNPS@SiO2) nanospheres/target DNA/avidin complexes as the sensing platform and allowed a limit of detection of 50 nM (Chen et al., 2022). Paper-based assays that allow the extraction and purification of RNA directly from human clinical nasopharyngeal specimens through a poly(ether sulfone) paper matrix and the *in situ* amplification of the RNA directly within the same paper matrix have been also reported (Kolluri et al., 2021; Rodriguez et al., 2015). Other versatile and cost-effective devices involve polydimethylsiloxane (PDMS)/paper hybrid microfluidic biosensors for the rapid, sensitive and instrument-free detection of the main meningitis-causing bacteria (Dou et al., 2014), Figure 8.3C, or paper-origami device with multiplexing through five channels to enable the detection of several target pathogens/virus at the same time (Yang et al., 2018), Figure 8.3D. However, the biggest limitation of PCR in practical biosensing applications is the temperature control, which must be cycled between approx. 95 and 63–65°C in a fast and reproducible manner. Besides, most of the publications that deal with the development of new analytical biosensing strategies and assays for the identification of

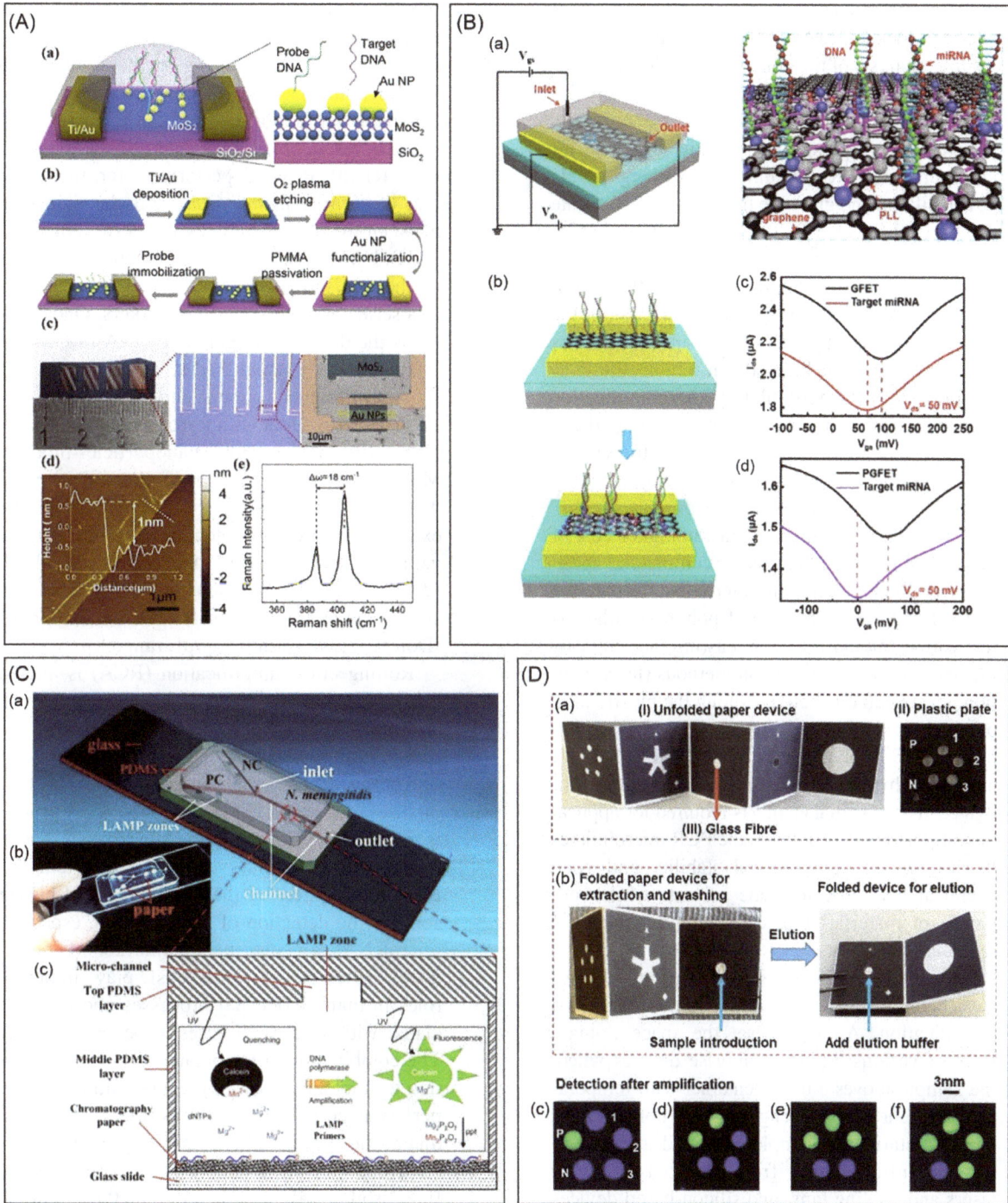

FIGURE 8.3 (A) Material characterization and device fabrication of the MoS₂ FET-based biosensor. (a) The structure diagram and (b) fabrication process flow diagram. (c) A typical fabricated chip. (d) The AFM image of MoS2 indicates that the height of MoS₂ is ca. 0.8 nm. (e) Raman spectroscopy of pristine MoS₂ with peaks at 386.1 and 404.5 cm⁻¹, tested at a 488 nm excitation wavelength. (Adapted with permission from Liu et al., 2019. Copyright American Chemical Society.) (B) (a) PGFET biosensor schematic. (b) Schematic principles of GFET and PGFET for miRNA detection. Comparison of miRNA detection results between GFET (c) and PGFET (d) biosensor. (Reprint with permission from Gao et al., 2022. Copyright American Chemical Society.) (C) Layout of a PDMS/paper hybrid microfluidic device. (a) 3D illustration of the schematic of the chip layout. (b) A photograph of the hybrid microfluidic device for infectious disease diagnosis. (c) A cross-section view of the LAMP zone illustrating the principle of the LAMP detection. (Reprint with permission from Dou et al., 2014. Copyright American Chemical Society.) (D) Design of the paper device for the detection of three targets with internal positive (P) and negative control (N). (a) The three components of the device. (b) The sample was introduced into the glass fibre and DNA was extracted, followed by washing and eluting to the reaction chamber of the plastic plate for LAMP reaction. (c–f) The green emission occurs in the presence of pyrophosphate under UV excitation: (c) blank sample; 1: Leptospira; 2: Brucella; 3: BoHV-1. (Adapted with permission from Yang et al., 2018. Copyright American Chemical Society.)

PCR amplified nucleic acids rely on classic PCR heating blocks usually followed by the *ex-situ* or *off-line* analysis of the collected sample.

Microelectromechanical (MEMS) technologies and semiconductor industry are currently integrated to develop microfluidic systems for miniaturized PCR instruments (Ahrberg et al., 2016). Silicon and glass are the most used substrate materials of the chip-PCR reaction due to the convenience in using standard photolithography and chemical etching techniques to produce the microfluidic channels, reservoirs and valves. Besides the benefits provided by the silicon substrate, it has been shown that the bare silicon inhibits the PCR reaction by reducing/inhibiting the amplification efficiency, while thermal insulation of the silicon substrate is usually needed to reduce the energy losses. All these disadvantages were eliminated by using glass substrates, which become an alternative for PCR microfluidics, but the higher cost of fabrication hindered their use in commercial applications. Nonetheless, an alternative fabrication method is given by the use of polymers either by replication (mould injection, casting, hot embossing, etc.) or direct fabrication methods (laser ablation, plasma etching, etc.). Among the mostly used polymers, polydimethylsiloxane (PDMS), polycarbonate (PC), polymethylmethacrylate (PMMA), and polyethylene terephathalate (PET) are highlighted. The essential elements required for application of polymer substrates to the PCR microfluidics fabrication are high thermal stability and good chemical resistance to some solvents while consisting of materials that cannot inhibit the PCR reaction.

Microfluidic PCR devices of varying designs have been developed for effective and fast DNA amplification. Among these, the space domain PCR of various designs comprise devices where the sample moves along a channel with temperature gradient along its length or time domain PCR where a stationary sample is heated and cooled through infrared light from tungsten lamps or lasers and cooling fans, and silicon-based devices through Peltier and Joule effect. The miniaturization of PCR devices offers the opportunity to shorten the amplification times, enhance sample throughput and minimize human intervention widening the perspective of POC diagnostic tools. The elimination of the temperature cycles during PCR represents a significant simplification by removing the requirements of heating and cooling systems. This can be achieved by isothermal nucleic acid amplification methods.

Loop-mediated isothermal amplification (LAMP) technique was developed to increase amplification efficacy in terms of sensitivity and specificity (Dou et al., 2014; Foo et al., 2020) by using a higher number of primers when compared to PCR, while the working temperature for the amplification reaction is fixed at approx. 65°C. Comparative investigations of LAMP assays with six primers coupled to a lateral-flow biosensor with RT-qPCR were performed for the detection of *Bordetella pertussis*, and the agreement between the two techniques was 98% (Sun et al., 2022). LAMP technique was combined with electrochemical biosensing systems for the detection of a series of nucleic acid biomarkers. One example is the detection of genomic DNA of Group B *Streptococci* by using LAMP with a set of four specially designed primers, and a dual-signal electrochemical readout with ferrocene and methylene blue indicators on gold nanoparticles-modified MoS_2 electrode surface, which allowed a limit of detection of 0.23 fg/μL (Fu et al., 2020). Other examples involve the detection of genes of oncogenic human papillomavirus (Izadi et al., 2021), *Mycobacterium avium* subspecies paratuberculosis (Chand et al., 2018) or *Neisseria meningitides* (Dou et al., 2014), in the fg/μL range.

Rolling circle amplification (RCA) is another isothermal process that is usually carried out at approx. 30°C and where a short nucleic acid primer is amplified to a long single-stranded DNA or RNA using a circular template and special polymerases (Ali et al., 2014). Thus, the RCA product is a concatemer that contains hundreds of tandem repeats complementary to the circular template. RCA was applied for the development of biosensors for the detection of nucleic acids, especially microRNAs, among others biomarkers (Gu et al., 2018). MicroRNAs (miRNAs) play important roles in many biological processes and are associated with various diseases, especially cancers. The combination of nanomaterials and amplification technologies showed great potential for high-performance detection of miRNAs in molecular diagnostic systems. RCA was integrated with the catalysed hairpin assembly (CHA), an enzyme-free, isothermal amplification method, into an electrochemical biosensor for the detection of microRNA-21 (S. Wang et al., 2019). This system involves a hairpin DNA probe capable to hybridize with the microRNA-21 target, which competes with a second hairpin structure that contains a G-quadruplex sequence amplified by the RCA method. In this report, hemin is used as an electroactive reporter molecule, which binds to the G-quadruplex structure allowing a detection limit of 13.5 fM microRNA-21.

The development of a programmable and universal sensing platform for the detection of clinically relevant molecules for disease monitoring and diagnosis remains a challenge but research toward this desiderate led to the combination of RCA with CRISPR)/Cas systems. An immobilization-free electrochemical biosensing platform for sensitive and specific detection of disease-related nucleic acids was developed (Qing et al., 2021). In this strategy, a universal blocker probe (BP) is designed as the cleavage substrate for the Cas12a enzymes. A universal reporter probe (RP) labelled with methylene blue is designed to hybridize with the BP. Typically, in the absence of targets, Cas12a enzymes are silent, and the duplex between BP and RP is electrochemically inactive at the surface of a reduced graphene oxide-modified electrode (rGO/GCE). Contrary, upon recognition of target amplified through RCA, the cleavage activity of Cas12a enzymes will be activated, resulting in the digestion of the BP. In these conditions, the free single-stranded RP labelled with methylene blue is adsorbed at the electrode surface leading to an increased electrochemical signal. Nonetheless, by tuning the sequence for target recognition in RCA components, this strategy was applied to the detection of microRNAs and DNA of *parvovirus* B19 with limits of detection of 0.83 aM and 0.52 aM, respectively (Qing et al., 2021).

8.2.3 Proteins

Proteomics is the research domain that involves the analysis of the entire protein complement of a cell, tissue or organism under a specific set of conditions (Yu et al., 2010). Proteomics is a complex research domain since the protein profile differs from cell to cell and even from time to time. For example, through a variety of mechanisms, cancer cells provide the protein biomarker material for their own detection. Protein biomarkers are molecular indicators of biological status and can be assayed to evaluate the presence of cancer and therapeutic interventions. Protein biomarkers

may be detectable in the blood, other body fluids, or tissues. In cancer diagnostics, patients can be tested for individual protein biomarkers that detect cancers at very early stages and be used to monitor cancer progression or remission during therapy. Table 8.1 summarizes the main protein biomarkers used for cancer detection and monitoring.

There are multiple analysis methods (mass spectrometry, electrophoresis, immunoassays among others) for investigating proteins, but their detection and identification in complex biological samples are probably one of the most important and difficult issues. Among different biorecognition elements, aptamers (see Section 8.2.1) can be employed as high-affinity tools for biomarker detection in a series of medical anomalies including cancer. Aptasensing of cancer can detect tumour biomarkers by measuring the changes of biomarkers level in the tissue histogenesis and blood. Also, it gives complementary information such as cell differentiation, cell functionalization and the emergence of cancer cells by the qualitative and quantitative study (Solhi and Hasanzadeh, 2020).

Label-free electrochemical aptasensors are analytical tools that use an aptamer as the biorecognition factor combined with an electrochemical transducer surface to create measured electrical signals due to the modification of the interfacial characteristics of the aptasensor surface upon aptamer-analyte binding. On the other hand, the labelled aptasensors are usually in a sandwich-type configuration, where the target molecule is captured between two aptamers, or between an aptamer and an antibody that binds different regions of the target. In this case, the second recognition element is labelled. The first electrochemical aptasensor was reported in 2004 (Ikebukuro et al., 2004), and since then, aptamer-based electrochemical devices to detect cancer biomarkers are in a continuous development. They can be classified depending on the targeted biomarker, transduction element and field of application, among others (Díaz-Fernández et al., 2020).

Most aptasensors use the thiol-Au chemistry to anchor the aptamer on the bulk metal/metal oxide transducer surface

TABLE 8.1

Protein Biomarkers Used in Cancer Detection and Monitoring (Solhi and Hasanzadeh, 2020)

Cancer Type	General Biomarkers Used for Cancer Detection
Breast	BRCA1, BRCA2, CA 15-3, CA 125, CA 27.29, CEA, NY-BR-1, ING-1, HER2/NEU, ER/PR
Prostate	PSA
Ovarian	CA 125, HCG, p53, CEA, CA 549, CASA, CA 19-9, CA 15-3, MCA, MOV-1, TAG72
Liver	AFP, CEA
Lung	CEA, CA 19-9, SCC, NSE, NY-ESO-1
Colon	CEA, EGFR, p53
Melanoma	Tyrosinase, NY-ESO-1
Gastric carcinoma	CA72-4, CEA, CA19-9
Oesophagus carcinoma	SCC
Trophoblastic	SCC, hCG
Bladder	BTA, BAT, FDP, NMP22, HA, HAase, BLCA-4, CYFRA21-1
Leukaemia	BCR, ABL, PML, BCL1, BCL2, ETO

(Rahi et al., 2016), but alternative immobilization strategies on carbon (Tertis et al., 2019) or paper-based materials (Y. Wang et al., 2019) are also developed through the use of metal nanoparticles (Paniagua et al., 2019). For example, a highly sensitive and selective aptasensor for quantitative detection of interleukin-6 was developed by using a glassy carbon electrode modified with p-aminobenzoic acid, p-aminothiophenol and gold nanoparticles. A thio-terminated aptamer specific for interleukin-6 was immobilized on the surface of the modified electrode via the formation of gold-sulphur bonds allowing a detection limit of 1.6 pg/mL, within the range of physiological concentration of the protein in blood samples collected from patients suffering of colorectal cancer (Tertis et al., 2019).

Simultaneous detection of multiple tumour biomarkers is desired to facilitate early diagnosis of cancer, hence providing a scientific reference for clinical treatment (Y. Wang et al., 2019). A multi-parameter paper-based electrochemical aptasensor for simultaneous detection of carcinoembryonic antigen (CEA) and neuron-specific enolase (NSE) was developed and tested in clinical samples (Y. Wang et al., 2019). The device was fabricated through wax and screen-printing and made use of amino functional graphene (NG)-thionin (THI)-gold nanoparticles (AuNPs) and prussian blue (PB)-poly (3,4-ethylenedioxythiophene) (PEDOT)-AuNPs nanocomposites to modify the working electrodes for immobilization of the CEA and NSE aptamers. The limits of detection were 2 pg/mL for CEA and 10 pg/mL for NSE, and these devices were evaluated in clinical samples. Also, a paper-based electrochemical biosensor with metal-organic framework and aptamer functionalization was developed for the detection of exosome, as noninvasive biomarker of cancer.

Magnetic nanomaterials have been utilized in synergy with aptamers for the oriented immobilization of proteins (Zhu et al., 2022) and also for their detection (Kiplagat et al., 2020). For example, avidin-modified $Fe_3O_4@SiO_2$ NanoCaptors® were used (Paniagua et al., 2019). In the sensors' design, horseradish peroxidase was immobilized onto the SiO_2 side of the Au-SiO_2 Janus nanoparticles while an anti-CEA aptamer with thiol and biotin modifications onto the Au side. Upon interaction with CEA, the complex was captured by the avidin-modified magnetic $Fe_3O_4@SiO_2$ NanoCaptors® allowing further magnetic deposition on carbon screen-printed electrodes for the amperometric detection of the cancer biomarker with a detection limit of 210 pg/mL (1.2 pM).

Electrochemistry was combined with surface plasmon resonance (SPR) to examine the aptamer-based detection kinetics of CEA with the aid of silver nanoclusters embedded in zirconium metal-organic framework (Guo et al., 2017) which allowed detection limits of 4.93 pg/mL and 0.3 ng/mL, respectively.

Electrochemical aptasensors for FDA-approved cancer biomarkers detection are reviewed in (Díaz-Fernández et al., 2020) and monitor biomarkers such as alpha-fetoprotein (AFP), PSA, carcinoembryonic antigen (CEA) and carbohydrate antigen 125 (CA125), among others.

Optical aptasensors can be classified based on their luminescence changes and light absorption as a result of interaction with different analytes and can be classified based on the different optical detection methods: fluorescence, chemiluminescence, SPR, surface-enhanced Raman scattering (SERS), colorimetric, etc. (Zahra et al., 2021). Various optical aptamers for cancer biomarkers detection have been reviewed in (Zahra et al., 2021) and include the detection of PSA, MCF-7, CA125, CEA, HER2 and MUC1 among others.

SERS has emerged as a promising and sensitive spectroscopic technique for biomarkers detection. An aptamer-based SERS sensor was developed with DNA linkers, which were complementary to the aptamer sequence, immobilized onto the surface of gold-silver-silver core-shell-shell nanostructures (GSSNT) (Ning et al., 2020). The capture probes were prepared by immobilizing specific aptamers of the target biomarker on magnetic beads (MBs). In the absence of the target biomarker, SERS detection probes were coupled with MBs via specific linker DNA-aptamer hybridization. In the presence of the target biomarker, the aptamer specifically recognized it, and GSSNTs were subsequently released into the supernatant leading to a decreased SERS signal. This sensor was applied for the simultaneous detection of protein biomarkers of multiple cancer-related exosomes such as PSMA, Her2 and AFP proteins of prostate cancer cell line (LNCaP), breast cancer cell line (SKBR3) and hepatocellular cancer cell line (HepG2).

Colorimetric aptasensors are desired for the detection of disease biomarkers, due to their simplicity, ease of use, accessibility and possibility to be integrated into POC devices due to the detection principle which consists in a simple visual colour change (Zhang and Liu, 2021). Apart from cancer biomarkers, aptasensing was largely used for the detection of neurological disorders (Erkmen et al., 2022). A gold nanoparticle-based label-free homogeneous phase colorimetric bioassay was developed for the detection of $A\beta_{1-40}$ oligomers ($A\beta O$), which are important biomarkers for diagnosis and monitoring progression of Alzheimer's disease (AD). The $A\beta O$-aptamer decorated AuNPs aggregated in a solution with high concentrations of salt and displayed a purple colour, characteristic to aggregated AuNPs. In the presence of $A\beta O$, the aptamer folded and stabilized AuNPs towards the salt-induced aggregation, and a hypsochromic shift was achieved. The proposed aptasensor was successfully applied for the detection of $A\beta O$ with a dynamic range of 1–600 nM and a low detection limit of 0.56 nM.

Another biomarker of AD is the p-tau231, which has emerged as a highly specific pathological biomarker of AD that remains normal in other dementia diseases. A dual, fluorescent and colorimetric biosensor for p-tau231 biomarker using fluorescent nitrogen-doped carbon dots (NCDs), gold nanoparticle (AuNPs) and specific aptamer of p-tau231 was proposed (Phan and Cho, 2022). The aptamer is immobilized on the surface of NCDs, causing fluorescence quenching. In the presence of p-tau231, the aptamer specifically binds to p-tau231 and dissociates from the NCDs, leading to the recovery of NCD fluorescence, which could be used for the quantification of p-tau231. For the visual detection of p-tau231, a colorimetric Cu-enhanced-Au aptablot has been exploited to improve the colorimetric intensity through the enlargement of the size of AuNPs and the transformation of the particle shape, becoming visible to the naked eye for the detection of p-tau231.

8.3 PROTEIN-BASED BIOSENSORS

Most protein-based biosensors rely either on enzymes for investigation of their activity and detection of small molecules and inhibitors or on antibodies for the detection of target protein biomarkers. An immunosensor is an affinity biosensor based on the specific interactions between an antigen and an antibody, with the latter immobilized on a transducer surface as the biorecognition element. Antibodies are protein components of the immune system with high affinity for antigens, which are foreign molecules in the body, e.g. xenobiotic, virus, bacteria, etc. Generally, the synthesis of antibodies requests a host organism inoculated with the target antigen, extraction and purification. All these procedures are laborious and came with a high cost that will be found in the final price of the specific antibody. However, while for different xenobiotic small molecules there are other appropriate detection methods, for protein detection the antibody-based biosensors are the most efficient. There are different configurations of the immunosensing systems, but the most common are the forward and sandwich-type immunoassays, Figure 8.4.

Forward immunoassays offer the greatest detection sensitivity for protein biomarkers and the lowest limits of detection (Landegren and Hammond, 2021), being at the same time the simplest configuration which corresponds to the most economic option. In this configuration, the immunosensor is incubated with a mixed solution of a known amount of labelled protein biomarker and an unknown sample. The recorded signal from the labelled protein biomarker is inversely proportional to the sample analyte amount. Due to the use of the limited amount of antibody, this format is known as *"limited reagent assay"*. On the other hand, in a sandwich-type immunoassay, the immobilized antibody captures the target protein biomarker from the solution, which then will be recognized by a second labelled antibody. In contrast to assays using single affinity reagents, sandwich immune assays require pairs of antibodies to recognize the target protein, enhancing specificity.

The majority of immunosensors designed for protein biomarker detection are either electrochemical or optical and makes use of labels derived from nanostructured materials such as metal nanoparticles, quantum dots or even biomolecules, especially enzymes, facilitating the detection of binding event (Nemčeková and Labuda, 2021; Pan et al., 2017). Nonetheless, nonlabelled or label-free immunosensors are designed so that the immunocomplex (i.e., the antigen–antibody complex) is directly determined by measuring the physical changes induced by the formation of the complex (Filik and Avan, 2019). In this case, the most employed measuring methods are potentiometry, reflectometry, elipsometry, SPR and SERS.

The majority of immunosensors designed to be used as POC for protein biomarker detection found applications in the detection of biomarkers related to cancer, brain injury, cardiovascular and neurodegenerative diseases or for detection of viruses. Different cancer biomarkers (Yáñez-Sedeño et al., 2017) have been investigated, and electrochemical (Mollarasouli et al., 2019) or optical (Pirzada and Altintas, 2020) immunosensors for their detection were reported.

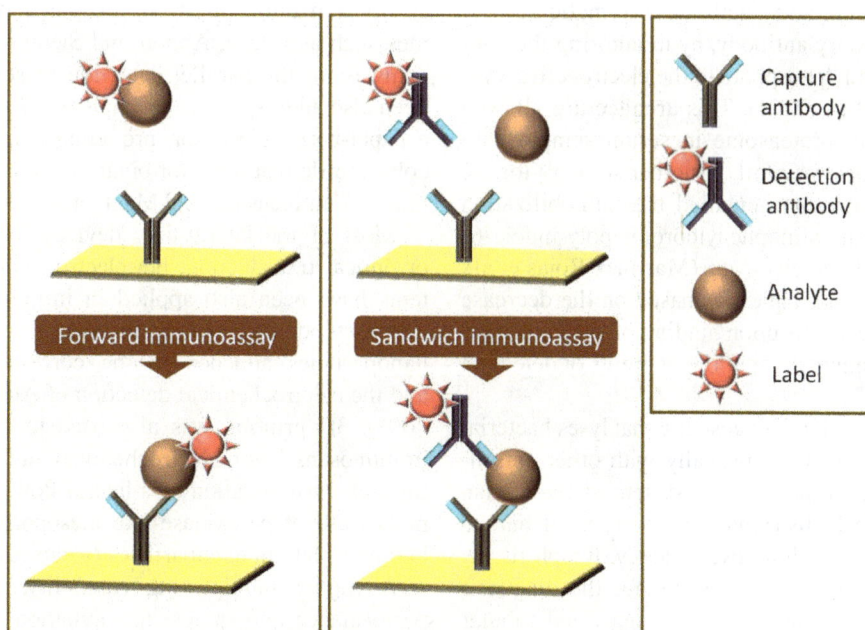

FIGURE 8.4 Schematic representation of immunosensors configuration.

Determination of brain injury biomarkers related to glial cell, axonal and neuronal injuries or due to immunological and inflammatory response (Yokobori et al., 2013) has been carried out in cerebrospinal fluid, serum and plasma through electrochemical (Pankratova et al., 2021), optical (Krausz et al., 2021) or electronic transduction by ion-selective FET (Moser et al., 2020). On their turn, myoglobin (Mb), cardiac troponins (cTn), creatine kinase MB (CK-MB) and myeloperoxidase (MPO), which are indicators of elevated risk of myocardial infarction, have been reported (Leva-Bueno et al., 2020). Alzheimer's disease (AD), Parkinson's disease (PD), transmissible spongiform encephalopathy (TSE), Huntington's disease (HD) and multiple sclerosis (MS) are mostly related to misfolded proteins in the central nervous system, such as amyloid-ß (Aß), tau, α-synuclein, huntingtin, prion and tau proteins (Karki et al., 2021; Li et al., 2020).

The ubiquitin-proteasome system (UPS) is one of the major protein degradation pathways, where abnormal UPS function has been observed in cancer and neurological diseases (Zheng et al., 2016). The circulating proteasome is a large enzyme complex released into the bloodstream upon disease progression, being approved as a protein biomarker. Its level can be correlated with the disease state while its enzymatic activity with the evolution of the drug treatment. An amperometric biosensor for the detection of proteasome was developed based on the immobilization of a capture antibody (Abβ) on Au electrodes functionalized with 4-mercaptophenylboronic acid, Figure 8.5A. The immunosensor can detect levels of proteasome in two methodologies: (1) the correlation of 20S concentration with its proteolytic activity when bound to Abβ, by employing a peptide substrate (Barsan and Diculescu, 2021; Henriques de Jesus et al., 2019), which upon proteolysis releases an electroactive marker; and (2) the correlation of 20S concentration with the enzymatic activity AlkP-marked secondary antibody, by monitoring the conversion of aminophenylphosphate to the electroactive aminophenol (Barsan et al., 2021). This architecture allowed the detection of 20S proteasome in serum sample with limits of detection below μg/mL. Another strategy for the detection of the proteasome entailed the immobilization of the antibody on an aminophenylboronic/poly-indole-6-carboxylic acid-modified electrode (Martínez-Rojas et al., 2020). The detection principle was based on the decrease of the polymer redox peaks upon binding of each constitution layer of the immunosensor allowing limits of detection of 6 ng/mL.

Lysozyme is a hydrolytic glycosidase that lyses bacterial cell membranes, acting synergistically with other antimicrobial polypeptides in the immune system of the human body. The lysozyme is distributed in a variety of human tissues and secretions such as liver, kidney, lymph tissue, tears, and saliva being a key indicator for the diagnosis and prognosis of some diseases, including renal tubular disease and acute monocytic leukaemia. A magnetoelastic immunosensor for lysozyme detection was employed by monitoring the changes in the resonance frequency under a magnetostrictive effect. The detection system is composed of a magnetoelastic chip with an immobilized lysozyme antibody, a solenoid coil and a vector network analyser, Figure 8.5B. The ME sensor was ultrasensitive to mass change, thus the frequency offset caused by mass change was used as an analytical signal to detect the content of lysozyme with LoD = 1.26 ng/mL (Huang et al., 2021).

Most immunoassays require long sample incubation/processing time, expensive test kits, an invasive acquisition process, demanding laboratory instrumentation and experienced personnel to operate and maintain the diagnostic system, which limits their accessibility and availability as frontline diagnostic tools, especially for patients in lower-income regions and countries (Gil Rosa et al., 2022). Miniaturized devices for POC immunosensing to comply with the sample rigour and detection limit still present a challenge.

Lateral-flow devices are the most well-established paper-based POC platforms. A relevant example is the actual pandemic and serological tests for COVID-19 are either based on the detection of SARS-CoV-2 immunoglobulin G and M (IgG and IgM) antibodies using enzyme-linked immunosorbent assay (ELISA) or lateral flow immunoassays (LFAs) but with limited utility for early-stage disease detection, or on antigen tests that enable the detection of viral nucleocapsid (N) and spike (S1) in the plasma of COVID-19 patients which are useful for accurate and early detection of COVID-19. There are different types of lateral-flow devices for SARS-CoV-2 virus detection (Pickering et al., 2021) based on optical, colorimetric or electrochemical detection and ingenious devices that integrate new nanostructured materials are still developed (Jia et al., 2021).

There are many commercially available POC analysis tools manufactured by renowned diagnostic companies such as Roche, Abbott and Siemens. But multiplexed analysis for the parallel detection of several markers has been also addressed. An example is a 7-segment display on a paper-based biosensor, providing compact and intuitive colorimetric read-outs for binary coded multiplexed detection of 7 antigens (Li and MacDonald, 2016).

Most of the lateral-flow devices rely on colorimetric or optical transduction, but electrochemical read-out systems have been also applied in immunosensing devices. The methodology usually involves the dissolution of Au nanoparticles attached to the corresponding biomarker and the electrochemical detection of Au^{2+} ions (Mao et al., 2008). 3D printing was also used to design a multiplex immunosensor for electrochemical multiplexed detection through customization of a lateral flow immunoassay that makes use of peroxidase-like mesoporous core-shell palladium@platium nanoparticles (Ruan et al., 2021).

A higher multiplexing capability, improved sensing performance and sample manipulation functions, such as liquid mixing, splitting and/or biological sample filtration,

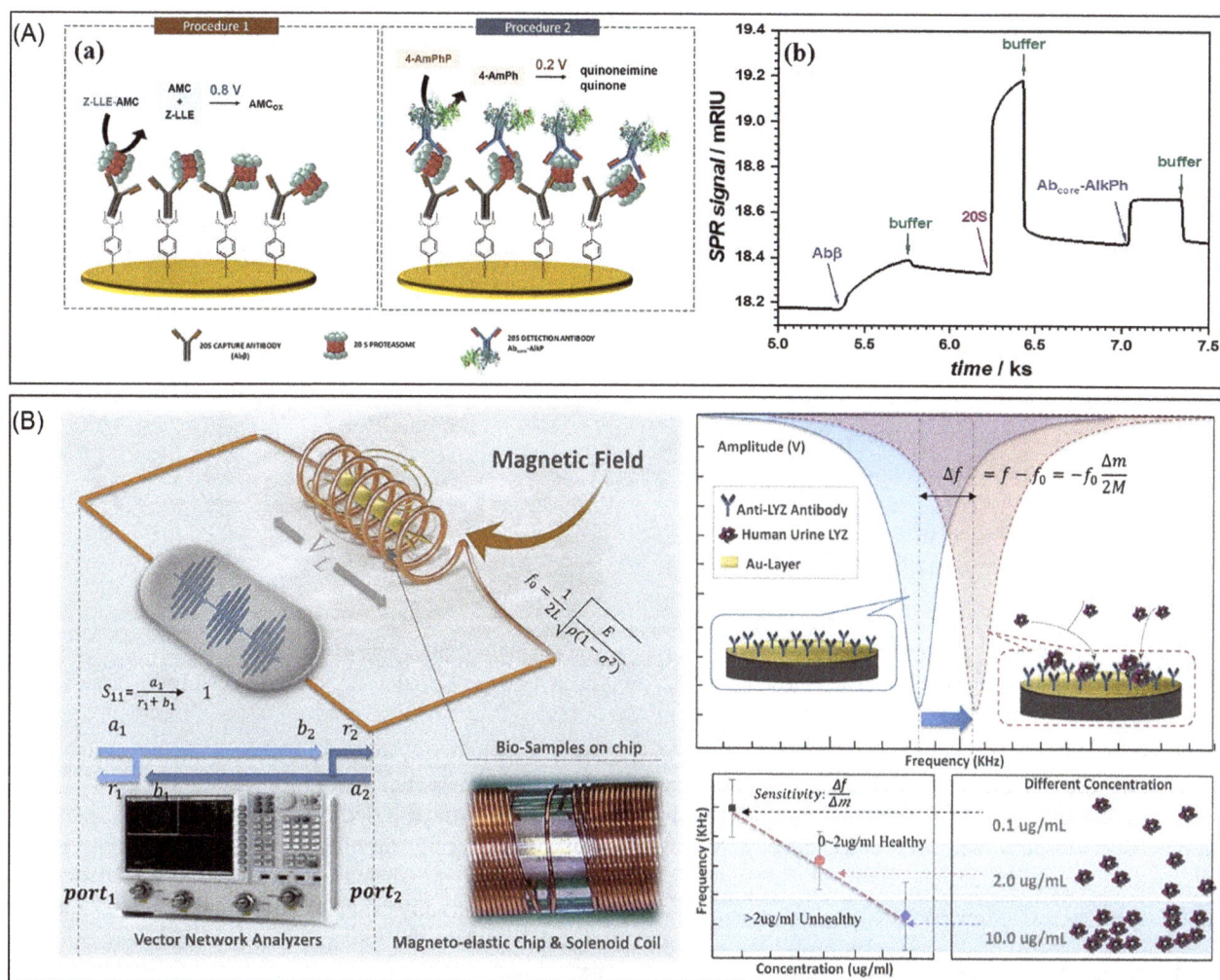

FIGURE 8.5 (A) (a) Schematic representation of the 20S proteasome detection principle through: Procedure 1 – the correlation of 20S concentration with its proteolytic activity when bound to Aβ, by employing a peptide substrate, which upon proteolysis releases an electroactive marker; and Procedure 2 – the correlation of 20S concentration with the enzymatic activity of AlkP-marked secondary antibody, by monitoring the conversion of aminophenylphosphate to the electroactive aminophenol; (b) SPR signal recorded at the Au/4-MPBA electrode after consecutive injections of Aβ antibody, 10 μg/mL 20S and Ab$_{core}$-AlkP. (Reprint from Barsan et al., 2021, under Creative Commons Attribution 4.0 International License.) (B) Schematic diagram of the lysozyme detection system and the magneto-elastic immunosensor sensing mechanism. (Reprint with permission from Huang et al., 2021. Copyright American Chemical Society.)

was achieved by using cellulose paper-based immunosensors. By using wax printing processes for microfluidic channels and reservoirs, the sample can be distributed in specific directions. The screen-printing process was used to design and produce the electrodes on paper support, which were functionalized with antibodies. Upon the addition of sample solution and (Ru(bpy)$_3^{2+}$)-labelled signal antibodies, (Ru(bpy)$_3^{2+}$)-tri-n-propylamine (TPA) electrochemiluminescence was used for the detection of all four tumour biomarkers.

8.4 CELL-BASED BIOSENSORS

The technological advance of the last decades has led to the development of a new class of biosensors, which address living cells as sensing elements. The fabrication process of the cell-based biosensors employs the immobilization/growth

of living cells in culture plates or onto transducers surfaces and are particularly designed for the detection of the intra- and extra-cellular events resulting from various stimuli.

At present, cell-based biosensors have gained interest as an alternative method of sensing because they have several advantages over traditional methods, including long-term recording in noninvasive ways, and have shown great potential for evaluating the cell interaction with pathogens and toxins, screening bioactive molecules for drug discovery or monitoring environments, as well as tools for physiological analysis of the cell. The main feature of the cell-analyte interaction is that the cellular processes initiated by the analyte lead to a series of biochemical reactions, thus allowing detection of the interaction by monitoring a single step of the reaction chain, Figure 8.6.

The detection method of cell-based biosensors differs from case to case, but the literature reports over the last

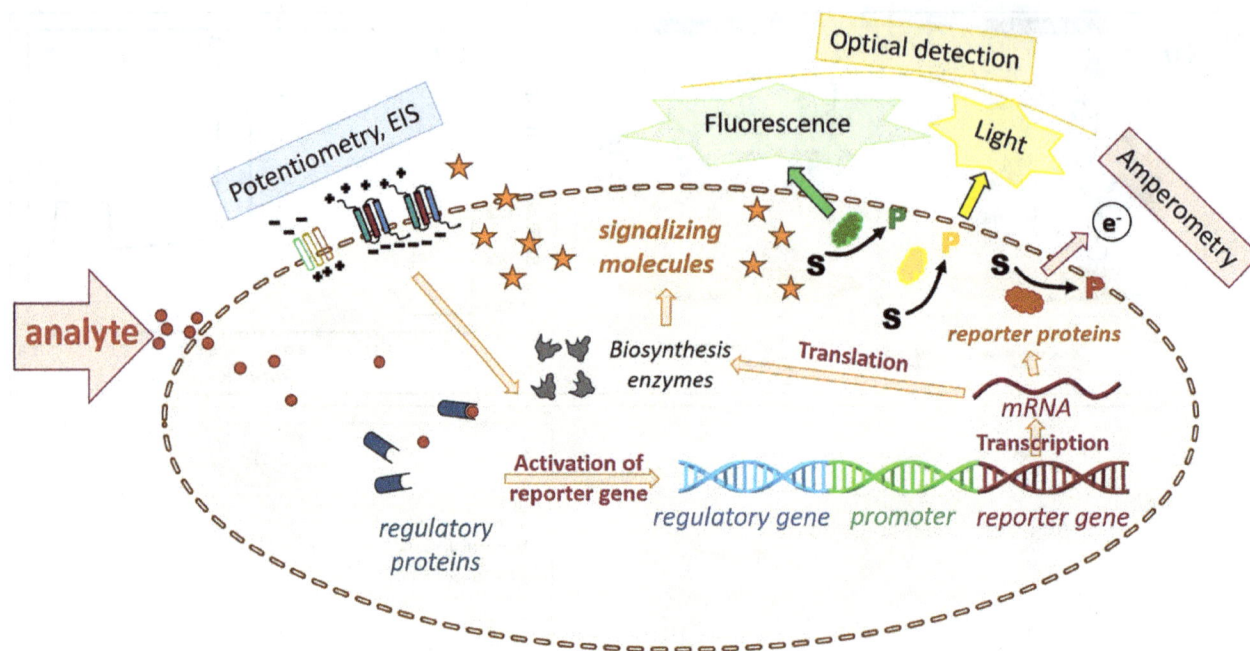

FIGURE 8.6　Scheme schematic representation of the cellular events that drive the detection mechanism on cell-based biosensors.

five years show that fluorescence detection is preferred, although the use of electrochemical or electronic biosensor is increasing. Cell-based biosensors are currently used for the detection of small molecules and their effects on the cells, pathogens and the evaluation of cellular events.

8.4.1　SMALL MOLECULES

Small molecules play different roles and affect organisms in different ways; hence, methods to sense these molecules within living cells have a wide range of applications in biology, biotechnology or medicine. The general approach of biosensing small molecules using cell-based biosensors is to take advantage of ligand-binding domains of cell membrane which bind specifically to such molecules. The applications include, but are not limited to, metabolic pathway regulation, analytical determination of metabolite concentration, environmental toxin detection or therapeutic response. Most of the biosensing strategies are based on optical detection, but electrochemical or electronic transductions are also used.

Using fluorescent proteins labelled to different intra- or extra-cellular receptors, the interaction of small molecules with living cells can be quantified by direct measurement of fluorescence intensity and correlated with a physiologic event. Changes in the cytosolic Ca^{2+} concentration allowed monitoring the relative changes in the glucose levels (Vinchhi et al., 2019). In this study, a cell-based biosensor was developed by inducing a FRET expressing gene in the INS-1E pancreatic β-cell line. Fluorescent proteins were also used to sense inflammation processes for screening various molecules as anti-inflammatory agents (Ye et al., 2019). Another cell-based biosensing strategy proposed the acquisition of the cell response to an active molecule as a

fluorescence image, followed by an image processing technique (Sukekawa et al., 2021, 2019).

Electrochemical-based biosensors, such as potentiometric, measuring the change in electrical properties, or impedimetric, measuring electrochemical impedance, are promising alternatives to laborious analysis methodologies, especially when the biological recognition between cells and analyte occurred. These strategies can be applied for a large class of small molecules. An example is the detection of cypermethrin, a voltage-gated sodium channel blocker in the central nervous system that leads to depolarization and hyper-excitation of the neurons, by a neuroblastoma cell-based bioelectric sensor in which neuroblastoma cell calcium uptake after treatment with nicotine and cypermethrin was fluorometrically determined (Apostolou et al., 2019). On the other hand, an electrochemical biosensor based on 3D cell modification of a glassy carbon electrode functionalized with L-cysteine/gold nanoparticle has been developed to evaluate the antioxidant effect of phloretin, where H_2O_2 was used as oxidative stress initiator (Ye et al., 2018).

8.4.2　PATHOGENS

Usually, the cell-analyte interaction is initiated by the binding of the analyte to the cellular receptors. In this way, membrane receptors interact with the surface proteins of pathogens, which leads to the activation of specific intracellular signalling pathways that can be translated into a measurable response specific to the analyte in question, mostly by optical or electrical means.

Cell-based biosensors allow the detection of bacteria mainly through components of the outer membrane such as the lipopolysaccharides for Gram-negative, and peptidoglycans for Gram-positive type. For example, a 293/hTLR4A-MD2-CD14 cell-based fluorescent biosensor to detect and identify lipopolysaccharides (Sun et al., 2018) was reported. For this, the promoter sequence of the critical signalling pathway gene ZC3H12A (encoding MCPIP1 protein) and enhanced green fluorescence protein (EGFP) were combined to construct a recombinant plasmid, which was transferred into 293/hTLR4A-MD2-CD14 cells. The principle is based on the interaction between lipopolysaccharides and TLR4 which induces a signalling pathway that results in green fluorescent protein expression. This cell-based biosensor was able to detect lipopolysaccharides with the limit of 0.075 μg/mL and was tested in foodstuff and biological samples.

Adenoviruses constitute a major healthcare burden but, when engineered, represent a useful tool for vectors in vaccination or gene and oncolytic therapies. A genetically encoded switch-on fluorescent biosensor consisting of a cyclized green fluorescent protein with an adenoviral protease cleavable site as a switch was used for adenovirus identification but also for the detection of HIV-1 protease activity (Guerreiro et al., 2019). The structurally distorted modified green fluorescent protein was expressed in mammalian cells, and the biosensor used for live-cell monitoring of adenovirus infection as the cleavage of the protease cleavable region led to the relief of the distortion in the green fluorescent protein and subsequent fluorescence emission.

Membrane-engineering is a generic methodology for increasing the selectivity of a cell biosensor against a target molecule, by electroinserting target-specific receptor molecules on the cell surface (Moschopoulou et al., 2011). A biosensor for the ultra-rapid (3 min) and sensitive (fg/mL level) detection of the SARS-CoV-2 virus was developed based on mammalian Vero cells, which were engineered by electroinserting the human chimeric spike S1 antibody (Mavrikou et al., 2020). The attachment of the protein to the membrane-bound antibodies resulted in a selective and considerable change in the cellular bioelectric properties, which were measured by means of a *Bioelectric Recognition Assay* (Kintzios et al., 2001), and no cross-reactivity was observed against the SARS-CoV-2 nucleocapsid protein. This assay was coupled with a customized portable read-out device, which was operated via smartphone/tablet, demonstrating the feasibility of applying the novel biosensor for the mass screening of SARS-CoV-2.

8.4.3 Cell Cytotoxicity

For the detection of toxic effect induced by different chemical agents, cell-based biosensors present unique advantages to other analytical methods given by the fact that cells respond to the toxic exposures in the manner related to actual physiologic response, which employs complex chain reaction mechanisms.

Either through optical, potentiometric, impedance or electronic detection, the complexity of chain reaction mechanisms allows the investigation of cytotoxicity for a large class of chemical agents including metal ions (Guo et al., 2021), toxins (Su et al., 2018) and chemotherapeutics (Luo et al., 2018; Mavrikou et al., 2019; Pan et al., 2019; Wei et al., 2019) or the study of the bioelectric signals of cells, relevant to cell biology in applications of drug development, ion channel studies and disease model establishment (Pulikkathodi et al., 2018).

8.5 CONCLUSIONS

DNA-, protein- and cell-based biosensors have shown to be successfully applied in the medical field for various applications, mainly for the detection of biomarkers for health management, rapid and early disease diagnosis, as well as for monitoring the treatment/therapy of individual patients. The detection of small molecules, nucleic acids sequences and proteins related to different diseases can be carried out with DNA biosensors that integrate chemistry and biology for achieving control and designing DNA sequences that can fold into specific structures with high affinity against their target (bio)molecules, with materials physics and engineering for platform development and electronic control of devices. Microfluidic systems on different materials such as paper or other polymeric synthetic supports are also prone for proteins and antibodies integration into POC devices and immunosensors for the identification of disease biomarkers and enzyme inhibitors with therapeutic action. The newest emerging cell-based biosensors are also used for the detection of pathogens and cellular cytotoxicity monitoring, enlarging their applicability in various analyte detection for POC testing.

REFERENCES

Ahrberg, C.D., Manz, A., Chung, B.G., 2016. Polymerase chain reaction in microfluidic devices. *Lab Chip* 16, 3866–3884. https://doi.org/10.1039/C6LC00984K.

Ali, M.M., Li, F., Zhang, Z., Zhang, K., Kang, D.K., Ankrum, J.A., Le, X.C., Zhao, W., 2014. Rolling circle amplification: A versatile tool for chemical biology, materials science and medicine. *Chem. Soc. Rev.* 43, 3324–3341. https://doi.org/10.1039/C3CS60439J.

Alkhamis, O., Canoura, J., Yu, H., Liu, Y., Xiao, Y., 2019. Innovative engineering and sensing strategies for aptamer-based small-molecule detection. *TrAC Trends Anal. Chem.* 121, 115698. https://doi.org/10.1016/J.TRAC.2018.115699.

Amor-Gutiérrez, O., Giulia, S.M., Fernández-Abedul, T., De La, A., Muñiz, E., Marrazza, G., 2020. Folding-based electrochemical aptasensor for the determination of β-lactoglobulin on poly-L-lysine modified graphite electrodes. *Sensors* 20, 2349. https://doi.org/10.3390/S20082349.

Apostolou, T., Mavrikou, S., Denaxa, N.K., Paivana, G., Roussos, P.A., Kintzios, S., 2019. Assessment of cypermethrin residues in tobacco by a bioelectric recognition assay (BERA) neuroblastoma cell-based biosensor. *Chemosensors* 7, 58. https://doi.org/10.3390/CHEMOSENSORS7040058.

Barsan, M.M., Diculescu, V.C., 2021. An antibody-based amperometric biosensor for 20S proteasome activity and inhibitor screening. *Analyst* 146, 3216–3224. https://doi.org/10.1039/d0an02426k.

Barsan, M.M., Sanz, C.G., Onea, M., Diculescu, V.C., 2021. Immobilized antibodies on mercaptophenylboronic acid monolayers for dual-strategy detection of 20s proteasome. *Sensors* 21, 16–18. https://doi.org/10.3390/s21082702

Bruch, R., Baaske, J., Chatelle, C., Meirich, M., Madlener, S., Weber, W., Dincer, C., Urban, G.A., 2019. CRISPR/Cas13a-powered electrochemical microfluidic biosensor for nucleic acid amplification-free miRNA diagnostics. *Adv. Mater.* 31, 1905311. https://doi.org/10.1002/ADMA.201905311.

Cai, C., Guo, Z., Cao, Y., Zhang, W., Chen, Y., 2018. A dual biomarker detection platform for quantitating circulating tumor DNA (ctDNA). *Nanotheranostics* 2, 12. https://doi.org/10.7150/NTNO.22419.

Canoura, J., Wang, Z., Yu, H., Alkhamis, O., Fu, F., Xiao, Y., 2018. No structure-switching required: A generalizable exonuclease-mediated aptamer-based assay for small-molecule detection. *J. Am. Chem. Soc.* 140, 9961–9971. https://doi.org/10.1021/JACS.8B04975/SUPPL_FILE/JA8B04975_SI_001.PDF.

Chand, R., Wang, Y.L., Kelton, D., Neethirajan, S., 2018. Isothermal DNA amplification with functionalized graphene and nanoparticle assisted electroanalysis for rapid detection of Johne's disease. *Sens. Actuators B Chem.* 261, 31–37. https://doi.org/10.1016/J.SNB.2018.01.140.

Chatterjee, N., Walker, G.C., 2017. Mechanisms of DNA damage, repair and mutagenesis. *Environ. Mol. Mutagen.* 58, 235. https://doi.org/10.1002/EM.22087.

Chen, H., Lin, S., Wang, Y., Fu, S., Ma, Y., Xia, Q., Lin, Y., 2022. Paper-based detection of Epstein-Barr virus using asymmetric polymerase chain reaction and gold silicon particles. *Anal. Chim. Acta* 1197, 339514. https://doi.org/10.1016/J.ACA.2022.339514.

Dai, Y., Somoza, R.A., Wang, L., Welter, J.F., Li, Y., Caplan, A.I., Liu, C.C., 2019. Exploring the trans-cleavage activity of CRISPR-Cas12a (cpf1) for the development of a universal electrochemical biosensor. *Angew. Chemie Int. Ed.* 58, 17399–17405. https://doi.org/10.1002/ANIE.201910772.

Davis, V.W., Bathe, O.F., Schiller, D.E., Slupsky, C.M., Sawyer, M.B., 2011. Metabolomics and surgical oncology: Potential role for small molecule biomarkers. *J. Surg. Oncol.* 103, 451–458. https://doi.org/10.1002/JSO.21831.

Deore, P.S., Gray, M.D., Chung, A.J., Manderville, R.A., 2019. Ligand-Induced G-Quadruplex Polymorphism: A DNA nanodevice for label-free aptasensor platforms. *J. Am. Chem. Soc.* 141, 14288–14297. https://doi.org/10.1021/JACS.9B06533/SUPPL_FILE/JA9B06533_SI_001.PDF.

Díaz-Fernández, A., Lorenzo-Gómez, R., Miranda-Castro, R., de-los-Santos-Álvarez, N., Lobo-Castañón, M.J., 2020. Electrochemical aptasensors for cancer diagnosis in biological fluids – A review. *Anal. Chim. Acta* 1124, 1–18. https://doi.org/10.1016/J.ACA.2020.04.022.

Dou, M., Dominguez, D.C., Li, X., Sanchez, J., Scott, G., 2014. A versatile PDMS/paper hybrid microfluidic platform for sensitive infectious disease diagnosis. *Anal. Chem.* 86, 7978–7986. https://doi.org/10.1021/AC5021694/SUPPL_FILE/AC5021694_SI_001.PDF.

Dunn, M.R., Jimenez, R.M., Chaput, J.C., 2017. Analysis of aptamer discovery and technology. *Nat. Rev. Chem.* 1(10), 1–6. https://doi.org/10.1038/s41570-017-0076.

Erkmen, C., Aydoğdu Tığ, G., Marrazza, G., Uslu, B., 2022. Design strategies, current applications and future perspective of aptasensors for neurological disease biomarkers.

TrAC Trends Anal. Chem. 154, 116675. https://doi.org/10.1016/J.TRAC.2022.116675.

Filik, H., Avan, A.A., 2019. Nanostructures for nonlabeled and labeled electrochemical immunosensors: Simultaneous electrochemical detection of cancer markers: A review. *Talanta* 205, 120153. https://doi.org/10.1016/J.TALANTA.2018.120153.

Foo, P.C., Nurul Najian, A.B., Muhamad, N.A., Ahamad, M., Mohamed, M., Yean Yean, C., Lim, B.H., 2020. Loop-mediated isothermal amplification (LAMP) reaction as viable PCR substitute for diagnostic applications: A comparative analysis study of LAMP, conventional PCR, nested PCR (nPCR) and real-time PCR (qPCR) based on Entamoeba histolytica DNA derived from. *BMC Biotechnol.* 20, 1–15. https://doi.org/10.1186/S12896-020-00629-8/FIGURES/8.

Fu, Y., Zhou, X., Duan, X., Liu, C., Huang, J., Zhang, T., Ding, S., Min, X., 2020. A LAMP-based ratiometric electrochemical sensing for ultrasensitive detection of Group B Streptococci with improved stability and accuracy. *Sens. Actuators B Chem.* 321, 128502. https://doi.org/10.1016/J.SNB.2020.128502.

Gao, J., Wang, C., Wang, C., Chu, Y., Wang, S., Sun, M.Y., Ji, H., Gao, Y., Wang, Y., Han, Y., Song, F., Liu, H., Zhang, Y., Han, L., 2022. Poly-l-Lysine-modified graphene field-effect transistor biosensors for ultrasensitive breast cancer miRNAs and SARS-CoV-2 RNA detection. *Anal. Chem.* 94, 1626–1636. https://doi.org/10.1021/ACS.ANALCHEM.1C03786/ASSET/IMAGES/LARGE/AC1C03786_0008.JPEG.

Gil Rosa, B., Akingbade, O.E., Guo, X., Gonzalez-Macia, L., Crone, M.A., Cameron, L.P., Freemont, P., Choy, K.L., Güder, F., Yeatman, E., Sharp, D.J., Li, B., 2022. Multiplexed immunosensors for point-of-care diagnostic applications. *Biosens. Bioelectron.* 203, 114050. https://doi.org/10.1016/J.BIOS.2022.114050.

Gilboa, T., Garden, P.M., Cohen, L., 2020. Single-molecule analysis of nucleic acid biomarkers – A review. *Anal. Chim. Acta* 1115, 61–85. https://doi.org/10.1016/J.ACA.2020.03.001.

Gu, L., Yan, W., Liu, L., Wang, S., Zhang, X., Lyu, M., 2018. Research progress on rolling circle amplification (RCA)-based biomedical sensing. *Pharmaceuticals* 11, 35. https://doi.org/10.3390/PH11020035.

Guerreiro, M.R., Freitas, D.F., Alves, P.M., Coroadinha, A.S., 2019. Detection and quantification of label-free infectious adenovirus using a switch-on cell-based fluorescent biosensor. *ACS Sensors* 4, 1654–1661. https://doi.org/10.1021/ACSSENSORS.9B00489/SUPPL_FILE/SE9B00489_SI_001.PDF.

Guo, C., Su, F., Song, Y., Hu, B., Wang, M., He, L., Peng, D., Zhang, Z., 2017. Aptamer-templated silver nanoclusters embedded in zirconium metal-organic framework for bifunctional electrochemical and SPR aptasensors toward carcinoembryonic antigen. *ACS Appl. Mater. Interfaces* 9, 41188–41198. https://doi.org/10.1021/ACSAMI.7B14952/ASSET/IMAGES/LARGE/AM-2017-149528_0006.JPEG.

Guo, H., Ji, J., Sun, J., Zhang, Y., Sun, X., 2021. Development of a living mammalian cell-based biosensor for the monitoring and evaluation of synergetic toxicity of cadmium and deoxynivalenol. *Sci. Total Environ.* 771. https://doi.org/10.1016/J.SCITOTENV.2020.144823.

Han, S., Liu, W., Zheng, M., Wang, R., 2020. Label-free and ultrasensitive electrochemical DNA biosensor based on urchinlike carbon nanotube-gold nanoparticle nanoclusters. *Anal. Chem.* 92, 4780–4787. https://doi.

org/10.1021/ACS.ANALCHEM.9B03520/SUPPL_FILE/
AC9B03520_SI_001.PDF.

Hashem, A., Hossain, M.A.M., Marlinda, A.R., Mamun, M.
Al, Sagadevan, S., Shahnavaz, Z., Simarani, K., Johan,
M.R., 2021. Nucleic acid-based electrochemical biosensors for rapid clinical diagnosis: Advances, challenges,
and opportunities. *Crit. Rev. Clin. Lab. Sci.* https://doi.
org/10.1080/10408363.2021.1997898.

Henriques de Jesus, C.S., Chiorcea-Paquim, A.M., Barsan, M.M.,
Diculescu, V.C., 2019. Electrochemical assay for 20S pro-
teasome activity and inhibition with anti-cancer drugs.
Talanta 198. https://doi.org/10.1016/j.talanta.2018.02.052

Hu, W., Huang, Y., Chen, C., Liu, Y., Guo, T., Guan, B.O., 2018.
Highly sensitive detection of dopamine using a graphene
functionalized plasmonic fiber-optic sensor with aptamer
conformational amplification. *Sens. Actuators B Chem.* 264,
440–447. https://doi.org/10.1016/J.SNB.2018.03.005.

Hua, Y., Ma, J., Li, D., Wang, R., 2022. DNA-based biosensors
for the biochemical analysis: A review. *Biosensor* 12, 183.
https://doi.org/10.3390/BIOS12030183.

Huang, X., Sang, S., Yuan, Z., Duan, Q., Guo, X., Zhang, H.,
Zhao, C., 2021. Magnetoelastic immunosensor via antibody
immobilization for the specific detection of lysozymes.
ACS Sensors 6, 3933–3938. https://doi.org/10.1021/
ACSSENSORS.1C00802/SUPPL_FILE/SE1C00802_
SI_004.MP4.

Hwang, M. T., Park, I., Heiranian, M., Taqieddin, A., You, S.,
Faramarzi, V., Pak, A. A., Van Der Zande, A. M., Aluru, N.
R., Bashir, R., 2021. Ultrasensitive detection of dopamine,
IL-6 and SARS-CoV-2 proteins on crumpled graphene FET
biosensor. *Adv. Mater. Technol.* 6, 2100712. https://doi.
org/10.1002/ADMT.202100712.

Ikebukuro, K., Kiyohara, C., Sode, K., 2004. Electrochemical
detection of protein using a double aptamer sandwich. *Anal. Letters.* 37, 2901–2909. https://doi.
org/10.1081/AL-200035778.

Izadi, N., Sebuyoya, R., Moranova, L., Hrstka, R., Anton, M.,
Bartosik, M., 2021. Electrochemical bioassay coupled to
LAMP reaction for determination of high-risk HPV infection in crude lysates. *Anal. Chim. Acta* 1187, 339145.
https://doi.org/10.1016/J.ACA.2021.339145.

Jia, Y., Sun, H., Tian, J., Song, Q., Zhang, W., 2021.
Paper-based point-of-care testing of SARS-CoV-
2. *Front. Bioeng. Biotechnol.* 9, 1197. https://doi.
org/10.3389/FBIOE.2021.773304/BIBTEX.

Jin, H., Zhao, C., Gui, R., Gao, X., Wang, Z., 2018a. Reduced graphene oxide/nile blue/gold nanoparticles complex-modified
glassy carbon electrode used as a sensitive and label-free
aptasensor for ratiometric electrochemical sensing of dopamine. *Anal. Chim. Acta* 1025, 154–162. https://doi.org/
10.1016/J.ACA.2018.03.036.

Jin, H., Zhao, C., Gui, R., Gao, X., Wang, Z., 2018b. Reduced graphene oxide/nile blue/gold nanoparticles complex-modified
glassy carbon electrode used as a sensitive and label-free
aptasensor for ratiometric electrochemical sensing of dopamine. *Anal. Chim. Acta* 1025, 154–162. https://doi.org/
10.1016/J.ACA.2018.03.036.

Karki, H.P., Jang, Y., Jung, J., Oh, J., 2021. Advances in the
development paradigm of biosample-based biosensors for early ultrasensitive detection of Alzheimer's
disease. *J. Nanobiotechnol.* 19, 1–24. https://doi.
org/10.1186/S12951-021-00814-7/FIGURES/8.

Kaur, B., Malecka, K., Cristaldi, D.A., Chay, C.S., Mames, I.,
Radecka, H., Radecki, J., Stulz, E., 2018. Approaching

single DNA molecule detection with an ultrasensitive electrochemical genosensor based on gold nanoparticles and
cobalt-porphyrin DNA conjugates. *Chem. Commun.* 54,
11108–11111. https://doi.org/10.1039/C8CC05362F.

Kim, D., Yoo, S., 2021. Aptamer-conjugated quantum dot optical
biosensors: Strategies and applications. *Chemosensors* 9,
318. https://doi.org/10.3390/CHEMOSENSORS9110318.

Kintzios, S., Pistola, E., Panagiotopoulos, P., Bomsel, M.,
Alexandropoulos, N., Bem, F., Ekonomou, G., Biselis,
J., Levin, R., 2001. Bioelectric recognition assay
(BERA). *Biosens. Bioelectron.* 16, 325–336. https://doi.
org/10.1016/S0956-5663(01)00127-0.

Kiplagat, A., Martin, D.R., Onani, M.O., Meyer, M., 2020.
Aptamer-conjugated magnetic nanoparticles for the effi-
cient capture of cancer biomarker proteins. *J. Magn.
Magn. Mater.* 497, 166063. https://doi.org/10.1016/J.
JMMM.2018.166063.

Kolluri, N., Kamath, S., Lally, P., Zanna, M., Galagan, J., Gitaka,
J., Kamita, M., Cabodi, M., Lolabattu, S.R., Klapperich,
C.M., 2021. Development and clinical validation of Iso-
IMRS: A novel diagnostic assay for *P. falciparum* Malaria.
Anal. Chem. 93, 2097–2105. https://doi.org/10.1021/ACS.
ANALCHEM.0C03847/ASSET/IMAGES/LARGE/
AC0C03847_0006.JPEG.

Kolpashchikov, D.M., Gerasimova, Y. V. (Eds.), 2013. *Nucleic Acid
Detection, Methods in Molecular Biology.* Humana Press,
Totowa, NJ. https://doi.org/10.1007/978-1-62703-535-4.

Krausz, A.D., Korley, F.K., Burns, M.A., 2021. The current state
of traumatic brain injury biomarker measurement methods.
Biosensors 11, 319. https://doi.org/10.3390/BIOS11090319.

Krishnan, S.K., Singh, E., Singh, P., Meyyappan, M., Nalwa, H.S.,
2019. A review on graphene-based nanocomposites for electrochemical and fluorescent biosensors. *RSC Adv.* 9, 8778–
8881. https://doi.org/10.1039/C8RA09577A.

Landegren, U., Hammond, M., 2021. Cancer diagnostics based
on plasma protein biomarkers: Hard times but great
expectations. *Mol. Oncol.* 15, 1715–1726. https://doi.
org/10.1002/1878-0261.12809.

Leva-Bueno, J., Peyman, S.A., Millner, P.A., 2020. A review on
impedimetric immunosensors for pathogen and biomarker
detection. *Med. Microbiol. Immunol.* 209, 343–362. https://
doi.org/10.1007/S00430-020-00668-0/TABLES/1.

Li, B., Tan, H., Jenkins, D., Srinivasa Raghavan, V., Rosa, B.G.,
Güder, F., Pan, G., Yeatman, E., Sharp, D.J., 2020. Clinical
detection of neurodegenerative blood biomarkers using graphene immunosensor. *Carbon N. Y.* 168, 144–162. https://
doi.org/10.1016/J.CARBON.2020.06.048.

Li, J., MacDonald, J., 2016. Multiplex lateral flow detection
and binary encoding enables a molecular colorimetric
7-segment display. *Lab Chip* 16, 242–245. https://doi.
org/10.1039/C5LC01323B.

Liang, M., Li, Z., Wang, W., Liu, J., Liu, L., Zhu, G., Karthik, L.,
Wang, M., Wang, K.F., Wang, Z., Yu, Jing, Shuai, Y., Yu, J.,
Zhang, L., Yang, Z., Li, C., Zhang, Q., Shi, T., Zhou, L.,
Xie, F., Dai, H., Liu, X., Zhang, J., Liu, G., Zhuo, Y., Zhang,
B., Liu, C., Li, S., Xia, X., Tong, Y., Liu, Y., Alterovitz, G.,
Tan, G.Y., Zhang, L.X., 2019. A CRISPR-Cas12a-derived
biosensing platform for the highly sensitive detection of
diverse small molecules. *Nat. Commun.* 10(1), 1–8. https://
doi.org/10.1038/s41467-019-11648-1.

Lin, M., Song, P., Zhou, G., Zuo, X., Aldalbahi, A., Lou,
X., Shi, J., Fan, C., 2016. Electrochemical detection
of nucleic acids, proteins, small molecules and cells

using a DNA-nanostructure-based universal biosensing platform. *Nat. Protoc.* 11(7), 1244–1263. https://doi.org/10.1038/nprot.2016.071.

Liu, J., Chen, X., Wang, Q., Xiao, M., Zhong, D., Sun, W., Zhang, G., Zhang, Z., 2019. Ultrasensitive monolayer MoS 2 field-effect transistor based DNA sensors for screening of down syndrome. *Nano Lett.* 19, 1437–1444. https://doi.org/10.1021/ACS.NANOLETT.8B03818.

Luo, S., Zhou, L., Dai, Y., Lu, Y., Liu, Q., 2018. Analysis of cytotoxic effects of chemotherapeutic agents for gastrointestinal cancer with cell-based impedance biosensor. *Sens. Mater.* 30, 1977–1987. https://doi.org/10.18494/SAM.2018.1910.

Mahas, A., Wang, Q., Marsic, T., Mahfouz, M.M., 2022. Development of Cas12a-based cell-free small-molecule biosensors via allosteric regulation of CRISPR array expression. *Anal. Chem.* 94, 4617–4626. https://doi.org/10.1021/ACS.ANALCHEM.1C04332/ASSET/IMAGES/LARGE/AC1C04332_0005.JPEG.

Mao, X., Baloda, M., Gurung, A.S., Lin, Y., Liu, G., 2008. Multiplex electrochemical immunoassay using gold nanoparticle probes and immunochromatographic strips. *Electrochem. Commun.* 10, 1636–1640. https://doi.org/10.1016/J.ELECOM.2008.08.032.

Martínez-Rojas, F., Diculescu, V.C., Armijo, F., 2020. Electrochemical immunosensing platform for the determination of the 20S proteasome using an aminophenylboronic/poly-indole-6-carboxylic acid-modified electrode. *ACS Appl. Bio Mater.* 3, 4941–4948. https://doi.org/10.1021/ACSABM.0C00478.

Mavrikou, S., Moschopoulou, G., Tsekouras, V., Kintzios, S., 2020. Development of a portable, ultra-rapid and ultrasensitive cell-based biosensor for the direct detection of the SARS-CoV-2 S1 spike protein antigen. *Sensors (Basel)* 20. https://doi.org/10.3390/S20113121.

Mavrikou, S., Tsekouras, V., Karageorgou, M.A., Moschopoulou, G., Kintzios, S., 2019. Detection of superoxide alterations induced by 5-fluorouracil on HeLa cells with a cell-based biosensor. *Biosensors* 8. https://doi.org/10.3390/BIOS9040126.

Melnychuk, N., Klymchenko, A.S., 2018. DNA-functionalized dye-loaded polymeric nanoparticles: Ultrabright FRET platform for amplified detection of nucleic acids. *J. Am. Chem. Soc.* 140, 10856–10865. https://doi.org/10.1021/JACS.8B05840/ASSET/IMAGES/LARGE/JA-2018-05840P_0007.JPEG.

Moschopoulou, G., Valero, T., Kintzios, S., 2011. Molecular identification through membrane engineering as a revolutionary concept for the construction of cell sensors with customized target recognition properties: The example of superoxide detection. *Procedia Eng.* 25, 1541–1544. https://doi.org/10.1016/J.PROENG.2011.12.381.

Moser, N., Leong, C.L., Hu, Y., Cicatiello, C., Gowers, S., Boutelle, M., Georgiou, P., 2020. Complementary metal-oxide-semiconductor potentiometric field-effect transistor array platform using sensor learning for multi-ion imaging. *Anal. Chem.* 92, 5276–5285. https://doi.org/10.1021/ACS.ANALCHEM.9B05836/ASSET/IMAGES/LARGE/AC9B05836_0003.JPEG.

Nagel, B., Dellweg, H., Gierasch, L.M., 1992. Glossary for chemists of terms used in biotechnology. *Pure Appl. Chem.* 64, 143–168. https://doi.org/10.1351/PAC199264010143/MACHINEREADABLECITATION/RIS.

Nemčeková, K., Labuda, J., 2021. Advanced materials-integrated electrochemical sensors as promising medical diagnostics tools: A review. *Mater. Sci. Eng. C* 120, 111751. https://doi.org/10.1016/J.MSEC.2020.111751.

Nicholson, J.K., Lindon, J.C., Holmes, E., 2008. "Metabonomics": Understanding the metabolic responses of living systems to pathophysiological stimuli via multivariate statistical analysis of biological NMR spectroscopic data. 29, 1181–1188. https://doi.org/10.1080/004982599238047.

Ning, C.F., Wang, L., Tian, Y.F., Yin, B.C., Ye, B.C., 2020. Multiple and sensitive SERS detection of cancer-related exosomes based on gold–silver bimetallic nanotrepangs. *Analyst* 145, 2795–2804. https://doi.org/10.1039/C9AN02180A.

Niu, C., Wang, C., Li, F., Zheng, X., Xing, X., Zhang, C., 2021. Aptamer assisted CRISPR-Cas12a strategy for small molecule diagnostics. *Biosens. Bioelectron.* 183, 113196. https://doi.org/10.1016/J.BIOS.2021.113196.

Pan, M., Gu, Y., Yun, Y., Li, M., Jin, X., Wang, S., 2017. Nanomaterials for electrochemical immunosensing. *Sensors* 17, 1041. https://doi.org/10.3390/S17051041.

Pan, Y., Hu, N., Wei, X., Gong, L., Zhang, B., Wan, H., Wang, P., 2019. 3D cell-based biosensor for cell viability and drug assessment by 3D electric cell/matrigel-substrate impedance sensing. *Biosens. Bioelectron.* 130, 344–351. https://doi.org/10.1016/J.BIOS.2018.08.046.

Paniagua, G., Villalonga, A., Eguílaz, M., Vegas, B., Parrado, C., Rivas, G., Díez, P., Villalonga, R., 2019. Amperometric aptasensor for carcinoembryonic antigen based on the use of bifunctionalized Janus nanoparticles as biorecognition-signaling element. *Anal. Chim. Acta* 1061, 84–91. https://doi.org/10.1016/J.ACA.2018.02.015.

Pankratova, N., Jovic, M., Pfeifer, M.E., 2021. Electrochemical sensing of blood proteins for mild traumatic brain injury (mTBI) diagnostics and prognostics: Towards a point-of-care application. *RSC Adv.* 11, 17301–17318. https://doi.org/10.1039/D1RA00589H.

Pellestor, F., Paulasova, P., 2004. The peptide nucleic acids (PNAs), powerful tools for molecular genetics and cytogenetics. *Eur. J. Hum. Genet.* 12(9), 694–700. https://doi.org/10.1038/sj.ejhg.5201226.

Phan, L.M.T., Cho, S., 2022. Fluorescent aptasensor and colorimetric aptablot for p-tau231 detection: Toward early diagnosis of Alzheimer's disease. *Biomedicines* 10. https://doi.org/10.3390/BIOMEDICINES10010093/S1.

Pickering, S., Batra, R., Merrick, B., Snell, L.B., Nebbia, G., Douthwaite, S., Reid, F., Patel, A., Kia Ik, M.T., Patel, B., Charalampous, T., Alcolea-Medina, A., Lista, M.J., Cliff, P.R., Cunningham, E., Mullen, J., Doores, K.J., Edgeworth, J.D., Malim, M.H., Neil, S.J.D., Galão, R.P., 2021. Comparative performance of SARS-CoV-2 lateral flow antigen tests and association with detection of infectious virus in clinical specimens: A single-centre laboratory evaluation study. *The Lancet Microbe* 2, e461–e471. https://doi.org/10.1016/S2666-5247(21)00143-9/ATTACHMENT/9062137E-9E22-459D-A34A-A70FC4107309/MMC1.PDF.

Pirzada, M., Altintas, Z., 2020. Recent progress in optical sensors for biomedical diagnostics. *Micromachines* 11. https://doi.org/10.3390/MI11040356.

Pulikkathodi, A.K., Sarangadharan, I., Chen, Y.-H., Lee, G.-Y., Chyi, J.-I., Lee, G.-B., Wang, Y.-L., 2018. A comprehensive model for whole cell sensing and transmembrane potential measurement using FET biosensors. *ECS J. Solid State Sci. Technol.* 7, Q3001–Q3008. https://doi.org/10.1149/2.0011807JSS/XML.

Qing, M., Chen, S.L., Sun, Z., Fan, Y., Luo, H.Q., Li, N.B., 2021. Universal and programmable rolling circle amplification-CRISPR/Cas12a-mediated immobilization-free electrochemical biosensor. *Anal. Chem.* 93, 7499–7507. https://doi.org/10.1021/ACS.ANALCHEM.1C00805/ASSET/IMAGES/LARGE/AC1C00805_0007.JPEG.

Qiu, G., Gai, Z., Tao, Y., Schmitt, J., Kullak-Ublick, G.A., Wang, J., 2020. Dual-functional plasmonic photothermal biosensors for highly accurate severe acute respiratory syndrome coronavirus 2 detection. *ACS Nano* 14, 5268–5277. https://doi.org/10.1021/ACSNANO.0C02439/ASSET/IMAGES/LARGE/NN0C02439_0005.JPEG.

Rahat, B., Ali, T., Sapehia, D., Mahajan, A., Kaur, J., 2020. Circulating cell-free nucleic acids as epigenetic biomarkers in precision medicine. *Front. Genet.* 11, 844. https://doi.org/10.3389/FGENE.2020.00844/BIBTEX.

Rahi, A., Sattarahmady, N., Heli, H., 2016. Label-free electrochemical aptasensing of the human prostate-specific antigen using gold nanospears. *Talanta* 156–157, 218–224. https://doi.org/10.1016/J.TALANTA.2016.05.029.

Rodriguez, N.M., Linnes, J.C., Fan, A., Ellenson, C.K., Pollock, N.R., Klapperich, C.M., 2015. Paper-based RNA extraction, in situ isothermal amplification, and lateral flow detection for low-cost, rapid diagnosis of influenza A (H1N1) from clinical specimens. *Anal. Chem.* 87, 7872–7878. https://doi.org/10.1021/ACS.ANALCHEM.5B01594/SUPPL_FILE/AC5B01594_SI_001.PDF.

Ruan, X., Wang, Y., Kwon, E.Y., Wang, L., Cheng, N., Niu, X., Ding, S., Van Wie, B.J., Lin, Y., Du, D., 2021. Nanomaterial-enhanced 3D-printed sensor platform for simultaneous detection of atrazine and acetochlor. *Biosens. Bioelectron.* 184. https://doi.org/10.1016/J.BIOS.2021.113238.

Soares, J.C., Melendez, M.E., Soares, A.C., Arantes, L.M.R.B., Da Cruz Rodrigues, V., Carvalho, A.L., Reis, R.M., Mattoso, L.H.C., Oliveira, O.N., 2020. Detection of HPV16 in cell lines deriving from cervical and head and neck cancer using a genosensor made with a DNA probe on a layer-by-layer matrix. *Mater. Chem. Front.* 4, 3258–3266. https://doi.org/10.1039/D0QM00530D.

Solhi, E., Hasanzadeh, M., 2020. Critical role of biosensing on the efficient monitoring of cancer proteins/biomarkers using label-free aptamer based bioassay. *Biomed. Pharmacother.* 132, 110848. https://doi.org/10.1016/J.BIOPHA.2020.110849.

Solis-Marcano, N.E., Morales-Cruz, M., Vega-Hernández, G., Gómez-Moreno, R., Binder, C., Baerga-Ortiz, A., Priest, C., Cabrera, C.R., 2021. PCR-assisted impedimetric biosensor for colibactin-encoding pks genomic island detection in *E. coli* samples. *Anal. Bioanal. Chem.* 413, 4673–4680. https://doi.org/10.1007/S00216-021-03404-6/FIGURES/7.

Su, K., Zhong, L., Pan, Y., Fang, J., Zou, Q., Wan, Z., Wang, P., 2018. Novel research on okadaic acid field-based detection using cell viability biosensor and Bionic e-Eye. *Sen. Actuators B Chem.* 256, 448–456. https://doi.org/10.1016/J.SNB.2017.08.097.

Sui, Y., Qi, L., Wu, J.K., Wen, X.P., Tang, X.X., Ma, Z.J., Wu, X.C., Zhang, K., Kokoska, R.J., Zheng, D.Q., Petes, T.D., 2020. Genome-wide mapping of spontaneous genetic alterations in diploid yeast cells. *Proc. Natl. Acad. Sci. U. S. A.* 117, 28191–28200. https://doi.org/10.1073/PNAS.2018633117/SUPPL_FILE/PNAS.2018633117.SD07.XLSX.

Sukekawa, Y., Mitsuno, H., Kanzaki, R., Nakamoto, T., 2019. Odor discrimination using cell-based odor biosensor system with fluorescent image processing. *IEEE Sens. J.* 19, 7192–7200. https://doi.org/10.1109/JSEN.2018.2916377.

Sukekawa, Y., Mitsuno, H., Kanzaki, R., Nakamoto, T., 2021. Binary mixture quantification using cell-based odor biosensor system with active sensing. *Biosens. Bioelectron.* 179, 113053. https://doi.org/10.1016/J.BIOS.2021.113053.

Sun, C., Xiao, F., Fu, J., Huang, X., Jia, N., Xu, Z., Wang, Y., Cui, X., 2022. Loop-mediated isothermal amplification coupled with nanoparticle-based lateral biosensor for rapid, sensitive, and specific detection of *Bordetella pertussis*. *Front. Bioeng. Biotechnol.* 9, 1471. https://doi.org/10.3389/FBIOE.2021.797957/BIBTEX.

Sun, J., Zhu, P., Wang, X., Ji, J., Habimana, J.D.D., Shao, J., Lei, H., Zhang, Y., Sun, X., 2018. Cell based-green fluorescent biosensor using cytotoxic pathway for bacterial lipopolysaccharide recognition. *J. Agric. Food Chem.* 66, 6869–6876. https://doi.org/10.1021/ACS.JAFC.8B01542/ASSET/IMAGES/LARGE/JF-2018-015428_0005.JPEG.

Tertis, M., Leva, P.I., Bogdan, D., Suciu, M., Graur, F., Cristea, C., 2019. Impedimetric aptasensor for the label-free and selective detection of Interleukin-6 for colorectal cancer screening. *Biosens. Bioelectron.* 137, 123–132. https://doi.org/10.1016/J.BIOS.2018.05.012.

Vinchhi, B., Boss, C., Hermant, A., Bouche, N., de Marchi, U., Dehollain, C., 2019. Optical pancreatic beta cell based biosensor, applications and glucose monitoring. *Proceedings of IEEE Sensors*, 2019-October. https://doi.org/10.1109/SENSORS43011.2018.8956793.

Wang, Q., Wang, J., Huang, Y., Du, Y., Zhang, Y., Cui, Y., Kong, D. Ming, 2022. Development of the DNA-based biosensors for high performance in detection of molecular biomarkers: More rapid, sensitive, and universal. *Biosens. Bioelectron.* 197, 113738. https://doi.org/10.1016/J.BIOS.2021.113739.

Wang, S., Lu, S., Zhao, J., Ye, J., Huang, J., Yang, X., 2019. An electric potential modulated cascade of catalyzed hairpin assembly and rolling chain amplification for microRNA detection. *Biosens. Bioelectron.* 126, 565–571. https://doi.org/10.1016/J.BIOS.2018.08.088.

Wang, Y., Luo, J., Liu, J., Sun, S., Xiong, Y., Ma, Y., Yan, S., Yang, Y., Yin, H., Cai, X., 2019. Label-free microfluidic paper-based electrochemical aptasensor for ultrasensitive and simultaneous multiplexed detection of cancer biomarkers. *Biosens. Bioelectron.* 136, 84–90. https://doi.org/10.1016/J.BIOS.2018.04.032.

Wei, X., Gu, C., Li, H., Pan, Y., Zhang, B., Zhuang, L., Wan, H., Hu, N., Wang, P., 2019. Efficacy and cardiotoxicity integrated assessment of anticancer drugs by a dual functional cell-based biosensor. *Sens. Actuators B Chem.* 283, 881–888. https://doi.org/10.1016/J.SNB.2018.12.085.

Whiley, L., Legido-Quigley, C., 2011. Current strategies in the discovery of small-molecule biomarkers for Alzheimer's disease. 3, 1121–1142. https://doi.org/10.4155/BIO.11.62.

Xiong, Y., Zhang, J., Yang, Z., Mou, Q., Ma, Y., Xiong, Y., Lu, Y., 2020. Functional DNA regulated CRISPR-Cas12a sensors for point-of-care diagnostics of non-nucleic-acid targets. *J. Am. Chem. Soc.* 142, 207–213. https://doi.org/10.1021/JACS.9B09211/SUPPL_FILE/JA9B09211_SI_001.PDF.

Yang, Z., Xu, G., Reboud, J., Ali, S.A., Kaur, G., McGiven, J., Boby, N., Gupta, P.K., Chaudhuri, P., Cooper, J.M., 2018. Rapid veterinary diagnosis of bovine reproductive infectious diseases from semen using paper-origami DNA microfluidics. *ACS Sens.* 3, 403–408. https://doi.org/10.1021/ACSSENSORS.7B00825/ASSET/IMAGES/LARGE/SE-2017-008259_0003.JPEG.

Ye, Y., Ji, J., Pi, F., Yang, H., Liu, J., Zhang, Y., Xia, S., Wang, J., Xu, D., Sun, X., 2018. A novel electrochemical biosensor for antioxidant evaluation of phloretin based on cell-alginate/L-cysteine/gold nanoparticle-modified glassy carbon electrode. *Biosens. Bioelectron.* 119, 119–125. https://doi.org/10.1016/J.BIOS.2018.07.051.

Ye, Y., Liu, K., Geng, S., Ji, J., Sun, J., Zhang, Y., Pi, F., Sun, X., 2019. A novel fluorescent recombinant cell-based biosensor for screening NLRP3 inflammasome inhibitors. *Sens. Actuators B Chem.* 301, 126864. https://doi.org/10.1016/J.SNB.2018.126864.

Yokobori, S., Hosein, K., Burks, S., Sharma, I., Gajavelli, S., Bullock, R., 2013. Biomarkers for the clinical differential diagnosis in traumatic brain injury-A systematic review. *CNS Neurosci. Ther.* https://doi.org/10.1111/cns.12127.

Yu, L.R., Stewart, N.A., Veenstra, T.D., 2010. Proteomics: The deciphering of the functional genome. *Essentials Genomic Pers. Med.* 89–96. https://doi.org/10.1016/B978-0-12-374934-5.00008-8.

Zahra, Q. ul ain, Khan, Q. A., Luo, Z., 2021. Advances in optical aptasensors for early detection and diagnosis of various cancer types. *Front. Oncol.* 11, 18. https://doi.org/10.3389/FONC.2021.632165/BIBTEX.

Zhang, F., Liu, J., 2021. Label-free colorimetric biosensors based on aptamers and gold nanoparticles: A critical review. *Anal. Sens.* 1, 30–43. https://doi.org/10.1002/ANSE.202000023.

Zhang, Y., Figueroa-Miranda, G., Lyu, Z., Zafiu, C., Willbold, D., Offenhäusser, A., Mayer, D., 2019. Monitoring amyloid-β proteins aggregation based on label-free aptasensor. *Sens. Actuators B Chem.* 288, 535–542. https://doi.org/10.1016/J.SNB.2018.03.049.

Zhao, X., Dai, X., Zhao, S., Cui, X., Gong, T., Song, Z., Meng, H., Zhang, X., Yu, B., 2021. Aptamer-based fluorescent sensors for the detection of cancer biomarkers. *Spectrochim. Acta Part A Mol. Biomol. Spectrosc.* 247, 119038. https://doi.org/10.1016/J.SAA.2020.119038.

Zheng, Q., Huang, T., Zhang, L., Zhou, Y., Luo, H., Xu, H., Wang, X., 2016. Dysregulation of ubiquitin-proteasome system in neurodegenerative diseases. *Front. Aging Neurosci.* 8, 303. https://doi.org/10.3389/FNAGI.2016.00303/BIBTEX.

Zhu, H., Lu, Y., Xia, J., Liu, Y., Chen, J., Lee, J., Koh, K., Chen, H., 2022. Aptamer-assisted protein orientation on silver magnetic nanoparticles: Application to sensitive leukocyte cell-derived chemotaxin 2 surface Plasmon resonance sensors. *Anal. Chem.* 94, 2109–2118. https://doi.org/10.1021/ACS.ANALCHEM.1C04448/ASSET/IMAGES/LARGE/AC1C04448_0007.JPEG.

9 Modelling Dissolving Microneedles Mediated Drug Delivery for COVID-19 Treatment

Charlotte R. Haigh, Prateek R. Yadav, and Diganta B. Das
Loughborough University

CONTENTS

9.1 INTRODUCTION

Since the emergence of SARS-CoV-2 and coronavirus disease (COVID-19), there has been a focus on improving patient treatments decreasing mortality rates and severity of symptoms due to the disease. Although there are currently no specific treatments for COVID-19, existing antiviral medications have been repurposed to aid patients in their recovery. The FDA-approved antiviral COVID-19 treatment involving the delivery of Remdesivir has been shown to inhibit SARS-CoV-2 *in-vitro* and shorten the time a patient takes to recover (Humeniuk et al., 2020). However, the most common method of delivering Remdesivir is intravenous, which has many associated disadvantages, such as poor patient compliance and blood-borne pathogen transmission (Sabri et al., 2020).

In recent years, there has been an increased interest in microneedles (MNs) due to their ability to deliver drugs in a painless and non-invasive way. The principles behind the MN use involve the creation of micron-sized pores in the skin, which penetrate the outermost layer of skin, the *stratum corneum* (SC). This allows molecules to bypass the major barrier to mass transfer within the skin (Kim et al., 2015). These MNs are categorised into solid, hollow, coated, dissolving, and hydrogel-forming MNs (Bhatnagar et al., 2019; Yadav et al., 2020). Among these, dissolving microneedles (DMNs) are of particular interest as they are completely biocompatible and dissolve when inserted into the skin to release the drug encapsulated within the DMN (Hassan et al., 2022). They, therefore, result in no production of sharp waste (Chu et al., 2016). DMNs provide advantages over solid MNs, as they can maintain controlled drug release over a longer time frame, by controlling the dissolution rate of the polymeric matrix (Guillot et al., 2020). Furthermore, their application only involves a single step as the DMN array is not to be removed after insertion as in other cases (Guillot et al., 2020; Waghule et al., 2019).

DMNs currently have a variety of applications, including uses in nicotine replacement therapy and to deliver local anaesthetics (Panda et al., 2021). They can be produced

DOI: 10.1201/9781003224464-9

from biodegradable polymers such as hyaluronic acid (HA), polyvinyl alcohol (PVA) and carboxymethyl cellulose (CMC), which are nontoxic, biodegradable and biocompatible (Leone et al., 2021; Yadav et al., 2021). The release profile of each DMN system is complex as it depends on the properties of its polymeric matrix, such as its molecular weight, chemistry and morphology (Yadav et al., 2020).

For an HA-based system, the DMN is initially rigid, allowing permeation through the upper layers of the skin. After insertion, MN imbibes water present in the skin, leading to pore formation within the MN. Adsorption then causes the polymer to swell, which creates pores large enough in the DMN for drug transport. A hydrolysis reaction occurs between steric bonds of the polymer, which results in a reduction of the polymer molecular weight and causes degradation of the MN array (Yadav et al., 2020). The production of a mathematical model that simulates the mechanisms involved during the transdermal drug delivery (TDD) is essential for the healthcare industry. The development of such a model will allow key parameters such as MN height, shape and patch size to be optimised in a faster and more cost-effective way than by running laboratory experiments. Currently, there are few publications involving the modelling and optimisation of DMN arrays when compared with solid or hollow MNs. However, key studies to date on DMNs will be discussed below as they provide some context to previous MN research. Kim et al. (2015) developed a mathematical model to predict the amount of fentanyl delivered into the skin by a DMN. The approach varied from previous approaches to modelling DMNs at the time of publication due to the inclusion of a biological membrane. The model found that the drug concentration within the skin was proportional to its mass fraction in the DMN. A non-linear relationship was found to occur between MN pitch and skin permeation (Kim et al., 2015). A similar numerical approach was followed by Ronnander et al. (2020) for modelling the transport of sumatriptan from pyramidal-shaped DMNs.

Chavoshi et al. (2019) developed a mathematical model on the dissolution of a Poly(lactic-co-glycolic acid) (PLGA) DMN containing aspirin and albumin, following the same principles as Ronnander et al. (2020). However, the model developed by Chavoshi et al. (2019) incorporated the autocatalytic effects within the PLGA polymer. The results obtained from the model were compared with experimental data. Differences were found in the release profiles from the experimentation and mathematical model. Chavoshi et al. (2019) concluded that this was due to difficulties when estimating some of the parameters required from the model, as it is difficult to find data in the literature for exactly the same experimental conditions.

Although there have been mathematical models created for other drugs, the existence of a mathematical model of the TDD of Remdesivir through a DMN does not exist. In addressing this point, the current chapter aims to present a modelling framework for DMNs and demonstrate its use to model the TDD of Remdesivir. To describe the mechanisms involved, the following physics were considered:

- Dissolution of polymer in the skin in the presence of interstitial skin fluid (ISF)
- Imbibition of skin interstitial fluid into the DMN
- Diffusive drug transport from DMN into the skin
- Drug pharmacokinetics (PKs) in blood

This chapter aims to investigate the influence of altering important DMN parameters like MN length and skin pH on Remdesivir release profiles and improving DMN efficiency.

9.2 METHODOLOGY

9.2.1 MATHEMATICAL MODELLING STRATEGIES

The algorithm chosen to use in the model was finite element analysis (FEA) as it allows the user to alter design variables with ease and can consider complex geometries and boundary conditions. After research into FEA software, COMSOL Multiphysics 5.4 was determined as a suitable software with various MN models already created. For example, Podder et al. (2011) used COMSOL software to simulate silicon hollow MNs to optimise the cross-sectional area of the MN bores.

The model developed in COMSOL Multiphysics 5.4 was created using 2D axial symmetry because the needles are concentric through the centre. HA with a molecular weight of 20 kDa was used to model the DMNs as it has been found to produce robust needles which dissolve quickly in the skin (Leone et al., 2021). The initial geometry of the needles was modelled as a cone, with a base radius of 125 μm and a height of 260 μm. The MNs are defined to be present in three layers of skin, the SC, viable epidermis (VE) and dermis (D). Therefore, each layer of skin was modelled as a porous medium with a different porosity where the key mass transport mechanism is governed by passive diffusion.

As the simulation was time-dependent, an adaptive mesh refinement approach was used to improve the accuracy of calculations and to reduce the computational cost (Dias et al., 2019). A study was computed to determine the optimal mesh size to compute an accurate result that minimised computational time.

Sections 9.2.2–9.2.6 describe the mechanisms and physics considered within the model. Table 9.1, found in the nomenclature, displays all parameter values used in the model.

9.2.2 ESTIMATION OF DIFFUSION COEFFICIENT FOR MODELLING

To consider the TDD of Remdesivir, it was necessary to first determine the drug diffusion coefficient. There is a lack of data available on the diffusion of Remdesivir. Therefore, an approach derived by Vadovic and Colver (1973) was used to calculate the diffusion coefficient (D_D), as shown in equation 9.1. This approach was able to correlate experimental data for 20 liquids with an average deviation of 6% (Vadovic and Colver, 1973).

$$D_D = \frac{0.216 \times 10^{-15} \left(v_m\right)^{\left(\frac{2}{3}\right)} \rho T}{\mu M_D} \qquad (9.1)$$

where v_m is the molar volume of Remdesivir at its boiling point, ρ is the density of Remdesivir injection formulation, T is temperature, μ is the viscosity of Remdesivir and M_D is the molecular weight of Remdesivir.

9.2.3 DISSOLUTION OF DMN IN THE SKIN

During the DMN dissolution process, the drug encapsulated in the MN was transferred into the skin based on the concentration gradient between the MN and the skin. The rate of MN degradation depends on the value of the mass transfer coefficient (K_D), which was estimated by considering the DMN design factors and medium properties (Zoudani and Soltani, 2020). The molar flux of drug from the MN to the skin was defined using equation 9.2.

$$J_{0,D} = \sum K_D \left(C_{D\_skin,\,i} - K\left(C_{D_{MN}}\right)\right) \qquad (9.2)$$

where $C_{D\_skin,\,i}$ is the concentration of Remdesivir in the skin within domain i, K is the partition coefficient and $C_{D_{MN}}$ is the concentration of Remdesivir in the MN.

To simplify the dissolution process of the MN, it was assumed that the ratio of the MN base radius to height ratio remains constant over time (Kim *et al.*, 2015). Equation 9.3 was therefore used to calculate the half-angle at the MN apex.

$$\tan\theta = \frac{R(t)}{h(t)} = \frac{R_0}{h_0} = constant \qquad (9.3)$$

where R_0 and h_0 are the initial MN base radius and height, respectively.

During the dissolution process, the MN volume is defined to decrease with time. Using results from a study produced by Leone et al. (2021) on the dissolution of a 20 kDa HA DMN, a 1200 s was defined as the time for the polymeric MN to dissolve completely. By assuming a constant mesh velocity at the dissolving MN surface, the change in the shape of the DMN was modelled over time in COMSOL by using the deformed geometry module and defining the prescribed mesh velocity in both the R and Z directions. For simplification, it was defined that the gap width between the MN and the skin layers is negligible (Al-Qallaf and Das, 2009). Therefore, the rate of skin pore closure is equal to the rate of MN degradation (Ogunjimi et al., 2020).

Once the mesh velocity was defined, the polymer dissolution kinetics were considered. In the skin, HA is degraded into fragments of various sizes, causing a decrease in polymer molecular weight (M_P). Acid hydrolysis of HA is a random process that follows first-order kinetics (Tømmeraas and Melander, 2008). Therefore, equation 9.4 was used to define the molecular weight of the polymeric scaffold over time. Values for polymeric degradation rate constant

(k_h) were obtained from experimental data produced by Tømmeraas and Melander (2008).

$$\frac{1}{M_P} = k_h t - \frac{1}{M_{P,\,t=0}} \qquad (9.4)$$

where $M_{P,\,t=0}$ is the initial polymer molecular weight.

The porosity of the matrix increases with time as the polymer degrades. By accounting for a first-order degradation rate of the HA polymer matrix, the following relationship was used to define the time-dependent porosity term (Rothstein et al., 2008).

$$\varepsilon_{MN}(t) = \frac{1}{2}\left[erf\left(\frac{t - t_{mean}}{\sqrt{2\sigma^2}}\right) + 1\right] \qquad (9.5)$$

The values for mean time for pore formation (t_{mean}) and variance in the time required to form pores (σ^2) were estimated from experimental data produced by Leone et al. (2021). However, these parameters will be further investigated within the study.

9.2.4 DIFFUSION OF SKIN INTERSTITIAL FLUID INTO THE DMN

When a DMN is inserted into the skin, interstitial fluid is initially adsorbed at a faster rate than drug release (Kamaly et al., 2016). The diffusion of interstitial fluid from the skin into the DMN was modelled by considering the transport process as a diluted species in a porous medium. For this physics module, the domain was bound to all three layers of skin. The initial concentration of drug and water was defined for each layer of skin. Millington and Quirk's model was used to calculate the tortuosity of each layer, as shown in equation 9.6.

$$\tau_i = \varepsilon_{P,\,i}^{-\frac{1}{3}} \qquad (9.6)$$

Once the effective porosity and the fluid diffusion coefficients were known, equation 9.7 was used to calculate the effective diffusion coefficient for each species in each domain.

$$D_{e,\,i} = \frac{\varepsilon_{P,\,i}}{\tau_i} D_{F,\,i} \qquad (9.7)$$

After all initial conditions and transport properties were defined in the model, the boundaries in which flux of water can and cannot occur were defined. Fick's first law was used to calculate the total initial flux of water across all domains bordering the needle to skin boundary.

$$J_{0,w} = \sum K_D \left(C_{w\_skin,\,i} - K\left(C_{w_{MN}}\right)\right) \qquad (9.8)$$

where $C_{w\_skin,\,i}$ is the concentration of water in the skin within domain i and $C_{w_{MN}}$ is the concentration of water within the MN.

All other boundaries except the interface between the layers of skin were bound to no flux conditions. By defining these parameters and boundary conditions, the mathematical model can solve the instantaneous flux of local moisture from the skin into the needle. It simultaneously solves Fick's first law equations 9.9 and 9.10.

$$J_W = \sum -D_{e,i} \times \nabla C_{W,i} \qquad (9.9)$$

$$J_{0,w} = -n \times J_w \qquad (9.10)$$

9.2.5 Diffusive Drug Release

Initially, drug molecules are tethered to the polymeric scaffold. However, as the ester bonds begin to break, molecules of drug will be released. As polymer degradation occurs, the drug diffusion coefficient increases as the MN becomes more porous (Yadav et al., 2020). The diffusive drug release may be modelled using Fick's law, where the effective diffusion coefficient for the drug in the MN can be calculated using the time-dependent porosity term previously defined in equation 9.5.

$$D_{D\_MN} = D_D \varepsilon_{MN}(t) \qquad (9.11)$$

The diffusion of Remdesivir was modelled using the transport of a diluted species module bound to the needle's domain. Initial conditions for the MN's water and Remdesivir concentrations were defined in addition to boundaries whereby flux is not able to occur. Flux conditions were allowed across the needle-skin boundary and the interfaces between layers of skin. However, all other boundaries had no flux conditions applied. Fick's first law was used to define the initial flux of drug across the needle-skin boundary.

$$J_{0,D} = \sum -K_D \left(C_{D\_skin,i} - K(C_{D_{MN}}) \right) \qquad (9.12)$$

Through implementing equations 9.11 and 9.12, the model can calculate the instantaneous flux of drug using a time-dependent study, which solves equations 9.13 and 9.14 simultaneously.

$$J_D = \sum -D_{D\_MN} \times \nabla C_{D,i} \qquad (9.13)$$

$$J_{0,D} = -n \times J_D \qquad (9.14)$$

9.2.6 Drug Pharmacokinetics

Once the drug encapsulated within the DMN diffuses into the skin, it is defined to be absorbed into the microcirculation after which it is eliminated from the body. A one-compartment model was determined suitable to describe Remdesivir PK when delivered through a DMN array because TDD bypasses adsorption in the stomach

and intestine (Olatunji et al., 2012). The metabolism of Remdesivir was described by equation 9.15, which was modelled in COMSOL using the global ODEs and DAEs module (Olatunji et al., 2012; Yadav et al., 2022).

$$V_d \frac{dC_{d\_blood}}{dt} = QS_a - K_e V_d C_{d\_blood} \qquad (9.15)$$

Here, V_d is the volume of distribution in the blood, C_{d_blood} is the Remdesivir concentration in blood at any time t, K_e is the elimination rate constant, Q is the Remdesivir flux coming out of the unit cross-sectional area of the skin, and S_a is the area of skin-blood interface. The data for volume of distribution and elimination rate constant for Remdesivir were taken from a phase I, first-in-human study by Humeniuk et al. (2020).

9.3 RESULTS AND DISCUSSION

9.3.1 Geometry Mesh

An adaptive mesh refinement study was completed where the model was resolved using successively finer and finer meshes, increasing the number of mesh elements (Figure 9.1).

As the element size decreases towards zero, the model size (number of mesh elements) increases towards infinity and the solution becomes more exact (COMSOL, 2016). After comparing the results for Remdesivir concentration within the skin for successively finer meshes, the changes in the output over time between meshes became smaller as shown in Figure 9.1a. Hence it was determined that the changes in output for a mesh containing more than 1520 elements were so small that they could be ignored. Therefore, a mesh with 1520 elements was chosen as the optimum to produce an accurate result with a fast computational time. This is identified in Figure 9.1a using an arrow sign. Figure 9.1b displays the element mesh used. The mesh displayed shows half of one DMN as a unit. It is worth noting that the MN patch consists of a 10×10 array; therefore, there are 200 of these (half) units in a single patch.

9.3.2 Concentration Distribution Profiles

Once Sections 9.2.2–9.2.6 were incorporated, the model's ability to produce concentration distribution profiles describing the transport of skin interstitial fluid and Remdesivir was assessed. The results from the first study, which assessed the model's ability to simulate the diffusive drug release from the DMN into the skin are seen in Figure 9.2. This shows a comparison of the simulation results of the diffusion of Remdesivir from the needle to the skin at time 0–400 s.

With this initial study confirming the presence of physics simulating the transport of Remdesivir, it was necessary to also confirm the diffusion of skin interstitial fluid. Figure 9.3 shows the simulation results for the concentration of water in the skin for times of 0–400 s.

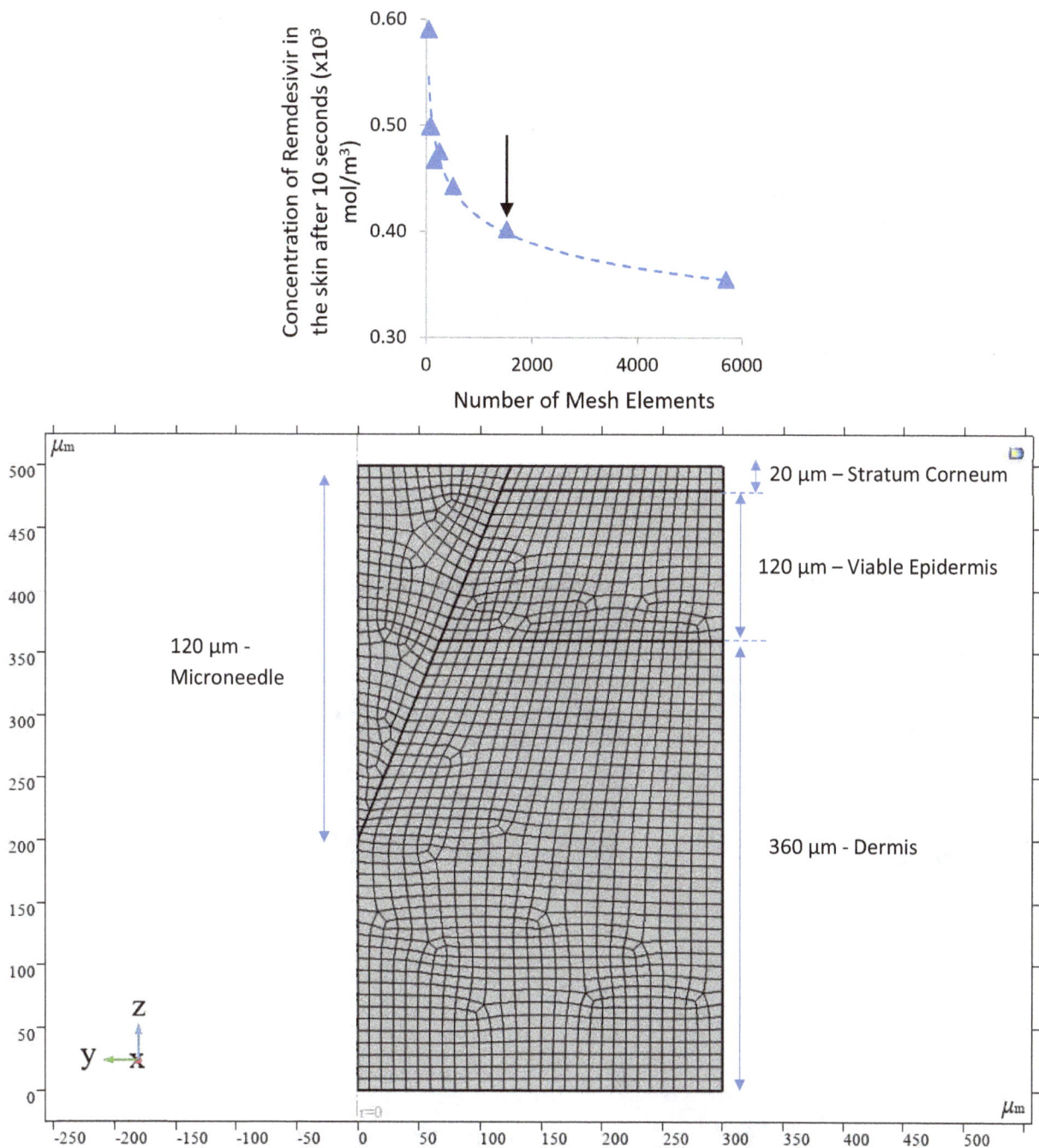

FIGURE 9.1 (a) Concentration of Remdesivir in skin for varying number of mesh elements. (b) A diagram of the extra fine COMSOL mesh containing 1520 elements showing different skin layers.

Figure 9.2 shows that there is a concentration gradient for Remdesivir in the skin, therefore confirming that there is a flux of drug through the skin layers. It also shows that the MN decreases in volume over time. Hence confirming the implementation of polymer dissolution kinetics in the model. Figure 9.2 is comparable to the surface concentration profile produced by Zoudani and Soltani (2020), describing the dissolution of a conical PVP MN containing fluorescein isothiocyanate-dextran. Figure 9.3 demonstrates a concentration gradient for water in the skin, causing a flux of water into the MN. This is further confirmed by the presence of

streamlines, shown in black, which describe the direction of the flux of water into the needle. As both figures show that the physics for the transport of species and MN dissolution have been successfully implemented in the model, studies of further detail could be issued.

9.3.3 EFFECT OF POLYMER BIODEGRADATION KINETICS

The porosity of the polymer present within the DMN changes with the time that the needle is present within the skin and depends on both the mean time for pore formation

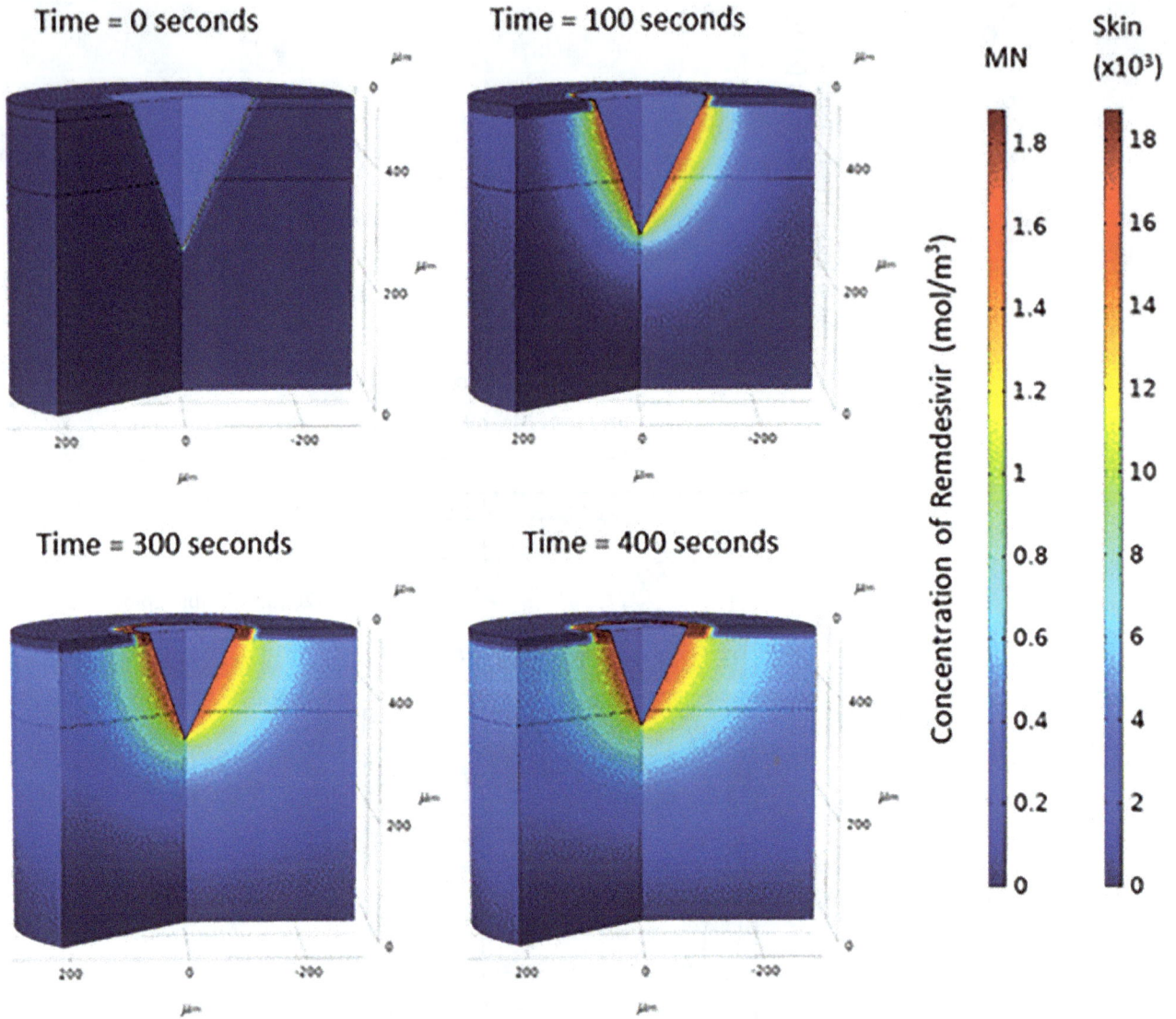

FIGURE 9.2 Remdesivir concentration profiles for DMN in the skin between 0 and 400 s.

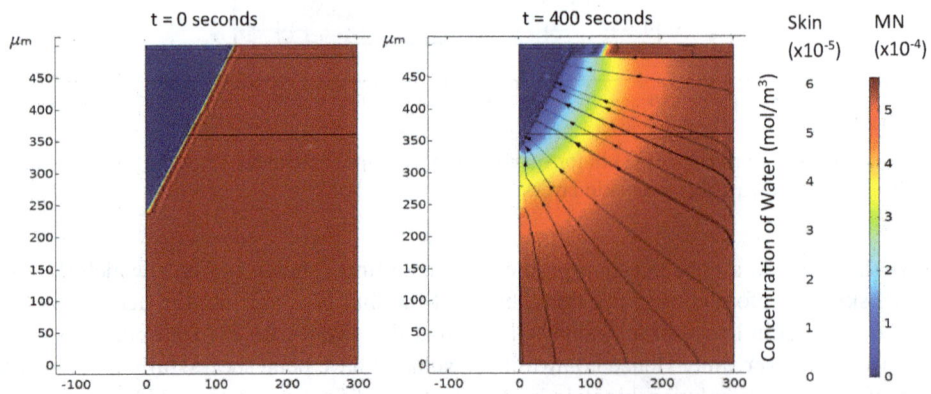

FIGURE 9.3 Water concentration and flux distribution for DMN in the skin at 0 and 400 s.

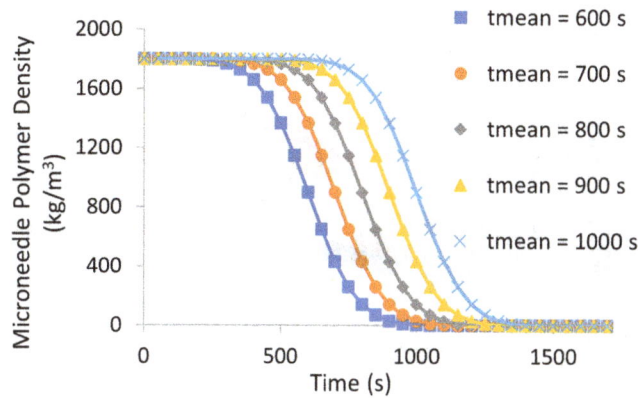

FIGURE 9.4 Relationship between the density of the DMN and the time the needle is present within the skin.

and variance in the time required to form pores. Both these constants depend on the polymer used for MN fabrication and the properties of the encapsulated drug. As no data is available for Remdesivir encapsulated in a 20 kDa HA DMN, the MN density over time was simulated for different values of the mean time for pore formation, as shown in Figure 9.4. The mean time for pore formation was estimated to be between 600 and 1000 s (Leone et al., 2021).

Figure 9.4 shows an S-shaped curve with an initial MN density of 1800 kg/m³ which decreases to zero. This shape can be explained through considering the mechanism of degradation. Initially, the MN density decreases slowly as surface erosion place. This is followed by a steady rate of decrease in MN density as controlled degradation occurs due to diffusion and hydrolysis of the polymeric MN. This is followed by bulk erosion of the MN. Figure 9.4 shows that as the value of mean time for pore formation increases from 600 to 1000 s, the time for the MN density to reach zero also increases by 400 s. This is because, when less pores are formed in the polymeric scaffold, there is less surface area for acid hydrolysis of HA to occur across. Therefore, the degradation is slower. The S-shape curve displayed in Figure 9.4 is expected, as the majority of drug release profiles from polymeric drug delivery systems result in a triphasic profile (Kamaly et al., 2016).

9.3.4 EFFECT OF SKIN pH

The mechanism of degradation of a polymeric MN depends on both the external environment, the payload property, and the skin property, such as skin pH. To determine the effect of altering skin pH on the rate of degradation of the HA DMN, the simulation was run for different values of polymeric degradation rate constant, which corresponded to different skin pH values. These values were extracted from a study conducted by Tømmeraas and Melander (2008) on the kinetics of HA hydrolysis in acidic solution at various pH values. This study found that over a 24-h period, there was negligible change in polymer molecular weight between skin pH values of 3.5–5.0. Therefore, no graph of

these results has been displayed. This was despite the value of polymeric degradation rate constant varying between 7.0×10^{-8} and $6.0 \times 10^{-9} h^{-1}$, respectively.

9.3.5 EFFECT OF DISSOLVING MICRONEEDLE LENGTH AND CENTRE-TO-CENTRE SPACING

To determine the effects of DMN length and centre-to-centre spacing on the concentration of Remdesivir in the skin over time, two studies were computed. In the first, five needle geometries were created with a radius of 100 μm and lengths of between 150 and 350 μm. The geometries were cylindrical so that the surface area to volume ratio of the needle remains constant between varying needle lengths. The simulation was run for each length, and the results were balanced for the initial volume of each needle. In the second study, the DMN geometry was kept the same, and the skin volume was varied for centre-to-centre spacings of between 400 and 1200 μm. Figure 9.5 shows the release profiles generated.

Figure 9.5a suggests that increasing the needle length increases the amount of Remdesivir delivered into the skin over time, even when the results are balanced for the initial needle volume. As the needle length increases from 150 to 350 μm, the Remdesivir concentration in the skin increases by $2.9 \times 10^{-6} mol/m^3$ (49%) after 100 s. This can be explained through considering the MN volume present in the VE and D layers of skin, which have increased transport properties when compared to the SC (Andrews et al., 2013). As the needle length increases, the percentage of needle volume present in the VE and D increases, leading to an increased release of Remdesivir. Chen et al. (2020) similarly found that polymeric MNs with a longer length presented higher TDD efficiency. The drug used in this study was insulin, and the MN lengths were varied between 124 and 445 μm. It was noted that using longer polymeric MNs may not be optimal, as shorter lengths would significantly reduce the pain due to skin piercing. Gomaa et al. (2010) also found that longer MNs may require a greater insertion force for their use to be effective.

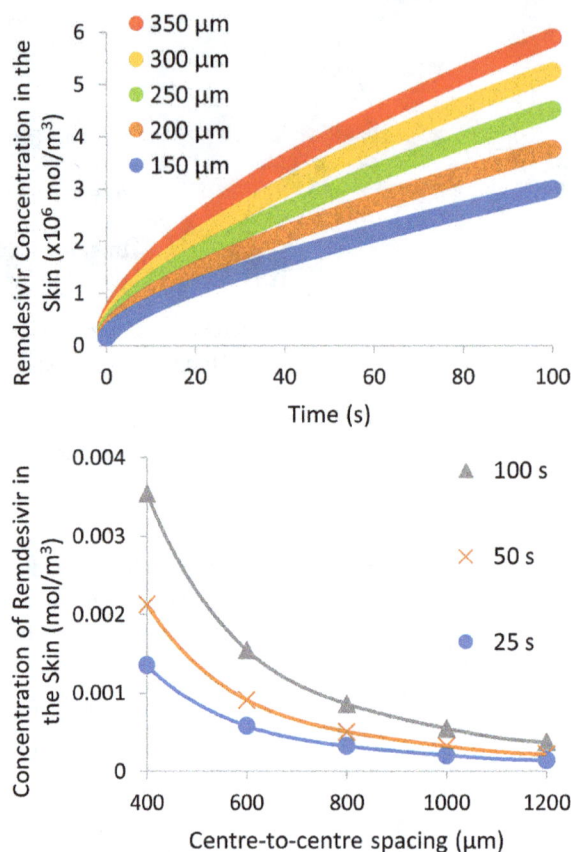

FIGURE 9.5   (a) Remdesivir concentration in the skin for the first 100 s after needle insertion for DMN lengths between 150 and 350 μm. (b) Remdesivir concentration in the skin for varying centre-to-centre spacings between DMNs, at times of 25, 50 and 100 s.

Figure 9.5b shows a non-linear relationship between the centre-to-centre spacing between DMNs and the concentration of Remdesivir in the skin. As the centre-to-centre spacing between DMNs increases from 400 to 1200 μm, the concentration of Remdesivir in the skin decreases by 3.17×10^{-3} mol/m³ (89%), at 100 s after DMN insertion. These results are comparable to Kim et al. (2015) who found a non-linear relationship between the centre-to-centre spacing between DMNs and the concentration of fentanyl in the skin. A larger amount of fentanyl was found to be delivered by DMN arrays with a smaller centre-to-centre spacing. However, neither results in Figure 9.5b nor produced by Kim et al. (2015) consider the effect of centre-to-centre spacing on the deformation of the skin. If a small centre-to-centre spacing is used, more force is required to insert the DMN array into the skin. The indentation of the skins surface will also be affected, which may lead to decreased diffusivity, insufficient insertion and leakage of drug (Olatunji et al., 2013).

9.3.6 Effect of Skin Thickness

Skin thickness varies between individuals depending on their race, sex, age and the location on their body (Ya-Xian

et al., 1999). To determine the effect of skin thickness on the concentration of Remdesivir in the blood, two studies were computed. In the first, only the SC thickness was varied between 16.7 and 22.3 μm (Ya-Xian et al., 1999). The total thickness of the skin was kept constant at 500 μm. In the second study, the thickness of the VE varied between 70 and 120 μm (Koehler et al., 2010). In this study, the total skin thickness varied with the VE thickness. Figure 9.6 shows the PK results for Remdesivir.

Figure 9.6 shows the combined effect of the adsorption and elimination of Remdesivir on the concentration of drug in the blood, when delivered through the DMN array. It represents a typical drug concentration in the blood vs time profile and is comparable to the simulated human PKs data produced by Hanafin et al. (2021) on a single 100 mg intraveneous dose of Remdesivir. Figure 9.5 displays the maximum plasma concentration (C_{max}) is reached at a time (T_{max}) of 2500–2700 s. It suggests that no drug reaches the blood before 100 s.

Figure 9.6a suggests that varying SC thickness between 16.7 and 22.3 μm has a negligible effect on Remdesivir concentration in the blood over time. Each SC thickness shows a maximum Remdesivir plasma concentration of 2.1×10^{-12} mol/m³ at 2600 s. This can be explained by considering the volume percentage of MN in the SC, VE

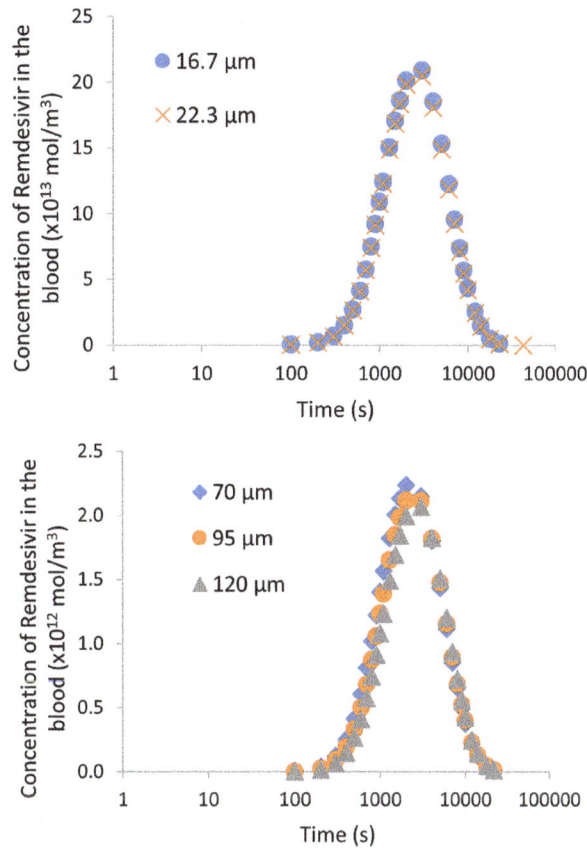

FIGURE 9.6   Remdesivir concentration in the blood over time for different skin thickness (a) For SC thicknesses of 16.7 and 22.3 μm. (b) For VE thicknesses of 70, 95 and 120 μm.

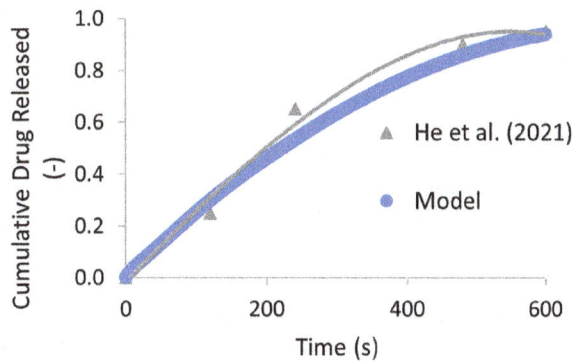

FIGURE 9.7   Modelling and experimentally determined data comparison for cumulative drug release from a DMN over time (He et al., 2021).

and D. For both SC thicknesses of 16.7 and 22.3 μm, most of the MN volume lies in the VE and D layers of skin (82% and 76%, respectively). As these skin layers have increased transport properties when compared to the SC, small changes in SC thickness have a negligible impact on the Remdesivir concentration in the blood over time (Andrews et al., 2013).

The results shown in Figure 9.6b suggest that as the VE thickness decreases by 42% (120–70 μm), the maximum Remdesivir concentration increases by 1.7×10^{-13} mol/m³

(8%). This is because, as the VE thickness decreases, the total skin thickness decreases, leading to a smaller distance for the drug molecules to diffuse across to reach the bloodstream.

9.3.7   COMPARISON TO LITERATURE DATA

The cumulative release data for Remdesivir over time was compared with experimental data produced by He et al. (2021) on the dissolution of obelisk-shaped DMNs containing propranolol hydrochloride. For this comparison, the model was run with an obelisk geometry, with the same dimensions. Figure 9.7 shows the results of this comparison.

Figure 9.7 shows similar release profiles for the experimental data and data produced through the model. The cumulative drug release was calculated by comparing the moles of drug released from the DMN to the number of moles initially encapsulated. Despite the similarity in the results displayed, He et al. (2021) used a HA polymer with a lower molecular weight of 10 kDa when compared to the HA polymer used in the model (20 kDa). Any slight difference in release rates may be because higher molecular weight HA DMNs exhibit a longer application time for complete dissolution into the skin compared with lower molecular weight HA DMNs (Leone et al., 2021). Also, differences may arise because of the drugs used in each study. Propranolol hydrochloride has a smaller molecular weight than Remdesivir (295.80 g/mol compared to 602.59 g/mol). Therefore, the drug diffusion coefficients will be different for each study.

9.4 MODEL LIMITATIONS

As this mathematical model was created with no access to laboratory experiments, constants such as the polymer mean time for pore formation, the variance in time required to form pores and the mass transfer coefficient were estimated using experimental data from published literature. As it is difficult to find data for matching experimental conditions used within the model, there will be inaccuracies in these parameters. There is also limited data available on the properties of Remdesivir, such as density and viscosity. Therefore, there may be inaccuracies in calculated values, such as the drug diffusion coefficient.

9.5 CONCLUSIONS

This study developed a mathematical model to describe the TDD of Remdesivir, an antiviral medication repurposed to treat COVID-19 symptoms, using a DMN system. A framework of governing equations was used to describe the dissolution of the MN in the skin in the presence of interstitial fluid, the diffusion of skin interstitial fluid into the DMN and the diffusive drug release from DMN into the skin. The model was constructed using COMSOL Multiphysics 5.4 with the geometries drawn using 2D axial symmetry. HA with a molecular weight of 20 kDa was used to model the

DMNs. The results, predicted by the theoretical model, agree well with published experimental data for similar systems.

Simulations showed that the polymer mean time for pore formation had a significant effect on the time for the DMN density to reach zero. As the value of mean time for pore formation increased from 600 to 1000 s, the time for the MN density to reach zero also increased by 400 s. These findings highlight the importance of polymer choice on the biodegradation kinetics of the DMN system. It was also found that varying skin pH between 3.5 and 5.0 had a negligible effect on the change in polymer molecular weight over a 24-h period. The model found that by increasing the needle length from 150 to 350 μm led to a 49% increase in Remdesivir concentration in the skin 100 s after DMN insertion. A non-linear relationship was found to exist between the centre-to-centre spacing between DMNs and the concentration of Remdesivir in the skin. As the centre-to-centre spacing increased from 400 to 1200 μm, the concentration of Remdesivir in the skin decreased by 3.17×10^{-3} mol/m^3 (89%), 100 s after DMN insertion.

Through the creation of a one-compartmental PK model of Remdesivir, it was found that changing the SC thickness between 16.7 and 22.3 μm had a negligible effect on the concentration of Remdesivir in the blood over time. Both SC thicknesses showed maximum Remdesivir plasma concentrations of 2.1×10^{-12} mol/m^3 at 2600 s. It was also found that as the VE thickness decreased by 42% (120–70 μm), the maximum Remdesivir concentration increased by 1.7×10^{-13} mol/m^3 (8%).

9.6 RECOMMENDATIONS FOR FUTURE WORK

Future work should aim to collect experimental data to allow the use of accurate parameters within the model. This would also help to determine any drug-polymer interactions that have not been considered so far. Experimental data would allow statistical analysis to be completed on the model's ability to accurately simulate the transport mechanisms occurring. Further efforts to improve the model could work on accounting for the differences in the rates of MN degradation and skin pore closure, through considering the gap present between the MN and skin. The model could also be changed to use a 3D geometry analysis instead of 2D axial symmetry. However, this will require more computing power and time to execute the simulations.

9.7 NOMENCLATURE

TABLE 9.1

List of All the Parameters Used within the Model

Symbol	Description	Unit	Value
v_m	Remdesivir molar volume at boiling point	m³/kmol	0.6
ρ	Density of Remdesivir injection formulation	kg/m³	1103
T	Temperature	K	310.5
μ	Viscosity of Remdesivir	Pas	0.0012
M_D	Drug molecular weight	kg/kmol	602.585
$M_{P, t=0}$	Initial polymer molecular weight	kg/kmol	20,000
M_P	Polymer molecular weight	kg/kmol	Time-dependant variable
K_D	Mass transfer coefficient	m/s	1×10^7
K	Partition coefficient	-	0.06306
C_{D_i}	Concentration of Remdesivir in domain i	mol/m³	SC: 0, VE: 0, D: 0, MN: 0.2987
C_{W_i}	Concentration of water in domain i	mol/m³	SC: 5.417×10^4, VE: 5.417×10^4, D: 5.417×10^4, MN: 0.01
J_x	Instantaneous flux of species x	mol/(m²s)	Time-Dependant Variable
R_0	Initial MN base radius	m	0.000125
h_0	Initial MN height	m	0.00026
k_h	Polymer degradation rate constant	s⁻¹	1.67×10^{-12}
ϵ_i	Porosity of domain i	-	SC: 0.1, VE: 0.3, D: 0.3, MN: Time-Dependant Variable
t_{mean}	Mean time for pore formation	s	800
σ^2	Variance in time required to form pores	s²	20,000
D_{x_MN}	Effective diffusion coefficient of species x in MN	m²/s	Time-Dependant Variable
$D_{F\_D,i}$	Fluid diffusion coefficient of Remdesivir in domain i	m²/s	SC: 5×10^{-14}, VE: 1×10^{-10}, D: 1×10^{-10}
$D_{F\_W,i}$	Fluid diffusion coefficient of water in domain i	m²/s	SC: 1×10^{-14}, VE: 1×10^{-11}, D: 1×10^{-11}
V_d	Remdesivir volume of distribution	m³	0.064
dQ/dt	Penetration rate of Remdesivir through the skin	mol/(m³s)	Time-Dependant Variable
S_a	Area of skin-blood interface	m³	0.0000162
K_e	Remdesivir elimination rate constant	s⁻¹	2.78×10^{-4}

Key: Stratum corneum (SC), Viable epidermis (VE), Dermis (D), Dissolving microneedle (MN).

REFERENCES

Al-Qallaf, Barrak, and Diganta Das. "Optimizing microneedle arrays to increase skin permeability for transdermal drug delivery." *Annals of the New York Academy of Sciences* 1161 (2009): 83–94.

Andrews, Samantha, Eunhye Jeong, and Mark Prausnitz. "Transdermal delivery of molecules is limited by full epidermis, not just stratum corneum." *Pharmaceutical Research* 30 (2013): 1099–1109.

Bhatnagar, Shubhmita, Pradeeptha Reddy Gadeela, Pranathi Thathireddy et al. "Microneedle-based drug delivery: Materials of construction." *Journal of Chemical Sciences* 131 (2019): 1–28.

Chavoshi, Sarvenaz, Mohammed Rabiee, Mehdi Rafizadeh et al. "Mathematical modeling of drug release from biodegradable polymeric microneedles." *Bio-Design and Manufacturing* 2 (2019): 96–107.

Chen, Shoukai, Daming Wu, Ying Liu et al. "Optimal scaling analysis of polymeric microneedle length and its effect on transdermal insulin delivery." *Journal of Drug Delivery Science and Technology* 56 (2020): 101547.

Chu, Leonard, Ling Ye, Ke Dong et al. "Enhanced stability of inactivated influenza vaccine encapsulated in dissolving microneedle patches." *Pharmaceutical Research* 33 (2016): 868–878.

COMSOL. "Finite Element Mesh Refinement Definition and Techniques." January 6, 2016. Accessed May 16, 2021. https://www.comsol.com/multiphysics/mesh-refinement.

Dias dos Santos, Thiago, Mathieu Morlighem, Hélène Seroussi et al. "Implementation and performance of adaptive mesh refinement in the Ice Sheet System Model (ISSM v4.14)." *Geoscientific Model Development* 12 (2019): 215–232.

Gomaa, Yasmine, Desmond Morrow, Martin Garland et al. "Effects of microneedle length, density, insertion time and multiple applications on human skin barrier function: Assessments by transepidermal water loss." *Toxicology in Vitro* 24 (2010): 1971–1978.

Guillot, Antonio, Ana Cordeiro, Ryan Donnelly et al. "Microneedle-based delivery: An overview of current applications and trends." *Pharmaceutics* 12 (2020): 12060569.

Hanafin, Patrick, Brian Jermain, Anthony Hickey et al. "A mechanism-based pharmacokinetic model of remdesivir leveraging interspecies scaling to simulate COVID-19 treatment in humans." *CPT: Pharmacometrics and Systems Pharmacology* 10 (2021): 89–99.

Hassan, Jasmin, Charlotte Haigh, Ahmed Tanvir et al. "Potential of microneedle systems for Covid-19 vaccination: Current trends and challenges." *Pharmaceutics* 14 (2022) 1066.

He, Jingjing, Zichen Zhang, Xianzi Zheng et al. "Design and evaluation of dissolving microneedles for enhanced dermal delivery of propranolol hydrochloride." *Pharmaceutics* 13 (2021): 579.

Humeniuk, Rita, Anita Mathias, Huyen Cao et al. "Safety, tolerability, and pharmacokinetics of remdesivir, an antiviral for treatment of COVID-19, in healthy subjects." *Clinical and Translational Science* 13 (2020): 896–906.

Kamaly, Nazila, Basit Yameen, Jun Wu et al. "Degradable controlled-release polymers and polymeric nanoparticles: Mechanisms of controlling drug release." *Chemical Reviews* 116 (2016): 2602–2663.

Kim, Kwang, Kevin Ita, and Laurent Simon. "Modelling of dissolving microneedles for transdermal drug delivery: Theoretical and experimental aspects." *European Journal of Pharmaceutical Sciences* 68 (2015): 137–143.

Koehler, Martin, Tanja Vogel, Peter Elsner et al. "In vivo measurement of the human epidermal thickness in different localizations by multiphoton laser tomography." *Skin Research and Technology* 16 (2010): 259–264.

Leone, Mara, Stefan Romeijn, Bram Slütter et al. "Hyaluronan molecular weight: Effects on dissolution time of dissolving microneedles in the skin and on immunogenicity of antigen." *European Journal of Pharmaceutical Sciences* 146 (2021): 105269.

Ogunjimi, Abayomi, Jamie Carr, Christine Lawson et al. "Micropore closure time is longer following microneedle application to skin of colour." *Scientific Reports* 10 (2020): 1–14.

Olatunji, Ololade, Diganta Das, and Vahid Nassehi et al. "Modelling transdermal drug delivery using microneedles: Effect of geometry on drug transport behaviour." *Journal of Pharmaceutical Sciences* 101 (2012): 164–175.

Olatunji, Ololade, Diganta Das, Martin Garland et al. "Influence of array interspacing on the force required for successful microneedle skin penetration: theoretical and practical approaches." *Journal of Pharmaceutical Sciences* 102 (2013): 1209–1221.

Panda, Apoorva, Purnendu Sharma, Nanjappa Shivakumar et al. "Nicotine loaded dissolving microneedles for nicotine replacement therapy." *Journal of Drug Delivery Science and Technology* 61 (2021): 102300.

Podder, Pranay, Dhiman Mallick, Dip Samajdar et al. "Design, simulation and study of MEMS based micro-needles and micro-pump for biomedical applications." *Proceedings of the 2011 COMSOL Conference in Bangalore*, 2011.

Ronnander, Paul, Laurent Simon, and Andreas Koch. "Experimental and mathematical study of the transdermal delivery of sumatriptan succinate from polyvinylpyrrolidone-based microneedles." *European Journal of Pharmaceutics and Biopharmaceutics* 146 (2020): 32–40.

Rothstein, Sam, William Federspiel, and Steven Little. "A simple model framework for the prediction of controlled release from bulk eroding polymer matrices." *Journal of Materials Chemistry* 16 (2008): 1873–1880.

Sabri, Akmal, Yujin Kim, Maria Marlow et al. "Intradermal and transdermal drug delivery using microneedles - Fabrication, performance evaluation and application to lymphatic delivery." *Advanced Drug Delivery Reviews* 153 (2020): 195–215.

Tømmeraas, Kristoffer, and Claes Melander. "Kinetics of hyaluronan hydrolysis in acidic solution at various pH values." *Biomacromolecules* 9 (2008): 1535–1540.

Vadovic, C. J., and C. P. Colver. "Infinite dilution diffusion coefficients in liquids." *AIChE Journal* 19 (1973): 546–551.

Waghule, Tejashree, Gautam Singhvi, Sunil Dubey et al. "Microneedles: A smart approach and increasing potential for transdermal drug delivery system." *Biomedicine & Pharmacotherapy* 109 (2019): 1249–1258.

Yadav, Prateek R., Lewis J. Dobson, Sudip K. Pattanayek et al. "Swellable microneedles based transdermal drug delivery: Mathematical model development and numerical experiments." *Chemical Engineering Science* 247 (2022): 117005.

Yadav, Prateek R., Tao Han, Ololade Olatunji et al. "Mathematical modelling, simulation and optimisation of microneedles for transdermal drug delivery: Trends and progress." *Pharmaceutics* 12 (2020): 1–31.

Yadav, Prateek R., Monika N. Munni, Lauryn Campbell et al. "Translation of polymeric microneedles for treatment of human diseases: Recent trends, progress, and challenges." *Pharmaceutics* 13 (2021): 13081132.

Ya-Xian, Zhen, Takaki Suetake, and Hachiro Tagami. "Number of cell layers of the stratum corneum in normal skin relationship to the anatomical location an the body, age, sex and physical parameters." *Archives of Dermatological Research* 291 (1999): 555–559.

Zoudani, Elham Lori and Madjid Soltani. "A new computational method of modeling and evaluation of dissolving microneedle for drug delivery applications: Extension to theoretical modeling of a novel design of microneedle (array in array) for efficient drug delivery." *European Journal of Pharmaceutical Sciences* 150 (2020): 105339.

10 Engineering Approaches for Cellular Therapeutics and Diagnostics

Hamdi Torun, Prashant Agrawal, Christopher Markwell,
YongQing Fu, Stephen Todryk, and Sterghios A. Moschos
Northumbria University

CONTENTS

10.1 INTRODUCTION

Recent advances in microsystems and nanotechnology make it feasible to engineer structures and devices to interact with biological entities at single-molecule and single-cell levels for various applications in therapeutics and diagnostics [1–3]. Technologies for probing cellular behaviour can be broadly divided into two categories: contact and non-contact methods (Figure 10.1). In this chapter, we discuss these technologies and their application towards the manipulation of biological entities for applications in cellular therapeutics and diagnostics.

10.2 CONTACT-BASED APPROACHES FOR CELLULAR APPLICATIONS

10.2.1 NANOPARTICLE-BASED PROBES

Functional nanoparticles such as gold nanoparticles, magnetic nanoparticles and quantum dots have been used as labelling and sensitivity-enhancing agents for the detection of molecules and cells [4–6]. These particles can be mixed with biofluids, cells, biopsy preparations and molecules extracted from the body for *in vitro* diagnostics using biosensing and microfluidic platforms. It is also possible to introduce these particles into the living body, and their interaction with target cells and molecules can be detected *in vivo* using external detectors and imagers [7,8]. In addition to diagnostics, untargeted lipid nanoparticle complexes are used in the clinic for both therapeutic and vaccination formulations [9] and can be functionalized with drug

molecules for site-specific drug delivery. Biodegradable polymeric nanoparticles have been developed for these applications [10,11] building the paradigm of targeting nanoparticles that assemble passively after dosing, increasing targeted tissue loading by ~60x [12]. Elsewhere, some of the particles such as gold nanoparticles can be used for therapeutics as they provide the means for the selective destruction of labelled cells [13,14].

Because many large biological proteins assume 3D conformations in the low nanometre scale, engineered nanoparticles down to the size of individual molecules present obvious opportunities for diagnostic and therapeutic agents as the rapid progress in this area is an attestation. Figure 10.2 shows the progress for three generations of nanoparticles used for cellular applications [15]. Moreover, microsystems technology is an enabler for devices that can interact with single cells for diagnostics and therapeutics. Several such engineering approaches are reviewed in this chapter in this area.

10.2.2 SCANNING PROBE MICROSCOPY FOR CELLULAR APPLICATIONS

Scanning probe microscopy is based on scanning a microscale probe with a nanoscale tip over a sample where the interaction of the tip and the sample is captured during scanning. The technology has several variations where the probe and the method of detection are varied based on the needs of different applications. One such variation is atomic force microscopy (AFM) where a microcantilever integrated

DOI: 10.1201/9781003224464-10

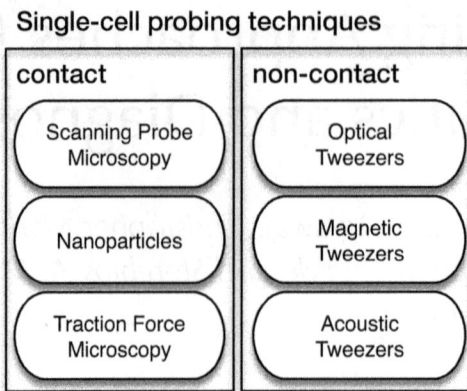

FIGURE 10.1   Different methods of single-cell level measurements and manipulations.

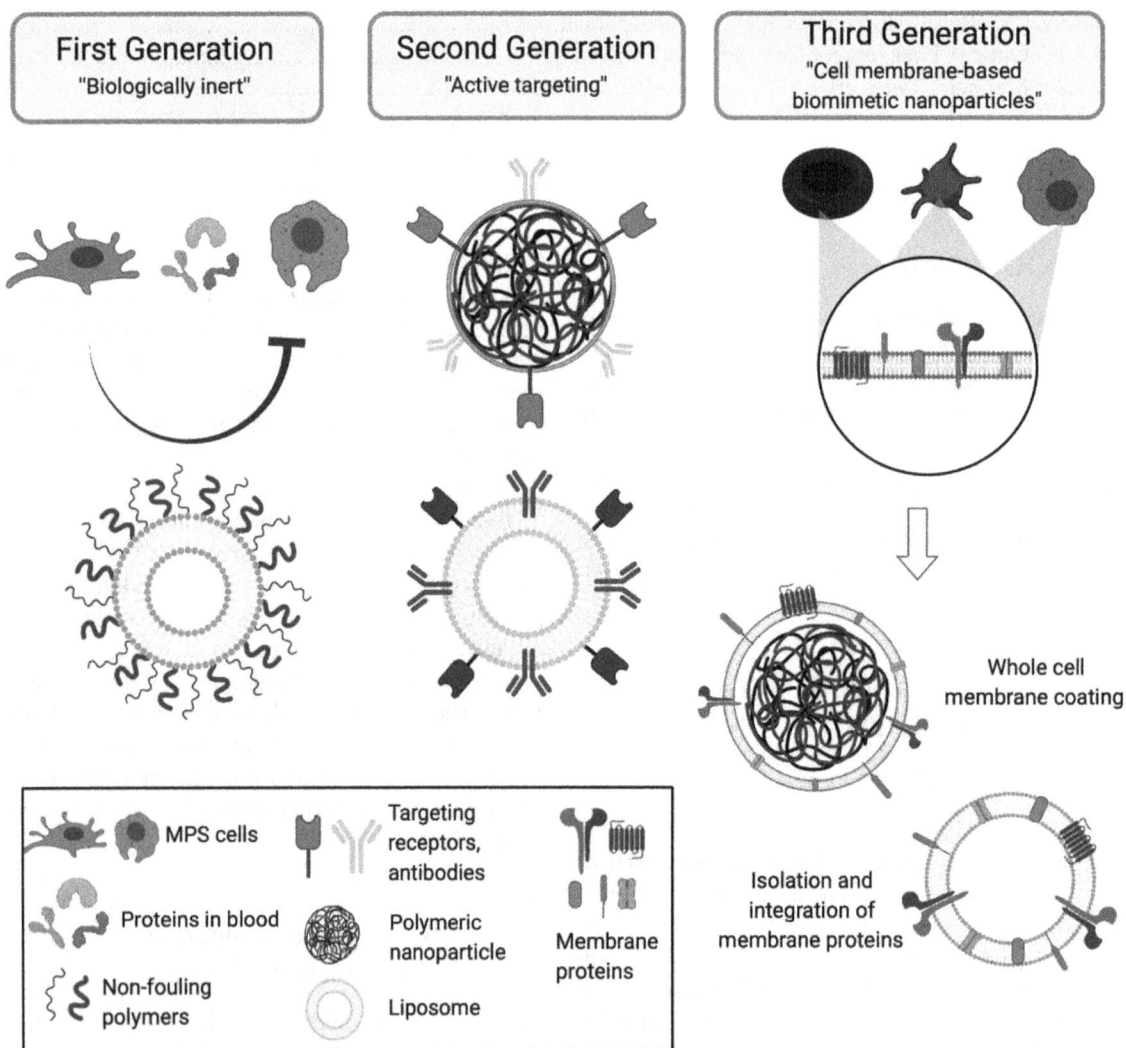

FIGURE 10.2   Nanoparticle-based probes and therapeutics are classified into three generations. The first generation of nanoparticles were inert and often featured polymer coatings such as polyethylene glycol that rendered the nanoparticles "stealth" to the immune system. The second generation of nanoparticles were decorated with targeting molecules to interact with target cells and target molecules. The third generation of nanoparticles are cell membrane-based, either natively produced in culture (e.g. exosomes) and labelled/loaded post-purification, or manufactured chemically and mechanically to render them biomimetic. (Adapted from Ref. [15].)

with a nanoscale sharp tip is used. The interaction of the cantilever and a sample results in deflection of the cantilever that is measured optically using a method called optical lever detection. The deflection of the cantilever can be interpreted as an interaction force when the spring constant of the cantilever is known. Using this method, interaction forces can be measured with a resolution of piconewtons while the distance between the base of the cantilever and the sample can be controlled with a resolution of nanometres. Crucial for biomolecular applications, AFM measurements can be performed in solution on live cells. Dynamic control of the distance while measuring the force precisely allows determination of viscoelastic properties of cells. There are excellent review papers summarizing this method [16,17].

AFM was originally developed as a topographical imaging tool, and it is possible to capture cellular, particulate and an even macromolecular (e.g. DNA) morphology using AFM. Differences in cell morphology can be used to differentiate normal and malignant cancer cells. Statistical differences in surface roughness between normal and leukaemic cells have been demonstrated in a small cohort study where cells were extracted using bone marrow collected from healthy individuals and patients [18]. The technology offers spatial resolution in sub-nanometre and nanometre scales, in out-of-plane and in-plane, respectively. The temporal resolution has been improving with high-speed AFM implementations where the cell images can be acquired at video rates. Different methods such as infrared spectroscopy have been combined with AFM to assess the morphological differences in conjunction with chemical heterogeneity for label-free identifying subcellular structures [19].

Morphological changes have also been reported as a method of assessing drug treatments, especially on cancer cells. The treatment of HeLa and Ishikawa cells using Paclitaxel, a chemotherapeutic drug, results in quantitative morphological changes as reported by Kim et al. [20]. In that study, AFM images reveal hole formation on the membranes of both cells after the treatment with the drug. Holes with depths of ~0.6 μm were imaged for cells with heights of ~4 μm. The interaction of drug molecules with single cells has also been investigated using a combined method of AFM and infrared spectroscopy [21]. Researchers investigated erlotinib, a tyrosine kinase inhibitor, adsorption on metallic nanoparticles using this system where erlotinib was measured spectroscopically while the metallic nanoparticles were imaged using AFM techniques. In another study, morphological differences between normal and cancer cells and how these are correlated with a tumour suppressor protein have been shown [22]. These approaches are promising in developing clinical interventions including drug discovery and assessing their efficacies in single-cell level.

Morphological changes are usually results of viscoelastic properties of cells, and the correlation between morphology and elasticity has been reported [23]. The elasticity of cells can be measured using AFM in liquid near-physiological conditions. This approach revealed that elasticity can be used to differentiate cancer and normal cells even when they are morphologically indistinguishable [24]. Pioneering work in this field demonstrated that the elasticity of cancer cells, especially metastatic cells, is significantly smaller than that of normal ones for breast and prostate cancer [24–27]. For example, the stiffness of metastatic cancer cells was measured 70% lower than normal mesothelial cells, both types of which were extracted from cancer patients with lung, breast and pancreatic cancers [27]. In addition, the researchers identified that the stiffness of tumour cells is narrowly distributed, whereas benign mesothelial cells displayed a broader distribution. After the framework was established, various research groups have used similar methods to characterize cells and to further categorize cancer cells in specific cancer stages. Investigation of cancer cells in different stages revealed a more complicated picture where cancerous, non-metastatic and metastatic cells can have different signatures [28]. A later work showed that the spread of elasticity values of a population of cells in a tissue sample can be used for diagnostics, where in early stages elasticity spectrum of a population of cells exhibits bimodal distribution with a peak in softened cells and broadening of the distribution towards higher stiffness values [29]. Obviously, cells are mechanically heterogenous and single values of elasticity do not capture the local variations. This was demonstrated in an earlier work with fibroblast cells where a localized map of elasticity over a single cell shows the presence of stiff cytoskeletal filaments including actin filaments, microtubules and intermediate filaments [30]. A further consideration is how the experiments are performed. Similar to any other experimental method, the measurements should be repeated to obtain statistically significant data. That implies the cells will be probed consecutively by the AFM cantilever, which can result in cellular cytoskeleton remodelling in response to the mechanical stimuli, that alters the elasticity values [31,32]. Experimental protocols should be developed to avoid this effect, and the measured elasticity values should be assessed accordingly.

A distinct feature of probe microscopy is the capability of measuring the interaction of biomolecules with cells at a single-molecule level. Towards this direction, AFM cantilevers were functionalized with antibodies, and their interaction with cancer cells was studied. Li et al. studied the interaction of CD20, a membrane protein present on B-cell lymphomas, with rituximab, an approved and widely clinically used monoclonal antibody therapeutic targeting CD20 [33]. They presented the interaction forces and the changes in cell morphologies due to the interactions. Specifically, the distribution of CD20 on the surface of cancer cells was obtained over an area of $500 \times 500 \, nm^2$ where interaction forces of 20–50 pN were measured. This approach where the spatial distribution of therapeutic agents over cell surfaces can complement conventional assays for drug design and evaluating therapeutic efficacy. For example, Guo et al. presented the interaction of an antibody against ICAM1, an intercellular adhesion molecule on cell surface, and proposed a metric to evaluate in vivo tumour targeting activity of a nanomedicine [34]. This molecule is considered as a target for immunotherapies for triple-negative breast cancer.

Combining the capabilities of AFM in single-molecule measurements and viscoelasticity characterization, Knoops et al. characterized the interaction between peroxiredoxins and their Toll-like receptors on the cell membrane [35]. Toll-like receptors are a family of eukayotic pattern recognition receptors that recognise pathogen-associated molecular patterns; their activation by pathogenic compounds, such as bacterial lipopolysaccharide, initiate and direct the nature of the innate immune response. Peroxiredoxins cellular anti-oxidative factors are released during cell damage and are specifically recognised by Toll-like receptor 4 (TLR4), serving thus as damage associated molecular patterns. The interaction forces between peroxiredoxins and Toll-like receptor were measured, the localized variations were shown, and the stiffening of cells after the interactions were quantified [35]. Specifically, in addition to showing that extracellular peroxiredoxins trigger a proinflammatory response, the research revealed that peroxiredoxin binding induces a cellular mechanoresponse resulting in cell stiffening. This is a direct evidence showing how cellular stiffening can be triggered during inflammation by a direct specific binding to TLR4. The combination of the AFM methods helped elucidate the inflammatory response in that research.

Being sensitive to surfaces, AFM can be used to probe molecules on the cell membranes such as G protein-coupled receptors (GPCR). In an interesting work, Pfreundschuh et al. introduced a method to map the interactions of a GPCR, a human protease-activated receptor, with two different ligands functionalized on an AFM cantilever while generating high-resolution AFM images [36].

Having the capability of generating high-resolution images of cell membranes and measuring interaction of molecules on the surface of cell membranes makes AFM methods also pivotal in understanding the mechanisms of antibiotics. In two different studies, mechanisms of pore formation on the cell membranes have been investigated as a result of antibiotics [37,38]. Zuition et al. investigated the action of daptomycin, a lipopeptide antibiotic, on living bacteria (*Bacillus subtilis* strains) [37]. This research revealed the formation of toroidal-shaped pores on bacteria membrane. Also, bacterial resistance to daptomycin was studied, and how the increased content of cardiolipin in the membrane can lead to resistance was explained based on the interaction of cardiolipin and daptomycin.

Figure 10.3 summarizes the use of AFM for single-cell applications including the characterization of viscoelastic

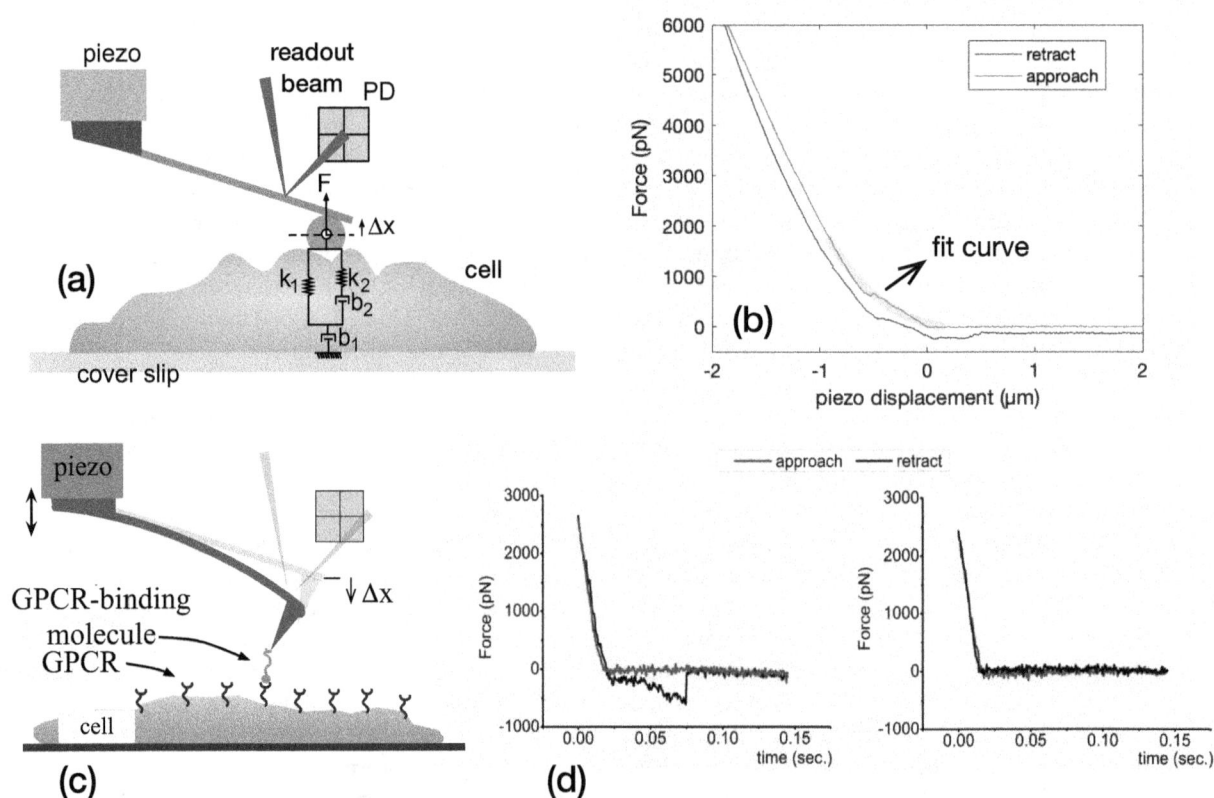

FIGURE 10.3 Probe microscopy for single-cell measurements: (a) measurement of viscoelastic properties of cells using a cantilever with a predefined tip interacting with a cell. The deflection of the tip Δx is measured using a photodetector (PD) to be represented as the interaction force shown in panel (b). The interaction force as a function of piezo displacement can be analysed to reveal the viscoelastic properties of the cells. Here the point where the cantilever makes contact with the cell is labelled as the reference displacement. The negative displacement region indicates the cantilever is pressed into the cell membrane. (c) The cantilevers can be functionalized with molecules that can bind on, e.g., GPCR to measure the interaction forces between receptor and ligand. Panel (d) shows example force curves that indicate the presence (left) and absence (right) of single-molecular interaction forces.

properties of cells and the measurement of molecules on cell membranes.

10.2.3 Traction Force Microscopy

Cells adhere to surfaces to survive and proliferate. The traction forces generated by cells on surfaces allow them to probe surrounding environments and organize extracellular matrices. Traction force microscopy (TFM) is a technique to measure these traction forces. One of the first studies visualized and measured these forces through the morphology of wrinkles created by cells on a soft silicone rubber [39]. The traction forces generated by cells on thin-sheets of rubber are visualized through elastic distortions (or wrinkles) on the surface. The size of the wrinkles are measured via phase contrast microscopy to provide qualitative estimates of the traction force magnitude and direction. Later developments explored the use of fluorescent beads or markers embedded in a soft substrate such as polyacrylamide hydrogels or silicon-based gels for more quantitative force measurements. Positions of the markers are tracked before and after the cell adherence to determine a traction force map from their displacement [40–42]. These markers can be beads embedded at regular spacings [43] or fluorescent photoresist markers lithographically printed on the substrate [44]. Polyacrylamide and silicon-based gels are widely used for 2D TFM as they undergo linear and elastic deformation across several orders of magnitude and are mechanically stable under repeated usage or cyclic testing [45,46].

From analysing cells on 2D substrates, TFM has evolved to assess cell traction forces in more natural 3D environments, with initial studies characterizing forces in multicellular clusters [47], on substrates with varying stiffness [48] and employing confocal microscopy for out-of-focal-plane force measurements [49]. Although 3D environments are more natural and result in significantly different cell behaviours compared to 2D cultures [50], 3D TFM poses several challenges including a technical one on obtaining sub-micrometre scale force resolution. There are also practical difficulties such as controlling mechanical properties of the 3D ECM due to reorganization and degradation as cells proliferate and challenges arising due to non-linear material properties of fibrous ECM's [51]. For 3D TFM, collagen type I hydrogel is the mostly commonly used material for mimicking ECM [52]. Due to the non-linear force-displacement relationship of the material, computational techniques are used to model the collagen ECM [50] and quantify the cell traction forces. Synthetic hydrogels of polyethylene glycol gels which demonstrate linear elastic deformation have also been used to quantify small deformations of single cells [53,54]. However, quantifying forces by cell clusters as cells proliferate and incorporating the ensuing material non-linearity is still a challenge in these synthetic hydrogels.

Microbead-based TFM provides a subcellular resolution for mapping traction stresses which assists in understanding several processes such as adhesion of fibroblasts [55],

differentiation between fibroblasts and myofibroblasts [56], mechanical response of cell clusters [57], migration of cell clusters [47], wound healing [56], inflammation and metastasis [58–61].

Another variation of TFM is a micropillar-based TFM, where deformation of micropillars on a substrate due to cell adhesion is measured to determine traction forces. For small displacements, the pillars act as a spring where the force is directly proportional to the deformation. As a result, this method avoids the computation of substrate properties and the assumption of linear response and provides displacement directly from the pillar displacement.

The micropillar devices can be either silicon-based or soft polymer and gel-based [62]. On silicon-based devices, pliant pillars and cantilevers are used to measure changes in electrical response of the devices [63–65]. Conversely, these devices can also be used to actuate cells to assess their response to external stimuli [66,67], while additional force sensors can also be integrated [68,69]. While incompatibility with the integration of electronics in soft polymer poses a challenge in direct measurement of forces, biocompatibility, ease of manufacture, optical transparency and tunable mechanical and chemical properties provide significant advantages to these devices for TFM [62]. Soft substrates can also be mechanically actuated to study cell response to stimuli via stretching [70,71] or magnetic actuation of pillars [72]. Micropillar-based TFM's have been used to study traction properties of various types of cells, such as stem cells, epithelial cells, neutrophills, fibroblasts, platelets and T-cells [73–80]. Pillar sizes are crucial to ensure appropriate cell adhesion and deformation of pillars. Traction force sensitivity can be increased by reducing the stiffness of the pillars, which implies that pillars undergo large deformations with small forces. This stiffness reduction can be achieved through pillars with high-aspect ratio or by reducing the pillar diameter. Techniques such as deep reactive ion-etching (DRIE) and subsequent polymer moulding have been used to fabricate high-aspect ratio and sub-micron-sized pillars for increased sensitivity and cell adhesion [72,74,75,81–83].

Although micropillar TFM's provide a simple way to measure traction forces, the substrate stiffness is limited compared to smooth (non-pillared) surfaces [84]. This also leads to challenges in estimating an effective stiffness of pillared surfaces for comparison with natural physiological systems [85]. The linear spring equivalent model is predominantly used, but it is limited to small deformations of <10% [86]. Therefore, integration of large deflection theory [87] in analytical models, accounting for substrate deformation [88], effect of substrate viscoelasticity on periodic excitation [89] and modelling non-linear deformation at large forces [90] are significant challenges in adopting this technology for cell traction force measurement.

Practically, manufacturing of these surfaces also poses a significant challenge in the wide-scale adoption of this technique. These surfaces are manufactured using a typically two-step process. The first step is fabrication of a master structure or a master mould, which serves as the negative of the required features. The second step is moulding the polymer liquid in the master mould and eventual peel off after setting it to obtain the desired surfaces. Fabrication of the master mould relies on complex fabrication procedures such as DRIE and photolithography, which are not readily accessible and pose challenges with relatability.

10.3 NON-CONTACT-BASED APPROACHES FOR CELLULAR APPLICATIONS

Although contact-based force measurement approaches are advantageous in various applications, non-contact force-based methods are preferred for sorting and trapping applications.

10.3.1 OPTICAL TWEEZERS FOR CELLULAR APPLICATIONS

The optical tweezer method is a non-contact force-based approach where a highly focused laser beam is used to generate scattering and gradient forces on particles that can be trapped within the focal region of the laser beam. Arthur Ashkin invented the method and developed the theory of optical tweezing [91,92]. Typically, a biological cell is larger than the wavelengths of light (500–1100 nm) used in this method, so the trapping mechanism is usually in ray optics regime where the diffraction does not have significant influence. Since the cells do not have to be anchored on a surface, optical tweezers can be used on cells in motion, which is particularly useful for cell sorting based on the differences in size and refractive indices [93].

Red blood cells (RBC) have been at the focus of this application as they are usually not adhered to a surface [94]. Using an infrared light source, in vivo trapping of RBCs has also been reported within living mice [95]. In addition to sorting and trapping, cells can be deformed by controlling the light source for rheological measurements. A method of investigating three-dimensional dynamic deformation of RBC membrane was presented using a dual-trap optical tweezers where two parallel trapping beams are used [96]. In this study, RBC's were elongated by 20%, and the method is suggested to investigate the effects of drug treatment on RBC elasticity.

An inherent advantage of the optical tweezers is the use of lasers that can also be utilized for spectroscopy. In an interesting work, Raman spectroscopy is integrated to a dual-beam optical tweezers to map the distribution of haemoglobin in trapped RBCs [97]. The integration of Raman spectroscopy as a diagnostic tool with optical tweezers is known as Raman tweezers [98]. Raman spectroscopy is based on measuring scattered light from a specimen illuminated with coherent light to determine vibrational modes of molecules. A combination of optical tweezers and Raman spectroscopy usually requires the addition of another laser source to excite the specimens. Raman tweezers have been used to investigate various cells including bacteria and eukaryotic cells, and even organelles as reported by Tang et al. where the researchers investigated Ca^{2+}-induced damage of mitochondria [99]. It has been demonstrated that Raman tweezers can also be used to differentiate between populations of different cells at single-cell level based on feature extraction and principle component analysis. Examples include differentiation between two strains of bacteria cells for their capability of biofilm formation [100], as well as between prostate and bladder cells in urine relevant to prostate cancer diagnostics [101]. Spectral information allowed the researchers to differentiate the cell types rather than cell size which is crucial for diagnostics [101].

Despite the advantages and capabilities of the optical tweezers, the fundamental trapping mechanism by irradiating the targets with focused light can be harmful. The detrimental effects of optical trapping on the cells and biologically relevant molecules have been observed especially in the visible and UV light spectra where the molecules often exhibit strong absorption. The damage mechanisms have been explained by photothermal damage due to one-photon and two-photon processes and thermal damage due to the elevation of temperatures of the targets and their surroundings [102]. Near-infrared bands have been suggested to reduce the damaging effects with 700–1100 nm band coined as the "biological window" outside of the absorption bands for O_2-binding haemoglobin and water [103]. However, physiological damage to cells has been reported also in this band [104]. Figure 10.4 summarizes different implementations of optical tweezers.

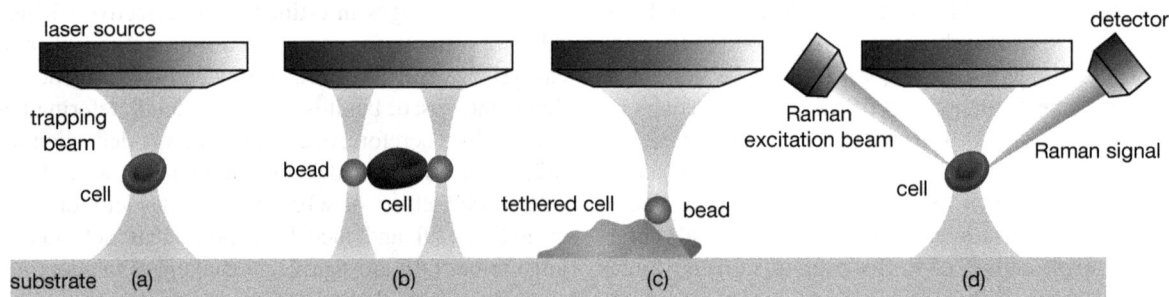

FIGURE 10.4 Optical tweezers can be used in different configurations to probe cells either (a) directly by illuminating the cells or (b) via microbeads that can be trapped. The microbeads are attached to the cells and are used as anchors. This configuration has advantages as the cells are not directly illuminated with laser beams that can be harmful for the cells especially at high power and in prolonged operation. The cells can be untethered or (c) tethered on a substrate. (d) Raman spectroscopy can be integrated with optical tweezers to investigate trapped cells.

10.3.2 Magnetic Tweezers

Magnetic tweezing is a non-contact method which uses magnetic beads controlled through a magnetic field gradient to measure forces on single molecules or cells (both extra- and intracellular). By tracking the motion of the particles through an optical microscope, magnetic tweezing can measure forces between pico- and nano-Newtons [105].

The magnetic field can be generated using permanent magnets; however, electromagnets provide significant flexibility to control the magnetic force by changing its magnitude, field direction and number of poles. The nature of magnetic field interaction with biological samples can also be altered by using different magnetic particles. For example, weak ferromagnetic and superparamagnetic particles are used for their zero residual magnetization and high susceptibility. Ferromagnetic nanoparticles are used with weak magnetic fields due to their high saturation magnetization [106]. More recently, particles with different shapes such as cylindrical rods and helices, and compositions of magnetic materials have widened the performance and application scope of magnetic tweezers [107,108]. For example, anisotropically shaped magnetic particles can be aligned in a specific direction on the action of a homogeneous magnetic field to control their orientation and self-assembly for sensing and diagnostics as bio-probes. Metallic and alloy-based nanorods have enhanced torque measurement sensitivity without significantly altering the stretching force [108–110]. Active magneto-optic materials widen applications for integration of optical control and measurements. For example, addition of a gold segment to a nanorod made it possible to detect the orientation of the rods with nanometre resolution [110].

Magnetic tweezers have been used to study the response of single cells and large cellular populations to torque and stretching forces. In this method, the application of torque is relatively straightforward and can be used to investigate helical structures such as DNA or molecular motors. Several approaches have been developed to decouple the effect of stretching forces on torque measurements and expand the range of force and torque measurements [111–115]. In a few of the early studies, torque on arginine-glycine-aspartic acid peptide-coated magnetic beads were used to measure cytoskeleton stiffening properties of endothelial cells, observing the actin cytoskeleton dependent stiffening response [116]. The study highlighted the effect of different transmembrane molecules (e.g. integrin subunits, cadherins and cell surface proteoglycans) in transferring mechanical stimuli to the cytoskeleton filaments, leading to the potential usage of this technique to study intracellular signalling mechanisms. Stretching forces via collagen-coated beads over single and large cell populations have been used to observe changes in intracellular calcium concentrations [117,118], activation of RhoA molecules and their role in determining signalling pathways via tension [119,120]. Interestingly, magnetic tweezers allowed the observation of retraction of integrin-bound beads in a direction opposite to that of the applied stress that required integrin activation, Rho signalling and cytoskeletal tension generation. Saphirstein et al. also demonstrated the role of vascular smooth muscle cells in determining the stiffness of aortic cells which is linked to cardiovascular diseases [121]. This study identified focal adhesions of vascular muscle cells as a significant regulator of aortic stiffness. This result highlights the focal adhesion as a potential therapeutic target as increases in aortic stiffness are linked to cardiovascular diseases. Magnetic tweezers can also be used to measure viscoelastic properties of cell cytoplasm and cell membrane stiffnesses and to study cellular organelles [122].

The use of external magnetic fields also allows magnetic tweezers to be coupled with different technologies for diagnosis and added functionality. For example, integration with microscopic imaging can increase spatial and temporal sensitivity [123,124]. The in-plane position of beads are obtained by real-time correlation of the bead images with a resolution of a few nanometres, whereas the out-of-plane position of the beads are obtained using diffraction effects with an accuracy of 10 nm. On similar lines, optical magnetic twisting cytometry has been used to study the effect of periodic forces on cell rigidity [125–127]. Fluorescent resonant energy transfer (FRET) coupled with magnetic tweezers has also been used to study live signalling in cells via fluorescence microscopy [128,129]. For example, Poh et al. measured changes in a Rac (guanosine phosphatase molecule) activity in airway muscle cells using FRET in response to mechanical stimuli applied through magnetic fields [129].

Magnetic tweezers can also be coupled with other active and passive force probe techniques to expand functionality for different applications [124,130]. The use of deformable substrates (TFM) along with magnetic tweezers has been used to measure contraction force and torque on bovine pulmonary artery cells on a bed of elastic micropillars [109]. This study shows that the response of the cells to periodic stimuli is frequency-dependent. The cells show enhanced reinforcement to periodic stimuli with a frequency range of 0.5–1 Hz. The precise mechanism was not clear, but the method revealed the frequency-dependent response of the cells *in vivo*. Magnetic micropillars on a soft elastomer have been actuated to mechanically stimulate cells on a surface [131]. In this method, the cells are not contaminated with magnetic particles, while still allowing the continuous simultaneous stimulation of cells through tension and compression, and force measurement with TFM.

As discussed above, magnetic tweezers offer a versatile method to study cellular interaction with external stimuli as wells as their signalling response [132–134]. Due to the ease of applying magnetic fields, magnetic tweezers are extremely versatile in measuring stretching forces (perpendicular to the substrate) as well as torque or twisting forces (parallel to the substrate) [113,135–137]. Although the equipment is simple to use and highly customizable, magnetic tweezers are not widely commercialized and standardized. Although a non-contact method,

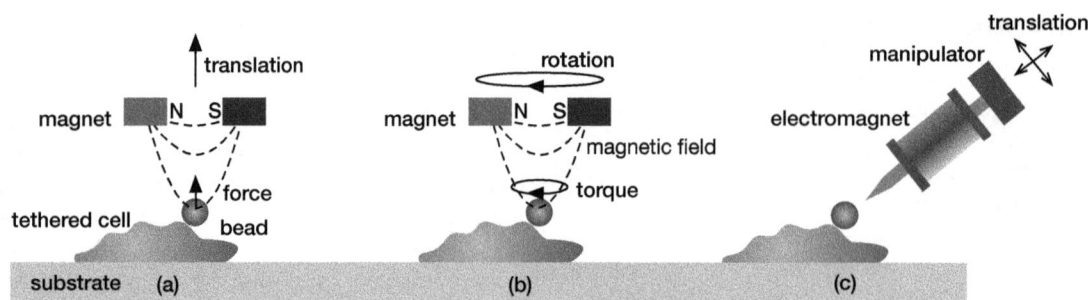

FIGURE 10.5 Magnetic tweezers can be implemented using (a and b) permanent magnets or (c) electromagnets to probe cells. Translational movement of the permanent magnets or electromagnets can be used to (a,c) apply force, whereas rotational motion of the permanent magnets can be used to (b) apply torque.

the technique involves inserting magnetic material in the sample which can alter responses and is a source of contamination. Biomolecules can also demonstrate hysteresis with the applied magnetic field, when the orientation of the magnetic domains within the structure does not perfectly follow the external magnetic field but retain part of their alignment. In this case, cyclic change in the direction of external magnetic field can result in heating of the sample [138]. Figure 10.5 depicts the magnetic tweezers to apply force and torque on single cells.

10.3.3 ACOUSTIC TWEEZERS

Acoustic tweezers are a versatile technology using ultrasonic waves to trap and manipulate particles. The most common type of acoustic tweezers use piezoelectric transducers to generate a focused ultrasound beam creating an acoustic radiation pressure field with a stable potential well where particles and cells can be trapped and held stably [139]. The position and number of these stable points can be altered by changing the frequency of the wave or moving the transducers; the trapped particles undergo a displacement as well with the movement of the potential well. One of the first studies demonstrated the trapping of frog eggs in a focused ultrasound beam [140]. A second type of acoustic tweezers utilize travelling acoustic waves interacting with other waves [141–143] and phononic structures (usually periodic structures with varying mass densities and acoustomechanical properties to control the reflection, refraction and transmission of waves) [144–146] to generate arbitrary pressure fields. The pressure fields generated from these interacting waves and structures can be modulated and controlled for dynamic trapping of cells. A third type of tweezers utilize acoustic streaming flows to trap particles in vortices. Acoustic streaming is a second-order steady flow which is generated by the absorption of acoustic energy by the liquid [147,148]. Cells can be trapped in these flows generated next to an oscillating pillar [149,150] or a bubble [151] and can be sorted based on their size and densities for diagnostics.

The flexibility to generate acoustic fields in a 3D space allows acoustic tweezers to manipulate particles and cells in all three spatial directions for sorting [152,153], separating

[154] and focusing [155]. Cells can be manipulated in a user-defined path in a 2D plane [156] and position controlled out of plane [157] towards applications in complex dynamic printing of cells [157,158]. Recently, some studies with in-vivo manipulation of cells in blood vessels have also been reported [155]. Introducing gradient acoustic forces, it was possible to exceed the drag forces of flow at a blood velocity of 5–10 mm/s in 50–70-μm-diameter blood vessels. Apart from translation, acoustic tweezers can also non-invasively rotate cells [142,159–162], as well as larger organisms, like zebrafish embryos [163] with negligible effects to their cellular properties when the field strengths are controlled.

The label-free, non-invasive nature of acoustic tweezers has been used for manipulating a wide range of cell types for different applications. For instance, size-dependent manipulation by acoustic tweezers have been used for collecting and analysing blood platelets [164], circulating tumour cells (CTCs) [165] and primary human trophoblasts (PHT) derived exosomes [166]. Acoustic waves have also been used to trap and pattern cell clusters in 2D arrays to study cell-cell communication and interaction [158,167]. For example, Feng et al. qualitatively studied the gap junctional intercellular communication via the exchange of dyes between cells trapped using acoustic tweezers [158] as a demonstration of the technology. Single-cell trapping is still a challenge as higher frequency devices (requiring smaller length scales) are necessary; however, recent studies with gigahertz frequencies have demonstrated individual trapping of *Plasmodium falciparum*–infected RBC were observed after 2D patterning [167].

Acoustic tweezers have also been used to study the bond strength and conformational properties of single molecules [168]. Termed as acoustic field spectroscopy, molecules are attached to glass microspheres and tethered to a glass chamber, and a standing acoustic wave is generated which displaces the glass microsphere. The acoustic field pressure and microsphere displacement provide a measure of the molecule bond strength. The key advantage of acoustic field spectroscopy over other methods is that such measurements can be performed for several samples at a time, increasing throughput.

Acoustic tweezers offer several advantages over other non-contact-based techniques [169,170]. For instance, the radiation forces acting on the biological specimens are significantly lower than methods such as optical tweezers [171,172]. As a result, cell response, characteristics, function and viability are maintained during operation and measurement. Cells do not need to be tagged for measurement, as opposed to tagging by magnetic particles in magnetic tweezers, and can also be manipulated in their preferred cell culture medium. Acoustic tweezers do not affect or alter physical properties such as size, density and shape and other native properties like refractive index and charge. The setup and operation are also simple [157] allowing integration in microfluidic devices as surface acoustic waves [149,173,174].

Despite the versatility of acoustic tweezers, their adoption by medical and scientific communities faces several challenges. For instance, these tweezers have limited spatial resolution. High frequency devices can solve this problem but face challenges in device fabrication and complex operation. The biocompatibility and versatility of the tweezers is also a drawback as a lot of studies are applicable to specific systems and are not standardized. This limits the adoption of this technology by non-technical users in the biomedical community.

10.4 FUTURE PROSPECTS

The measurement capabilities of contact and non-contact methods summarized in this chapter for single-cell level measurements have been improving over the last a couple of decades. The spatial and temporal resolution of the measurements have reached to the level where it is possible to observe biomolecular processes near-physiological conditions. It is expected that further improvements will be demonstrated in near future. The application of these methods for diagnostics is anticipated to grow following these improvements. Interpretation of the basic measurements into diagnostics requires understanding of fundamental processes where the development in these techniques will be pivotal. These techniques not only allow us to answer some of the key questions in cell biology, but they also lead to new questions.

There are still some areas for improvement to utilize these methods for a wider range of applications. The throughput of the measurements is still limited in applications where statistically significant data require large sample sizes. An obvious solution can be either parallel measurements where arrays of devices operate simultaneously, or serial measurements are made very fast within a reasonable time scale. Both alternatives are challenging, and future developments are expected along these lines.

An important trend in the field is the multimodal functionalities brought together by integration of different techniques in a single setup. Inherent advantages of individual techniques are important in this aspect. For example, it is relatively easy to integrate a Raman spectroscopy and confocal microscopy to optical tweezers for multimodal measurements that can lead to better diagnostics; however, the established position of these analytical techniques in research and diagnostic laboratories suggests that

manipulation tools are more likely to become commercially adopted if delivered as add-ons to existing analytical platforms. Likewise, scanning probe microscopy methods are complemented with optical measurements.

Another exciting area for development is *in vivo* applications of the techniques covered in this chapter. Implementation of single-cell measurement techniques *in vivo* is severely limited presently due to technical difficulties and safety, although real-time niche applications in conditionally accessible biofluids such as kidney dialysis and transplantation surgery offer opportunities for near term technique and application development. However, long-time scale measurements *in vivo* can lead to exciting applications in therapeutics in addition to diagnostics. Further miniaturization is required for this prospect.

REFERENCES

1. Li, Y., Zhao, L., Yao, Y. & Guo, X. Single-molecule nanotechnologies: An evolution in biological dynamics detection. *ACS Appl. Bio Mater.* **3**, 68–85 (2020).
2. Zhang, Y., Li, M., Gao, X., Chen, Y. & Liu, T. Nanotechnology in cancer diagnosis: Progress, challenges and opportunities. *J. Hematol. Oncol.* **12**, 137 (2019).
3. Heath, J. R. Nanotechnologies for biomedical science and translational medicine. *Proc. Natl. Acad. Sci.* **112**, 14436–14443 (2015).
4. Wilson, R. The use of gold nanoparticles in diagnostics and detection. *Chem. Soc. Rev.* **37**, 2028 (2008).
5. Wu, Y., Ali, M. R. K., Chen, K., Fang, N. & El-Sayed, M. A. Gold nanoparticles in biological optical imaging. *Nano Today* **24**, 120–140 (2019).
6. Wagner, A. M., Knipe, J. M., Orive, G. & Peppas, N. A. Quantum dots in biomedical applications. *Acta Biomater.* **94**, 44–63 (2019).
7. Bouché, M. et al. Recent advances in molecular imaging with gold nanoparticles. *Bioconjug. Chem.* **31**, 303–314 (2020).
8. Pons, T. et al. In vivo imaging of single tumor cells in fast-flowing bloodstream using near-infrared quantum dots and time-gated imaging. *ACS Nano* **13**, 3125–3131 (2019).
9. Suzuki, Y. & Ishihara, H. Difference in the lipid nanoparticle technology employed in three approved siRNA (Patisiran) and mRNA (COVID-19 vaccine) drugs. *Drug Metab. Pharmacokinet. J.* **41**, 100424 (2020).
10. Karlsson, J., Vaughan, H. J. & Green, J. J. Biodegradable polymeric nanoparticles for therapeutic cancer treatments. *Annu. Rev. Chem. Biomol. Eng.* **9**, 105–127 (2018).
11. Ahlawat, J., Henriquez, G. & Narayan, M. Enhancing the delivery of chemotherapeutics: Role of biodegradable polymeric nanoparticles. *Molecules* **23**, 2157 (2018).
12. Debacker, A. J., Voutila, J., Catley, M., Blakey, D. & Habib, N. Delivery of oligonucleotides to the liver with GalNAc: From research to registered therapeutic drug. *Mol. Ther.* **28**, 1759–1771 (2020).
13. Patino, T. et al. Multifunctional gold nanorods for selective plasmonic photothermal therapy in pancreatic cancer cells using ultra-short pulse near-infrared laser irradiation. *Nanoscale* **7**, 5328–5337 (2015).
14. Pissuwan, D., Valenzuela, S. M., Killingsworth, M. C., Xu, X. & Cortie, M. B. Targeted destruction of murine macrophage cells with bioconjugated gold nanorods. *J. Nanoparticle Res.* **9**, 1109–1124 (2007).

15. Sushnitha, M., Evangelopoulos, M., Tasciotti, E. & Taraballi, F. Cell membrane-based biomimetic nanoparticles and the immune system: Immunomodulatory interactions to therapeutic applications. *Front. Bioeng. Biotechnol.* **8**, 627 (2020).

16. Garcia, R. Nanomechanical mapping of soft materials with the atomic force microscope: Methods, theory and applications. *Chem. Soc. Rev.* **49**, 5850–5884 (2020).

17. Krieg, M. et al. Atomic force microscopy-based mechanobiology. *Nat. Rev. Phys.* **1**, 41–57 (2019).

18. Gaman, A., Osiac, E., Rotaru, I. & Taisescu, C. Surface morphology of leukemic cells from chronic myeloid leukemia under atomic force microscopy. *Curr. Heal. Sci. J.* **39**, 45–47 (2013).

19. Kennedy, E., Al-Majmaie, R., Al-Rubeai, M., Zerulla, D. & Rice, J. H. Quantifying nanoscale biochemical heterogeneity in human epithelial cancer cells using combined AFM and PTIR absorption nanoimaging. *J. Biophotonics* **8**, 133–141 (2015).

20. Kim, K. S. et al. AFM-detected apoptotic changes in morphology and biophysical property caused by paclitaxel in ishikawa and HeLa cells. *PLoS One* **7**, e30066 (2012).

21. Piergies, N. et al. Nanoscale image of the drug/metal monolayer interaction: Tapping AFM-IR investigations. *Nano Res.* **13**, 1020–1028 (2020).

22. Kaul-Ghanekar, R. et al. Tumor suppressor protein SMAR1 modulates the roughness of cell surface: Combined AFM and SEM study. *BMC Cancer* **9**, 350 (2009).

23. Guo, Q., Xia, Y., Sandig, M. & Yang, J. Characterization of cell elasticity correlated with cell morphology by atomic force microscope. *J. Biomech.* **45**, 304–309 (2012).

24. Cross, S. E. et al. AFM-based analysis of human metastatic cancer cells. *Nanotechnology* **19**, 384003 (2008).

25. Li, Q. S., Lee, G. Y. H., Ong, C. N. & Lim, C. T. AFM indentation study of breast cancer cells. *Biochem. Biophys. Res. Commun.* **374**, 609–613 (2008).

26. Lekka, M. et al. Cancer cell detection in tissue sections using AFM. *Arch. Biochem. Biophys.* **518**, 151–156 (2012).

27. Cross, S. E., Jin, Y.-S., Rao, J. & Gimzewski, J. K. Nanomechanical analysis of cells from cancer patients. *Nat. Nanotechnol.* **2**, 780–783 (2007).

28. Faria, E. C. et al. Measurement of elastic properties of prostate cancer cells using AFM. *Analyst* **133**, 1498 (2008).

29. Plodinec, M. et al. The nanomechanical signature of breast cancer. *Nat. Nanotechnol.* **7**, 757–765 (2012).

30. Haga, H. et al. Elasticity mapping of living fibroblasts by AFM and immunofluorescence observation of the cytoskeleton. *Ultramicroscopy* **82**, 253–258 (2000).

31. Lekka, M. & Laidler, P. Applicability of AFM in cancer detection. *Nat. Nanotechnol.* **4**, 72 (2009).

32. Cross, S. E., Jin, Y.-S., Rao, J. & Gimzewski, J. K. Applicability of AFM in cancer detection. *Nat. Nanotechnol.* **4**, 72–73 (2009).

33. Li, M. et al. Nanoscale mapping and organization analysis of target proteins on cancer cells from B-cell lymphoma patients. *Exp. Cell Res.* **319**, 2812–2821 (2013).

34. Guo, P. et al. Using atomic force microscopy to predict tumor specificity of ICAM1 antibody-directed nanomedicines. *Nano Lett.* **18**, 2254–2262 (2018).

35. Knoops, B. et al. Specific interactions measured by AFM on living cells between peroxiredoxin-5 and TLR4: Relevance for mechanisms of innate immunity. *Cell Chem. Biol.* **25**, 550–559 (2018).

36. Pfreundschuh, M. et al. Identifying and quantifying two ligand-binding sites while imaging native human membrane receptors by AFM. *Nat. Commun.* **6**, 8857 (2015).

37. Zuttion, F. et al. High-speed atomic force microscopy highlights new molecular mechanism of daptomycin action. *Nat. Commun.* **11**, 6312 (2020).

38. Maity, S., Melcrová, A. & Roos, W. H. Surface active antibiotics in action: A real-time study of the killing mechanism using high speed-atomic force microscopy. *Biophys. J.* **121**, 220 (2022).

39. Harris, A. K., Wild, P. & Stopak, D. Silicone rubber substrata: A new wrinkle in the study of cell locomotion. *Science (80-.).* **208**, 177–179 (1980).

40. Franck, C., Maskarinec, S. A., Tirrell, D. A. & Ravichandran, G. Three-dimensional traction force microscopy: A new tool for quantifying cell-matrix interactions. *PLoS One* **6**, e17833 (2011).

41. Butler, J. P., Toli-Nørrelykke, I. M., Fabry, B. & Fredberg, J. J. Traction fields, moments, and strain energy that cells exert on their surroundings. *Am. J. Physiol. - Cell Physiol.* **282**, 595–605 (2002).

42. Munoz, J. J. Non-regularised inverse finite element analysis for 3D traction force microscopy. *Int. J. Numer. Anal. Model.* **13**, 763–781 (2016).

43. Bergert, M. et al. Confocal reference free traction force microscopy. *Nat. Commun.* **7**, 1–10 (2016).

44. Balaban, N. Q. et al. Force and focal adhesion assembly: A close relationship studied using elastic micropatterned substrates. *Nat. Cell Biol.* **3**, 466–472 (2001).

45. Roca-Cusachs, P., Conte, V. & Trepat, X. Quantifying forces in cell biology. *Nat. Cell Biol.* **19**, 742–751 (2017).

46. Khetan, S. et al. Degradation-mediated cellular traction directs stem cell fate in covalently crosslinked three-dimensional hydrogels. *Nat. Mater.* **12**, 458–465 (2013).

47. Trepat, X. et al. Physical forces during collective cell migration. *Nat. Phys.* **5**, 426–430 (2009).

48. Sunyer, R. et al. Collective cell durotaxis emerges from long-range intercellular force transmission. *Science (80-.).* **353**, 1157–1161 (2016).

49. Bastounis, E. et al. Both contractile axial and lateral traction force dynamics drive amoeboid cell motility. *J. Cell Biol.* **204**, 1045–1061 (2014).

50. Steinwachs, J. et al. Three-dimensional force microscopy of cells in biopolymer networks. *Nat. Methods* **13**, 171–176 (2016).

51. Hall, M. S. et al. Toward single cell traction microscopy within 3D collagen matrices. *Exp. Cell Res.* **319**, 2396–2408 (2013).

52. Bloom, R. J., George, J. P., Celedon, A., Sun, S. X. & Wirtz, D. Mapping local matrix remodeling induced by a migrating tumor cell using three-dimensional multiple-particle tracking. *Biophys. J.* **95**, 4077–4088 (2008).

53. Legant, W. R. et al. Measurement of mechanical tractions exerted by cells in three-dimensional matrices. *Nat. Methods* **7**, 969–971 (2010).

54. Palacio, J. et al. Numerical estimation of 3D mechanical forces exerted by cells on non-linear materials. *J. Biomech.* **46**, 50–55 (2013).

55. Yang, Z., Lin, J. S., Chen, J. & Wang, J. H. C. Determining substrate displacement and cell traction fields-a new approach. *J. Theor. Biol.* **242**, 607–616 (2006).

56. Li, B. & Wang, J. H. C. Fibroblasts and myofibroblasts in wound healing: Force generation and measurement. *J. Tissue Viability* **20**, 108–120 (2011).

Engineering Approaches for Cellular Therapeutics and Diagnostics **167**

57. Li, B., Li, F., Puskar, K. M. & Wang, J. H. C. Spatial patterning of cell proliferation and differentiation depends on mechanical stress magnitude. *J. Biomech.* **42**, 1622–1627 (2009).

58. Engler, A. J., Sen, S., Sweeney, H. L. & Discher, D. E. Matrix Elasticity Directs Stem Cell Lineage Specification. *Cell* **126**, 677–689 (2006).

59. Indra, I. et al. An in vitro correlation of mechanical forces and metastatic capacity. *Phys. Biol.* **8**, 015015 (2011).

60. Jannat, R. A., Dembo, M. & Hammer, D. A. Traction forces of neutrophils migrating on compliant substrates. *Biophys. J.* **101**, 575–584 (2011).

61. Malandrino, A., Kamm, R. D. & Moeendarbary, E. In vitro modeling of mechanics in cancer metastasis. *ACS Biomater. Sci. Eng.* **4**, 294–301 (2018).

62. Rajagopalan, J. & Saif, M. T. A. MEMS sensors and microsystems for cell mechanobiology. *J. Micromech. Microeng.* **21**, 054002 (2011).

63. Polacheck, W. J. & Chen, C. S. Measuring cell-generated forces: A guide to the available tools. *Nat. Methods* **13**, 415–423 (2016).

64. Matsudaira, K. et al. MEMS piezoresistive cantilever for the direct measurement of cardiomyocyte contractile force. *J. Micromech. Microeng.* **27**, 105005 (2017).

65. Takahashi, H. et al. Rigid two-axis MEMS force plate for measuring cellular traction force. *J. Micromech. Microeng.* **26**, 105006 (2016).

66. Scuor, N. et al. Design of a novel MEMS platform for the biaxial stimulation of living cells. *Biomed. Microdevices* **8**, 239–246 (2006).

67. Antoniolli, F., Maggiolino, S., Scuor, N., Gallina, P. & Sbaizero, O. A novel MEMS device for the multidirectional mechanical stimulation of single cells: Preliminary results. *Mech. Mach. Theory* **78**, 131–140 (2014).

68. Fior, R., Maggiolino, S., Codan, B., Lazzarino, M. & Sbaizero, O. A study on the cellular structure during stress solicitation induced by BioMEMS. *Proceedings of Annual International Conference of the IEEE Engineering in Medicine and Biology Society (EMBS)*, pp. 2455–2458 (2011) doi:10.1109/IEMBS.2011.6090682.

69. Zhang, L. & Dong, J. Design, fabrication, and testing of a SOI-MEMS-based active microprobe for potential cellular force sensing applications. *Adv. Mech. Eng.* **2012**, 785798 (2012).

70. Mann, J. M., Lam, R. H. W., Weng, S., Sun, Y. & Fu, J. A silicone-based stretchable micropost array membrane for monitoring live-cell subcellular cytoskeletal response. *Lab Chip* **12**, 731–740 (2012).

71. Lam, R. H. W., Sun, Y., Chen, W. & Fu, J. Elastomeric microposts integrated into microfluidics for flow-mediated endothelial mechanotransduction analysis. *Lab Chip* **12**, 1865–1873 (2012).

72. Sniadecki, N. J. et al. Magnetic microposts as an approach to apply forces to living cells. *Proc. Natl. Acad. Sci. U. S. A.* **104**, 14553–14558 (2007).

73. Nelson, C. M. et al. Emergent patterns of growth controlled by multicellular form and mechanics. *Proc. Natl. Acad. Sci. U. S. A.* **102**, 11594–11599 (2005).

74. Fu, J. et al. Mechanical regulation of cell function with geometrically modulated elastomeric substrates. *Nat. Methods* **7**, 733–736 (2010).

75. Ghassemi, S. et al. Cells test substrate rigidity by local contractions on submicrometer pillars. *Proc. Natl. Acad. Sci. U. S. A.* **109**, 5328–5333 (2012).

76. Yang, M. T., Fu, J., Wang, Y.-K., Desai, R. A. & Chen, C. S. Assaying stem cell mechanobiology on microfabricated elastomeric substrates with geometrically modulated rigidity. *Nat. Protoc.* **6**, 187–213 (2011).

77. Gupta, M. et al. Chapter 16: Micropillar substrates: A tool for studying cell mechanobiology. in *Biophysical Methods in Cell Biology* (ed. Paluch, E. K.) vol. 125, pp. 289–308 (Cambridge, MA: Academic Press, 2015).

78. Boudou, T. et al. A microfabricated platform to measure and manipulate the mechanics of engineered cardiac microtissues. *Tissue Eng. Part A* **18**, 910–919 (2012).

79. Serrao, G. W. et al. Myocyte-depleted engineered cardiac tissues support therapeutic potential of mesenchymal stem cells. *Tissue Eng. Part A* **18**, 1322–1333 (2012).

80. Hinson, J. T. et al. Titin mutations in iPS cells define sarcomere insufficiency as a cause of dilated cardiomyopathy. *Science (80-.).* **349**, 982–986 (2015).

81. Du Roure, O. et al. Force mapping in epithelial cell migration. *Proc. Natl. Acad. Sci. U. S. A.* **102**, 2390–2395 (2005).

82. Kim, D. H., Pak, K. W., Park, J., Levchenko, A. & Sun, Y. Microengineered platforms for cell mechanobiology. *Annu. Rev. Biomed. Eng.* **11**, 203–233 (2009).

83. Shiu, J.-Y., Aires, L., Lin, Z. & Vogel, V. Nanopillar force measurements reveal actin-cap-mediated YAP mechanotransduction. *Nat. Cell Biol.* **20**, 262–271 (2018).

84. Miroshnikova, Y. A. et al. Adhesion forces and cortical tension couple cell proliferation and differentiation to drive epidermal stratification. *Nat. Cell Biol.* **20**, 69–80 (2018).

85. Ghibaudo, M. et al. Traction forces and rigidity sensing regulate cell functions. *Soft Matter* **4**, 1836–1843 (2008).

86. Chen, J., Li, H., SundarRaj, N. & Wang, J. H. C. Alpha-smooth muscle actin expression enhances cell traction force. *Cell Motil. Cytoskeleton* **64**, 248–257 (2007).

87. Xiang, Y. & LaVan, D. A. Analysis of soft cantilevers as force transducers. *Appl. Phys. Lett.* **90**, 1–4 (2007).

88. Zhao, Y., Lim, C. C., Sawyer, D. B., Liao, R. & Zhang, X. Cellular force measurements using single-spaced polymeric microstructures: Isolating cells from base substrate. *J. Micromechanics Microengineering* **15**, 1649–1656 (2005).

89. Lin, I. K. et al. Viscoelastic mechanical behavior of soft microcantilever-based force sensors. *Appl. Phys. Lett.* **93**, 1–4 (2008).

90. Zheng, X. R. & Zhang, X. Microsystems for cellular force measurement: A review. *J. Micromechanics Microengineering* **21**, 054003 (2011).

91. Ashkin, A. Acceleration and trapping of particles by radiation pressure. *Phys. Rev. Lett.* **24**, 156–159 (1970).

92. Ashkin, A. Optical trapping and manipulation of neutral particles using lasers. *Proc. Natl. Acad. Sci.* **94**, 4853–4860 (1997).

93. MacDonald, M. P., Spalding, G. C. & Dholakia, K. Microfluidic sorting in an optical lattice. *Nature* **426**, 421–424 (2003).

94. Grover, S. C., Skirtach, A. G., Gauthier, R. C. & Grover, C. P. Automated single-cell sorting system based on optical trapping. *J. Biomed. Opt.* **6**, 14 (2001).

95. Zhong, M.-C., Wei, X.-B., Zhou, J.-H., Wang, Z.-Q. & Li, Y.-M. Trapping red blood cells in living animals using optical tweezers. *Nat. Commun.* **4**, 1768 (2013).

96. Rancourt-Grenier, S. et al. Dynamic deformation of red blood cell in dual-trap optical tweezers. *Opt. Express* **18**, 10462 (2010).

97. Rusciano, G., De Luca, A., Pesce, G. & Sasso, A. Raman tweezers as a diagnostic tool of hemoglobin-related blood disorders. *Sensors* **8**, 7818–7832 (2008).

98. Snook, R. D., Harvey, T. J., Correia Faria, E. & Gardner, P. Raman tweezers and their application to the study of singly trapped eukaryotic cells. *Integr. Biol.* **1**, 43–52 (2009).

99. Tang, H. et al. NIR Raman spectroscopic investigation of single mitochondria trapped by optical tweezers. *Opt. Express* **15**, 12708 (2007).

100. Samek, O. et al. Identification of individual biofilm-forming bacterial cells using Raman tweezers. *J. Biomed. Opt.* **20**, 51038 (2015).

101. Harvey, T. J. et al. Spectral discrimination of live prostate and bladder cancer cell lines using Raman optical tweezers. *J. Biomed. Opt.* **13**, 64004 (2008).

102. Peterman, E. J. G., Gittes, F. & Schmidt, C. F. Laser-induced heating in optical traps. *Biophys. J.* **84**, 1308–1316 (2003).

103. Blázquez-Castro, A. Optical tweezers: Phototoxicity and thermal stress in cells and biomolecules. *Micromachines* **10**, 507 (2019).

104. Rasmussen, M. B., Oddershede, L. B. & Siegumfeldt, H. Optical tweezers cause physiological damage to Escherichia coli and Listeria bacteria. *Appl. Environ. Microbiol.* **74**, 2441–2446 (2008).

105. Xin, Q. et al. Analytical methods magnetic tweezers for the mechanical research of DNA at the single molecule level. *Anal. Methods* **9**, 5720–5730 (2017).

106. Mahony, J. J. O., Platt, M., Kilinc, D. & Lee, G. Synthesis of superparamagnetic particles with tunable morphologies: The role of nanoparticle–Nanoparticle interactions. *Langmuir* **29**, 2546–2553 (2013).

107. Tavacoli, J. W. et al. The fabrication and directed self-assembly of micron- sized superparamagnetic non-spherical particles. *Soft Matter* **9**, 9103–9110 (2013).

108. Zhang, Y. & Wang, Q. Magnetic-plasmonic dual modulated FePt-Au ternary heterostructured nanorods as a promising nano-bioprobe. *Adv. Mater.* **24**, 2485–2490 (2012).

109. Lin, Y., Kramer, C. M., Chen, C. S. & Reich, D. H. Probing cellular traction forces with magnetic nanowires and microfabricated force sensor arrays. *Nanotechnology* **23**, 075101 (2012).

110. Zhang, Y. et al. Magnetic manipulation and optical imaging of an active plasmonic single-particle Fe-Au nanorod. **1056**, 15292–15298 (2011).

111. Mosconi, F., Allemand, J. F. & Croquette, V. Soft magnetic tweezers: A proof of principle. *Rev. Sci. Instrum.* **82**, 34302 (2011).

112. Janssen, X. J. A. et al. Electromagnetic torque tweezers: A versatile approach for measurement of single-molecule twist and torque. *Nano Lett.* **12**, 3634–3639 (2012).

113. Oberstrass, F. C., Fernandes, L. E. & Bryant, Z. Torque measurements reveal sequence-specific cooperative transitions in supercoiled DNA. *Proc. Natl. Acad. Sci.* **109**, 6106–6111 (2012).

114. Lipfert, J., Wiggin, M., Kerssemakers, J. W. J., Pedaci, F. & Dekker, N. H. Freely orbiting magnetic tweezers to directly monitor changes in the twist of nucleic acids. *Nat. Commun.* **2**, 439 (2011).

115. Lipfert, J., Kerssemakers, J. W. J., Jager, T. & Dekker, N. H. Magnetic torque tweezers: Measuring torsional stiffness in DNA and RecA-DNA filaments. *Nat. Methods* **7**, 977–980 (2010).

116. Wang, N., Butler, J. P. & Ingber, D. E. Mechanotransduction across the cell surface and through the cytoskeleton. *Science* **260**, 1124–1128 (1993).

117. Glogauer, M., Ferrier, J. & McCulloch, C. A. Magnetic fields applied to collagen-coated ferric oxide beads induce stretch-activated Ca^{2+} flux in fibroblasts. *Am. J. Physiol. Physiol.* **269**, C1093–C1104 (1995).

118. Glogauer, M. et al. Calcium ions and tyrosine phosphorylation interact coordinately with actin to regulate cytoprotective responses to stretching. *J. Cell Sci.* **21**, 11–21 (1997).

119. Zhao, X. et al. Force activates smooth muscle alpha-actin promoter activity through the Rho signaling pathway. *J. Cell Sci.* **120**, 1801–1809 (2007).

120. Matthews, B. D., Overby, D. R., Mannix, R. & Ingber, D. E. Cellular adaptation to mechanical stress : Role of integrins, Rho, cytoskeletal tension and mechanosensitive ion channels. *J. Cell Sci.* **119**, 508–518 (2006).

121. Saphirstein, R. J. et al. The focal adhesion: A regulated component of aortic stiffness. *PLoS One* **8**, e62461 (2013).

122. Bausch, A. R., Mo, W. & Sackmann, E. Measurement of local viscoelasticity and forces in living cells by magnetic tweezers. *Biophys. J.* **76**, 573–579 (1999).

123. Gosse, C. & Croquette, V. Magnetic tweezers: Micromanipulation and force measurement at the molecular level. *Biophys. J.* **82**, 3314–3329 (2002).

124. Kilinc, D. & Lee, G. U. Integrative Biology molecule and cell biophysics. *Integr. Biol.* **6**, 27–34 (2014).

125. Fabry, B. et al. Scaling the microrheology of living cells. *Phys. Rev. Lett.* **87**, 148102 (2001).

126. Trepat, X. et al. Universal physical responses to stretch in the living cell. *Nature* **447**, 592–595 (2007).

127. Puig-de-morales, M. et al. Cytoskeletal mechanics in adherent human airway smooth muscle cells: Probe specificity and scaling of protein-protein dynamics. *Am. J. Physiol. Physiol.* **287**, C643–C654 (2004).

128. Na, S. & Wang, N. Application of fluorescence resonance energy transfer and magnetic twisting cytometry to quantify mechanochemical signaling activities in a living cell. *Sci. Signal.* **1**, pl1–pl1 (2008).

129. Poh, Y., Na, S., Chowdhury, F., Ouyang, M. & Wang, Y. Rapid activation of Rac GTPase in living cells by force is independent of Src. *PLoS One* **4**, 1–7 (2009).

130. Tanase, M., Biais, N. & Sheetz, M. Magnetic tweezers in cell biology. in *Cell Mechanics* vol. 83, pp. 473–493 (Academic Press, 2007).

131. Bidan, C. M. et al. Magneto-active substrates for local mechanical stimulation of living cells. *Sci. Rep.* **8**, 1464 (2018).

132. De Vlaminck, I. & Dekker, C. Recent advances in magnetic tweezers. *Annu. Rev. Biophys.* **41**, 453–472 (2012).

133. Oddershede, L. B. Force probing of individual molecules inside the living cell is now a reality. *Nat. Chem. Biol.* **8**, 879–886 (2012).

134. Monachino, E., Spenkelink, L. M. & Van. Oijen, A. M. Watching cellular machinery in action, one molecule at a time. *J. Cell Biol.* **216**, 41–51 (2017).

135. Kilinc, D., Blasiak, A., Mahony, J. J. O., Suter, D. M. & Lee, G. U. Magnetic tweezers-based force clamp reveals mechanically distinct apCAM domain interactions. *Biophys. J.* **103**, 1120–1129 (2012).

136. Tabdili, H. et al. Cadherin-dependent mechanotransduction depends on ligand identity but not affinity. *J. Cell Sci.* **125**, 4362–4371 (2012).

137. Shang, H. & Lee, G. U. Magnetic tweezers measurement of the bond lifetime-force behavior of the IgG-Protein A specific molecular interaction. *Am. Chem. Soc.* **13**, 6640–6646 (2007).

138. Rocha, M. S. Integrative biology extracting physical chemistry from mechanics: A new approach to investigate DNA

interactions with drugs and proteins in single molecule experiments. *Integr. Biol.* **7**, 967–986 (2015).

139. Wu, J. & Du, G. Acoustic radiation force on a small compressible sphere in a focused beam. *J. Acoust. Soc. Am.* **87**, 997–1003 (1990).

140. Wu, J. Acoustical tweezers. *J. Acoust. Soc. Am.* **89**, 2140–2143 (1991).

141. Marzo, A. et al. Holographic acoustic elements for manipulation of levitated objects. *Nat. Commun.* **6**, 1–7 (2015).

142. Foresti, D. & Poulikakos, D. Acoustophoretic contactless elevation, orbital transport and spinning of matter in air. *Phys. Rev. Lett.* **112**, 24301 (2014).

143. Démoré, C. E. M. et al. Acoustic tractor beam. *Phys. Rev. Lett.* **112**, 174302 (2014).

144. Melde, K., Mark, A. G., Qiu, T. & Fischer, P. Holograms for acoustics. *Nature* **537**, 518–522 (2016).

145. Cummer, S. A., Christensen, J. & Alù, A. Controlling sound with acoustic metamaterials. *Nat. Rev. Mater.* **1**, 16001 (2016).

146. Memoli, G. et al. Metamaterial bricks and quantization of meta-surfaces. *Nat. Commun.* **8**, 1–8 (2017).

147. Sadhal, S. S. Acoustofluidics 13: Analysis of acoustic streaming by perturbation methods. *Lab Chip* **12**, 2292–2300 (2012).

148. Sadhal, S. S. Acoustofluidics 16: Acoustics streaming near liquid-gas interfaces: Drops and bubbles. *Lab Chip* **12**, 2771–2781 (2012).

149. Ding, X. et al. Cell separation using tilted-angle standing surface acoustic waves. *Proc. Natl. Acad. Sci. U. S. A.* **111**, 12992–12997 (2014).

150. Van Phan, H. et al. Vibrating membrane with discontinuities for rapid and efficient microfluidic mixing. *Lab Chip* **15**, 4206–4216 (2015).

151. Hashmi, A., Yu, G., Reilly-Collette, M., Heiman, G. & Xu, J. Oscillating bubbles: A versatile tool for lab on a chip applications. *Lab Chip* **12**, 4216–4227 (2012).

152. Schmid, L., Weitz, D. A. & Franke, T. Sorting drops and cells with acoustics: Acoustic microfluidic fluorescence-activated cell sorter. *Lab Chip* **14**, 3710–3718 (2014).

153. Ren, L. et al. A high-throughput acoustic cell sorter. *Lab Chip* **15**, 3870–3879 (2015).

154. Augustsson, P., Karlsen, J. T., Su, H. W., Bruus, H. & Voldman, J. Iso-acoustic focusing of cells for size-insensitive acousto-mechanical phenotyping. *Nat. Commun.* **7**, 1–9 (2016).

155. Galanzha, E. I. et al. In vivo acoustic and photoacoustic focusing of circulating cells. *Sci. Rep.* **6**, 1–15 (2016).

156. Shi, J. et al. Acoustic tweezers: Patterning cells and microparticles using standing surface acoustic waves (SSAW). *Lab Chip* **9**, 2890–2895 (2009).

157. Guo, F. et al. Three-dimensional manipulation of single cells using surface acoustic waves. *Proc. Natl. Acad. Sci. U. S. A.* **113**, 1522–1527 (2016).

158. Guo, F. et al. Controlling cell-cell interactions using surface acoustic waves. *Proc. Natl. Acad. Sci. U. S. A.* **112**, 43–48 (2015).

159. Ahmed, D. et al. Rotational manipulation of single cells and organisms using acoustic waves. *Nat. Commun.* **7**, 11085 (2016).

160. Reboud, J. et al. Shaping acoustic fields as a toolset for microfluidic manipulations in diagnostic technologies. *Proc. Natl. Acad. Sci. U. S. A.* **109**, 15162–15167 (2012).

161. Bernard, I. et al. Controlled rotation and translation of spherical particles or living cells by surface acoustic waves. *Lab Chip* **17**, 2470–2480 (2017).

162. Hahn, P., Lamprecht, A. & Dual, J. Numerical simulation of micro-particle rotation by the acoustic viscous torque. *Lab Chip* **16**, 4581–4594 (2016).

163. Sundvik, M., Nieminen, H. J., Salmi, A., Panula, P. & Hæggström, E. Effects of acoustic levitation on the development of zebrafish, Danio rerio, embryos. *Sci. Rep.* **5**, 1–11 (2015).

164. Chen, Y. et al. High-throughput acoustic separation of platelets from whole blood. *Lab Chip* **16**, 3466–3472 (2016).

165. Dao, M. et al. Acoustic separation of circulating tumor cells. *Proc. Natl. Acad. Sci. U. S. A.* **112**, 4970–4975 (2015).

166. Wu, M. et al. Isolation of exosomes from whole blood by integrating acoustics and microfluidics. *Proc. Natl. Acad. Sci. U. S. A.* **114**, 10584–10589 (2017).

167. Collins, D. J. et al. Two-dimensional single-cell patterning with one cell per well driven by surface acoustic waves. *Nat. Commun.* **6**, 1–11 (2015).

168. Sitters, G. et al. Acoustic force spectroscopy. *Nat. Methods* **12**, 47–50 (2015).

169. Friend, J. & Yeo, L. Y. Microscale acoustofluidics: Microfluidics driven via acoustics and ultrasonics. *Rev. Mod. Phys.* **83**, 647–704 (2011).

170. Ding, X. et al. Surface acoustic wave microfluidics. *Lab Chip* **13**, 3626 (2013).

171. Carovac, A., Smajlovic, F. & Junuzovic, D. Application of Ultrasound in Medicine. *Acta Inform. Medica* **19**, 168 (2011).

172. Wiklund, M. Acoustofluidics 12: Biocompatibility and cell viability in microfluidic acoustic resonators. *Lab Chip* **12**, 2018–2028 (2012).

173. Voiculescu, I. & Nordin, A. N. Acoustic wave based MEMS devices for biosensing applications. *Biosens. Bioelectron.* **33**, 1–9 (2012).

174. Nguyen, V. H., Kaulen, C., Simon, U. & Schnakenberg, U. Single interdigital transducer approach for gravimetrical SAW sensor applications in liquid environments. *Sensors (Switzerland)* **17**, 2931 (2017).

11 Drug Delivery Using Cold Plasma

Sotirios I. Ekonomou, Saliha Saad, Aniko Varadi, and Alexandros Ch. Stratakos
University of the West of England Bristol

CONTENTS

11.1 INTRODUCTION: FUNDAMENTALS OF CP AND DRUG DELIVERY

Over the past few decades, the trend of replacing traditional techniques in drug delivery has been intensively investigated. Since 1879 when plasma was first discovered as the fourth state of matter by Sir William Crookes and named in 1927 by Irving Langmuir (Gates, 2018), it has been significantly optimised to substitute traditional treatments within various fields with remarkable efficacy. Plasma treatment can be divided into thermal and cold/non-thermal atmospheric plasma. CP or cold atmospheric pressure plasma (CAPP) can be conveniently generated in an atmospheric environment with a gas temperature as low as room temperature, while higher power and pressures need to obtain thermal plasma. CPs can be generated using a wide range of adjustable temperatures, energy and power input, type, pressure and gas composition.

In most cases, CPs are generated by using a range of single gases or gas mixtures, such as oxygen (O_2), nitrogen (N_2), carbon dioxide (CO_2), helium (He) and argon (Ar). From the physical and chemical reactions that are taking place, a non-equilibrium state between electrons and neutral ions and free radicals is achieved (Ma et al., 2022). Additionally, these reactions may lead to the production of numerous stable reactive atomic and molecular molecules and atoms, e.g., reactive oxygen species (ROS), reactive nitrogen species (RNS) and finally, the generation of highly energetic UV photons.

In the past, CP has been used for sterilisation and decontamination of various surfaces and, more recently, as a novel processing technology in the food industry. The application of CPs has been gaining increasing importance in inactivating many types of common pathogenic bacteria on different substrates (Asimakopoulou et al., 2022; Ekonomou and Boziaris, 2021; González-González et al., 2021a) as well as inactivating endogenous enzymes (Sonawane & Patil, 2020). Moreover, the broad field of technological applications of CPs and the continuous research for innovative cold plasma

applications in the new century resulted in the use of CPs in medical technology, biotechnology and pharmacy. This ground-breaking and emerging field is called plasma medicine, and the medical applications of CPs can be branched into (1) "Indirect" and (2) "Direct" cold plasma-based techniques. In this way, indirect plasma techniques can be applied to treat various materials, coatings and surfaces, while the direct techniques are focused on the direct application of CPs in the human or animal body and living tissue (von Woedtke et al., 2013). CPs can be generated by a range of devices, with dielectric barrier discharge (DBD) being the most widely used indirect plasma source (Figure 12.1a). Indirect plasmas are produced between two electrodes and then transported to the desired area via the gas flow. Indirect plasmas are preferable for the treatment of living cells and tissues. However, the more versatile direct jet plasma (Figure 12.1b) can be applied to perform *in vitro* and *in vivo* treatments using plasma needle or torch devices (O'Connor et al., 2014). Hybrid plasmas are less commonly used and combine the plasma production technique of DBD with the properties of jet plasmas (Heinlin et al., 2011). A broad spectrum of direct, indirect and hybrid plasma sources offered for biomedical applications has been reported recently, atmospheric pressure plasma plume, jet, glow discharge torch (APGD-t), CP brush, CP needle, floating-electrode DBD (FEDBD), CP jets, micro-plasma jets and considerably more. However, not all CP sources have been proven valuable tools for biomedical applications, and further biological characterisation is needed to prove their potential.

Despite CP technology being a new principle in the medical field, its applications in drug delivery have experienced considerable growth. Drugs are currently being used as effective means to improve health and increase longevity. Drug delivery can be described as the transportation process of a therapeutic agent with appropriate pharmacokinetics to achieve the desired effect. Medications can be administered into the body by swallowing (through the

(a)

(b)

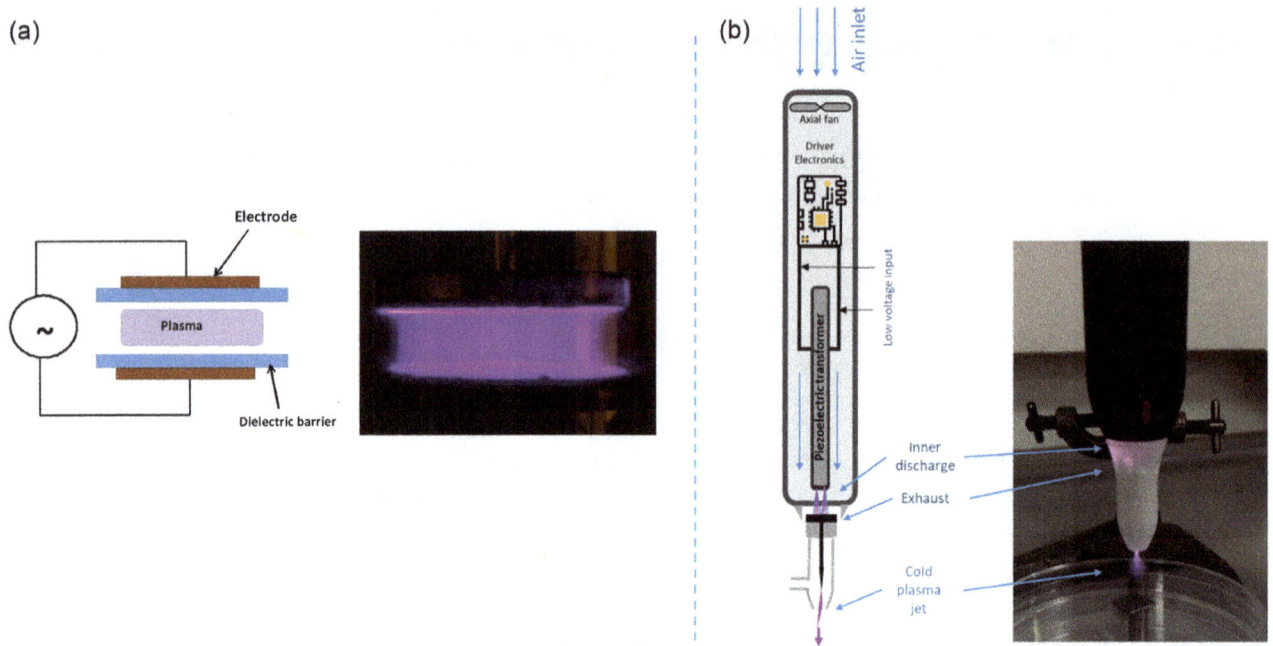

FIGURE 11.1   Schematic view (left) and photograph (right) of (a) an atmospheric pressure diffuse plasma generated by a dielectric barrier discharge (DBD) adopted from Laroussi (2018) and a schematic view (left) and a photograph (right) of (b) a handheld CP jet nozzle system with attached nozzle against a nonconductive surface adapted from González-González et al. (2021b).

gastrointestinal tract), inhalation, absorption through the skin or intravenous injection (Chamundeeswari et al., 2018).

However, some of the existing methods for drug delivery pose significant drawbacks. For example, drug administration by swallowing leads to reduced absorption and bioavailability of the drug as it moves through the gastrointestinal tract. More recently, direct drug administration to infected or at risk body areas has been attempted. However, local administration of drugs with sustained release can be achieved through nanostructures acting as the drug's delivery vehicles at the desired site (Ekonomou et al., 2022) and, more recently, by localised, non-invasive plasma treatment, which offers the possibility of controlled drug release at the molecular level (Heinlin et al., 2011).

Over the last two decades, many researchers in the field of plasma medicine demonstrated that CPs could be applied for clinical and biomedical applications (beyond decontamination) to improve drug delivery and surpass the limitations of pain, electric shock, patient discomfort, skin deformation and irritation observed with the application of traditional methods. Plasma medicine is being extensively investigated for various therapeutic applications in dermatology, dentistry, infection control and oncology. This chapter focuses on the direct or indirect application of CPs for enhancing drug delivery.

11.2   COLD ATMOSPHERIC PLASMA APPLICATIONS FOR TRANSDERMAL DRUG DELIVERY

The transdermal delivery system of drugs is an attractive method for painless, non-invasive delivery and sustainable

release through the skin to the blood circulation. CP techniques have proven their potential as transdermal drug delivery approaches to improve the absorption rate of various drugs.

CP treatments only affect the surface of different types of materials without causing alteration of the physical, chemical, mechanical, electrical and optical properties of their interior (Tabares and Junkar, 2021). Thus, several research studies aimed to treat dermatological problems, such as skin wounds and infections, by plasma which is relatively easy to achieve by adjusting the discharge and plasma parameters (Gan et al., 2021; Mai-Prochnow et al., 2014; O'Connor et al., 2014). However, efficient transdermal drug delivery remains a challenge mainly due to the skin's barrier properties, namely *stratum corneum* (SC), the highly lipophilic outermost layer of the epidermis. CP applications have been focused on their effects on the skin barrier as an approach for drug delivery in clinical and biomedical applications beyond a disease treatment. For instance, it was found that a 9-channel plasma jet array treatment promoted transdermal delivery of patent blue V, which is a synthetic triphenylmethane hydrophilic dye with a molecular weight (MW) of 1159.427 Da used in medicine as a dye to colour lymph vessels (Lv et al., 2021). The authors investigated the effect of the jet array under 7 kV, 7 kHz and 9 kV, 9 kHz treatment for 3 and 6 min on porcine ear skin using He or a mix of He with 0.5% oxygen (O_2) working gas composition. Their results revealed the potential of using CP jet array treatment for enhancing transdermal drug delivery with drug penetration across the skin being enhanced between 2 and 110 times after the plasma treatment. It also increased with treatment time as well as when the working gas was a mixture of He/O$_2$. A limitation when a high applied voltage was used (9 kV with pulse frequency at 9 kHz, 6 min) is that it caused an increased heating of the skin. However, lower voltage treatments did not have the same undesirable effect. The authors concluded that the increased density of the ROS flux observed with treatment time and the 0.5% O$_2$ incorporation into the working gas enhanced the penetration efficiency of the drug. A potential mechanism for this could be the intracellular lipid layer's oxidation, causing instability of the SC's structure due to the presence of unsaturated lipids and cholesterol. Moreover, further studies showed a beneficial transdermal delivery of topical anaesthetic cream with lidocaine (MW: 234.34 Da) and prilocaine (MW: 220.316 Da) as a pre-treatment prior to CO$_2$ laser treatment for post-acne scars (Xin et al., 2021a,b) and galantamine hydrobromide (MW: 368.3 Da) used as a drug for Alzheimer's disease treatment (Shimizu et al., 2016). Even though the exact mechanism for increasing skin permeability upon CP treatment is yet to be identified, Van der Paal et al. (2019) proposed that CP-generated reactive species induce lipid oxidation of the intracellular lipid layers. This effect leads to cross-linkages between the SC's anchored lipids and subsequent formation of nanopores, thus facilitating the increased transdermal permeation of drug molecules.

One of the three pathways that allow a drug to penetrate the SC is the intracellular route – together with the intercellular route and follicular route – where the drug must diffuse across the keratinocytes found in the lipid matrix and then be transported straight to the dermis. Recently, Lee et al. (2021) used an atmospheric pressure, Ar-plasma jet device to treat keratinocytes for 10, 30 and 60 s and proved that CP treatment significantly increased the transfer of high-MW fluorescein-dextrans (70 and 150 kDa). The increased transmission observed was due to the plasma-induced nitric oxides into the Ar-treated HaCaT cells that can regulate the junctions anchoring the cells of the second skin barrier and lead to increased permeability. Moreover, the fluorescent signals obtained showed a significantly increased intensity of the high-MW molecules in the Ar-treated HaCaT cells compared with the untreated cells, suggesting that Ar-plasma increases the permeation of molecules over 1 kDa that are difficult to penetrate the skin. As we have already discussed, the first barrier of the skin is SC, while the second skin barrier, located below SC, mainly consists of a group of proteins able to form a strong barrier that can inhibit penetration of external agents (Elias, 2005). In another study, it was demonstrated that Ar-plasma treatment for 5 min using a DBD plasma jet on HaCaT cells modulated the function of E-cadherin which is involved in cell-to-cell interactions, and its modulation can affect the skin's permeability (Lee et al., 2018). This is in line with the Schmidt et al. (2020) study that investigated the effect of CP treatment both *in vitro* on keratinocytes and *in vivo* on murine skin. Plasma treatment was carried out using an atmospheric pressure Ar-plasma jet kINPen Med at a frequency of 1 MHz and a constant distance of 8 mm. In

vitro experiments revealed how the plasma-derived ROS of hydrogen peroxide, nitrite, nitrate or hypochlorous acid act on skin cells, while it is known that short-lived species (such as hydroxyl radicals, superoxide anion, etc.) deteriorate in the culture media (Bekeschus et al., 2017). The authors found that plasma-derived reactive species modify the junctional network by affecting expression levels of transmembrane proteins, which promotes tissue oxygenation and oxidation of SC-lipids. Following this, *in vivo* plasma treatment caused histological changes to the murine skin leading to increased levels of curcumin (MW: 368.38) within the SC, but plasma treatment did not show to affect curcumin permeation in the epidermal region.

11.3 COLD ATMOSPHERIC PLASMA APPLICATIONS IN ONCOLOGY – CANCER TREATMENT

In this section, special consideration will be given to the most recent approaches of CP as an efficient method for promoting drug delivery against numerous carcinogenic cells as a new growing field in plasma medicine called "plasma oncology". Although this is a relatively new field, with the first results reported by Kieft et al. (2004), the increasing number of research articles showing the successful promotion of drugs in treating a broad spectrum of different tumour cells has fuelled the hope for CP to be used as a new therapy for cancer. Several studies outlined in this section show that CP can be used as a novel physical drug delivery tool against tumour cells *in vitro*.

11.3.1 *IN VITRO* CP APPLICATIONS FOR DRUG DELIVERY IN CANCER CELL LINES

The development of efficient and safe drug delivery methods to substitute the well-established techniques of electroporation and sonoporation remains challenging. CP has been identified as a promising method to substitute these techniques since the anticancer effects of plasma seem to be uniform and are not restricted to a particular type of tumour.

It has been shown that direct and indirect CP treatment can be used as a new physical drug delivery tool for low and high-MW molecules into human cervical cancer HeLa cells and murine breast carcinoma 4T1 cells (Vijayarangan et al., 2020). In the work of Vijayarangan et al. (2020), a He-plasma jet DBD reactor was employed to treat cells directly with a constant capillary-to-cells distance set at 11 mm or indirectly with plasma-activated media (PAM) generated at high voltage pulses (14 kV, 100 Hz pulse frequency) after 100 s. Interestingly, the uptake efficiency in 4T1 cells after CP treatment increased. Furthermore, the internalisation efficiency revealed a size-dependent uptake into 4T1 cells with the highest molecular weight FITC-Dextran of 150 kDa showing a successful but lower delivery into the cells than the same agent, at lower MW of 4 and 70 kDa.

These results revealed that CP-induced uptake could be more challenging for higher MW compounds, while similar observations were made in a previous study of the same research group for HeLa cells (Vijayarangan et al., 2018). In addition, Vijayarangan et al. (2020) confirmed the successful delivery of doxorubicin (MW: 543.4 Da) into 4T1 cells with direct CP treatment, an agent used as chemotherapy medication to treat various types of cancer. However, in the same study, when cells were treated indirectly, with PAM alone or in combination with electric fields, no cellular uptake was observed compared to direct CP treatment. The increased drug uptake observed after direct CP treatment was due to the disruption of the transient plasma membrane by the reactive oxygen and nitrogen species (RONS) produced which lasted for tens of minutes, allowing the efficient uptake of small and high-MW substances into HeLa and 4T1 cells.

The work of Vijayarangan et al. (2020) revealed the great potential of CP for increased cell uptake and sparked new opportunities for combined protocols. Xu et al. (2016) investigated the biological effects of CP treatment on multiple myeloma cells (LP-1 MM). They found that plasma treatment could be applied as a new strategy to overcome the resistance of myeloma cells to chemotherapy medications, which is a considerable challenge. Following CP treatment using a plasma jet device at 10 kHz pulse frequency and a fixed distance of 2.0 cm for 30 and 40 s, myeloma cells revealed an increased sensitivity in bortezomib (MW: 384.237 Da) first-line drug in myeloma chemotherapy. The combined treatment of CP and bortezomib on LP-1 cells showed a significantly decreased cell viability compared with each treatment used alone, indicating the synergistic effect and the higher delivery of the chemotherapeutic agent into the cells after plasma treatment. The higher sensitivity of myeloma cells was previously found by Xu et al. (2012) that could be due to the downregulation of the *CY1A1* gene expression that accelerates the bortezomib metabolism in myeloma cells. Further investigation is needed to prove if CP treatment could decrease the CYP1A1 expression leading to improved sensitivity to bortezomib. Direct or indirect CP treatment can demonstrate different sensitivity in different cancer cell lines. Glioblastoma (GBM) is the most common type of malignant brain tumour in adults. To elucidate the effect of CP treatment, Köritzer et al. (2013) analysed glioblastoma cell lines LN18, LN229 and U87MG. They used a CP device based on surface micro discharge technology where the electrode for plasma production was placed at the top inside a closed box, and plasma was produced at 8.5 kV voltage and 1 kHz pulse frequency in ambient air. Remarkably, CP treatment for 30 to 180 s restored the responsiveness of resistant LN18 glioma cells towards therapy with the first-line chemotherapeutic drug temozolomide (TMZ) compared to treatment with TMZ (MW: 194.151 Da) alone. This synergistic effect indicates a high likelihood of CP treatment in improving the delivery of chemotherapeutic drugs across the blood-brain barrier to reach and treat tumour cells, revealing the high potential of CP in cancer

therapy. In another interesting study, a He-CP jet device with two electrodes was used as a promising technique to promote immune checkpoint blockade (ICB) therapy against cancer by treating a patch with hollow-structured microneedles acting as microchannels to promote the release of tumour-associated antigens and CP reactive species (Chen et al., 2020). The authors demonstrated that 4 min of CP treatment led to a synergistic effect of CP treatment and ICB therapy integrated with the hollow microneedles relying on the transdermal delivery of CP's reactive species and immune checkpoint inhibitors as anti-programmed death-ligand 1 antibody (aPDL1) into the target tumour cells. Furthermore, it is known that CP can allow gene transfer since 2005 when Ogawa et al. used an atmospheric DBD plasma device and successfully transferred plasmid DNA encoding green fluorescence protein (GFP) into neuronal and HeLa cells. The study of Chen et al. (2020) provided a suitable minimally invasive technique for cancer treatment.

In essence, the above results demonstrate the enhanced effect of CP and regular chemotherapeutic agents to minimise drug resistance in cancer cell lines and increase drug delivery on site. However, further *in vivo* experiments should be carried out to tweeze out the beneficial effects of CP treatment on different tumour cell types and pave the way for medical applications.

11.3.2 ANTICANCER EFFECT OF CP IN COMBINATION WITH NANOPARTICLES (NPs) FOR DELIVERING DRUGS

Another novel approach to controlling drug delivery in modern medicine is using NPs as drug delivery carriers to overcome the drawbacks of commercial delivery systems. NPs have numerous advantages, such as improving hydrophobic drug delivery, reducing drug degradation in the gastrointestinal tract, sustaining and triggering the release and many more. Therefore, the application of CP for delivering anticancer drugs using nano-vehicles has revolutionised cancer treatment and is of high interest (Xu et al., 2012). In addition to traditional chemotherapy and radiotherapy, the synergistic action of NPs loaded with chemotherapeutic agents together with CP technology has shown their potential in cancer therapy (Cheng et al., 2014; Zhu et al., 2016). It has been exhibited that CP treatment coupled with polymeric NPs led to an *in vitro* synergistic inhibition of breast cancer cell growth and downregulation of metastasis-related gene expression (VEGF, MMP9, MMP2, MTDH), which are involved in minimising drug resistance and can promote drug uptake (Zhu et al., 2016). Gold nanoparticles (Au-NPs) can be used as drug delivery carriers due to their low toxicity to normal cells and selective toxicity to specific cancer cell lines (Connor et al., 2005; Patra et al., 2007). Indeed, He et al. (2018) observed that when Au-NPs were used alone, in agreement with other reports, low cytotoxicity was revealed against cancer cells (Connor et al., 2005). Although when Au-NPs were used in combination with a DBD plasma treatment for 30 s at high voltage (75 kV), increased synergistic cytotoxicity and enhanced uptake of Au-NPs on U373MG glioblastoma cells were revealed. The authors indicated that the long-lived reactive species did not play a major role in the enhanced uptake of AuNP, suggesting that physical effects play a minor role, while chemical effects induced by direct and indirect exposure to CAP appeared as the primary mediator due to increased endocytosis observed. Kong et al. (2011), in their review of the interaction of CP and drug-loaded NPs with cells, documented that NPs may favourably deposit near cancerous cells instead of healthy cells due to their different mechanical properties (Iyer et al., 2009). More recently, the same effect has been observed by Manaloto et al. (2020) by visualising the glioma cell morphology using spectral imaging. Images of *in vitro* brain cancer cells U373MG after the combined treatment of CP and Ag-NPs demonstrated morphological changes (losing the astrocyte shape) with a higher distribution of Ag-NPs in cancerous cells. Even though the exact mechanism of the combined effect of CP and NPs is as of yet poorly understood, the existing findings in the field show an enhanced selective permeability through the induction of membrane disruption of CP species leading to facilitated intracellular diffusion of NPs towards diseased sites within a tissue (Kong et al., 2011).

To date, exciting progress has been made in plasma oncology, but many challenges remain for the successful development of cancer therapy for different types of cancer. A synopsis of the main findings of the existing CP-mediated cancer drug delivery studies can be found in Table 11.1.

11.4 DEVELOPMENT OF CONTROLLED RELEASING SURFACES BY COLD PLASMA MODIFICATION FOR DRUG DELIVERY

CP has the capacity to change the surface characteristics of different materials and create surfaces able to absorb therapeutic agents and release them in a controlled way to patients. The use of plasma treatment to control and improve the rate of drug delivery of bioactive agents has created new opportunities for the development of drug delivery systems.

In terms of drug delivery, surface modification of materials for biomedical use by cold plasma has become a field of high scientific interest due to the numerous advantages and the selectivity of this technique, exhibiting great potential for various applications. CP can be efficiently applied to enhance the surface properties of these biomaterials without the drawbacks of the traditional surface modification techniques, e.g. machining, grinding and chemical grafting, and can modify the surface characteristics only in a few nanometres depth without affecting the bulk attributes of the materials (Reyna-Martínez et al., 2018). The desired alterations obtained by CP surface modification on various biomaterials range from improving surface adhesion and wettability, achieving sterilisation, as well as biocompatibility and bioactivity by changing the surface's chemical composition to allow the immobilisation or controlled release of drugs and bioactive molecules (Yoshida et al., 2013). In the present section, CP treatments of different biomaterials will be discussed to define their potential use for the successful administration of drugs or bioactive molecules.

Polymers are widely used in various biomedical applications such as medical devices, surgical implants and prosthetic biomaterials and are among the new drug carriers for effective drug delivery. Interestingly, Labay et al. (2015) used a corona discharge CP device at ambient air to load polypropylene (PP) hernia meshes with ampicillin as a new method to treat possible post-surgery infections. Open hernia repair is a very common surgical operation where the hernia is pushed back into the abdomen, and the abdominal wall becomes fortified using stitches or a synthetic mesh. A surgical mesh is a medical implant that supports damaged tissue around hernias as it heals, and significant physio-chemical differences among available meshes can be found. A biomedical implant refers to an artificial functional organ that can fully restore the injured natural organ or tissue of the body without causing any adverse effects (Stloukal et al., 2017). One of the main advantages of CP is that it can be used to manipulate the surface properties of PP meshes-implants and improve their wettability, limiting the efficient drug loading. In this approach, Labay et al. (2015) demonstrated that after 3.5 s of CP functionalisation, the wettability of the PP meshes significantly increased, leading to a 3-fold higher loading (59.5%) capacity and 84.6% total release of ampicillin, which could be useful apropos of local treatment. After plasma functionalisation, a progressive increase in surface roughness was revealed that did not affect the fibroblast (CRL-1658) adhesion following the findings of Pandiyaraj et al. (2009) on PP films after long CP treatment for 2–10 min. This is an important finding as the microstructure of the biomedical material is important in promoting the initial attachment to the surface and the subsequent proliferation of the cells. It is generally accepted that cell adhesion is greater on rough surfaces, but the adhesion of cells may vary depending on the cell line. More recently, Zahedi et al. (2021) experimented with developing PP meshes loaded with betaine hydrochloride that can be exploited as wound dressings for the controlled drug delivery in diabetic wounds. In particular, the PP meshes were functionalised using direct plasma treatment at ambient air conditions and then CP was applied for 30 s to allow the polymerisation of the polyethylene glycol (PEG) using an Argon-plasma bubble reactor to fix and delay drug release. The amount of loaded betaine on the plasma-treated meshes reached almost 80%, while HPLC analysis revealed an *in vitro* drug release of up to 10%. The *in vivo* results of this study presented the wound healing potential of the CP-treated PP meshes on rat skin, where the modified meshes induced faster tissue regeneration and accelerated wound closure compared to the control group.

Zhu et al. (2015) demonstrated that a DBD CP system using He (working gas) could be applied to modify the surface of electrospun scaffolds to increase the adsorption of vitronectin and poly(lactic-co-glycolic) acid (PLGA) microspheres loaded with bovine serum albumin (BSA) as a bioactive compound. Their results revealed a

TABLE 11.1

List of Research Papers Presenting the *In Vitro* Effects of CP Treatment for Promoting Drug Delivery in Various Cancer Cell Lines

Plasma Source and Gas	Plasma Treatment	Cell Types (Lines)	Results	Reference
Plasma jet (He)	Direct	Breast cancer cells (MDA-MB-231)	Synergistic anticancer effect of 60s CP and drug-loaded NPs against breast cancer cell growth due to increased cellular uptake of drug-loaded NPs	Zhu et al. (2016)
			Downregulation of metastasis-related gene expression (VEGF, MMP9, MMP2, MTDH) led to decreased drug resistance	
Plasma jet gun based on DBD (He)	Direct (hybrid)	Human cervical cancer cells (HeLa)	Permeabilisation of propidium iodide in the cells was up to seven times higher after CP treatment with 1,000, 10,000 and 100,000 pulses at 10 Hz for 100 s	Vijayarangan et al. (2018)
			Drug delivery observed through the formation of 6.5 nm diameter pores	
			Plasma-induced permeabilisation was dependent on endocytosis	
Plasma jet gun based on DBD (He)	Direct (hybrid)	Human cervical cancer (HeLa) and murine breast carcinoma cells (4T1)	High uptake levels in both HeLa and 4T1 cells (100s of CP treatment)	Vijayarangan et al. (2020)
			High MW molecules of FITC-Dextran were successfully delivered into 4T1 cells after 100s of CP treatment	
	Indirect		Increased doxorubicin uptake and more efficient delivery when the drug was added in PAM after CP treatment	
Plasma jet (He)	Direct	Human cervical cancer cells (HeLa)	Dextrans with MW of 3 and 10 kDa were delivered in HeLa cells only in the plasma-treated area	Leduc et al. (2009)
High-voltage DBD atmospheric plasma system (ambient air)	Direct	Human glioblastoma multiforme cells (U373MG)	Increased cellular uptake of Ag-NPs after the synergistic treatment with a low dose of 0.07 µg/mL Ag-NP in combination with 25s CAP at 75 kV	Manaloto et al. (2020)
High-voltage DBD atmospheric plasma system (ambient air)	Direct and Indirect	Human brain glioblastoma cancer cell (U373MG-CD14)	Increased uptake and accumulation into U373MG cells after 30s of CP treatment (direct and indirect)	He et al. (2018)
			Chemical effects and endocytosis were the major uptake mechanisms	

significantly higher cell proliferation on CP-treated scaffolds with embedded BSA-loaded microspheres after 7 days. Furthermore, CP treatment decreased the hydrophobicity of the scaffolds leading to high adsorption of vitronectin while displaying interconnected porous topography with homogenous distribution of PLGA microspheres that released the encapsulated bioactive factor and promoted chondrogenesis. There is a high demand for novel strategies to improve the articular cartilage's poor tissue regeneration, and this study provided a new strategy through the synergistic effect of CP and bioactive compound-loaded microspheres and electrospinning for tissue regeneration. Numerous studies presented the prospect of using CP treatment to control the release kinetics of the drug for various applications, including CP-treated biodegradable porous silicon microparticles (pSi MPs) loaded with camptothecin, an anticancer agent (McInnes et al., 2016), electrospun poly(e-caprolactone) (PCL) mats loaded with dopamine to form polydopamine coatings (Xie et al., 2012), partially dissolvable CP-treated polymer microneedles (MNs) for the efficient delivery of drugs and vaccines (Lee et al., 2015; Nair et al., 2015) and calcium phosphate (CaP) ceramics loaded with antibiotics to prevent infections (Canal et al., 2016b) or coated with a biodegradable copolymer to control the subsequent drug delivery of simvastatin (Canal et al., 2016a).

Plasma surface modification has been extensively used as an effective technique to promote drug delivery and controlled drug release for numerous applications in plasma medicine. However, some issues remain. In order to achieve successful surface modification by CP treatment, it generally requires an in-depth knowledge of physics, chemistry and engineering of the surfaces and a robust biological background to explain the effects of plasma modification prior to *in vivo* practical application.

11.5 CONCLUSIONS

This chapter introduces the fundamentals of CP, state of the art and primary challenges in the field of CP-mediated drug delivery. Given that at present drug delivery has many limitations, CP could be a suitable alternative method to circumvent at least some of these limitations in a number of diseases (e.g. skin infections, cancer). However, even though various atmospheric CP sources for biomedical applications have been described in the literature, most of them have only been characterised by *in vitro* cell biology, and as of yet the number of *in vivo* studies is limited. In the future, CP-based therapies are anticipated to become common practice in medicine as plasma treatment proves to be a simple, rapid, cost-effective and substrate-independent and can significantly reduce the level of the required drugs for therapy by improving their overall drug delivery. Although to ensure this, it is necessary to develop flexible and modular plasma devices that can be employed in plasma medicine to treat variable target areas at clinical settings or even at home. To overcome these challenges and meet the requirements, CP research should continue to evolve and help translate the *in vitro* results into clinical applications by gaining more in-depth insights into the mechanisms involved in plasma-induced effects in tissues.

REFERENCES

Asimakopoulou, E., Ekonomou, S., Papakonstantinou, P., Doran, O., Stratakos, A.C., 2022. Inhibition of corrosion causing Pseudomonas aeruginosa using plasma-activated water. *J. Appl. Microbiol.* 132, 2781–2794. https://doi.org/10.1111/jam.15391.

Bekeschus, S., Schmidt, A., Niessner, F., Gerling, T., Weltmann, K.D., Wende, K., 2017. Basic research in plasma medicine - a throughput approach from liquids to cells. *J. Vis. Exp.* 2017, 1–9. https://doi.org/10.3791/56331.

Canal, C., Khurana, K., Gallinetti, S., Bhatt, S., Pulpytel, J., Arefi-Khonsari, F., Ginebra, M.P., 2016a. Design of calcium phosphate scaffolds with controlled simvastatin release by plasma polymerisation. *Polymer (Guildf)* 92, 170–178. https://doi.org/10.1016/j.polymer.2016.03.069.

Canal, C., Modic, M., Cvelbar, U., Ginebra, M.P., 2016b. Regulating the antibiotic drug release from β-tricalcium phosphate ceramics by atmospheric plasma surface engineering. *Biomater. Sci.* 4, 1454–1461. https://doi.org/10.1039/c6bm00411c.

Chamundeeswari, M., Jeslin, J., Verma, M.L., 2018. Nanocarriers for drug delivery applications. *Environ. Chem. Lett.* 17(2), 849–865. https://doi.org/10.1007/S10311-018-00841-1.

Chen, G., Chen, Z., Wen, D., Wang, Z., Li, H., Zeng, Y., … Gu, Z., 2020. Transdermal cold atmospheric plasma-mediated immune checkpoint blockade therapy. *Proc. Natl. Acad. Sci. U.S.A.* 117(7), 3687–3692. https://doi.org/10.1073/pnas.1917891117.

Cheng, X., Murphy, W., Recek, N., Yan, D., Cvelbar, U., Vesel, A., Mozetič, M., Canady, J., Keidar, M., Sherman, J.H., 2014. Synergistic effect of gold nanoparticles and cold plasma on glioblastoma cancer therapy. *J. Phys. D. Appl. Phys.* 47, 335402. https://doi.org/10.1088/0022-3727/47/33/335402.

Connor, E.E., Mwamuka, J., Gole, A., Murphy, C.J., Wyatt, M.D., 2005. Gold nanoparticles are taken up by human cells but do not cause acute cytotoxicity. *Small* 1, 325–327. https://doi.org/10.1002/SMLL.200400093.

Ekonomou, S.I., Boziaris, I.S., 2021. Non-thermal methods for ensuring the microbiological quality and safety of seafood. *Appl. Sci.* 11, 1–30. https://doi.org/10.3390/app11020833.

Ekonomou, S. I., Akshay Thanekar, P., Lamprou, D. A., Weaver, E., Doran, O., & Stratakos, A. C. 2022. Development of Geraniol-Loaded Liposomal Nanoformulations against Salmonella Colonization in the Pig Gut. *J. Agric. Food Chem.* 70, 7004–7014. https://doi.org/10.1021/acs.jafc.2c00910.

Elias, P.M., 2005. Stratum corneum defensive functions: An integrated view. *J. Invest. Dermatol.* 125(2), 183–200.

Gan, L., Jiang, J., Duan, J.W., Wu, X.J.Z., Zhang, S., Duan, X.R., Song, J.Q., Chen, H.X., 2021. Cold atmospheric plasma applications in dermatology: A systematic review. *J. Biophotonics* 14, 1–9. https://doi.org/10.1002/jbio.202000415.

Gates, D., 2018. Plasma: An international open access journal for all of plasma science. *Plasma* 1, 45–46. https://doi.org/10.3390/plasma1010004.

González-González, C.R., Labo-Popoola, O., Delgado-Pando, G., Theodoridou, K., Doran, O., Stratakos, A.C., 2021a. The effect of cold atmospheric plasma and linalool nanoemulsions against Escherichia coli O157:H7 and Salmonella on ready-to-eat chicken meat. *LWT* 149, 111898. https://doi.org/10.1016/J.LWT.2021.111898.

González-González, C.R., Hindle, B.J., Saad, S., Stratakos, A.C., 2021b. Inactivation of listeria monocytogenes and salmonella on stainless steel by a piezoelectric cold atmospheric plasma generator. *Appl. Sci.* 11. https://doi.org/10.3390/app11083567.

He, Z., Liu, K., Manaloto, E., Casey, A., Cribaro, G.P., Byrne, H.J., Tian, F., Barcia, C., Conway, G.E., Cullen, P.J., Curtin, J.F., 2018. Cold atmospheric plasma induces ATP-dependent endocytosis of nanoparticles and synergistic U373MG cancer cell death. *Sci. Rep.* 8, 1–11. https://doi.org/10.1038/s41598-018-23262-0.

Heinlin, J., Isbary, G., Stolz, W., Morfill, G., Landthaler, M., Shimizu, T., Steffes, B., Nosenko, T., Zimmermann, J.L., Karrer, S., 2011. Plasma applications in medicine with a special focus on dermatology. *J. Eur. Acad. Dermatology Venereol.* 25, 1–11. https://doi.org/10.1111/j.1468–3083.2010.03702.x.

Iyer, S., Gaikwad, R.M., Subba-Rao, V., Woodworth, C.D., Sokolov, I., 2009. Atomic force microscopy detects differences in the surface brush of normal and cancerous cells. *Nat. Nanotechnol.* 4(6), 389–393. https://doi.org/10.1038/nnano.2009.77.

Kieft, I.E., Broers, J.L.V., Caubet-Hilloutou, V., Slaaf, D.W., Ramaekers, F.C.S., Stoffels, E., 2004. Electric discharge plasmas influence attachment of cultured CHO K1 cells. *Bioelectromagnetics* 25, 362–368. https://doi.org/10.1002/bem.20005.

Kong, M.G., Keidar, M., Ostrikov, K., 2011. Plasmas meet nanoparticles—where synergies can advance the frontier of medicine. *J. Phys. D. Appl. Phys.* 44, 174018. https://doi.org/10.1088/0022-3727/44/17/174018.

Köritzer, J., Boxhammer, V., Schäfer, A., Shimizu, T., Klämpfl, T.G., Li, Y.F., Welz, C., Schwenk-Zieger, S., Morfill, G.E.,

Zimmermann, J.L., Schlegel, J., 2013. Restoration of sensitivity in chemo - Resistant glioma cells by cold atmospheric plasma. *PLoS One* 8. https://doi.org/10.1371/journal.pone.0064498.

Labay, C., Canal, J.M., Modic, M., Cvelbar, U., Quiles, M., Armengol, M., Arbos, M.A., Gil, F.J., Canal, C., 2015. Antibiotic-loaded polypropylene surgical meshes with suitable biological behaviour by plasma functionalisation and polymerisation. *Biomaterials* 71, 132–144. https://doi.org/10.1016/j.biomaterials.2015.08.023.

Laroussi, M., 2018. Plasma medicine: A brief introduction. *Plasma* 1, 47–60. https://doi.org/10.3390/plasma1010005.

Leduc, M., Guay, D., Leask, R.L., Coulombe, S., 2009. Cell permeabilisation using a non-thermal plasma. *New J. Phys.* 11. https://doi.org/10.1088/1367-2630/11/11/115021.

Lee, H.Y., Choi, J.H., Hong, J.W., Kim, G.C., Lee, H.J., 2018. Comparative study of the Ar and He atmospheric pressure plasmas on E-cadherin protein regulation for plasma-mediated transdermal drug delivery. *J. Phys. D. Appl. Phys.* 51. https://doi.org/10.1088/1361-6463/aabd8c.

Lee, I.C., He, J.S., Tsai, M.T., Lin, K.C., 2015. Fabrication of a novel partially dissolving polymer microneedle patch for transdermal drug delivery. *J. Mater. Chem. B* 3, 276–285. https://doi.org/10.1039/c4tb01555j.

Lee, S., Choi, J., Kim, J., Jang, Y., Lim, T.H., 2021. Atmospheric pressure plasma irradiation facilitates transdermal permeability of aniline blue on porcine skin and the cellular permeability of keratinocytes with the production of nitric oxide. *Appl. Sci.* 11. https://doi.org/10.3390/app11052390.

Lv, Y., Nie, L., Duan, J., Li, Z., Lu, X., 2021. Cold atmospheric plasma jet array for transdermal drug delivery. *Plasma Process. Polym.* 18, 1–9. https://doi.org/10.1002/ppap.202000180.

Ma, C., Nikiforov, A., De Geyter, N., Morent, R., Ostrikov, K. (Ken), 2022. Plasma for biomedical decontamination: From plasma-engineered to plasma-active antimicrobial surfaces. *Curr. Opin. Chem. Eng.* 36, 100764. https://doi.org/10.1016/j.coche.2021.100764.

Mai-Prochnow, A., Murphy, A.B., McLean, K.M., Kong, M.G., Ostrikov, K., 2014. Atmospheric pressure plasmas: Infection control and bacterial responses. *Int. J. Antimicrob. Agents* 43, 508–517. https://doi.org/10.1016/j.ijantimicag.2014.01.025.

Manaloto, E., Gowen, A.A., Lesniak, A., He, Z., Casey, A., Cullen, P.J., Curtin, J.F., 2020. Cold atmospheric plasma induces silver nanoparticle uptake, oxidative dissolution and enhanced cytotoxicity in glioblastoma multiforme cells. *Arch. Biochem. Biophys.* 689, 108462. https://doi.org/10.1016/j.abb.2020.108462.

McInnes, S.J.P., Michl, T.D., Delalat, B., Al-Bataineh, S.A., Coad, B.R., Vasilev, K., Griesser, H.J., Voelcker, N.H., 2016. "Thunderstruck": Plasma-polymer-coated porous silicon microparticles as a controlled drug delivery system. *ACS Appl. Mater. Interfaces* 8, 4467–4476. https://doi.org/10.1021/acsami.5b12433.

Nair, K., Whiteside, B., Grant, C., Patel, R., Tuinea-Bobe, C., Norris, K., Paradkar, A., 2015. Investigation of plasma treatment on micro-injection moulded microneedle for drug delivery. *Pharmaceutics* 7, 471–485. https://doi.org/10.3390/pharmaceutics7040471.

O'Connor, N., Cahill, O., Daniels, S., Galvin, S., Humphreys, H., 2014. Cold atmospheric pressure plasma and decontamination. Can it contribute to preventing hospital-acquired infections? *J. Hosp. Infect.* 88, 59–65. https://doi.org/10.1016/j.jhin.2014.06.015.

Pandiyaraj, K.N., Selvarajan, V., Deshmukh, R.R., Gao, C., 2009. Modification of surface properties of polypropylene (PP) film using DC glow discharge air plasma. *Appl. Surf. Sci.* 255, 3965–3971. https://doi.org/10.1016/J.APSUSC.2008.10.090.

Patra, H.K., Banerjee, S., Chaudhuri, U., Lahiri, P., Dasgupta, A.K., 2007. Cell selective response to gold nanoparticles. *Nanomed. Nanotechnol. Biol. Med.* 3, 111–119. https://doi.org/10.1016/J.NANO.2007.03.005.

Reyna-Martínez, R., Céspedes, R.I.N., Alonso, M.C.I., Acosta, Y.K.R., 2018. Use of cold plasma technology in biomaterials and their potential utilization in controlled administration of active substances. *J. Mater. Sci.* 4, 0–9. https://doi.org/10.19080/JOJMS.2018.04.555649.

Schmidt, A., Liebelt, G., Striesow, J., Freund, E., von Woedtke, T., Wende, K., Bekeschus, S., 2020. The molecular and physiological consequences of cold plasma treatment in murine skin and its barrier function. *Free Radic. Biol. Med.* 161, 32–49. https://doi.org/10.1016/j.freeradbiomed.2020.09.026.

Shimizu, K., Tran, A. N., Kristof, J., & Blajan, M. (2016, June). Investigation of atmospheric microplasma for improving skin permeability. In: *Proceedings of the 2016 Electrostatics joint conference* (pp. 13–18).

Sonawane, S.K., T, M., Patil, S., 2020. Non-thermal plasma: An advanced technology for food industry. *Food Sci. Technol. Int.* 26, 727–740. https://doi.org/10.1177/1082013220929474.

Stloukal, P., Novák, I., Mičušík, M., Procházka, M., Kucharczyk, P., Chodák, I., … & Sedlařík, V., 2018. Effect of plasma treatment on the release kinetics of a chemotherapy drug from biodegradable polyester films and polyester urethane films. *Int. J. Polym. Mater.* 67(3), 161–173. https://doi.org/10.1080/00914037.2017.1309543.

Tabares, F.L., Junkar, I., 2021. Cold plasma systems and their application in surface treatments for medicine. *Molecules* 26. https://doi.org/10.3390/molecules26071903.

Van der Paal, J., Fridman, G., Bogaerts, A., 2019. Ceramide cross-linking leads to pore formation: Potential mechanism behind CAP enhancement of transdermal drug delivery. *Plasma Process. Polym.* 16, 1900122. https://doi.org/10.1002/PPAP.201900122.

Vijayarangan, V., Delalande, A., Dozias, S., Pouvesle, J.M., Pichon, C., Robert, E., 2018. Cold atmospheric plasma parameters investigation for efficient drug delivery in HeLa cells. *IEEE Trans. Radiat. Plasma Med. Sci.* 2, 109–115. https://doi.org/10.1109/TRPMS.2017.2759322.

Vijayarangan, V., Delalande, A., Dozias, S., Pouvesle, J.M., Robert, E., Pichon, C., 2020. New insights on molecular internalisation and drug delivery following plasma jet exposures. *Int. J. Pharm.* 589, 119874. https://doi.org/10.1016/j.ijpharm.2020.119874.

von Woedtke, T., Reuter, S., Masur, K., Weltmann, K.D., 2013. Plasmas for medicine. *Phys. Rep.* 530, 291–320. https://doi.org/10.1016/j.physrep.2013.05.005.

Xie, J., Michael, P.L., Zhong, S., Ma, B., MacEwan, M.R., Lim, C.T., 2012. Mussel inspired protein-mediated surface modification to electrospun fibers and their potential biomedical applications. *J. Biomed. Mater. Res. - Part A* 100 A, 929–938. https://doi.org/10.1002/jbm.a.34030.

Xin, Y., Wen, X., Hamblin, M.R., Jiang, X., 2021a. Transdermal delivery of topical lidocaine in a mouse model is enhanced by treatment with cold atmospheric plasma. *J. Cosmet. Dermatol.* 20, 626–635. https://doi.org/10.1111/jocd.13581.

Xin, Y., Wen, X., Jiang, X., 2021b. Analgesic effect of topical lidocaine is enhanced by cold atmospheric plasma pretreatment in facial CO_2 laser treatments. *J. Cosmet. Dermatol.* 20, 2794–2799. https://doi.org/10.1111/jocd.13983.

Xu, D., Hu, J., De Bruyne, E., Menu, E., Schots, R., Vanderkerken, K., Van Valckenborgh, E., 2012. Dll1/Notch activation contributes to bortezomib resistance by upregulating CYP1A1 in multiple myeloma. *Biochem. Biophys. Res. Commun.* 428, 518–524. https://doi.org/10.1016/j.bbrc.2012.10.071.

Xu, D., Luo, X., Xu, Y., Cui, Q., Yang, Y., Liu, D., Chen, H., Kong, M.G., 2016. The effects of cold atmospheric plasma on cell adhesion, differentiation, migration, apoptosis and drug sensitivity of multiple myeloma. *Biochem. Biophys. Res. Commun.* 473, 1125–1132. https://doi.org/10.1016/j.bbrc.2016.04.027.

Yoshida, S., Hagiwara, K., Hasebe, T., Hotta, A., 2013. Surface modification of polymers by plasma treatments for the enhancement of biocompatibility and controlled drug release. *Surf. Coatings Technol.* 233, 99–107. https://doi.org/10.1016/j.surfcoat.2013.02.042.

Zahedi, L., Ghourchi Beigi, P., Shafiee, M., Zare, F., Mahdikia, H., Abdouss, M., Abdollahifar, M.A., Shokri, B., 2021. Development of plasma functionalised polypropylene wound dressing for betaine hydrochloride controlled drug delivery on diabetic wounds. *Sci. Rep.* 11, 1–18. https://doi.org/10.1038/s41598-021-89105-7.

Zhu, W., Castro, N.J., Cheng, X., Keidar, M., Zhang, L.G., 2015. Cold atmospheric plasma modified electrospun scaffolds with embedded microspheres for improved cartilage regeneration. *PLoS One* 10, 1–18. https://doi.org/10.1371/journal.pone.0134729.

Zhu, W., Lee, S.J., Castro, N.J., Yan, D., Keidar, M., Zhang, L.G., 2016. Synergistic effect of cold atmospheric plasma and drug loaded core-shell nanoparticles on inhibiting breast cancer cell growth. *Sci. Rep.* 6, 1–11. https://doi.org/10.1038/srep21974.

12 Ultrasound-Mediated Delivery of Therapeutics

Sophie V. Morse, Tiffany G. Chan, and Javier Cudeiro-Blanco
Imperial College London

Antonios N. Pouliopoulos
King's College London

CONTENTS

12.1 INTRODUCTION

The effective treatment of brain diseases remains an area of high unmet need, with therapies suffering from particularly high attrition rates compared to other indications. Between 2002 and 2012, a total of 244 compounds were assessed in clinical trials for the treatment of Alzheimer's disease, but only one of these agents was approved (99.6% attrition) [1]. The main reason for this high attrition rate is the presence of the blood-brain barrier (BBB), a highly selective semi-permeable biological barrier, which limits the movement of substances from the blood to the brain parenchyma and vice versa. It has been claimed that over 98% of conventional small molecule drugs are unable to cross this barrier and reach the brain, rendering them ineffective [2].

The BBB is formed by the endothelial cells that line the cerebral vasculature, as well as the surrounding basement membrane, pericytes, astrocytes and perivascular macrophages. It plays an essential role in maintaining brain homeostasis and protects the brain parenchyma from blood-borne toxins and pathogens. In contrast to the vascular endothelial cells lining the blood vessels of the rest of the body, cerebral endothelial cells have distinct morphological, structural and functional properties, which act to severely limit both paracellular and transcellular transport [3–6].

Adjacent cerebral endothelial cells are connected by tight junctions and adherens junctions of less than 1 nm in diameter as shown in Figure 12.1. They are made up of proteins, such as occludins, claudins and cadherins, which form homophilic and heterophilic intercellular contacts [3–6]. The presence of these junctions prevents unregulated paracellular transport into the brain, allowing only very small ions and molecules across.

Transcellular transport across the BBB is also restricted. In addition to having few pores and vesicles, cerebral endothelial cells express efflux transporters, such as P-glycoprotein, which actively work to export compounds out of the endothelial cells and back into the bloodstream,

DOI: 10.1201/9781003224464-12

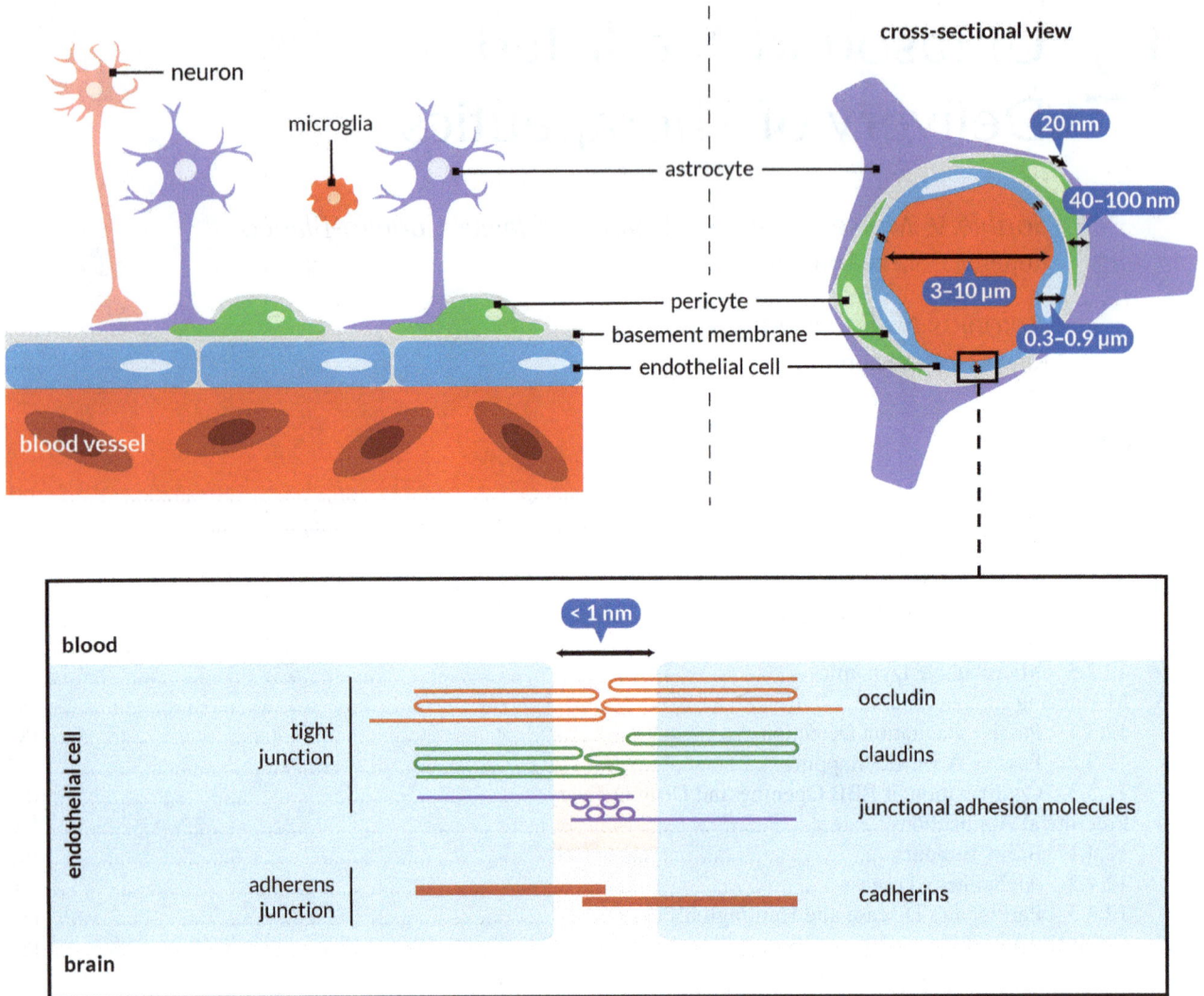

FIGURE 12.1   Structure of the blood-brain barrier (BBB) [4].

limiting brain exposure [3–6]. As a result, brain capillaries are stated to be 50–100 times tighter than peripheral microvessels [3].

While a healthy and intact BBB is crucial to maintain brain homeostasis, it represents a major obstacle for the treatment of brain diseases. For years, researchers have tried to pinpoint the exact physiochemical properties required for therapeutic agents to efficiently cross this barrier. Characteristics such as low molecular weight, low polarity, moderate lipophilicity and a low propensity for hydrogen bond formation have all been suggested to lead to improved BBB penetration. However, designing effective therapeutics that can successfully fulfil these criteria is non-trivial and remains a great challenge to the pharmaceutical industry [7–9].

As an alternative, there have been efforts to utilise a 'Trojan horse' approach and hijack existing transcellular transport pathways to enter the brain. Multiple studies have attempted to target endogenous receptors involved in receptor-mediated transcytosis across the BBB, such as the transferrin receptor [10]. While promising results have been shown using this strategy, high efficiency is difficult to

achieve and agents cannot be localised to a specific region, resulting in low therapeutic concentrations being delivered to target locations within the brain.

The injection of hyperosmotic solutions has also been investigated as a way to increase the permeability of the BBB [11]. Solutions such as mannitol can increase the osmotic pressure inside blood vessels, leading to the shrinking of endothelial cells, widening of junctions and increased vascular permeability. This technique, however, is not as localised as one would want and can lead to off-target side effects, such as seizures [12,13].

Complementary to the above approaches, there has been considerable interest in the design of methods that can completely bypass the BBB in a localised way. The use of focused ultrasound (FUS) and microbubbles has emerged as a promising method to disrupt the BBB in a non-invasive, targeted and transient manner [14]. Microbubbles are small particles (~0.5–10 μm in diameter; similar in size to red blood cells) composed of a gas core surrounded by a lipid or protein shell, which are commercially available and clinically approved as contrast agents for ultrasound imaging

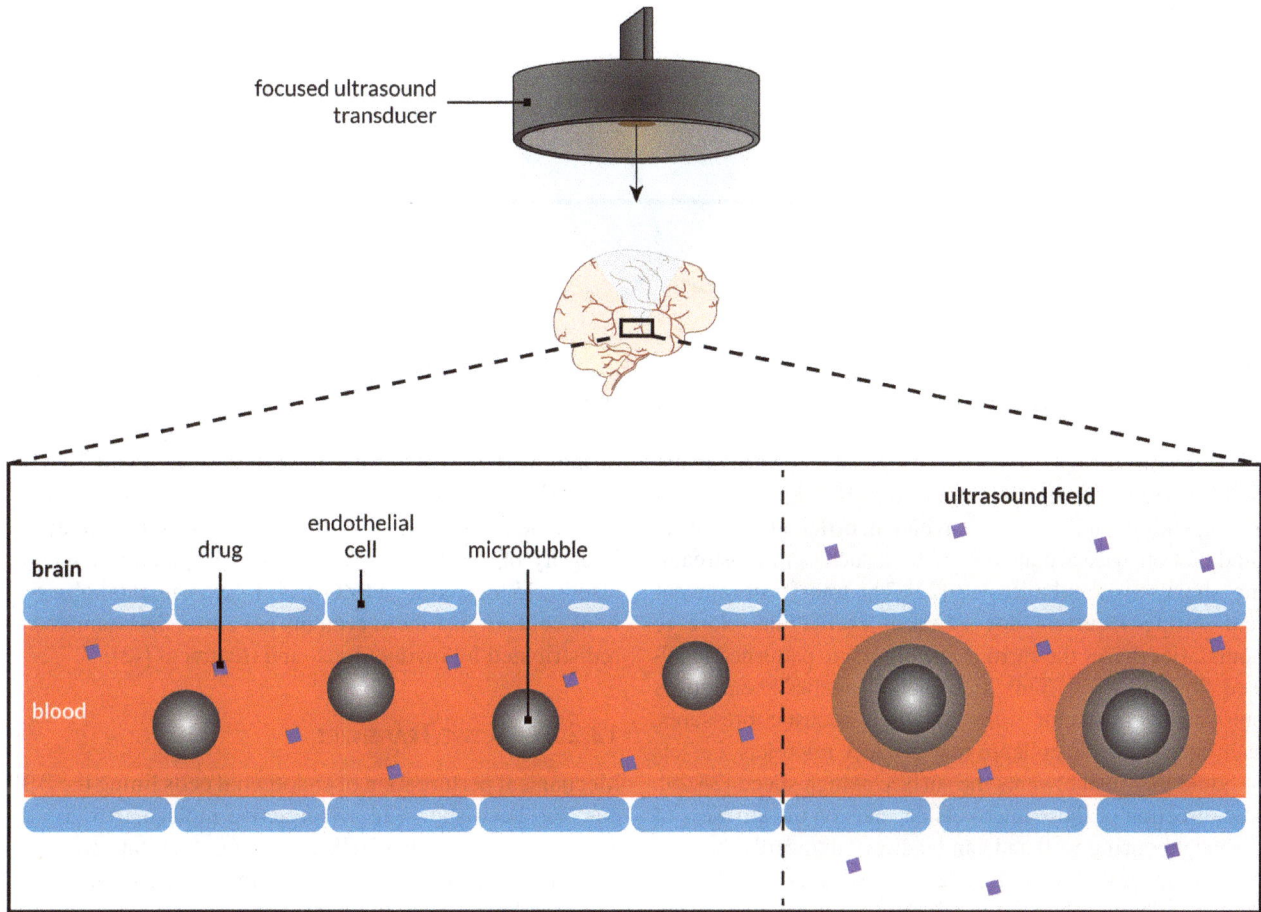

FIGURE 12.2 Illustration of the use of focused ultrasound and microbubbles to increase blood-brain barrier permeability and allow drug delivery to the brain.

[15,16]. Focused ultrasound-mediated drug delivery involves the co-injection of microbubbles and a drug of interest into the bloodstream. Ultrasound is focused onto the region of interest within the brain, causing microbubbles that enter the region to oscillate. This, in turn, increases the permeability of the BBB at this location and allows drugs to enter the brain in a localised manner as illustrated in Figure 12.2.

FUS-mediated drug delivery to the brain could revolutionise the treatment of brain diseases. To date, a wide variety of substances have been successfully delivered to the brain *in vivo* using ultrasound-mediated methods. These include small molecules, but also larger substances, which are traditionally challenging to deliver, such as liposomes and monoclonal antibodies.

In this chapter, we will describe the mechanisms of action of ultrasound-mediated drug delivery to the brain and provide examples of its use to date in both preclinical and clinical settings. While we have focused on brain applications in this chapter, it should be noted that FUS-mediated drug delivery is not limited to the brain; FUS can be used to deliver drugs to other organs, such as the liver and prostate [17–21], enhancing drug delivery across biological barriers, such as those imposed by cellular membranes and blood-tumour barriers.

12.2 MECHANISM OF ACTION

In this section, we will discuss the different microbubble behaviours that can be induced by ultrasound exposure and the biological outcomes that can be achieved from this.

12.2.1 ACOUSTIC CAVITATION

The complex behaviour of microbubbles during exposure to ultrasound is known as acoustic cavitation. Microbubbles oscillate volumetrically in response to the alternating phases of an ultrasound wave [22]. They expand during the rarefactional phase and contract during the compressional phase, as shown in Figure 12.3.

The oscillatory behaviour of each microbubble depends on the features of the driving acoustic field, the properties of the microbubble, the surrounding fluid and its interaction with neighbouring microbubbles. Therapeutic ultrasound exposure typically occurs at frequencies lower than 2 MHz, whereas imaging frequencies are generally higher. Below 2 MHz, microbubble oscillations are favoured allowing them to expand more. When these oscillations occur within microvessels or near the boundary of large vessels, they induce normal and shear stresses on the vascular walls.

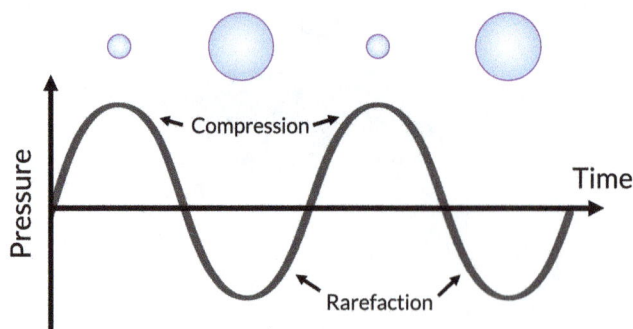

FIGURE 12.3 Volumetric oscillations of microbubbles in an ultrasound pressure field (compression and expansion of the bubbles during the compression and rarefactional phases of the ultrasound wave).

The force produced is linked to different modes of cavitation (for more detail please see Section 12.2.5).

Depending on the type of cavitation, different patterns of fluid motion arise around the bubbles, such as microstreaming, shockwaves and microjets [23]. The transfer of momentum during recurrent non-spherical oscillations close to boundaries drives the surrounding fluid into patterned localised flows [24–27]. This fluid motion applies shear stress onto nearby vascular walls. At high acoustic pressures, asymmetric bubble collapse may produce microjets, i.e. jets of liquid that rush towards the surface at high speed [28,29]. The direction of these microjets depends on the stiffness of the neighbouring wall and can be directed towards the wall or away from it. Violent bubble collapses may also lead to the formation of shockwaves, which occur when the velocity of the bubble wall exceeds the speed of sound.

The mechanical interactions described so far can produce three different bioeffects: sonoporation, tight junction disruption and upregulated active transport. In the following sections, each biological outcome will be briefly discussed.

12.2.2   Sonoporation

Microbubble activity in proximity to cells can produce pores in their cellular membrane. This is known as sonoporation, and its mechanism has been extensively studied *in vitro* [30]. Oscillating microbubbles poke individual cells, causing localised membrane perforations that are reversible [31]. The resulting pores differ in size [32] and are larger when microjetting is also observed [29]. In general, modifying the acoustic parameters and the microbubble population characteristics yields a wide spectrum of sonoporation effects [33–37]. Interestingly, opening a pore in a single cell can also affect neighbouring cells, activating mechanotransductive pathways and generating calcium waves [38–40].

12.2.3   Tight Junction Disruption

Tight junction proteins provide a tight physical seal that blocks the transcellular transport of most molecules that are larger than 400 Da [41,42]. Microbubble oscillations compromise the integrity of these junctions in cerebral microvessels, leading to enhanced paracellular transport

of administered drugs into the brain parenchyma [43,44]. Numerical simulations have shown that shockwaves induced by bubble implosion are sufficient to disrupt the molecular integrity of the proteins that form tight junctions, such as claudin-15, occludin and ZO-1 [45]. Experimental observations confirm that these proteins are either disassembled or redistributed following ultrasound treatment [43].

12.2.4   Active Transport

Mechanical perturbation of endothelial cells lining the BBB has also been shown to affect active molecular transport. There is evidence that FUS treatments facilitate caveolin-mediated transcytosis through the upregulation of caveolin-1, a membrane protein involved in vesicle formation [46]. The same effect has been observed in non-human primates [47]. Interestingly, caveolin-1 was only detected in animals treated at low pressures, suggesting that active transport is primarily affected by low stresses and non-violent bubble collapses.

12.2.5   Microbubble Dynamics

In the case of an isolated bubble embedded within an infinite liquid, one can discriminate two types of cavitation behaviour: non-inertial and inertial.

The motion of the bubble wall is described by the Rayleigh-Plesset equation [48]. The terms of this equation can be separated into a "pressure function" and an "inertial function". According to Flynn, an oscillating microbubble undergoes inertial cavitation if the inertial function is lower (i.e. more negative) than the pressure function [49]. When the applied pressure exceeds a specific threshold, the inertia of the surrounding fluid dominates the collapse of the bubble wall during the compression phase. This is called inertial cavitation. On the other hand, if the pressure function dominates, then we refer to it as non-inertial cavitation. Both non-inertial and inertial cavitation can be sustained over time for multiple acoustic cycles, and if this is the case, then we refer to it as stable cavitation. When microbubble oscillations cease, either due to fragmentation or due to gas dissolution, we refer to it as transient cavitation. It is worth noting that these regimes are not mutually exclusive, i.e. we can have stable inertial cavitation, when inertial collapses persist over time [50].

12.3 MONITORING

A unique feature of FUS brain therapies is the ability to monitor them in real time and confirm their efficacy using established imaging modalities. In this section, we will discuss the methods for passive cavitation detection (PCD) and passive acoustic mapping (PAM), which are used to passively monitor the cavitation activity of microbubbles. We will also discuss methods used to confirm BBB opening and assess drug delivery within the targeted regions.

12.3.1 PASSIVE CAVITATION DETECTION

Microbubbles exposed to therapeutic FUS produce their own acoustic signals that can be passively captured using a separate ultrasound transducer. This transducer, called a passive cavitation detector, is typically aligned concentrically or at an angle with the FUS transducer that is delivering the therapeutic beam.

The bandwidth and sensitivity of the two transducers are very different and must be chosen carefully to minimise interference between the therapeutic field and the received microbubble emissions. As mentioned before, therapeutic ultrasound uses frequencies below 2 MHz, whereas microbubble emissions cover a broader range of frequencies. For this reason, different fabrication materials are used depending on the application: FUS is delivered using resonant piezo-ceramic transducers, while microbubble emissions are detected using piezo-polymers that can capture broadband emissions.

The acoustic signal produced by a microbubble population is first detected with the PCD transducer, filtered with a high-pass filter, amplified and finally captured with an oscilloscope. The received signal is the integration of all the acoustic emissions emanating from within the focal volume and is therefore 1-dimensional and does not provide spatial information.

PCD signals are useful in determining the mixture of cavitation behaviours occurring *in vitro* and *in vivo*. It has been shown that these signals correlate with the induced bioeffects in rodents [51–53], rabbits [14,54,55], pigs [56] and non-human primates [57–59].

The key goal of PCD is to quantify acoustic emissions from microbubbles to produce a metric of cavitation dose that could be linked to different bioeffects. Frequency analysis of the captured time-domain signals provides a useful tool for the discrimination of cavitation activity within the microbubble population. Stable and recurrent oscillations give rise to harmonic and ultra-harmonic emissions. Harmonic emissions (f_h) are integer multiples of the fundamental frequency (f_c) ($f_h = nf_c$, $n = 2, 3, 4, \ldots$), whereas ultra-harmonic emissions (f_u) are only the odd halves of the fundamental frequency ($f_u = (n/2)f_c$, $n = 3, 5, 7, \ldots$). The amplitude of these distinct peaks can be averaged

throughout the bandwidth of the signal to derive a metric of stable harmonic and stable ultra-harmonic dose.

As previously described, inertial cavitation occurs when the applied pressure exceeds a specific threshold, and the inertia of the surrounding fluid dominates the bubble wall collapse [49]. This type of cavitation may be sustained over time [60], but it is also likely to lead to transient events such as microbubble fragmentation [61]. Violent collapses produce a sharp spike in the time-domain signal, which translates into a broadband acoustic signature. When a sufficient proportion of the microbubble population undergoes transient cavitation behaviour, the broadband floor of the emission spectrum increases [61–63].

High inertial cavitation doses have been correlated with both beneficial [62–64] and detrimental [57,65,66] bioeffects. Based on these observations, feedback-loop systems using PCD monitoring have been tested *in vivo* to control the degree of BBB opening and drug delivery [51,59,67–70]. These systems allow us to understand the landscape of cavitation behaviours within the entire focal volume in a simple and inexpensive manner. However, they do not provide spatial information which can be critical to accurately monitor brain therapies.

12.3.2 PASSIVE ACOUSTIC MAPPING

Although PCD is an invaluable tool for the study of bubble cavitation, it does not provide spatial information about the location of these emissions. It is essential to know the precise location of microbubble activity to confirm the accuracy of the treatment and minimise off-target effects.

To overcome this limitation, the use of imaging arrays was independently proposed by Gyöngy and Coussios [71] and Salgaonkar et al. [72]. PAM and passive cavitation imaging (PCI) have since become an indispensable tool in monitoring the spatial profile of cavitation activity in biomedical applications [23,73,74]. Both techniques require the passive detection of acoustic signals using multi-element arrays. The radiofrequency data obtained with these arrays is processed for every channel independently using delay-and-sum algorithms [71,75,76], and the space in front of the array is discretised into pixels. Finally, the total acoustic energy, along with useful spectral features, can be extracted from each pixel within a region of interest on the imaging plane.

The delay-and-sum approach has an acceptable lateral resolution and a poor axial resolution due to the so-called tail artefact. Tail artefacts appear due to the diffraction pattern of the imaging array and the constructive interference of signals from multiple sources within the region of interest. Multiple approaches have been proposed to improve the spatial resolution of PAM, including robust Capon beamforming [77], sum-of-harmonics [78] and higher-order statistics for data-adaptive beamforming [79]. The axial-to-lateral resolution ratio is poor for linear or phased imaging arrays but improves considerably for probes with a larger aperture size, as with hemispherical arrays [80–83].

Regardless of the imaging system, 2D and 3D PAM have shown good correlation with BBB opening and drug delivery in both preclinical [82,84,85] and clinical studies [86–88]. Current approaches focus on real-time implementation of PAM, which would permit adaptive control of exposure conditions during treatment [47,82,83,89–92].

12.3.3 CONFIRMATION OF BBB OPENING AND DRUG DELIVERY

Successful FUS-mediated BBB opening can be confirmed with a multitude of methods. The most widely used method to assess BBB opening in animals and humans is T_1-weighted contrast-enhanced MRI. Gadolinium-based contrast agents are impermeable to the intact BBB but can enter the brain parenchyma after FUS treatment. BBB opening in preclinical animal models can be observed as the area of enhanced contrast in either static or dynamic contrast-enhancement T1-weighted scans [14,93–97]. Other modalities, such as PET and SPECT imaging, have also been employed to confirm BBB opening and drug delivery. In this case, the drug of interest is labelled with an appropriate radioisotope to confirm its delivery [87,98]. The advantage of using established medical imaging modalities is their scalability to humans; in fact, MRI is being used to confirm BBB opening in ongoing clinical trials [87,99–103].

Apart from the aforementioned imaging contrast agents, preclinical FUS treatments have also been confirmed optically using fluorescent dyes, such as Evans Blue and fluorescently labelled dextran [104,105]. These markers have variable molecular size and have been used to identify the size threshold for FUS-mediated drug delivery [106–108]. Finally, in the field of gene therapy, confirmation of viral delivery can be achieved by incorporating a reporter gene in the plasmid that is being delivered [109–112]. Ongoing research focuses on tagging drugs or drug carriers, like liposomes, with MRI, PET, SPECT or optical agents to establish their delivery following FUS exposure [113–117].

12.4 PRECLINICAL APPLICATIONS

So far, we have described the physical mechanisms of FUS-mediated drug delivery, the biological effects induced on the BBB and the existing strategies to monitor this procedure. In this section, we will focus on the application of this drug delivery technology for the treatment of brain tumours, Alzheimer's disease, Parkinson's disease and Huntington's disease.

12.4.1 BRAIN TUMOURS

FUS has been used to deliver drugs of low and high molecular weight across the BBB to treat brain tumours, including glioblastoma and diffuse intrinsic pontine glioma. Examples of small molecule drugs successfully delivered for this purpose include temozolomide (molecular mass: 194 Da), doxorubicin (543 Da), carboplatin (371 Da), irinotecan (587 Da), carmustine (214 Da) and paclitaxel (853 Da) [69,118–123]. Larger substances delivered include bevacizumab (149 kDa) [124], humanised antihuman epidermal growth factor receptor (2HER2-cerbB2), monoclonal antibodies [97], IL-12 (70 kDa) [125] and dopamine D-4 receptor-targeting antibodies [126]. The use of FUS has also been explored for the treatment of brain metastases [124,127].

Focused ultrasound-mediated delivery can benefit drugs of all sizes. For example, while temozolomide, a small molecule drug approved for the treatment of gliomas, is able to cross the BBB by itself, animal studies have shown that the use of FUS can double its delivery to the brain, slow tumour growth rates and provide longer survival [119,121,128]. On the other hand, FUS can enhance the delivery of larger drugs, which have limited BBB penetration by themselves, by factors of 3–10.

In addition to enhancing drug delivery at the target location, FUS can be used at pressures that will trigger immune effects, which could be beneficial for the treatment of brain tumours. FUS has shown to recruit tumour-infiltrating lymphocytes at targeted tumour sites with or without the administration of anticancer drugs or specific immune triggering agents, such as IL-12 [125,127,129–131].

12.4.2 ALZHEIMER'S DISEASE

Given that many Alzheimer's disease clinical trials have failed due to poor drug delivery across the BBB, there has been much excitement about the potential of this non-invasive and localised ultrasound approach to revolutionise the Alzheimer's disease treatment landscape [1]. To date, both therapeutic and molecular imaging agents have been delivered with FUS to Alzheimer's disease animal models [132], including BAM-10 anti-beta-amyloid antibodies [133], RN2N tau-specific antibodies [134], GSK-3 (glycogen-synthase kinase-3) inhibitors and intravenous immunoglobulin [135,136]. These studies have observed neurogenesis and reduced amounts of amyloid plaques, which have been correlated with improved behavioural performance and memory [137,138].

12.4.3 PARKINSON'S DISEASE AND HUNTINGTON'S DISEASE

FUS has also been used to deliver therapeutic genomic material in models of Parkinson's disease and Huntington's disease. To enhance gene expression and neurotrophic factor delivery for Parkinson's disease, FUS has been used to deliver gene-encoding viral vectors, such as recombinant AAV-2 [139,140]; non-viral vectors, such as liposomal plasmids and microbubbles conjugated with genes and cationic plasmids; as well as neurotrophic factors, such as brain-derived neurotrophic factors (BDNF), glia-derived neurotrophic factors (GDNF) and neurturin [139–147]. By delivering *BDNF* and *GDNF* genes, neuroprotective effects have been observed, with lower astrocyte activation and apoptosis markers as well as improved behaviour.

For Huntington's disease, where brain cells progressively degenerate leading to a loss of motor skills and cognitive function, FUS has enhanced the delivery of small interfering RNA (siRNA) to decrease the expression of mutated Huntingtin protein [148]. The delivery of GDNF-plasmid-liposomes in mice with Huntington's disease has also shown advantageous effects, including reductions in oxidative stress and apoptosis, improved motor performance and neuronal survival [149].

12.5 CLINICAL APPLICATIONS

Following two decades of extensive preclinical testing, FUS-induced BBB opening has entered the phase of clinical implementation. As of April 2022, there are three approaches being tested for non-invasive drug delivery into the central nervous system. These approaches, along with examples of ongoing clinical trials, are given below.

12.5.1 Ongoing Clinical Trials

Clinical systems for performing FUS treatments in the human brain can be separated into three broad categories: MRI-guided transducers, implantable transducers and neuronavigation-guided transducers. Clinical trials using these systems are currently underway at multiple sites around the world.

The most widely used system is the ExAblate Neuro™ device from Insightec©. ExAblate is a hemispherical multi-element array, which is guided by MRI. Its 1024 elements allow for electronic steering of the ultrasonic beam within the treatment region. MRI guidance ensures precise targeting and in-line monitoring of the treatment. The ExAblate system has been used in multiple clinical trials, with an excellent safety and efficacy profile to date [99,100,150,151]. The first trial using this system was completed in 2018 by Lipsman et al. [99], where they showed that the BBB of Alzheimer's disease patients could be opened reversibly, with no contrast enhancement detected 1 day after treatment.

Another approach, which was the first to be tested in humans, uses an implantable transducer fixed within the skull following craniotomy. The SonoCloud™ device made by CarThera© allows for repeated treatment of the same brain region without the need for repositioning. Without the presence of the skull between the brain and the transducer, ultrasound can freely travel into the brain without significant beam distortions. This device originally comprised of a single transducer, but recent developments have included the use of multiple transducers implanted following tumour resection. SonoCloud™ has been successfully tested in glioblastoma patients, demonstrating a considerable increase in the delivery of chemotherapeutics within the tumour area [101,103,152]. Results from the first FUS trial using this device were published in 2016 by Carpentier et al. [101],

where they showed that the BBB of patients with glioblastoma, who were exposed to monthly FUS treatments before systemic chemotherapy, could be disrupted without detectable adverse effects on MRI or clinical examination.

The third approach uses a single-element FUS transducer guided by a neuronavigation system. Neuronavigation systems are based on an infrared camera that can optically detect a set of fiducial markers, which are reflecting spheres oriented in a unique way in space so they can be identified by the camera. These markers are placed on the transducer and the patient's head, permitting real-time tracking of their respective positions. A pre-loaded MRI image is used to guide the treatment. This approach was pioneered by NaviFUS©, showing evidence of safety and efficacy in glioblastoma patients [102]. A similar system has been developed by Delsona Therapeutics©/Columbia University, which incorporates real-time monitoring of the treatment using both PCD and PAM [57,102,153,154]. Neuronavigation-based systems are portable and cost-efficient, offering the capability of fast and convenient treatments across different patient populations. The first clinical study using neuronavigation guidance was published in 2021 [102]. They showed that with their system reversible BBB opening, within 24 h, could be achieved in patients with recurrent glioblastoma.

Due to the success of these studies, a multitude of clinical trials are ongoing (Table 12.1). These studies are aimed at a range of neurological diseases, such as Alzheimer's disease, Parkinson's disease, Parkinson's dementia, amyotrophic lateral sclerosis, GBM and diffuse intrinsic pontine glioma.

12.5.2 Challenges and Future Prospects

Despite the tremendous success of early-phase clinical trials, many challenges remain before the widespread implementation of FUS therapies. The first challenge is cost and time. FUS systems require expensive equipment and, in the case of MRI-guided systems, are performed within clinical MRI scanners. Scanning time for conducting and confirming FUS therapies induces a significant cost. Reducing the time required for each treatment will not only reduce the cost but also encourage reimbursement from the private and public sector.

The second challenge is treatment confirmation. As of 2022, BBB opening is verified in the clinic through contrast-enhanced T_1-weighted MRI. Although this is sufficient to prove efficacy, there are concerns regarding the potential toxicity of gadolinium-based contrast agents. Patients with compromised renal function are excluded from clinical trials involving MRI-guided systems and patients with normal renal function can be affected by gadolinium administration, particularly if repeated treatments are required. Therefore, there is a pressing need for alternatives to MRI-based confirmation. One potential solution is to use microbubble monitoring systems, like PCD or PAM, as alternative methods for confirming drug delivery. However, calibrating drug delivery with acoustic data poses

TABLE 12.1

Active Clinical Trials on FUS-Induced BBB Opening for Targeted Drug Delivery[a]

NCT Number	Device	Indication	Drug
NCT03714243	ExAblate	HER2-positive breast cancer brain metastases	N/A
NCT04440358	ExAblate	Recurrent glioblastoma	Carboplatin
NCT04417088	ExAblate	Recurrent glioblastoma	Carboplatin
NCT04526262	ExAblate	Alzheimer's disease	N/A
NCT03321487	ExAblate	Amyotrophic lateral sclerosis	N/A
NCT02343991	ExAblate	Brain tumours	Doxorubicin
NCT03608553	ExAblate	Parkinson's dementia	N/A
NCT04559685	ExAblate	Recurrent high-grade glioma	Aminolevulinic acid HCl (ALA)
NCT03551249	ExAblate	High-grade glioma	N/A
NCT03616860	ExAblate	Glioblastoma	N/A
NCT04998864	ExAblate	High-grade glioma	N/A
NCT03739905	ExAblate	Alzheimer's disease	N/A
NCT03744026	SonoCloud	Recurrent glioblastoma	Carboplatin
NCT04528680	SonoCloud	Recurrent glioblastoma	Paclitaxel
NCT04063514	BrainSonix	Low grade glioma	N/A
NCT04118764	Delsona Therapeutics	Alzheimer's disease	N/A
NCT04804709	Delsona Therapeutics	Diffuse midline glioma	Panobinostat
NCT04988750	NaviFUS	Recurrent glioblastoma	N/A
NCT04446416	NaviFUS	Recurrent glioblastoma	Bevacizumab

[a] Active FUS trials registered on clinicaltrials.gov in March 2022.

other challenges, stemming from the variability of imaging sensors and their sensitivity, and their low spatial and contrast resolution compared to MRI.

12.6 CONCLUSIONS

In this chapter, we have described how FUS combined with systemically administered microbubbles can alter the permeability of the BBB, providing a non-invasive, transient and targeted approach for drug delivery to the brain. This technology is extremely promising for the treatment of brain diseases, where the BBB presents a significant challenge, but it can also be used in other parts of the body. A good understanding of the types of microbubble activity, when they occur and what bioeffects they cause allows us to better design ultrasound sequences for specific disease targets to maximise therapeutic effects and minimise adverse effects. Over the past two decades, significant advances in the technology have been made in preclinical studies to confirm its safety and efficacy for the treatment of brain diseases. Due to its promising results, clinical trials are now underway with several devices across the world to treat patients with brain tumours, Alzheimer's disease, Parkinson's disease and amyotrophic lateral sclerosis.

REFERENCES

1. J. L. Cummings, T. Morstorf, and K. Zhong, "Alzheimer's drug development candidates failures," *Alzheimers. Res. Ther.*, vol. 6, no. 37, pp. 1–7, 2014.

2. W. M. Pardridge, "The blood-brain barrier: Bottleneck in brain drug development," *Neurotherapeutics*, 2005, doi: 10.1007/bf03206638.

3. N. Abbott, L. Ronnback, and E. Hansson, "Astrocyte-endothelial interactions at the blood-brian barrier," *Nat. Rev. Neurosci.*, vol. 7, no. 1, pp. 41–53, 2006.

4. C. D. Arvanitis, G. B. Ferraro, and R. K. Jain, "The blood–brain barrier and blood–tumour barrier in brain tumours and metastases," *Nat. Rev. Cancer*, vol. 20, no. 1. pp. 26–41, 2020, doi: 10.1038/s41568-019-0205-x.

5. J. Fenstermacher, P. Gross, N. Sposito, V. Acuff, S. Pettersen, and K. Gruber, "Structural and functional variations in capillary systems within the brain," *Ann. N. Y. Acad. Sci.*, vol. 529, no. 1, pp. 21–30, 1988, doi: 10.1111/j.1749-6632.1988. tb51416.x.

6. H. Kadry, B. Noorani, and L. Cucullo, "A blood–brain barrier overview on structure, function, impairment, and biomarkers of integrity," *Fluids Barriers CNS*, vol. 17, no. 1, pp. 1–24, 2020, doi: 10.1186/s12987-020-00230-3.

7. E. G. Chikhale, P. S. Burton, and R. T. Borchardt, "Hydrogen bonding potential as a determinant of the in vitro and in situ blood-brain barrier permeability of peptides," *Pharm. Res. An Off. J. Am. Assoc. Pharm. Sci.*, vol. 11, pp. 412–419, 1994.

8. V. A. Levin, "Relationship of octanol/water partition coefficient and molecular weight to rat brain capillary permeability," *J. Med. Chem.*, vol. 23, pp. 682–684, 1980.

9. C. A. Lipinski, F. Lombardo, B. W. Dominy, and P. J. Feeney, "Experimental and computational approaches to estimate solubility and permeability in drug discovery and development settings," *Adv. Drug Deliv. Rev.*, vol. 46, pp. 3–26, 2001.

10. V. Mishra et al., "Targeted brain delivery of AZT via trasferrin anchored pegylated albumin nanoparticles," *J. Drug Target.*, vol. 14, pp. 45–53, 2006.

11. S. I. Rapoport, M. Hori, and I. Klatzo, "Testing of a hypothesis for osmotic opening of the blood-brain barrier," *Am. J. Physiol.*, vol. 223, no. 2, pp. 323–331, 1972, doi: 10.1152/ajplegacy.1972.223.2.323.

12. N. Marchi et al., "Seizure-promoting effect of blood-brain barrier disruption," *Epilepsia*, vol. 48, no. 4, pp. 732–742, 2007, doi: 10.1111/j.1528-1167.2007.00988.x.

13. N. D. Doolittle et al., "Safety and efficacy of a multicenter study using intraarterial chemotherapy in conjunction with osmotic opening of the blood-brain barrier for the treatment of patients with malignant brain tumors," *Cancer*, vol. 88, no. 3, pp. 637–647, 2000, doi: 10.1002/(SICI)1097-0142(20000201)88:3<637::AID-CNCR22>3.0.CO;2-Y.

14. K. Hynynen, N. McDannold, N. Vykhodtseva, and F. A. Jolesz, "Noninvasive MR imaging-guided focal opening of the blood-brain barrier in rabbits," *Radiology*, vol. 220, no. 3, pp. 640–646, 2001, doi: 10.1148/radiol.2202001804.

15. C. P. Nolsøe and T. Lorentzen, "International guidelines for contrast-enhanced Ultrasonography: Ultrasound imaging in the new millennium," *Ultrasonography*, vol. 35, no. 2. pp. 89–103, 2016, doi: 10.14366/usg.15057.

16. M. Schneider, "Characteristics of SonoVue™," *Echocardiography*, vol. 16, no. 7, pp. 743–746, 1999.

17. K. Fischer, et al., "Renal ultrafiltration changes induced by focused ultrasound," *Radiology*, vol. 253, no. 3, pp. 697–705, 2009, doi: 10.1148/radiol.2532082100.

18. H. Chen and J. Hwang, "Ultrasound-targeted microbubble destruction for chemotherapeutic drug delivery to solid tumors," *J. Ther. Ultrasound*, vol. 1, no. 1, p. 10, 2013, doi: 10.1186/2050-5736-1-10.

19. S. Eggen et al., "Ultrasound-enhanced drug delivery in prostate cancer xenografts by nanoparticles stabilizing microbubbles," *J Control. release*, vol. 187, pp. 39–49, 2014.

20. S. Tinkov et al., "New doxorubicin-loaded phospholipid microbubbles for targeted tumor therapy: In vivo characterization," *J Control. Release*, vol. 148, pp. 368–72, 2010.

21. P. Li et al., "Ultrasound triggered drug release from 10-hydroxycamptothecin-loaded phospholipid microbubbles for targeted tumor therapy in mice," *J. Control. Release*, vol. 162, pp. 349–54, 2012.

22. T. G. Leighton and R. E. Apfel, "The acoustic bubble," *J. Acoust. Soc. Am.*, vol. 96, no. 4, pp. 2616–2616, 1994, doi: 10.1121/1.410082.

23. E. Stride and C. Coussios, "Nucleation, mapping and control of cavitation for drug delivery," *Nat. Rev. Phys.*, vol. 1, no. 8, pp. 495–509, 2019, doi: 10.1038/s42254-019-0074-y.

24. A. A. Doinikov and A. Bouakaz, "Effect of a distant rigid wall on microstreaming generated by an acoustically driven gas bubble," *J. Fluid Mech.*, vol. 742, no. 2, pp. 425–445, 2014, doi: 10.1017/jfm.2012.656.

25. A. A. Doinikov and A. Bouakaz, "Acoustic microstreaming around a gas bubble," *J. Acoust. Soc. Am.*, vol. 127, no. 2, p. 703, 2010, doi: 10.1121/1.3279793.

26. S. Elder and W. L. Nyborg, "Acoustic streaming resulting from a resonant bubble," *J. Acoust. Soc. Am.*, vol. 28, no. 1, p. 155, 2005, doi: 10.1121/1.1918085.

27. W. L. Nyborg, "Acoustic streaming near a boundary," *J. Acoust. Soc. Am.*, vol. 30, no. 4, pp. 329–339, 1958, doi: 10.1121/1.1909587.

28. F. Yuan, C. Yang, and P. Zhong, "Cell membrane deformation and bioeffects produced by tandem bubble-induced jetting flow," *Proc. Natl. Acad. Sci. U. S. A.*, vol. 112, no. 51, pp. E7039–E7047, 2015, doi: 10.1073/pnas.1518679112.

29. C.-D. Ohl et al., "Sonoporation from jetting cavitation bubbles," *Biophys. J.*, vol. 91, no. 11, pp. 4285–4295, 2006, doi: 10.1529/biophysj.105.075366.

30. B. Helfield, X. Chen, S. C. Watkins, and F. S. Villanueva, "Biophysical insight into mechanisms of sonoporation," *Proc. Natl. Acad. Sci. U. S. A.*, vol. 113, no. 36, pp. 9983–9988, 2016, doi: 10.1073/pnas.1606915112.

31. Y. Hu, J. M. F. Wan, and A. C. H. Yu, "Membrane perforation and recovery dynamics in microbubble-mediated sonoporation," *Ultrasound Med. Biol.*, vol. 39, no. 12, pp. 2393–2405, 2013, doi: 10.1016/j.ultrasmedbio.2012.08.003.

32. Y. Zhou, R. E. Kumon, J. Cui, and C. X. Deng, "The size of sonoporation pores on the cell membrane," *Ultrasound Med. Biol.*, vol. 35, no. 10, pp. 1756–1760, 2009, doi: 10.1016/j.ultrasmedbio.2009.05.012.

33. F. E. Shamout et al., "Enhancement of non-invasive transmembrane drug delivery using ultrasound and microbubbles during physiologically relevant flow," *Ultrasound Med. Biol.*, vol. 41, no. 9, pp. 2435–2448, 2015, doi: 10.1016/j.ultrasmedbio.2015.05.003.

34. Y. Qiu et al., "The correlation between acoustic cavitation and sonoporation involved in ultrasound-mediated DNA transfection with polyethylenimine (PEI) in vitro," *J. Control. Release*, vol. 145, no. 1, pp. 40–48, 2010, doi: 10.1016/j.jconrel.2010.04.010.

35. R. Karshafian, P. D. Bevan, R. Williams, S. Samac, and P. N. Burns, "Sonoporation by ultrasound-activated microbubble contrast agents: Effect of acoustic exposure parameters on cell membrane permeability and cell viability," *Ultrasound Med. Biol.*, vol. 35, no. 5, pp. 847–860, 2009, doi: 10.1016/j.ultrasmedbio.2008.10.012.

36. K. Kooiman, M. Foppen-Harteveld, A. F. W. van der Steen, and N. de Jong, "Sonoporation of endothelial cells by vibrating targeted microbubbles," *J. Control. Release*, vol. 154, no. 1, pp. 35–41, 2011, doi: 10.1016/j.jconrel.2011.04.008.

37. Z. Fan, D. Chen, and C. X. Deng, "Characterization of the dynamic activities of a population of microbubbles driven by pulsed ultrasound exposure in sonoporation," *Ultrasound Med. Biol.*, vol. 40, no. 6, pp. 1260–1272, 2014, doi: 10.1016/j.ultrasmedbio.2012.12.002.

38. Z. Fan, R. E. Kumon, J. Park, and C. X. Deng, "Intracellular delivery and calcium transients generated in sonoporation facilitated by microbubbles," *J. Control. Release*, vol. 142, no. 1, pp. 31–39, 2010, doi: 10.1016/j.jconrel.2009.09.031.

39. I. Beekers et al., "High-resolution imaging of intracellular calcium fluctuations caused by oscillating microbubbles," *Ultrasound Med. Biol.*, 2020, doi: 10.1016/j.ultrasmedbio.2020.03.029.

40. R. E. Kumon et al., "Spatiotemporal effects of sonoporation measured by real-time calcium imaging," *Ultrasound Med. Biol.*, vol. 35, no. 3, pp. 494–506, 2009, doi: 10.1016/j.ultrasmedbio.2008.09.003.

41. C. Zihni, C. Mills, K. Matter, and M. S. Balda, "Tight junctions: From simple barriers to multifunctional molecular gates," *Nat. Rev. Mol. Cell Biol.*, vol. 17, no. 9, pp. 564–580, 2016, doi: 10.1038/nrm.2016.80.

42. H. C. Bauer, I. A. Krizbai, H. Bauer, and A. Traweger, "'You shall not pass'-tight junctions of the blood brain barrier," *Front. Neurosci.*, vol. 8, no. DEC. Frontiers Media SA, p. 392, 2014, doi: 10.3389/fnins.2014.00392.

43. N. Sheikov, N. McDannold, S. Sharma, and K. Hynynen, "Effect of focused ultrasound applied with an ultrasound contrast agent on the tight junctional integrity of the brain microvascular endothelium," *Ultrasound Med.*

Biol., vol. 34, no. 7, pp. 1093–1104, 2008, doi: 10.1016/j.ultrasmedbio.2007.12.015.

44. X. Shang, P. Wang, Y. Liu, Z. Zhang, and Y. Xue, "Mechanism of low-frequency ultrasound in opening blood–tumor barrier by tight junction," *J. Mol. Neurosci.*, vol. 43, no. 3, pp. 364–369, 2011, doi: 10.1007/s12031-010-9451-9.

45. A. Goliaei, U. Adhikari, and M. L. Berkowitz, "Opening of the blood-brain barrier tight junction due to shock wave induced bubble collapse: A molecular dynamics simulation study," *ACS Chem. Neurosci.*, vol. 6, no. 8, pp. 1296–1301, 2015, doi: 10.1021/acschemneuro.5b00116.

46. R. Pandit, W. K. Koh, R. K. P. Sullivan, T. Palliyaguru, R. G. Parton, and J. Götz, "Role for caveolin-mediated transcytosis in facilitating transport of large cargoes into the brain via ultrasound," *J. Control. Release*, vol. 327, pp. 667–675, 2020, doi: 10.1016/j.jconrel.2020.09.015.

47. S. Bae, K. Liu, A. N. Pouliopoulos, and E. E. Konofagou, "Coherence-factor-based passive acoustic mapping for real-time transcranial cavitation monitoring with improved axial resolution," in *2021 IEEE International Ultrasonics Symposium*, pp. 1–4, 2021, doi: 10.1109/IUS52206.2021.9593643.

48. P. Marmottant et al., "A model for large amplitude oscillations of coated bubbles accounting for buckling and rupture," *J. Acoust. Soc. Am.*, vol. 118, no. 6, pp. 3499–3505, 2005, doi: 10.1121/1.2109427.

49. H. G. Flynn, "Cavitation dynamics: II. Free pulsations and models for cavitation bubbles," *J. Acoust. Soc. Am.*, vol. 58, no. 6, pp. 1160–1170, 1975, doi: 10.1121/1.380799.

50. C. C. Church and E. L. Carstensen, "'Stable' inertial cavitation," *Ultrasound Med. Biol.*, vol. 27, no. 10, pp. 1435–1437, 2001, doi: 10.1016/S0301-5629(01)00441-0.

51. M. O'Reilly and K. Hynynen, "Blood-brain barrier: Real-time feedback-controlled focused ultrasound disruption by using an acoustic emissions-based controller," *Radiology*, vol. 263, no. 1, pp. 96–106, 2012, doi: 10.1148/radiol.11111417.

52. T. Sun, G. Samiotaki, S. Wang, C. Acosta, C. C. Chen, and E. E. Konofagou, "Acoustic cavitation-based monitoring of the reversibility and permeability of ultrasound-induced blood-brain barrier opening," *Phys. Med. Biol.*, vol. 60, no. 23, pp. 9079–9094, 2015, doi: 10.1088/0031-9155/60/23/9079.

53. T. Sun et al., "Closed-loop control of targeted ultrasound drug delivery across the blood–brain/tumor barriers in a rat glioma model," *Proc. Natl. Acad. Sci. U. S. A.*, vol. 114, no. 48, pp. E10281–E10290, 2017, doi: 10.1073/pnas.1713328114.

54. N. J. McDannold, N. I. Vykhodtseva, and K. Hynynen, "Microbubble contrast agent with focused ultrasound to create brain lesions at low power levels: MR imaging and histologic study in rabbits," *Radiology*, vol. 241, 2006, doi: 10.1148/radiol.2411051170.

55. K. Hynynen, N. McDannold, H. Martin, F. A. Jolesz, and N. Vykhodtseva, "The threshold for brain damage in rabbits induced by bursts of ultrasound in the presence of an ultrasound contrast agent (Optison®)," *Ultrasound Med. Biol.*, vol. 29, no. 3, pp. 473–481, 2003, doi: 10.1016/S0301-5629(02)00741-X.

56. C. P. Pacia et al., "Feasibility and safety of focused ultrasound-enabled liquid biopsy in the brain of a porcine model," *Sci. Rep.*, vol. 10, no. 1, pp. 1–9, 2020, doi: 10.1038/s41598-020-64440-3.

57. A. N. Pouliopoulos et al., "Safety evaluation of a clinical focused ultrasound system for neuronavigation guided blood-brain barrier opening in non-human primates," *Sci. Rep.*, vol. 11, no. 1, p. 15043, 2021, doi: 10.1038/s41598-021-94188-3.

58. F. Marquet et al., "Real-time, transcranial monitoring of safe blood-brain barrier opening in non-human primates," *PLoS One*, vol. 9, no. 2, pp. 1–11, 2014, doi: 10.1371/journal.pone.0084310.

59. H. A. S. Kamimura et al., "Feedback control of microbubble cavitation for ultrasound-mediated blood–brain barrier disruption in non-human primates under magnetic resonance guidance," *J. Cereb. Blood Flow Metab.*, p. 0271678X1775351, 2018, doi: 10.1177/0271678X17753514.

60. C. C. Church and E. L. Carstensen, "Stable inertial cavitation," *Ultrasound Med. Biol.*, vol. 27, no. 10, pp. 1435–1437, 2001, doi: 10.1016/S0301-5629(01)00441-0.

61. W. T. Shi, F. Forsberg, A. Tornes, J. Østensen, and B. B. Goldberg, "Destruction of contrast microbubbles and the association with inertial cavitation," *Ultrasound Med. Biol.*, vol. 26, no. 6, pp. 1009–1019, 2000, doi: 10.1016/S0301-5629(00)00223-4.

62. W. S. Chen, A. A. Brayman, T. J. Matula, L. A. Crum, and M. W. Miller, "The pulse length-dependence of inertial cavitation dose and hemolysis," *Ultrasound Med. Biol.*, vol. 29, no. 5, pp. 739–748, 2003, doi: 10.1016/S0301-5629(03)00029-2.

63. W.-S. Chen, A. A. Brayman, T. J. Matula, and L. A. Crum, "Inertial cavitation dose and hemolysis produced in vitro with or without Optison," *Ultrasound Med. Biol.*, vol. 29, no. 5, pp. 725–737, 2003.

64. S. M. Graham et al., "Inertial cavitation to non-invasively trigger and monitor intratumoral release of drug from intravenously delivered liposomes," *J. Control. release*, vol. 178, pp. 101–107, 2014, doi: 10.1016/j.jconrel.2012.12.016.

65. S. Xu et al., "Correlation between brain tissue damage and inertial cavitation dose quantified using passive cavitation imaging," *Ultrasound Med. Biol.*, 2019, doi: 10.1016/J.ULTRASMEDBIO.2019.07.004.

66. J. H. Hwang, J. Tu, A. A. Brayman, T. J. Matula, and L. A. Crum, "Correlation between inertial cavitation dose and endothelial cell damage in vivo," *Ultrasound Med. Biol.*, vol. 32, no. 10, pp. 1611–1619, 2006, doi: 10.1016/j.ultrasmedbio.2006.07.016.

67. C. Bing et al., "Characterization of different bubble formulations for blood-brain barrier opening using a focused ultrasound system with acoustic feedback control," *Sci. Rep.*, vol. 8, no. 1, 2018, doi: 10.1038/s41598-018-26330-7.

68. B. Cheng, C. Bing, and R. Chopra, "The effect of transcranial focused ultrasound target location on the acoustic feedback control performance during blood-brain barrier opening with nanobubbles," *Sci. Rep.*, vol. 9, no. 1, p. 20020, 2019, doi: 10.1038/s41598-019-55629-2.

69. N. McDannold et al., "Acoustic feedback enables safe and reliable carboplatin delivery across the blood-brain barrier with a clinical focused ultrasound system and improves survival in a rat glioma model," *Theranostics*, vol. 9, no. 21, pp. 6284–6299, 2019, doi: 10.7150/thno.35892.

70. N. Hockham, C. C. Coussios, and M. Arora, "A real-time controller for sustaining thermally relevant acoustic cavitation during ultrasound therapy," *IEEE Trans. Ultrason. Ferroelectr. Freq. Control*, vol. 57, no. 12, pp. 2685–2694, 2010, doi: 10.1109/TUFFC.2010.1742.

71. M. Gyöngy and C.-C. Coussios, "Passive cavitation mapping for localization and tracking of bubble dynamics," *J. Acoust. Soc. Am.*, vol. 128, no. 4, pp. EL175–EL180, 2010, doi: 10.1121/1.3467491.

72. V. A. Salgaonkar, S. Datta, C. K. Holland, and T. D. Mast, "Passive cavitation imaging with ultrasound arrays," *J. Acoust. Soc. Am.*, 2009, doi: 10.1121/1.3238260.

73. K. J. Haworth et al., "Passive imaging with pulsed ultrasound insonations," *J. Acoust. Soc. Am.*, 2012, doi: 10.1121/1.4728230.

74. K. J. Haworth, K. B. Bader, K. T. Rich, C. K. Holland, and T. D. Mast, "Quantitative frequency-domain passive cavitation imaging," *IEEE Trans. Ultrason. Ferroelectr. Freq. Control*, vol. 64, no. 1, pp. 177–191, 2017, doi: 10.1109/TUFFC.2016.2620492.

75. C. R. Jensen, R. W. Ritchie, M. Gyöngy, J. R. T. Collin, T. Leslie, and C.-C. Coussios, "Spatiotemporal monitoring of high-intensity focused ultrasound therapy with passive acoustic mapping," *Radiology*, vol. 262, no. 1, pp. 252–261, 2012, doi: 10.1148/radiol.11110670.

76. M. Gyöngy and C.-C. Coussios, "Passive spatial mapping of inertial cavitation during HIFU exposure," *IEEE Trans. Biomed. Eng.*, vol. 57, no. 1, pp. 48–56, 2010, doi: 10.1109/TBME.2009.2026907.

77. C. Coviello et al., "Passive acoustic mapping utilizing optimal beamforming in ultrasound therapy monitoring," *J. Acoust. Soc. Am.*, vol. 137, no. 5, pp. 2573–2585, 2015, doi: 10.1121/1.4916694.

78. E. Lyka, C. Coviello, R. Kozick, and C.-C. Coussios, "Sum-of-harmonics method for improved narrowband and broadband signal quantification during passive monitoring of ultrasound therapies," *J. Acoust. Soc. Am.*, vol. 140, no. 1, pp. 741–754, 2016, doi: 10.1121/1.4958991.

79. E. Lyka, C. M. Coviello, C. Paverd, M. D. Gray, and C. C. Coussios, "Passive acoustic mapping using data-adaptive beamforming based on higher order statistics," *IEEE Trans. Med. Imaging*, 2018, doi: 10.1109/TMI.2018.2843291.

80. C. Crake, S. T. Brinker, C. M. Coviello, M. S. Livingstone, and N. J. McDannold, "A dual-mode hemispherical sparse array for 3D passive acoustic mapping and skull localization within a clinical MRI guided focused ultrasound device," *Phys. Med. Biol.*, vol. 63, no. 6, p. 65008, 2018, doi: 10.1088/1361-6560/aab0aa.

81. M. A. O'Reilly, R. Jones, and K. Hynynen, "Three-dimensional transcranial ultrasound imaging of microbubble clouds using a sparse hemispherical array," *IEEE Trans. Biomed. Eng.*, vol. 61, no. 4, pp. 1285–1294, 2014.

82. R. M. Jones, L. Deng, K. Leung, D. McMahon, M. A. O'Reilly, and K. Hynynen, "Three-dimensional transcranial microbubble imaging for guiding volumetric ultrasound-mediated blood-brain barrier opening," *Theranostics*, vol. 8, no. 11, pp. 2909–2926, 2018, doi: 10.7150/thno.24911.

83. R. M. Jones, D. McMahon, and K. Hynynen, "Ultrafast three-dimensional microbubble imaging in vivo predicts tissue damage volume distributions during nonthermal brain ablation," *Theranostics*, vol. 10, no. 16, pp. 7211–7230, 2020, doi: 10.7150/thno.47281.

84. Y. Yang et al., "Cavitation dose painting for focused ultrasound-induced blood-brain barrier disruption," *Sci. Rep.*, vol. 9, no. 1, 2019, doi: 10.1038/s41598-019-39090-9.

85. J. J. Choi, R. C. Carlisle, C. Coviello, L. Seymour, and C. C. Coussios, "Non-invasive and real-time passive acoustic mapping of ultrasound-mediated drug delivery," *Phys. Med. Biol.*, vol. 59, no. 17, pp. 4861–4877, 2014, doi: 10.1088/0031-9155/59/17/4861.

86. P. Anastasiadis et al., "Localized blood–brain barrier opening in infiltrating gliomas with MRI-guided acoustic emissions–controlled focused ultrasound," *Proc. Natl. Acad. Sci.*, vol. 118, no. 37, 2021, doi: 10.1073/PNAS.2103280118.

87. Y. Meng, C. B. Pople, S. Suppiah, and N. Lipsman, "MR-guided focused ultrasound for brain tumors," *Image-Guided Hypofractionated Stereotactic Radiosurgery*. CRC Press, pp. 415–427, 2021, doi: 10.1201/9781003037095-22.

88. R. M. Jones et al., "Echo-focusing in transcranial focused ultrasound thalamotomy for essential tremor: A feasibility study," *Mov. Disord.*, p. mds.28226, 2020, doi: 10.1002/mds.28226.

89. H. A. S. Kamimura et al., "Real-time passive acoustic mapping using sparse matrix multiplication," *IEEE Trans. Ultrason. Ferroelectr. Freq. Control*, vol. 68, no. 1, pp. 164–177, 2021, doi: 10.1109/TUFFC.2020.3001848.

90. S. Y. Wu et al., "Efficient blood-brain barrier opening in primates with neuronavigation-guided ultrasound and real-time acoustic mapping," *Sci. Rep.*, 2018, doi: 10.1038/s41598-018-25904-9.

91. C. Arvanitis, N. McDannold, and G. Clement, "Fast passive cavitation mapping with angular spectrum approach," *J. Acoust. Soc. Am.*, vol. 138, no. 3, p. 1845, 2015, doi: 10.1121/1.4933873.

92. A. Patel, S. J. Schoen, and C. D. Arvanitis, "Closed-loop spatial and temporal control of cavitation activity with passive acoustic mapping," *IEEE Trans. Biomed. Eng.*, vol. 66, no. 7, pp. 2022–2031, 2019, doi: 10.1109/TBME.2018.2882337.

93. F. Vlachos, Y.-S. S. Tung, and E. Konofagou, "Permeability dependence study of the focused ultrasound-induced blood-brain barrier opening at distinct pressures and microbubble diameters using DCE-MRI," *Magn. Reson. Med.*, vol. 66, no. 3, pp. 821–830, 2011, doi: 10.1002/mrm.22848.

94. F. Vlachos, Y.-S. Tung, and E. E. Konofagou, "Permeability assessment of the focused ultrasound-induced blood-brain barrier opening using dynamic contrast-enhanced MRI," *Phys. Med. Biol.*, vol. 55, no. 18, pp. 5451–5466, 2010, doi: 10.1088/0031-9155/55/18/012.

95. N. McDannold et al., "Blood-brain barrier disruption and delivery of irinotecan in a rat model using a clinical transcranial MRI-guided focused ultrasound system," *Sci. Rep.*, vol. 10, no. 1, p. 8766, 2020, doi: 10.1038/s41598-020-65617-6.

96. A. N. Pouliopoulos et al., "Temporal stability of lipid-shelled microbubbles during acoustically-mediated blood-brain barrier opening," *Front. Phys.*, vol. 8, p. 137, 2020, doi: 10.3389/fphy.2020.00137.

97. M. Kinoshita, N. McDannold, F. A. Jolesz, and K. Hynynen, "Targeted delivery of antibodies through the blood-brain barrier by MRI-guided focused ultrasound," *Biochem. Biophys. Res. Commun.*, vol. 340, no. 4, pp. 1085–1090, 2006, doi: 10.1016/j.bbrc.2005.12.112.

98. N. D. Sheybani et al., "ImmunoPET-informed sequence for focused ultrasound-targeted mCD47 blockade controls glioma," *J. Control. Release*, vol. 331, pp. 19–29, 2021, doi: 10.1016/j.jconrel.2021.01.023.

99. N. Lipsman et al., "Blood–brain barrier opening in Alzheimer's disease using MR-guided focused ultrasound," *Nat. Commun.*, vol. 9, no. 1, p. 2336, 2018, doi: 10.1038/s41467-018-04529-6.

100. A. Abrahao et al., "First-in-human trial of blood–brain barrier opening in amyotrophic lateral sclerosis using MR-guided focused ultrasound," *Nat. Commun.*, vol. 10, no. 1, p. 4373, 2019, doi: 10.1038/s41467-019-12426-9.

101. A. Carpentier et al., "Clinical trial of blood-brain barrier disruption by pulsed ultrasound," *Sci. Transl. Med.*, vol. 8, no. 343, p. 343re2, 2016, doi: 10.1126/scitranslmed.aaf6086.

102. K.-T. Chen et al., "Neuronavigation-guided focused ultrasound for transcranial blood-brain barrier opening and immunostimulation in brain tumors," *Sci. Adv.*, vol. 7, no. 6, p. eabd0772, 2021, doi: 10.1126/sciadv.abd0772.

103. A. Idbaih et al., "Safety and feasibility of repeated and transient blood-brain barrier disruption by pulsed ultrasound in patients with recurrent glioblastoma," *Clin. Cancer Res.*, vol. 25, no. 13, pp. 3793–3801, 2019, doi: 10.1158/1078-0432.CCR-18-3643.

104. S. N. Tabatabaei, M. S. Tabatabaei, H. Girouard, and S. Martel, "Hyperthermia of magnetic nanoparticles allows passage of sodium fluorescein and Evans blue dye across the blood–retinal barrier," *Int. J. Hyperth.*, vol. 32, no. 6, pp. 657–665, 2016, doi: 10.1080/02656736.2016.1193903.

105. Z. K. Englander et al., "Focused ultrasound mediated blood–brain barrier opening is safe and feasible in a murine pontine glioma model," *Sci. Rep.*, vol. 11, no. 1, p. 6521, 2021, doi: 10.1038/s41598-021-85180-y.

106. S. V. Morse et al., "Rapid short-pulse ultrasound delivers drugs uniformly across the murine blood-brain barrier with negligible disruption," *Radiology*, vol. 291, no. 2, pp. 459–466, 2019, doi: 10.1148/radiol.2019181625.

107. J. J. Choi, K. Selert, F. Vlachos, A. Wong, and E. E. Konofagou, "Noninvasive and localized neuronal delivery using short ultrasonic pulses and microbubbles," *Proc. Natl. Acad. Sci. U. S. A.*, vol. 108, no. 40, pp. 16539–16544, 2011, doi: 10.1073/pnas.1105116108.

108. J. J. Choi, S. Wang, Y. S. Tung, B. Morrison, and E. E. Konofagou, "Molecules of various pharmacologically-relevant sizes can cross the ultrasound-induced blood-brain barrier opening in vivo," *Ultrasound Med. Biol.*, vol. 36, no. 1, pp. 58–67, 2010, doi: 10.1016/j.ultrasmedbio.2009.08.006.

109. J. O. Szablowski and M. Harb, "Focused ultrasound induced blood-brain barrier opening for targeting brain structures and evaluating chemogenetic neuromodulation," *J. Vis. Exp.*, vol. 2020, no. 166, p. e61352, 2020, doi: 10.3791/61352.

110. J. O. Szablowski, A. Lee-Gosselin, B. Lue, D. Malounda, and M. G. Shapiro, "Acoustically targeted chemogenetics for the non-invasive control of neural circuits," *Nat. Biomed. Eng.*, vol. 2, no. 7, pp. 475–484, 2018, doi: 10.1038/s41551-018-0258-2.

111. S. Wang, O. O. Olumolade, T. Sun, G. Samiotaki, and E. E. Konofagou, "Noninvasive, neuron-specific gene therapy can be facilitated by focused ultrasound and recombinant adeno-associated virus," *Gene Ther.*, vol. 22, no. 1, pp. 104–110, 2015, doi: 10.1038/gt.2014.91.

112. S. Wang et al., "Non-invasive, focused ultrasound-facilitated gene delivery for optogenetics," *Sci. Rep.*, vol. 7, p. 39955, 2017, doi: 10.1038/srep39955.

113. P. Cressey, M. Amrahli, P. W. So, W. Gedroyc, M. Wright, and M. Thanou, "Image-guided thermosensitive liposomes for focused ultrasound enhanced co-delivery of carboplatin and SN-38 against triple negative breast cancer in mice," *Biomaterials*, vol. 271, p. 120758, 2021, doi: 10.1016/J.BIOMATERIALS.2021.120758.

114. M. Amrahli et al., "Mr-labelled liposomes and focused ultrasound for spatiotemporally controlled drug release in triple negative breast cancers in mice," *Nanotheranostics*, vol. 5, no. 2, pp. 125–142, 2021, doi: 10.7150/NTNO.52168.

115. M. Amate, J. Goldgewicht, B. Sellamuthu, J. Stagg, and F. T. H. Yu, "The effect of ultrasound pulse length on microbubble cavitation induced antibody accumulation and distribution in a mouse model of breast cancer," *Nanotheranostics*, vol. 4, no. 4, pp. 256–269, 2020, doi: 10.7150/ntno.46892.

116. M. Aryal, I. Papademetriou, Y. Z. Zhang, C. Power, N. McDannold, and T. Porter, "MRI monitoring and quantification of ultrasound-mediated delivery of liposomes dually labeled with gadolinium and fluorophore through the blood-brain barrier," *Ultrasound Med. Biol.*, vol. 45, no. 7, pp. 1733–1742, 2019, doi: 10.1016/j.ultrasmedbio.2019.02.024.

117. Y. S. Kang, U. Bickel, and W. M. Pardridge, "Pharmacokinetics and saturable blood-brain barrier transport of biotin bound to a conjugate of avidin and a monoclonal antibody to the transferrin receptor," *Drug Metab. Dispos.*, vol. 22, no. 1, pp. 99–105, 1994.

118. D. Y. Zhang et al., "Ultrasound-mediated delivery of paclitaxel for glioma: A comparative study of distribution, toxicity, and efficacy of albumin-bound versus cremophor formulations," *Clin. Cancer Res.*, vol. 26, pp. 477–486, 2020.

119. K. C. Wei et al., "Focused ultrasound-induced blood-brain barrier opening to enhance temozolomide delivery for glioblastoma treatment: A preclinical study," *PLoS One*, vol. 8, p. e58995, 2012.

120. L. H. Treat, N. McDannold, N. Vykhodtseva, Y. A. Zhang, K. Tam, and K. Hynynen, "Targeted delivery of doxorubicin to the rat brain at therapeutic levels using MRI-guided focused ultrasound," *Int. J. Cancer2*, vol. 121, pp. 901–907, 2007.

121. H. L. Liu, C. Y. Huang, J. Y. Chen, H. Y. J. Wang, P. Y. Chen, and K. C. Wei, "Pharmacodynamic and therapeutic investigation of focused ultrasound-induced blood-brain barrier opening for enhanced temozolomide delivery in glioma treatment," *PLoS One*, vol. 9, p. e114311, 2014.

122. K. Beccaria, et al., "Ultrasound-induced opening of the blood-brain barrier to enhance temozolomide and irinotecan delivery: An experimental study in rabbits," *J. Neurosurg.*, vol. 124, pp. 1602–1610, 2016.

123. H. L. Liu et al., "Blood-brain barrier disruption with focused ultrasound enhances delivery of chemotherapeutic drugs for glioblastoma treatment," *Radiology*, vol. 255, pp. 415–425, 2010.

124. H. L. Liu et al., "Focused ultrasound enhances central nervous system delivery of bevacizumab for malignant glioma treatment," *Radiology*, vol. 281, pp. 99–108, 2016.

125. P. Y. Chen, H. Y. Hsieh, C. Y. Huang, C. Y. Lin, K. C. Wei, and H. L. Liu, "Focused ultrasound-induced blood-brain barrier opening to enhance interleukin-12 delivery for brain tumor immunotherapy," *J. Transl. Med.*, vol. 13, p. 93, 2015.

126. M. Kinoshita, N. McDannold, F. A. Jolesz, and K. Hynynen, "Noninvasive localized delivery of Herceptin to the mouse brain by MIR-guided focused ultrasound-induced blood-brain barrier disruption," *PNAS*, vol. 103, pp. 11719–11723, 2006.

127. H. L. Liu et al., "In vivo assessment of macrophage CNS infiltration during disruption of the blood-brain barrier with focused ultrasound: A magnetic resonance imaging study," *J. Cereb. Blood Flow Metab.*, vol. 30, p. 674, 2010.

128. A. Papachristodoulou et al., "Chemotherapy sensitization of glioblastoma by focused ultrasound-mediated delivery of therapeutic liposomes," *J. Control. Release*, vol. 295, pp. 130–139, 2019, doi: 10.1016/j.jconrel.2018.12.009.

129. P. Y. Chen, K. C. Wei, and H. L. Liu, "Neural immune modulation and immunotehrapy assisted by focused ultrasound induced blood-brain barrier opening," *Hum. Vaccines Immunother.*, vol. 11, pp. 2682–2687, 2015.

130. H. L. Liu, H. Y. Hsieh, L. A. Lu, C. W. Kang, M. F. Wu, and C. Y. Lin, "Low-pressure pulsed focused ultrasound with microbubbles promotes an anticancer immunological response," *J. Transl. Med.*, vol. 10, p. 221, 2012.

131. E. J. Park, Y. Z. Zhang, N. Vykhodtseva, and N. McDannold, "Ultrasound-mediated blood-brain/blood-tumour barrier

disruption improves outcomes with trastuzumab in a breast cancer brain metastasis model," *J Control. release*, vol. 163, pp. 277–284, 2012.

132. S. B. Raymond, L. H. Treat, J. D. Dewey, N. J. McDannold, K. Hynynen, and B. J. Bacskai, "Ultrasound enhanced delivery of molecular imaging and therapeuitc agents in Alzheimer's Disease mouse model," *PLoS One*, vol. 3, p. e2175, 2008.

133. J. F. Jordão et al., "Antibodies targeted to the brain with image-guided focused ultrasound reduces amyloid-β plaque load in the TgCRND8 mouse model of Alzheimer's disease," *PLoS One*, vol. 5, no. 5, 2010, doi: 10.1371/journal.pone.0010549.

134. R. M. Nisbet, A. der Jeugd, G. Leinenga, H. T. Evans, P. W. Janowicz, and J. Götz, "Combined effects of scanning ultrasound and a tau-specific single chain antibody in a tau transgenic mouse model," *Brain*, vol. 140, no. 5, pp. 1220–1230, 2017.

135. P. H. Hsu et al., "Focused ultrasound-induced blood-brain barrier opening enhances GSK-3 inhibitor delivery for amyloid-beta plaque reduction," *Sci. Rep.*, vol. 8, no. 1, 2018, doi: 10.1038/s41598-018-31071-8.

136. S. Dubey et al., "Clinically approved IVIg delivered to the hippocampus with focused ultrasound promotes neurogenesis in a model of Alzheimer's disease," *Proc. Natl. Acad. Sci.*, vol. 117, no. 51, pp. 32691–32700, 2020.

137. A. Burgess et al., "Alzheimer disease in a mouse model: Mr imaging-guided focused ultrasound targeted to the hippocampus opens the blood-brain barrier and improves pathologic abnormalities and behavior," *Radiology*, vol. 273, no. 3, pp. 736–745, 2014, doi: 10.1148/radiol.14140245.

138. G. Leinenga and J. Götz, "Scanning ultrasound removes amyloid-b and restores memory in an Alzheimer's disease mouse model," *Sci. Transl. Med.*, vol. 7, no. 278, 2015, doi: 10.1126/scitranslmed.aaa2512.

139. P.-H. Hsu et al., "Noninvasive and targeted gene delivery into the brain using microbubble-facilitated focused ultrasound," *PLoS One*, vol. 8, no. 2, p. e57682, 2012.

140. Z. Noroozian et al., "MRI-guided focused ultrasound for targeted delivery of rAAV to the brain," *Methods Mol. Biol.*, vol. 1950, pp. 177–197, 2019.

141. B. Baseri et al., "Activation of signaling pathways following localized delivery of systemically administered neurotrophic factors across the blood-brain barrier using focused ultrasound and microbubbles," *Phys. Med. Biol.*, vol. 57, no. 7, pp. N65–N81, 2012, doi: 10.1088/0031-9155/57/7/N65.

142. F. Wang et al., "Targeted delivery of GDNF through the blood--brain barrier by MRI-guided focused ultrasound," *PLoS One*, vol. 7, no. 12, p. e52925, 2012.

143. C. Y. Lin et al., "Focused ultrasound-induced blood-brain barrier opening for non-viral, non-invasive, and targeted gene delivery," *J. Control. Release*, vol. 212, pp. 1–9, 2015, doi: 10.1016/j.jconrel.2015.06.010.

144. C.-Y. Lin et al., "Non-invasive, neuron-specific gene therapy by focused ultrasound-induced blood-brain barrier opening in Parkinson's disease mouse model," *J. Control. Release*, vol. 235, pp. 72–81, 2016.

145. C. H. Fan et al., "Noninvasive, targeted, and non-viral ultrasound-mediated GDNF-plasmid delivery for treatment of Parkinson's disease," *Sci. Rep.*, vol. 6, 2016, doi: 10.1038/srep19579.

146. L. Long et al., "Treatment of Parkinson's disease in rats by Nrf2 transfection using MRI-guided focused ultrasound delivery of nanomicrobubbles," *Biochem. Biophys. Res. Commun.*, vol. 482, no. 1, pp. 75–80, 2017.

147. C.-Y. Lin, Y.-C. Lin, C.-Y. Huang, S.-R. Wu, C.-M. Chen, and H.-L. Liu, "Ultrasound-responsive neurotrophic factor-loaded microbubble-liposome complex: Preclinical investigation for Parkinson's disease treatment," *J. Control. Release*, vol. 321, pp. 519–528, 2020.

148. A. Burgess, Y. Huang, W. Querbes, D. W. Sah, and K. Hynynen, "Focused ultrasound for targeted delivery of siRNA and efficient knockdown of Htt expression," *J. Control. release*, vol. 163, no. 2, pp. 125–129, 2012.

149. C. Y. Lin et al., "Focused ultrasound-induced blood brain-barrier opening enhanced vascular permeability for GDNF delivery in Huntington's disease mouse model," *Brain Stimul.*, vol. 12, no. 5, pp. 1143–1150, 2019, doi: 10.1016/j.brs.2019.04.011.

150. C. Gasca-Salas et al., "Blood-brain barrier opening with focused ultrasound in Parkinson's disease dementia," *Nat. Commun.*, vol. 12, no. 1, pp. 1–7, 2021, doi: 10.1038/s41467-021-21022-9.

151. T. Mainprize et al., "Blood-brain barrier opening in primary brain tumors with non-invasive MR-guided focused ultrasound: A clinical safety and feasibility study," *Sci. Rep.*, vol. 9, no. 1, 2019, doi: 10.1038/s41598-018-36340-0.

152. N. Asquier et al., "Blood-brain barrier disruption in humans using an implantable ultrasound device: Quantification with MR images and correlation with local acoustic pressure," *J. Neurosurg.*, pp. 1–9, 2019, doi: 10.3171/2018.9.JNS182001.

153. A. N. Pouliopoulos et al., "Neuronavigation-guided focused ultrasound for non-invasive blood-brain barrier opening in the prefrontal cortex of Alzheimer's disease patients with real-time cavitation monitoring," in *International Symposium on Therapeutic Ultrasound (ISTU)*, 2021.

154. A. N. Pouliopoulos, S. Y. Wu, M. T. Burgess, M. E. Karakatsani, H. A. S. Kamimura, and E. E. Konofagou, "A clinical system for non-invasive blood–brain barrier opening using a neuronavigation-guided single-element focused ultrasound transducer," *Ultrasound Med. Biol.*, vol. 46, no. 1, pp. 73–89, 2020, doi: 10.1016/j.ultrasmedbio.2019.09.010.

13 The Ongoing Emergence of Technology in Healthcare to Enhance Patient Outcomes

Dan J. Corbett, Maurice Hall, Lezley-Anne Hanna, and Heather E. Barry
Queen's University Belfast

CONTENTS

13.1 EMERGING TECHNOLOGIES IN HEALTHCARE – HOW DID WE GET HERE?

Over the last few decades, technology, particularly that related to telecommunications, has developed at an accelerated pace. Innovations, improved manufacturing processes and design improvements have all led to an exponential increase in the public's use and acceptance of technology in their daily lives. Indeed, the current ubiquity of technology within the lives of many is unparalleled and appears to be increasing (Edwards 2021). One need to only look at the current worldwide use of smartphones to appreciate the exceptionally commonplace nature of technology – at present, the number of smartphone subscriptions alone is reaching parity with global population, with the total number of devices in use set to rise to almost eight billion within the next 5 years (Statista 2022; Ramalingam et al. 2021). Likewise, the capability of such devices is increasing at a remarkable rate, due to a burgeoning software and application development market. For example, the number of applications available for use within the Android mobile device operating system alone has increased approximately 200-fold in just over 10 years (Li et al. 2022), illustrating the enormous potential which is offered by these technologies.

The rationale for the application of technology within healthcare more broadly, whilst multifactorial, is also clear. Above all else, the assistance which technology provides with respect to the handling of large volumes of data, patients' or otherwise, is massively advantageous, and whilst not a panacea, certainly provides opportunity for the lessening of strain on healthcare systems which are currently overburdened by ageing populations, the increased prevalence of multimorbidity, polypharmacy and ongoing threats and realisation of global pandemic. Moreover, the use of technology provides the opportunity for more effective provision of healthcare within low and low-middle income countries, enabling lifespan and quality of life improvements (Pan and Xu 2014). Whilst the impact of technology within more wide-ranging and operational aspects of healthcare provision is significant, the emergence of technology within healthcare is set to continue to offer a great deal of benefit at an individual end-user level. Again, this is the result of a range of contributing factors in addition to the immediate access which patients now have to useful technology, including cultural changes across populations which have seen patients wishing to take a greater level of responsibility for their own health and well-being, reflecting the manner in which modern medicine is practiced (Pipich 2018).

These drivers, and others related and in addition to them, have led to the development, growth and expansion of healthcare technology, which the World Health Organization has defined as the "application of organized knowledge and skills in the form of devices, medicines, vaccines, procedures, and systems developed to solve a health problem and improve quality of lives" (Keestra 2021), and which goes further to indicate both the breadth of the area, and ongoing and potential impacts that these technologies, if used correctly, can have on population health. The parallels which can be drawn between the unprecedented rate of innovation within the digital space and that which is taking place within the many areas of healthcare go further still to reinforce the role that technology has to play, and the synergistic effects to be gained from the concurrent development of these areas. A counterpoint to this, however, is the need to understand that the relationship between technology and the provision of healthcare is *truly* synergistic, and that these technologies alone will not lead to the enhancement of patient outcomes – thus, it is crucial that the role that these resources play in this arena is acutely understood, such that full benefit can be derived, and that the employment of these aspects to the detriment of outcome is avoided.

13.2 THE APPLICATION OF TECHNOLOGIES TO ENHANCE OUTCOMES

To better understand the value which can be derived from the use of emerging technologies within healthcare, it is useful to consider several key areas within which the use of technology has provided benefit, or indeed, which has led to the development of entirely new approaches for healthcare management. Areas such as telemedicine, and the sophisticated use of smart devices alone or in conjunction with wearable devices, offer significant insight into how technology can be leveraged to these ends, while also illustrating limitations and considerations which should be made to allow the potential benefits of these technologies to be realised.

13.2.1 TELEMEDICINE

Telemedicine, as a wider term, and which is synonymous with terms including "telehealth", "e-health" and "online health", and which is derived from the Greek term *tēle*, meaning "far off" or "from a distance", describes the use of digital communication technologies to facilitate clinical interactions between healthcare providers and end users, complementing or often replacing those which would have traditionally taken place in person (Wootton 2001). It should be noted, perhaps because of some linguistic ambiguity of the encapsulating term, that telemedicine is by no means restricted to the use of a telephone and, indeed, is facilitated by a range of telecommunications technologies. The advantage offered by such approaches are clear and are of benefit to patient and practitioner alike – access to healthcare is facilitated, particularly to those who may find it difficult to attend physical appointments for a range of reasons, including the lack of transport or availability during times when such consultations would traditionally take place, or indeed, as a result of suffering from conditions where mobilisation to clinical primary and secondary care settings may be challenging and potentially detrimental to the patient's health state. Moreover, the leveraging

of telemedicine permits healthcare providers to be more efficient with their time, enhancing overall effectiveness of service delivery.

In current times, the value of approaches which allow for the genuine interpersonal interaction between patient and healthcare provider, as well as the effective communication of patient-centric data in a remote way, has been brought into sharp focus as a result of the global COVID-19 pandemic, due to their role in the continuance of function of healthcare systems, including the management of coronavirus-infected patients, while simultaneously reducing or removing the risks which traditional, in-person approaches would present, particularly for those with morbidities which result in elevated risk (Portnoy, Waller, and Elliott 2020). Indeed, whilst the majority of telemedicine activity takes place within high and upper-middle income regions, the literature indicates that there are benefits in relation to the quality of healthcare offerings, providing a clear rationale for the extension of use of telemedicine in a more significant extent to low and low-middle income locales which lack extensive levels of healthcare provision, as it may play a key role in providing access to medicine where none may previously have existed, leading to marked enhancement in population health and development.

The fundamentals of telemedical practice are straightforward, incorporating the transmission of healthcare information across distances digitally, with this information allowing for a full range of medical activities to be delivered, from the creation and maintenance of patient medical records, through to the diagnosis of ailments and the education and training of other healthcare professionals. This transmission of data can occur both in real-time, or in an asynchronous manner, depending on the specific activity being undertaken. Key examples of real-time activities involve those which are mediated by teleconferencing platforms, facilitating conversations between patient and practitioner, or indeed, amongst interprofessional teams (Koch et al. 2018; Kern et al. 2020); asynchronous approaches involve the transmission of data of some sort which can be used for the purposes of diagnosis or patient monitoring (for example, the provision of test results to another clinician with expertise in a particular clinical area) (Craig and Petterson 2005; Brown et al. 2022; Ross Kerr 2020). Whilst the core activities which define telemedicine would likely be regarded as being rudimentary within the wider span of emerging technologies, the implementation of telemedicine and its impact are still significant areas of research – an interrogation of the MEDLINE database provides approximately 1600 indexed research articles authored over the most recent 5-year period, and indeed, as described earlier, the COVID-19 pandemic has provided added impetus for related research, with approximately 550 of these articles investigating the use of telemedicine-related technologies within this context. In addition, extensive randomised controlled trials supported by major funding bodies have been carried out which investigate the impact that these approaches may have (Salisbury et al. 2017). This significant context may well have acted as a catalyst for the increasingly commonplace use of telemedicine within healthcare settings, and importantly, the acceptance of its use, which is key to its success (Fisk, Livingstone, and Pit 2020; Hu et al. 1999). As such, there is a significant and potentially increasing role for telemedicine to play within healthcare, which in turn has led to an interest in the integration of other emerging technologies, for the purposes of enhancing related outcomes.

13.2.2 The Emergence of Connected Health

As described in the previous section, the activities which constitute telemedicine are multitudinous and involve the transfer of information in relation to the care of patients in a range of modalities. This very nature lends the approach to bring about paradigm shifts in core areas of healthcare practice, which may lead to the introduction of efficiencies within healthcare organisations, whilst simultaneously offering benefit to involved patients. One such approach is that of telemonitoring, which involves a more targeted set of goals, namely to make use of telecommunications for the purposes of monitoring a patient's disease status. Telemonitoring is included within the intermediary grouping of home-telecare, which seeks to support patients, particularly those with chronic conditions, within their own home (Pare, Jaana, and Sicotte 2007). Given the extent of population ageing, and the inter-related increasing prevalence in chronic conditions, with approximately one in every three adults now suffering from multimorbidity (Hajat and Stein 2018), the use of these remote approaches offers a potentially more humanistic and economically efficient way to manage these patient groups.

Whilst telemonitoring itself is arguably an evolved aspect of telemedicine, it also is undergoing its own evolutionary process, influenced by changing attitudes to healthcare, the empowerment of other professions within healthcare to take a larger role in the management of patient populations and, indeed, the development of technologies which can feed reliable clinical information into systems, further potentiating their value and more closely aligning them to orthodox in-person clinical consultations (Caulfield and Donnelly 2013; Frist 2014). This evolution has led to the emergence of *connected health*, which seeks to make use of telemonitoring practices and associated emerging technologies to encourage patients to take greater ownership over the monitoring and management of their conditions, whilst also enabling their collaboration with healthcare professionals to be less point-in-time and more reactive in nature, such that condition worsening can be at best avoided and, at worst, identified as quickly as possible (Caulfield and Donnelly 2013; Burmaoglu et al. 2017).

The characteristics of connected health strongly lend it to the management of chronic disease, and as such, it is unsurprising that its use has been investigated for the management of the most prevalent chronic morbidities including cardiovascular disease and diabetes (Colorafi 2016;

Wongvibulsin et al. 2019). This is especially promising given the enormity of the populations who suffer from one or both disease sets, and indeed, the extent of additional morbidities which can be caused by them when control is poor. Telemonitoring enables streamlining of patient interaction and care, and a focus to be placed on those who are experiencing deterioration, whilst those who are experiencing successful disease control can benefit from minimal intervention.

Despite the significant potential offered by the effective use of connected health approaches, barriers to more extensive use exist as a result of a combination of issues including the need for a more extensive evidence base, public awareness and acceptance, and the availability of supporting technologies which would allow for full value to be extracted. With respect to the evidence base, meta-analyses do point to the enhanced effectiveness of connected health for the control of major conditions such as hypertension; however, limitations are often reported with respect to the duration or follow up involved in the interventions used to gather this evidence, or in relation to the number of studies available, and/or the population sizes within them (Omboni 2019; Wongvibulsin et al. 2019). Whilst currently to the detriment of further success, the identification of public awareness and acceptance as challenges provides new avenues for the enhancement of rollout of connected health activities, enabling work to be undertaken to address these issues and, indeed, enhance the attractiveness of patients' own ability to play a major role within the management of their health. The extent of these issues should not be undervalued, with general patient awareness of connected health appearing to be low (Barr et al. 2014). This in turn may lead to a lack of engagement or the emergence of concerns around the use of the approach due to a lack of knowledge and related confidence, or indeed with respect to patient comfort in the sharing of their healthcare data electronically (Kuziemsky et al. 2018). Thus, education of student groups is key to the ongoing success and use of connected health and telemedical approaches more widely, which in turn necessitates the involvement of multidisciplinary care teams, who can provide explanations and encourage and reassure patients with respect to the processes themselves and the derived benefits. However, whilst the role of these professionals, particularly those in primary care, such as community pharmacists, would reasonably be expected to provide positive impact, this does appear to be under-investigated (Barr, McElnay, and Hughes 2012).

13.2.3 The Role of Connected Health in Treatment Adherence

Adherence can be very simply defined as the process of patients doing, to the fullest extent, what has been recommended or instructed by their healthcare providers (Vermeire et al. 2001). Achieving successful outcomes for patients depends heavily on their adherence to prescribed treatment or management strategies (Martin et al. 2005), with poor adherence having a clear and obvious link to worsening of patients' disease states, increased rates of co- and multi-morbidities and, ultimately in many cases, worsened mortality rates (Lehane and Mccarthy 2009). Further, if treatment adherence is related only to pharmacotherapy, i.e. the taking of provided medicines in the manner instructed by the prescriber, poor adherence can bring about additional issues including the development of treatment resistance (B. Nachega et al. 2011; Bangsberg, Kroetz, and Deeks 2007) and the heightened risk of suffering from adverse effects where medicines with particular pharmacodynamic characteristics are prescribed (Haddad 2001). Rates of non-adherence in more general terms are high, estimated to be around 25% (DiMatteo 2004) and linked to a myriad of causative factors, including the nature of the condition which is being treated, such as in cases where there is a reduction in cognitive function, or in patient motivation to manage their own care (Smith et al. 2017; Burns et al. 2013).

For the reasons outlined above, adherence is a major concern in public health from broad social and economic perspectives, and thus it is unsurprising that investigations centred on the understanding of adherence, and in relation to the development of technological approaches to improve it, are abundant. From a broader perspective, the core tenets of connected health indicate that its use may be assistive to those who are poorly adherent, and that innovation and the development of new technologies in this area may empower patients and healthcare teams to potentiate outcomes. There are many means by which enhancement of patient adherence can be brought about which fall under the umbrella of connected health, and which also vary in complexity and success. In many cases, the active role of the patient is necessitated, as activities involve the engagement and/or taking of some action or another by the individual, which can be grouped into the categories of "to read", "to do" or "to connect" (Agher et al. 2020). With reference to these domains, more widely known approaches involve the provision of condition and/or medicine-specific information in the form of training materials which can be accessed by the patient at home via their computer, etc., or indeed, the provision of access to coaching and support groups, which in turn may help them to improve knowledge of their own condition and its management, ultimately understanding the importance of adherence to management strategies (Kvedar, Coye, and Everett 2014). Despite the popularity of these approaches, and the measurable benefits which they appear to provide, there is a need for further optimisation work to be undertaken in order to allow these techniques to be regularly and reliably recommended (Conn et al. 2009).

The use of electronic reminders (which can be grouped within the co-domain of "to read" and "to do") also form another popular approach in leveraging connected health for the purposes of enhancing adherence. In brief, this technique involves the sending of electronic messages to a patient for the purposes of reminding them to take medications, or otherwise engage in activities related to the management of their condition, which given repeated use, are anticipated to bring about the reduction of non-adherent behaviours. These messages can take the form of short message service (SMS) reminders, those which are sent via an application installed on a mobile device, or indeed, which make use of a dedicated device (Tao et al. 2015). Whilst these approaches do appear to be elegant in their simplicity, and indeed, their use may understandably be automatically assumed to enhance outcomes, the reality appears to be more complex, with variability in outcome of such interventions being brought about by key aspects such as disease type, and the duration over which the intervention is used (Tao et al. 2015). Other approaches which leverage digital communication channels to contact the patient for the purposes of improving adherence, and which appear to offer similar success rates include the use of video-calling technology which can be automated or involve authentic human interaction with a professional – whilst the latter of these two approaches would appear to not introduce as much efficiency as automated approaches, it may offer alternative benefits, in that what is communicated to the patient can be reactively augmented, ultimately allowing the practitioner to focus more on the offering of support, rather than acting in a controlling manner (West et al. 2012).

So far, each of the interventions discussed in relation to the enhancement of patient adherence has required active involvement of the patient with respect to the actioning of the treatment regimen. Further, approaches necessitate manual outcome monitoring to ascertain impact, for example, via the monitoring dose count changes on inhaler devices, or by patient self-reporting (Strandbygaard, Thomsen, and Backer 2010; Cocosila et al. 2009). The latter approach here is of particularly dubious value given the nature of the problem which is being addressed, and as such, the value or accuracy of the data provided by the patient may lead to issues in evaluating the value of the intervention. Thus, opportunities have been identified to harness innovative digital/electronic solutions which may allow for more "true" indications of patient medication adherence to be observed. This may ultimately facilitate the creation of interconnected systems which can actively respond in instances where adherence is poor and, likewise, not engage with patients who are successfully managing their conditions.

Developments in microprocessor technology have undoubtedly played a role in the emergence of the Medication Event Monitoring System (MEMS), constituted of standard medicine containers that contain a sensor and processor within the cap, and which in turn are triggered by, and log, container openings, facilitating patient adherence to be ascertained from both the frequency and timing of package opening (Olivieri et al. 1991). The fundamental purposes of such a technology are to remove the need for patients to record information in relation to their own adherence and prevent the provision of inaccurate data resulting from poor record keeping. These attributes have rendered the use of MEMS particularly attractive in areas of research which focus on levels of patient medication adherence and indeed appear to have proven useful. However, perhaps unexpectedly, more extensive studies into the validity of this technology for adherence monitoring have indicated that the benefits may not be as large as expected, with MEMS-based records correlating moderately well with patients' own manual record keeping (Shi et al. 2010). The rationale for this is multivariate; however, key factors appear to be the lack of consensus or standardisation of the term adherence, and further, the acknowledgement that medication non-adherence comes about as a result of more complex aspects of behaviour (Hartman, Lems, and Boers 2019; Easthall 2019), in turn making the wider comparison and identification of trends across related studies exceptionally complex.

This situation described above provides a well-formed example of how certain healthcare practice areas may not have developed at a comparable rate to the technologies which are emerging within them, which at present may well be preventing the full realisation of the potential which these systems can offer. Thus, to leverage this potential, it is critical that relevant healthcare practices are considered in conjunction with the technologies that are available which could assist with their implementation, and with reference to the direction of travel in those areas, such that those practices can be appropriately future-proofed. This conclusion would indicate that healthcare development in the modern age is perhaps more multidisciplinary than ever and now includes professions such as electrical engineering, software engineering and human factors (Pan and Xu 2014).

13.2.4 Leveraging Mobile Technology for Next-Generation Telemedicine and Health

The introductory section of this chapter discusses the omnipresence of mobile device technology as an allegory for the wider spanning integration of digital technologies throughout healthcare. When considered as its own domain more specifically, it is clear that the ongoing emergence of mobile device technology and capability is set to revolutionise how healthcare professionals deliver their impact. Moreover, as discussed earlier, it will further facilitate individuals' ownership of their health, in the context of condition management. The accessibility of mobile technology, and perhaps more importantly, the rate at which powerful software and applications are being developed for those platforms, have unleashed an entirely new area of potential within healthcare; however, it is crucial that such approaches are designed thoughtfully and in a manner which brings about a real and positive impact on population health.

When mobile devices are considered as a discrete tool for use within healthcare, their potential to deliver significant impact often relates not to the devices themselves, but rather the applications which are developed for use via those devices, and which can effectively harness the inherent usability and connectedness of these devices to facilitate a range of functionality for the purposes of bringing about health-related enhancements. The potential for this

resource provision is reflected in the sheer range of health-focused applications available for the two largest mobile device platforms, Alphabet's Android and Apple's iOS, which run into the many tens of thousands, and further to this, the fact that these operating systems themselves are being evolved to contain "baked in" health-related functionality, which in turn are used as key marketing tools to enhance sales of devices to target audiences.

As alluded to previously, technology and mobile apps are also used to support education and professional practice, including enabling the provision of up-to-date, evidence-based information at the point of care and enhancing decision-making in patient management (Hanna and Hall 2018; Keyworth et al. 2018). However, it should be noted that while technology-based interventions may have theoretical benefits for healthcare professionals, this does not mean they will be readily implemented or successful in practice (Keyworth et al. 2018). Barriers preventing the adoption of technology-based interventions by healthcare professionals and patients are discussed later in this chapter.

The burgeoning popularity of mobile devices, coupled with the ongoing development of applications within the healthcare space, has led to the emergence of *mHealth*, which the National Institutes of Health define as "use of mobile and wireless devices … to improve health outcomes, health care services, and health research" (National Institutes of Health 2018). Importantly, and unlike other technological/digital approaches which have been discussed here and elsewhere, mHealth, in many instances, offers the end user a service which is intended to provide enhanced health outcomes in the absence of a healthcare professional. This presents a uniquely dichotomous situation whereby individuals can be effectively empowered to take ownership of their own health, or to be put at risk as a result of the use of resources which are not supported by clinical evidence (Bates, Landman, and Levine 2018). This risk is elevated due to security, safety and privacy issues. Indeed, many applications have been shown to provide inaccurate and potentially life-threatening information (Bates, Landman, and Levine 2018).

This landscape has logically led to a requirement for assurance of standards and an ongoing debate around the certification of mHealth approaches, and for standalone health-related applications more specifically. Despite the surface-level sensibility of this approach, difficulties arise due to the fast pace of technological developments in this space, juxtaposed with comparatively sluggish timelines involved in certification processes (Chan and Misra 2014). As such, whilst mHealth has the potential to be exceptionally lucrative from a population health perspective, there is a need for related bureaucratic processes to adapt to match the rate of progress seen within the area which requires this additional regulation. This is reinforced by the time, resource and evidence base required for healthcare academics to assess the value of these software packages via traditional means such as randomised controlled trial. Given the significant challenges in evaluating this area of digital health in terms of impact, here is a need for a new area of research to come to the fore which can track, evaluate and recommend resources robustly and at high speed (Nilsen et al. 2012).

Despite the challenges in mHealth, the area clearly offers multiple strands of opportunity for integration into existing healthcare infrastructures, including via the offering of condition-related data collection, support provision and the optimisation of adherence and management strategies (Becker et al. 2014). Particular value can be extracted in these areas when these opportunities are realised by teams led by those with expertise in the delivery and evaluation of healthcare activities, as previous learnings from related evidence bases can be integrated proactively, enhancing the likelihood of success.

Patient adherence is again an area of interest within the context of mHealth and has been researched extensively, allowing its use as an exemplar of the emergence of these technologies within healthcare. Mobile devices offer significant advantages in comparison to other approaches for the enhancement of adherence, including those which make use of other digital systems. Perhaps the most important of these benefits is the incorporation of multiple hardware components in a form factor, which is easily transportable and which can accompany a patient at all times (Pérez-Jover et al. 2019). For example, the provision of a camera, microphone and multiple connectivity technologies allows for the streamlining of multiple communication types, fully supporting comparatively simple data transfer tasks through to enabling real-time personal interaction at any time. Additionally, the ability for these communication strands to be packaged together via thoughtful application design can enable the provision of intuitive, easy-to-use resources for patients who may have ordinarily struggled with the use of less user-friendly alternatives.

A further potential exploitable benefit of mobile technologies in the domain of patient adherence is the heightened ability to incorporate aspects of social connection between peers, allowing for those within similar situations to support one another, thus removing the need for engagement with healthcare professionals to provide this support and encouragement. This has produced, for example, applications which facilitate the interaction between patients who meet with one another for the purposes of discussing their conditions, but importantly to verify each other's taking of medications via active engagement with the application interface. These verifications create a record which can be transmitted to healthcare professionals at specific points in time, providing them with reliable, yet entirely user-generated data. Such approaches have been shown to have some success in a range of areas, including patients' motivation to take medicines, and improved outlook on their condition due to the ability to discuss it with others, coupled with clinicians indicating that non-adherence issues could be quickly and confidently detected (Fujita et al. 2018). The use of social interaction, and indeed wider social media-type approaches, may well provide a new tool for the enhancement of patient outcomes. While the investigation of this area does appear to be accelerating, it is also currently prototypical, meaning that firm conclusions about its use in more general terms cannot be drawn. Initial reports do appear to be promising, but should be viewed with caution, given concerns over aspects such as data privacy and the quality and accuracy of information which can be provided both actively and passively via social interactions with non-experts. These challenges must be addressed to engender user confidence and widespread implementation of these approaches (Elnaggar et al. 2020).

Whilst the discussion here around the use of mobile technologies is far from exhaustive, the selected examples do clearly communicate that the wider area of mHealth offers enormous potential in bringing about improvement of individual and population health outcomes. Challenges exist with respect to identifying the most effective and clinically sound approaches within the area. This is largely due to the extent of baseline noise, which is increasing at a remarkably fast pace as a result of accelerated technology development, and which currently prevents such technologies from being confidently recommended, due to the availability of full understanding of related benefits and risks (Backes et al. 2021). This does, however, support the need for healthcare professionals to ensure that they have their place in the development and evaluation of new and reliable resources in healthcare technology within those cross-cutting and widening multidisciplinary teams discussed previously.

13.2.5 Combinatorial Approaches: The Emergence of Connected Health Devices

The ability for mobile devices, and more specifically, smartphones, to connect to the internet wirelessly via WiFi and/or cellular connection is a key advantage in their use within healthcare. Data, of all forms, can be efficiently transferred to patients or between patient and practitioner, with barriers relating to distance and the need for physical transfer almost being removed. This connectivity is one of the multiple approaches for wireless networking of which mobile devices are capable, providing enormous opportunities to bring about the connection of multiple peripheral devices to smart devices wirelessly, and by extension, the collection of health-related data, in ways which would have been unimaginable just a few decades ago. While the development and access to these technologies may have been initially slowed by the emergence of competing standards (Bing 2008), in more recent times, there has been the coalescence around several wireless local area network standards which have become a common parlance when mobile devices are discussed. In addition to WiFi standards, Bluetooth and near-field communication (NFC), which make use of their own particular frequency ranges (Cisco Systems 2015), are commonly referred to when the utilisation of wireless communication is discussed, and indeed, this has transferred through to the area of connected health, with these commonplace technological standards being harnessed for the transfer of a range of types of health-related data between devices.

The ongoing emergence of these technologies, particularly when considered together with the exponential increase in sophistication offered by software applications, has enabled the development of devices which can not only capitalise on the ability to connect to mobile devices wirelessly without the need for their own internet connection but can make synergistic use of state-of-the-art sensor technologies, etc. for the collection of data which are to be transferred to the device for a range of purposes. As would be expected, the health sector is catered to in a significant way and has subsequently seen the wider area of connected health being furnished with a range of "connected health devices" (El Amrani et al. 2017). In more general terms, the emergence of these devices often removes the need for manual steps in the reporting of health-related data points, ultimately removing the opportunities for these flows of data to fail or the provision of inaccurate records. Therefore, the necessitation for a patient to measure their blood pressure using a less "intelligent" sphygmomanometer, followed by the input of the produced reading into a software interface of some kind, for example, can be entirely removed, allowing for data collection to be considerably less onerous (Omboni 2019). The value of this is perhaps more clearly elucidated when particular population sets are considered along with the roles which a patient or user must play in order to bring about the success of the process. For example, the seemingly straightforward process of logging a blood glucose measurement requires abilities in psychomotility, device interaction, perception, cognitive interpretation and decision-making. Ironically, this may be a challenging task for those who require the application most, such as older adults (Harte et al. 2014). The enhancement of these approaches via the development of "smarter" systems may effectively remove proficiency requirements in most of these domains, providing a considerable amount of added benefit to the end user.

Given the role that such devices may be able to play in the monitoring of conditions and the exertion of positive effects on patients' management thereof, there appears to be a surprising paucity of available research which focuses on the use of such devices and the measurement of improvement in patient outcome. It also seems that practitioners have not yet been able, due to resource or choice, to avail of these technologies as part of their patient management plans (El Amrani et al. 2017). Reasons for lack of implementation in practice, and hence a lack of data being generated to ascertain value and impact, include concerns around patients' use of, and motivation to use, the devices themselves, the potential for technical problems to arise, the nature in which the data are presented and related difficulties in data processing (El Amrani et al. 2017; Haluza et al. 2016). Perhaps most importantly, practitioners have expressed concerns around the use of these technologies being to the detriment of communication and professional relationships with their patients (Nambiar, Reddy, and Dutta 2017).

Significant advances in the area of connected health devices continue to be made. Despite their use not being commonplace (for reasons outlined in the previous few paragraphs), and the likelihood of clinical underuse in the short-to-medium term, connected health devices may eventually form key elements of how next-generation modern healthcare is provided. Of those ongoing advances, the genesis of the Internet of Things (IoT) (Holler et al. 2014) and its substrata of wearable devices are perhaps two of the most notable. They have the potential to make monitoring of healthcare-related data optimally streamlined, as well as bring about full integration of this data collection into an end-user's day-to-day life, thus normalising a task which currently requires some level of active actioning.

The IoT can be defined as a "set of technologies, systems and design principles, associated with the emerging wave of internet-connected things within the physical environment" (Holler et al. 2014), and as a grouping, provides a range of means by which data can be collected from single or multiple inputs, and subsequently for that information to be shared out to the devices forming part of that interconnected system. The vast range of "things" which can constitute IoT systems further illustrates the potential power of the phenomenon. For example, the breadth of sensor systems and their associated flexibility in application are able to collect massive amounts of data points and types which can then feed processing and recording systems. This can ultimately result in positive outcomes at organisation level through to individual end-user levels. The wide-ranging benefits include enhanced efficiencies within healthcare organisation, improved patient medication adherence, reduction in diagnosis or treatment error and, importantly, the facilitation of activities which relate to preventative medicine (Thangam et al. 2022). Unsurprisingly, the growing usefulness of these technologies within health more widely is reflected in the growth of market size, which exceeded US$100 billion in 2019, and is growing at a rate of 18% (Businesswire 2020). Indeed, the healthcare industry is predicted to solidify its stake as one of the largest users of IoT technologies within the next decade, with over 40% of all IoT devices being used within this area (Moko Smart 2022). Moreover, these developments are not restricted to more basic aspects of health, with the creation of reliable, palm-size IoT systems for clinical tests such as electrocardiogram (ECG) and electroencephalography (EEG) being developed, tested and used currently (Bhagyashri and Hirekodi 2022; Laport et al. 2020). Thus, the IoT is unquestionably on track to become a pillar of modern healthcare.

A major area of value and usefulness of IoT is the home setting. Indeed, the ongoing emergence of truly interconnected connected health devices within the home will facilitate better integration of health monitoring with the user's day-to-day activities, although there is no direct communication with their healthcare team. However, despite the removal of a great deal of user responsibility, the need for active engagement persists, albeit at a low level. As a lack of such engagement may introduce failure at the point of data collection, it should be a key consideration within the design of any such solution.

The developments in IoT and interconnected health devices, in parallel with accomplishments in sensor electronics, etc., have led to the next evolutionary stage within health technology, namely wearable devices. As suggested by the nomenclature, wearable devices are worn on the user's body, allowing for the continual recording, logging and transmission of health-related data via wireless connections, rendering the user's role almost entirely passive, save for the initial wearing and minor maintenance of the device itself (for example, battery charging). The range of data which wearables are currently able to detect and process is impressive, spanning across the physiological, biochemical and spatial (Dunn, Runge, and Snyder 2018). Additionally, devices are available which can provide internal anatomical data via processes such as endoscopy (Alam et al. 2019), although strictly speaking, these probably fall outside of the definition of a wearable device.

In part, the continual emergence of wearable connected health devices is being driven by the consumer health market, consisting of users who, in the main, are in good health, but who wish to be empowered to monitor their own physiological state for a range of reasons, including the enhancement of their well-being (Yetisen et al. 2018). When the consumer wearables market is considered, it is typically in the form of wristwatch-type products. These products typically include Apple Watch, Samsung Gear and Garmin and Fitbit devices. Whilst these brands possess an enormity of the market, their devices also are noteworthy due to the wealth and sophistication of data which they can produce (Mück et al. 2019), the useability of the device themselves and their companion mobile applications and, thus, the ease of use of the systems overall for the end user. The projected ability for these devices to shape the future landscape of IoT-supported connected health is underpinned by ongoing research into the validity of the data that they collect and process, which may support their use as off-the-shelf solutions to enhance access to, and the successful collection of important patient diagnostic data.

Interest in the use of consumer-level wearable health devices has translated to greater levels of explorative investigation by healthcare researchers and shown promising results in many cases. For example, smartwatches offer potential in helping with the detection of conditions such as atrial fibrillation (AF) and could well offer significant value in patient management post-diagnosis and after treatment initiation (Inui et al. 2020; Pantelopoulos et al. 2017). Indeed, the use of these devices in the management of AF is currently under investigation through a large-scale clinical trial involving over 450,000 participants (Lubitz et al. 2021). Smartwatches have also shown promise in the monitoring of other markers of cardiovascular function; devices equipped with pulse oximeters show close parity with medical-grade equipment when used in non-extreme conditions (Lauterbach et al. 2021; Hermand et al. 2021; Schiefer et al. 2021). These examples are particularly encouraging

given the extent of the global population affected by cardiovascular conditions and illuminate the role that these devices can play more generally within the management of complex conditions.

In line with what has been discussed elsewhere in relation to the use of health technologies, wearables are not excluded from regulatory requirements when there is an intention to use them medically. The need for this is clear, both from more straightforward aspects of medical relevance and acceptable levels of functional accuracy, and the bringing about of patient benefit in comparison to control, among other reasons. As such, it is important that regulatory processes are introduced which can be afforded to these devices. This necessitates regulators ensuring there is ongoing modernisation and adaptability of their processes to permit beneficial developments at a rate which mirrors that of the base technologies themselves. Regulators do appear to be taking appropriate action, having issued guidance and implemented approval workflows for these devices, which draw many parallels to processes undertaken for the approval of more rudimentary devices and medicines. These will continue to be of importance to device manufacturers, particularly in cases where diagnostic data are collected which offer particular risk (Jiang, Mück, and Yetisen 2020). The implementation of these processes has already seen many wearable devices receiving approval in territories including the United States for the detection of clinical data related to conditions such as diabetes (Dunn, Runge, and Snyder 2018); however, many of these appear to relate to devices which carry out that monitoring as their exclusive task, rather than being integrated into a consumer appliance which offers a plethora of health and non-health functionality. This suggests it will be some time before these devices become reliable resources for the clinically relevant management of patients' health.

The future of wearables, and by extension, connected health devices, is a promising one for healthcare, with next-generation devices already emerging which may offer value in the areas of neurology, respiratory health, ophthalmology and even mental health (Avenga 2020). However, the pathway to their success is rugose and reliant on the progression of other areas in addition to those which relate exclusively to technological development. There is no doubt that a greater assimilation of expertise from the key loci in the overall area of connected health is fundamental to our ability to move forward and, ultimately, bring about marked enhancements in healthcare.

13.3 BARRIERS TO THE EMERGENCE OF HEALTHCARE TECHNOLOGIES

For many reasons, the rate and quantity of developments in the arena of healthcare technologies, and more specifically, in relation to connected health, would invariably bring about a renaissance in healthcare, were the requirements for such reliant only on technological advancement. However, the reality of bringing about step-change within the area of health is multifaceted and reliant on the satisfaction of many requirements, including those which are intrinsically related to regulation, as discussed elsewhere in this chapter.

As is the case with any element of healthcare development, economic considerations are crucial to the success of connected health. At a base level, technology, regardless of how unsophisticated, costs money, including that which is needed to pay for the assets which form the physical part of the system(s), and the associated requirements for implementation, including the design of new data handling architectures and the training of users at all points of data flow. These fiscal considerations are especially acute when a key motivating factor for the enhanced implementation of digital health is its ability to bring about bettered access to healthcare for those within LMICs and become a priority when testing feasibility (Chen and Liu 2020). The absence of financial support for the hardware and supporting infrastructure represents a significant barrier for technological advancements to bring about their intended effect. Thus, further to regulatory and certification requirements, governments must consider how a greater use of such technologies may bring about more affordable and accessible services and make necessary budgetary changes which will enable these advances to produce more resilient healthcare systems (O'Leary et al. 2015).

Regardless of the ability of technology to deliver desired requirements and the availability of funding to enable its use, the success of healthcare technologies is dependent on their acceptance by those individuals directly implicated in their use, namely patients and healthcare practitioners. The non-use of healthcare technologies is an area which is keenly investigated (Keyworth et al. 2018), highlighting key and perhaps oft-ignored barriers which commonly affect the very individuals targeted to benefit from these technologies. A considerable number of barriers to the use of healthcare technologies appear to be linked to a lack of education or digital skills, particularly within older adults. Such barriers can include a lack of awareness of the benefits that technology can bring, insufficient education in the area or proficiency in the use of the resources which are provided, and anxieties regarding the correct use of those systems (Rogers and Mead 2004; Delemere and Maguire 2021; Chen and Chan 2013). In addition, a literature review (2018) of 69 studies investigating what maximises the effectiveness and implementation of technology-based interventions to support healthcare professional practice revealed various facilitators and barriers. Facilitators included ensuring senior peer endorsement and integration into clinical workload. Barriers to implementation included organisational challenges and also the design, content and technical issues with the technology-based intervention (Keyworth et al. 2018). Further information in relation to the behavioural aspects of non-use may be gleaned from reference to work in salient areas which overlap with connected health technologies. These include the use of assistive technologies, further illustrating the diversity in the rationale for non-use, including those which are more obvious, such as the incompatibility of the technology with users' disabilities. Non-use of these technologies can also be related to their removal of concealment of an individual's health status, which is perceived to draw unwanted attention towards their health-related issues from family members and others (Söderström and Ytterhus 2010). Indeed, non-use of certain technologies has been linked to the improvement in patients' health which have been brought about in part *by that* technology, with the user erroneously believing that they no longer need to engage with that device/technology after a certain point in order to maintain their improved health state (Dijcks et al. 2006). This fallacy appears to be commonplace wherever long-term management strategies for the maintenance of positive health states are employed (Ulrik et al. 2006; Ashoorkhani et al. 2018).

The barriers outlined here are important but far from exhaustive; they, and others, must be key considerations of successfully designed healthcare technologies at large and in the field of connected health technologies of all kinds. Full credence of these barriers is necessary to produce successful interventions which result in tangible patient benefit, and whilst addressing these may lead to the reduction of what may idealistically be provided by related technologies, this is a compromise which will inarguably allow for maximum benefit to be extracted.

To end this section on a positive note, a brief overview of some potential solutions to the aforementioned key barriers has been included. As previously mentioned, a large part of the success of connected health technologies hinges on the perceptions of the end user, and their acceptance and desire to make use of those technologies. Whilst it is unlikely that any device or system could be designed which would satisfy all users, the consideration of certain characteristics may significantly enhance uptake. One such approach which has been widely employed in the consideration of how technologies can be more widely adopted is the technology acceptance model (TAM), introduced more than 30 years ago. This addresses the key components of ease of use and perceived usefulness, and their influence in users' technology-centric behaviours (Davis 1989; Marangunić and Granić 2015). The TAM has seen both adoption and modification for use in the design of various technologies, allowing it to address environments and user groups effectively, including professionals and patients. In doing so, it allows for both the evaluation and more considered design of systems which are thus more likely to bring about positive outcome on implementation (Attié and Meyer-Waarden 2022). Unfortunately, the divergence of more recently emerging acceptance models from the "original" TAM prevents the provision of a single model which is transferable in its application, and as such, the criteria used when designing and evaluating such technologies for acceptance require careful consideration by system designers (Rahimi et al. 2018; Orruño et al. 2011).

13.4 FUTURE DIRECTIONS

The domain of healthcare technology benefits from its intrinsic link to the rich vein of technological developments at large, given that the healthcare industry, and the global community of healthcare professionals, continually strive to work at the cutting edge and to capitalise wherever possible on developments which may translate effectively to patient care. The emergence of technologies in other areas and disciplines provide considerable levels of insight into the shape of healthcare delivery in years to come, and in some instances, breakthroughs are already being made which give a glimpse into the future.

One such technology, virtual reality (VR), currently most linked with applications in entertainment, is now increasingly of interest within healthcare, given its ability to immerse the user in realistic environments for a range of purposes. In short, approaches in VR involve the user entering "into" an environment simply by putting on a headset and ancillary devices which can provide stereoscopic video and 360° audio, and in many cases, also involves the provision of controllers or other functionality which allows the user to interact with and manipulate the environment around them, permitting them to undertake various tasks. The emergence of these technologies was quickly found to be of interest in the area of training for healthcare professionals, allowing practitioners to be brought into challenging, but "safe" environments where they could acquire and practice their own skills, and receive training and guidance, without the need for an equivalent physical space to exist (Mantovani et al. 2003). Additional to this, the use of VR technology enhances training efficiency, by negating resource availability, and related restrictions on the number of individuals who can be trained simultaneously. The ability for VR to offer benefit in patient-facing applications is another defined area of interest, particularly where the placing of a patient into an environment in a safe way can offer therapeutic benefit. This translates perhaps most clearly to the treatment of mental health conditions, where treatment strategies may already involve immersive approaches. To date, preliminary investigations into the use of VR in this way have included the development of treatment strategies for disorders including phobias, post-traumatic stress, addiction and psychosis – however, levels of data and success have varied widely (Gregg and Tarrier 2007; Emmelkamp and Meyerbröker 2021). Thus, further work is required to fully elucidate the role of this emerging technology within these areas of health and others. Ongoing developments within the area of VR include the inclusion of digital avatars driven by artificial intelligence. Digital avatars may allow for more authentic and responsive interactions between the user and the virtual environment in the absence of another "live" participant (O'Connor 2019). Furthermore, approaches which blur the lines between the VR and reality itself via the creation of "digital twins" may facilitate remote management and even surgical procedures (Laaki, Miche, and Tammi 2019).

Perhaps the most exciting avenue to be explored within healthcare in the coming years is that of artificial intelligence (AI), which offers opportunities to generate novel strategies for the management of health and augment those which currently exist, making them more powerful, accurate and significantly more valuable. AI, which incorporates elements such as machine learning and language processing, is a field which has progressed rapidly, and which already constitutes its own industry. AI applications are being created for many aspects of everyday life, with the intention of automating processes and enhancing efficiencies.

Within health, AI approaches can potentiate the use of many of the methods and technologies which have been discussed within this chapter, along with many others, from the diagnostic analysis of data provided by connected health devices, to the ability for advice to be provided to patients in response to the same. The ability for AI to benefit from "learning" from massive data sets, in manner not possible by humans due to the volume of data alone, may lead to these technologies making more accurate clinical decisions than healthcare professionals. Whilst this may seem far-fetched, IBM's "Watson for Oncology" is being used at the present time to make decisions around cancer treatments and has made clinical decisions of striking similarity to those suggested by trained and experienced multidisciplinary oncological treatment teams (Jie, Zhiying, and Li 2021). It is also worth noting that machine learning is increasingly being utilised in healthcare research, given the manual analysis of text-rich datasets can be error-prone and unscalable. For example, one of the authors of this chapter has demonstrated how machine learning facilitated the examination of over 3000 Fitness to Practise cases involving UK healthcare professionals. These cases were initially converted to text files, with other pre-processing steps implemented, before a topic analysis method (non-negative matrix factorisation; machine learning) was employed for data analysis (Hanna and Hanna 2019).

The potential impact which AI can exert on healthcare is clearly massive and may bring about wide-ranging improvements in outcomes more generally, as well as more granular enhancements in factors such as patients' involvement in their own health, care quality, productivity and medical discovery, etc., and all with greater efficiency and reduced operating costs (Lee and Yoon 2021). However, AI, regardless of its ability to learn, communicate and make decisions, is still a technological system, which will meet the same challenges and barriers that all other technologies face in healthcare, some of which have been outlined in this work, and which may limit adoption. Additionally, whilst outcomes from use cases are encouraging, these are perhaps more blinkered than reports suggest. For example, the ability to predict outcomes, treatment success, etc., in patients is dependent on data sets which have been used to "train" the system, essentially rendering the process useless in circumstances which do not match a well-defined, highly restricted, and even discriminatory set of criteria (Panch, Mattie, and Celi 2019). Conversations around AI are incomplete without some allusion to the ethical implications involved in the use of these systems, given factors such as the amount of data which are shared and the way that data are used (i.e. to potentially allow private enterprises to make use of patient data for the purposes of developing products for profit). There are many other elements which may require sociological and even philosophical consideration to be undertaken before solutions to these ethical concerns can be rolled out at a population level (Smallman 2022). Despite these challenges, and within the scope of the emerging nature of the technology, it would be surprising if AI fails to become a pillar of healthcare systems in years to come.

Notwithstanding the enormity of their projected impact, VR and AI are just two avenues on which we can expect healthcare to travel in years to come and, importantly, in the not-distant future. It is useful to note that many of the technologies which we can expect to see serving diverse roles within health already exist and are being used for other purposes, in various states of maturity. That has also been the case for each of the emerging technologies which have found their way into assisting patients so far. There is appetite to translate useful technologies into the healthcare sphere, with encouragement coming from both those professionals within health who have an awareness of what is being done in other areas and those who are involved in the development of such technologies, given the lucrative and persistent nature of this market.

An educated prediction may include the hypothesis that future directions being added to the healthcare map will be exceptionally combinatorial and may well tie in with technologies which have been discussed elsewhere within this chapter, in addition to others. Drastically enhanced computing power and the ability for systems to handle, process and draw conclusions from extraordinarily large data sets have led to the emergence of "big data", which in turn can centralise hugely useful information such as patient data sets, treatment outcomes and a host of other useful clinical, management-centric and professional performance data (Dash et al. 2019; Hanna and Hanna 2019) This, combined with technologies such as AI, labs-on-chip-based testing and enhancements in rapid drug and device manufacturing will accelerate the area of personalised medicine (Wu et al. 2018; Jhunjhunwala and Kapil 2022; Katakam, Adiki, and Satapathy 2022), potentially allowing us to move away from population-based empirical models for the treatment of disease and, instead, toward something which is maximally targeted. It is not long ago that the suggestion that someone who had fallen ill could be tested at their bedside, and their diagnostic test results instantly generated and fed into a data set which would allow both the selection of a medication that would cater specifically to that patient's needs, and its manufacture within hours or minutes within the care setting, would be touted as something straight from a science-fiction novel. However, the technologies which have been discussed here, as well as others within this publication, indicate how the building blocks of such a process are presently being refined. It would be wise to assume that these will soon start to be built together to allow maximal outcomes for those being cared for by the wider healthcare community.

Even with these developments on the horizon, it is already possible to take these suggestions further, which may allow us to see further into the future. 3D printing, considered at length by colleagues elsewhere within this book, is being explored for applications above and beyond the manufacture of medicines. Sophisticated versions of the technique are being used to create tissue scaffolds for the regeneration of cardiovascular tissues and even bone (Lan et al. 2022). Indeed, materials which have been incorporated with mammalian cells have also been produced which may ultimately allow for the regeneration of human tissues and the treatment and cure of conditions which may otherwise have required life-long treatment or permanent synthetic mechanical intervention (Chang et al. 2022). The explosion in the recent popularity of genomics provides another example of the notable impact of technology and utilisation of big data , whereby the ultrafast analysis (again benefitted by the integration of AI) of an individual's genetic material can reveal more about their health in addition to what has been mentioned above, which again benefits from quick processing, comparison against huge data sets, and which will provide practitioners with greater information about optimal disease management (Gulfidan, Beklen, and Arga 2021).

Perhaps ironically, one way in which we might predict health to change in the future does not involve the treatment of patients at all, or at least, may involve the marked increase in the role of preventative medicine within healthcare, and in turn, the treatment of patients for conditions they are yet to experience. The drastic improvement in healthcare over the last decades has seen global increases in life expectancy, and related to this, estimations that levels of disease globally could decrease by 40% by 2040, with over half of these gains in global health coming about as a result of preventative medicine and other contributing factors (McKinsey Global Institute 2020). Of course, many of these preventative measures relate to ones which are currently commonplace, including vaccination. However, the ability to again make use of technologies such as genomics, big data, AI and advanced drug discovery and manufacturing techniques may allow us to routinely identify patients' conditions and undertake highly effective prophylactic action, expanding both life span and quality.

13.5 CONCLUSIONS

The delivery of healthcare does not simply coexist with technology, deriving benefit from it where possible. Rather, these two entities are intertwined, serving each other in a range of ways, all of which benefit healthcare practitioners, who are enabled to do their jobs more efficiently and successfully, catering to an increasing and ageing population, and to patients and the population at large, who are experiencing life-enhancing benefits to their care, and to overall health outcomes. A single word which has been mentioned earlier in this chapter efficiently summarises the keystone role which technology plays within healthcare –

empowerment – for the reasons noted in the previous sentences, but also for its ability to permit individuals to take greater responsibility for their own health, as it now allows patients to learn more about the workings of their own body and its failings. This empowerment ultimately feeds through to individuals' and populations' education and knowledge about their health and ability to derive synergistic benefit where exogenous action may need to be taken to maintain the status of well-being.

This chapter, whilst a far from the exhaustive account of healthcare technology, is intended to indicate how technology fuels the survival and development of our healthcare systems, and healthcare practices more generally. There is not an action taken by a practitioner or patient which has, or will soon have some input from a technological system, with the hope being that in most cases, significant benefit will result.

This chapter has also reinforced that in many cases, we are not where we need to be yet, and whilst there is a rationale for the use of technologies within health, there is a concurrent need to apply scientific rigour to their use, such that they can be used in a manner which *actually* enacts change, rather than as a result of outcomes which are hypothesised, but never stringently tested. In many cases, this will require not the enhancement of modification of the technology itself, but rather the modernisation of the infrastructure or mode of practice within which it sits – the irony being that the pace of development of technology in healthcare may be slowed by the pace of development of the technology in question.

Few would disagree with the statement that the future role of technology within healthcare is bright, given the ongoing emergence of technologies which allow us to undertake activities at breath-taking levels of speed, accuracy and value. Again, the combination of processes which currently exist in novel ways has significant potential to be one of the most important key developments which we can expect to see within healthcare in the future.

REFERENCES

Agher, Dahbia, Karima Sedki, Rosy Tsopra, Sylvie Despres, and Marie-Christine Jaulent. 2020. "Influence of connected health interventions for adherence to cardiovascular disease prevention: A scoping review." *Applied Clinical Informatics* 11 (04): 544–555. https://doi.org/10.1055/s-0040-1715649.

Alam, Mohammad Wajih, Md Hanif Ali Sohag, Alimul H. Khan, Tanin Sultana, and Khan A. Wahid. 2019. "IoT-Based intelligent capsule endoscopy system: A technical review." In *Intelligent Data Analysis for Biomedical Applications*, pp. 1–20. Academic Press: UK.

Ashoorkhani, Mahnaz, Reza Majdzadeh, Jaleh Gholami, Hassan Eftekhar, and Ali Bozorgi. 2018. "Understanding non-adherence to treatment in hypertension: A qualitative study." *International Journal of Community Based Nursing and Midwifery* 6 (4): 314.

Attié, Elodie, and Lars Meyer-Waarden. 2022. "The acceptance and usage of smart connected objects according to adoption stages: An enhanced technology acceptance model integrating the diffusion of innovation, uses and gratification and privacy calculus theories." *Technological Forecasting and Social Change* 176: 121485.

Avenga. 2020. "20 Examples of Wearables and IoT Disrupting Healthcare." Accessed 07/08/2022. https://www.avenga.com/magazine/wearables-iot-healthcare/.

B. Nachega, Jean, Vincent C. Marconi, Gert U. Van Zyl, Edward M. Gardner, Wolfgang Preiser, Steven Y. Hong, Edward J. Mills, and Robert Gross. 2011. "HIV treatment adherence, drug resistance, virologic failure: Evolving concepts." *Infectious Disorders - Drug Targets* 11 (2): 167–174. https://doi.org/10.2174/187152611795589663.

Backes, Claudine, Carla Moyano, Camille Rimaud, Christine Bienvenu, and Marie P. Schneider. 2021. "Digital medication adherence support: Could healthcare providers recommend mobile health apps?" *Frontiers in medical technology* 2: 616242.

Bangsberg, David R., Deanna L. Kroetz, and Steven G. Deeks. 2007. "Adherence-resistance relationships to combination HIV antiretroviral therapy." *Current HIV/AIDS Reports* 4 (2): 65–72. https://doi.org/10.1007/s11904-007-0010-0.

Barr, Paul J., Shauna C. Brady, Carmel M. Hughes, and James C. McElnay. 2014. "Public knowledge and perceptions of connected health." *Journal of Evaluation in Clinical Practice* 20(3): 246–254.

Barr, Paul J., James C. McElnay, and Carmel M. Hughes. 2012. "Connected health care: The future of health care and the role of the pharmacist." *Journal of Evaluation in Clinical Practice* 18(1): 56–62.

Bates, David W., Adam Landman, and David M. Levine. 2018. "Health apps and health policy: What is needed?" *JAMA* 320 (19): 1975–1976. https://doi.org/10.1001/jama.2018.14378.

Becker, Stefan, Talya Miron-Shatz, Nikolaus Schumacher, Johann Krocza, Clarissa Diamantidis, and Urs-Vito Albrecht. 2014. "mHealth 2.0: Experiences, Possibilities, and Perspectives." *JMIR mHealth and uHealth* 2 (2): e24. https://doi.org/10.2196/mhealth.3328.

Bhagyashri, Pandurangi R., and Ashwini R. Hirekodi. 2022. "IoT devices for measuring pulse rates and ECG signals." In *Healthcare Systems and Health Informatics*, edited by Pawan Singh Mehra, Lalit Mohan Goyal, Arvind Dagur, & Anshu Kumar Dwivedi, pp. 17–32. CRC Press, Boca Raton, FL.

Bing, Benny. 2008. "Emerging technologies in wireless LANs: theory, design, and deployment."

Brown, Judith B., Sonja M. Reichert, Pauline Boeckxstaens, Moira Stewart, and Martin Fortin. 2022. "Responding to vulnerable patients with multimorbidity: An interprofessional team approach." *BMC Primary Care* 23 (1). https://doi.org/10.1186/s12875-022-01670-6.

Burmaoglu, Serhat, Ozcan Saritas, Levent Bekir Kıdak, and İpek Camuz Berber. 2017. "Evolution of connected health: A network perspective." *Scientometrics* 112 (3): 1419–1438. https://doi.org/10.1007/s11192-017-2431-x.

Burns, David, Henny Westra, Mickey Trockel, and Aaron Fisher. 2013. "Motivation and changes in depression." *Cognitive Therapy and Research* 37 (2): 368–379. https://doi.org/10.1007/s10608-012-9458-3.

Businesswire. 2020. "Global IoT in Healthcare Market 2020–2025: Demand Growing Exponentially, at a CAGR of 18%." Accessed 07/08/2022. https://www.businesswire.com/news/home/20200828005105/en/Global-%20IoT-in-Healthcare-Market-2020-2025-Demand-Growing-Exponentially-at-a-CAGR-%20of-18---ResearchAndMarkets.com.

Caulfield, Brian M., and Seamas C. Donnelly. 2013. "What is connected health and why will it change your practice?" *QJM* 106 (8): 703–707. https://doi.org/10.1093/qjmed/hct114. https://academic.oup.com/qjmed/article-pdf/106/8/703/4482740/hct114.pdf.

Chan, Steven R., and Satish Misra. 2014. "Certification of mobile apps for health care." *JAMA* 312 (11): 1155. https://doi.org/10.1001/jama.2014.9002.

Chang, Hyun Kyung, Dae Hyeok Yang, Mi Yeon Ha, Hyun Joo Kim, Chun Ho Kim, Sae Hyun Kim, Jae Won Choi, and Heung Jae Chun. 2022. "3D printing of cell-laden visible light curable glycol chitosan bioink for bone tissue engineering." *Carbohydrate Polymers* 287: 119328.

Chen, Ke, and Alan Chan. 2013. "Use or non-use of gerontechnology—A qualitative study." *International Journal of Environmental Research and Public Health* 10 (10): 4645–4666. https://doi.org/10.3390/ijerph10104645. http://www.mdpi.com/1660-4601/10/10/4645/pdf.

Chen, Sonia Chien-I., and Chenglian Liu. 2020. "Factors influencing the application of connected health in remote areas, Taiwan: A qualitative pilot study." *International Journal of Environmental Research and Public Health* 17 (4): 1282. https://doi.org/10.3390/ijerph17041282.

Cisco Systems. 2015. "Enterprise Mobility 8.5 Design Guide." Accessed 06/08/2022. https://www.cisco.com/c/en/us/td/docs/wireless/controller/8-5/Enterprise-Mobility-8-5-Design-Guide/Enterprise_Mobility_8-5_Deployment_Guide.pdf.

Cocosila, Mihail, Norm, Archer, R. Brian Haynes, and Yufei Yuan. 2009. "Can wireless text messaging improve adherence to preventive activities? Results of a randomised controlled trial." *International Journal of Medical Informatics* 78 (4): 230–238. https://doi.org/https://doi.org/10.1016/j.ijmedinf.2008.07.011. https://www.sciencedirect.com/science/article/pii/S1386505608001305.

Colorafi, Karen. 2016. "Connected health: A review of the literature." *mHealth* 2: 13–13. https://doi.org/10.21037/mhealth.2016.03.09. https://www.ncbi.nlm.nih.gov/pmc/articles/PMC5344146.

Conn, Vicki S., Adam R. Hafdahl, Pamela S. Cooper, Todd M. Ruppar, David R. Mehr, and Cynthia L. Russell. 2009. "Interventions to improve medication adherence among older adults: Meta-analysis of adherence outcomes among randomized controlled trials." *The Gerontologist* 49 (4): 447–462. https://doi.org/10.1093/geront/gnp037. https://academic.oup.com/gerontologist/article-pdf/49/4/447/1973592/gnp037.pdf.

Craig, John, and Victor Petterson. 2005. "Introduction to the practice of telemedicine." *Journal of Telemedicine and Telecare* 11 (1): 3–9. https://doi.org/10.1177/1357633X0501100102.

Dash, Sabyasachi, Sushil Kumar Shakyawar, Mohit Sharma, and Sandeep Kaushik. 2019. "Big data in healthcare: Management, analysis and future prospects." *Journal of Big Data* 6 (1). https://doi.org/10.1186/s40537-019-0217-0.

Davis, Fred D. 1989. "Perceived usefulness, perceived ease of use, and user acceptance of information technology." *MIS Quarterly* 13 (3): 319–340.

Delemere, Emma, and Rebecca Maguire. 2021. "Technology usage, eHealth literacy and attitude towards connected health in caregivers of paediatric cancer." *2021 IEEE International Symposium on Technology and Society (ISTAS)*, Ontario, 2021.

Dijcks, Beatrice. P. J., Luc. P. De Witte, G. Jan. Gelderblom, R. D. Wessels, and Mathijs. Soede. 2006. "Non-use of assistive technology in The Netherlands: A non-issue?" *Disability and Rehabilitation: Assistive Technology* 1 (1–2): 97–102. https://doi.org/10.1080/09638280500167548.

DiMatteo, M. Robin. 2004. "Variations in patients' adherence to medical recommendations: A quantitative review of 50 years of research." *Medical Care* 42 (3): 200–209. http://www.jstor.org/stable/4640729.

Dunn, Jessilyn, Ryan Runge, and Michael Snyder. 2018. "Wearables and the medical revolution." *Personalized Medicine* 15 (5): 429–448.

Easthall, Claire. 2019. "Medication nonadherence as a complex health behavior: There is more to it than just missed doses." *Journal of the American Geriatrics Society* 67 (12): 2439–2440. https://doi.org/10.1111/jgs.16143.

Edwards, Chris. 2021. "Moore's Law." *Communications of the ACM* 64 (2): 12–14. https://doi.org/10.1145/3440992.

El Amrani, Leila, Agnes Oude Engberink, Gregory Ninot, Maurice Hayot, and François Carbonnel. 2017. "Connected health devices for health care in French general medicine practice: Cross-sectional study." *JMIR mHealth and uHealth* 5 (12): e193. https://doi.org/10.2196/mhealth.7427. https://assetapi.jmir.pub/download?file=36781d5f981dfb914061bc74a0c59529.pdf&alt_file=7427-160681-8-SP.pdf.

Elnaggar, Abdelaziz, Van Ta Park, Sei J. Lee, Melinda Bender, Lee Anne Siegmund, and Linda G. Park. 2020. "Patients' use of social media for diabetes self-care: Systematic review." *Journal of Medical Internet Research* 22 (4): e14209. https://doi.org/10.2196/14209.

Emmelkamp, Paul M.G., and Katharina Meyerbröker. 2021. "Virtual reality therapy in mental health." *Annual Review of Clinical Psychology* 17 (1): 495–519. https://doi.org/10.1146/annurev-clinpsy-081219-115923.

Fisk, Malcolm, Anne Livingstone, and Sabrina Winona Pit. 2020. "Telehealth in the context of COVID-19: Changing perspectives in Australia, the United Kingdom, and the United States." *Journal of Medical Internet Research* 22 (6): e19264. https://doi.org/10.2196/19264.

Frist, William H. 2014. "Connected health and the rise of the patient-consumer." *Health Affairs* 33 (2): 191–193.

Fujita, Saki, Isaree Pitaktong, Graeme Vosit Steller, Victor Dadfar, Qinwen Huang, Sindhu Banerjee, Richard Guo, Hien Tan Nguyen, Robert Harry Allen, and Seth Shay Martin. 2018. "Pilot study of a smartphone application designed to socially motivate cardiovascular disease patients to improve medication adherence." *mHealth* 4: 1–1. https://doi.org/10.21037/mhealth.2017.11.01. http://europepmc.org/articles/pmc5803115?pdf=render.

Gregg, Lynsey, and Nicholas Tarrier. 2007. "Virtual reality in mental health." *Social Psychiatry and Psychiatric Epidemiology* 42 (5): 343–354. https://doi.org/10.1007/s00127-007-0173-4.

Gulfidan, Gizem, Hande Beklen, and Kazim Yalcin Arga. 2021. "Artificial intelligence as accelerator for genomic medicine and planetary health." *OMICS: A Journal of Integrative Biology* 25(12): 745–749.

Haddad, Peter M. 2001. "Antidepressant discontinuation syndromes." *Drug Safety* 24 (3): 183–197. https://doi.org/10.2165/00002018-200124030-00003.

Hajat, Cother, and Emma Stein. 2018. "The global burden of multiple chronic conditions: A narrative review." *Preventive Medicine Reports* 12: 284–293. https://doi.org/10.1016/j.pmedr.2018.10.008.

Haluza, Daniela, Marlene Naszay, Andreas Stockinger, and David Jungwirth. 2016. "Prevailing opinions on connected health in Austria: Results from an online survey." *International*

Journal of Environmental Research and Public Health 13 (8): 813. https://doi.org/10.3390/ijerph13080813. https://www.mdpi.com/1660-4601/13/8/813/pdf.

Hanna, Alan, and Lezley-Anne Hanna. 2019. "Topic analysis of UK fitness to practise cases: What lessons can be learnt?" *Pharmacy* 7 (3): 130. https://doi.org/10.3390/pharmacy7030130. https://pure.qub.ac.uk/ws/files/182713776/pharmacy_07_00130.pdf.

Hanna, Lezley-Anne, and Maurice Hall. 2018. "Launching and evaluating a mobile phone app to provide contemporary, evidence-based advice about self-treatable conditions." *European Journal for Person Centered Healthcare* 6 (3): 358–362.

Harte, Richard P., Liam G. Glynn, Barry J. Broderick, Alejandro Rodriguez-Molinero, Paul M. A. Baker, Bernadette McGuiness, Leonard Sullivan, Marta Diaz, Leo R. Quinlan, and Gearóid ÓLaighin. 2014. "Human centred design considerations for connected health devices for the older adult." *Journal of Personalized Medicine* 4 (2): 245–281. https://www.mdpi.com/2075-4426/4/2/245.

Hartman, Linda, Willem F. Lems, and Maarten Boers. 2019. "Outcome measures for adherence data from a medication event monitoring system: A literature review." *Journal of Clinical Pharmacy and Therapeutics* 44 (1): 1–5. https://doi.org/10.1111/jcpt.12757. https://research.vumc.nl/ws/files/8299462/Hartman_et_al._Outcome_measures_for_adherence_data_Journal_of_clinical_pharmacy_and_therapeutics.pdf.

Hermand, Eric, Clemence Coll, Jean-Paul Richalet, and Francois J. Lhuissier. 2021. "Accuracy and reliability of pulse o2 saturation measured by a wrist-worn oximeter." *International Journal of Sports Medicine* 42(14): 1268–1273.

Holler, Jan, Vlasios Tsiatsis, Catherine Mulligan, Stamatis Karnouskos, Stefan Avesand, and David Boyle. 2014. *Internet of Things*. Oxford: Academic Press.

Hu, Paul J., Patrick Y.K. Chau, Olivia R. Liu Sheng, and Kar Yan Tam. 1999. "Examining the technology acceptance model using physician acceptance of telemedicine technology." *Journal of Management Information Systems* 16 (2): 91–112. https://doi.org/10.1080/07421222.1999.11518247.

Inui, Tomohiko, Hiroki Kohno, Yohei Kawasaki, Kaoru Matsuura, Hideki Ueda, Yusaku Tamura, Michiko Watanabe, Yuichi Inage, Yasunori Yakita, Yutaka Wakabayashi, and Goro Matsumiya. 2020. "Use of a smart watch for early detection of paroxysmal atrial fibrillation: Validation study." *JMIR Cardio* 4 (1): e14857. https://doi.org/10.2196/14857.

Jhunjhunwala, Sanchit, and Sajan Kapil. 2022. "Rapid manufacturing of biomedical devices: Process alternatives, selection and planning." In *Advanced Micro-and Nano-Manufacturing Technologies*, edited by Joshi, Shrikrishna Nandkishor, and Pranjal Chandra, pp. 77–104. Singapore: Springer.

Jiang, Nan, Julia E. Mück, and Ali K. Yetisen. 2020. "The regulation of wearable medical devices." *Trends in Biotechnology* 38(2): 129–133.

Jie, Zhou, Zeng Zhiying, and Li Li. 2021. "A meta-analysis of Watson for oncology in clinical application." *Scientific Reports* 11 (1). https://doi.org/10.1038/s41598-021-84973-5.

Katakam, Prakash, Shanta Kumari Adiki, and Soumya Ranjan Satapathy. 2022. "Recent advancements of additive manufacturing for patient-specific drug delivery." In *Additive Manufacturing Processes in Biomedical Engineering*, edited by Babbar, Atul, Ankit Sharma, Vivek Jain, and Dheeraj Gupta, pp. 1–26. CRC Press, Boca Raton, FL.

Keestra, Sarai. 2021. "Structural violence and the biomedical innovation system: What responsibility do universities have in ensuring access to health technologies?" *BMJ Global Health* 6(5): e004916. https://doi.org/10.1136/bmjgh-2020-004916.

Kern, Christoph, Dun Jack Fu, Karsten Kortuem, Josef Huemer, David Barker, Alison Davis, Konstantinos Balaskas, Pearse A. Keane, Tom Mckinnon, and Dawn A. Sim. 2020. "Implementation of a cloud-based referral platform in ophthalmology: Making telemedicine services a reality in eye care." *British Journal of Ophthalmology* 104 (3): 312–317. https://doi.org/10.1136/bjophthalmol-2019-314161.

Keyworth, Chris, Jo Hart, Christopher.J. Armitage, and Mary P. Tully. 2018. "What maximizes the effectiveness and implementation of technology-based interventions to support healthcare professional practice? A systematic literature review." *BMC Medical Informatics and Decision Making* 18 (1). https://doi.org/10.1186/s12911-018-0661-3.

Koch, Roland, Andreas Polanc, Hannah Haumann, Gudula Kirtschig, Peter Martus, Christian Thies, Leonie Sundmacher, Carmen Gaa, Leonard Witkamp, and Stefanie Joos. 2018. "Improving cooperation between general practitioners and dermatologists via telemedicine: Study protocol of the cluster-randomized controlled TeleDerm study." *Trials* 19 (1). https://doi.org/10.1186/s13063-018-2955-2.

Kuziemsky, Craig, Shashi Gogia, Mowafa Househ, Carolyn Petersen, and Arindam Basu. 2018. "Balancing health information exchange and privacy governance from a patient-centred connected health and telehealth perspective." *Yearbook of Medical Informatics* 27 (01): 048–054. https://doi.org/10.1055/s-0038-1641195.

Kvedar, Joseph, Molly Joel Coye, and Wendy Everett. 2014. "Connected health: A review of technologies and strategies to improve patient care with telemedicine and telehealth." *Health affairs* 33(2): 194–199.

Laaki, Heikki, Yoan Miche, and Kari Tammi. 2019. "Prototyping a digital twin for real time remote control over mobile networks: Application of remote surgery." *IEEE Access* 7: 20325–20336. https://doi.org/10.1109/access.2019.2897018.

Lan, Weiwei, Xiaobo Huang, Di Huang, Xiaochun Wei, and Weiyi Chen. 2022. "Progress in 3D printing for bone tissue engineering: A review." *Journal of Materials Science* 57 (27): 12685–12709. https://doi.org/10.1007/s10853-022-07361-y.

Laport, Francisco, Adriana Dapena, Paula M. Castro, Francisco J. Vazquez-Araujo, and Daniel Iglesia. 2020. "A prototype of EEG system for IoT." *International Journal of Neural Systems* 30(07): 2050018.

Lauterbach, Claire J., Phebe A. Romano, Luke A. Greisler, Richard A. Brindle, Kevin R. Ford, and Matthew R. Kuennen. 2021. "Accuracy and reliability of commercial wrist-worn pulse oximeter during normobaric hypoxia exposure under resting conditions." *Research Quarterly for Exercise and Sport* 92 (3): 549–558. https://doi.org/10.1080/02701367.2020.1759768.

Lee, Donhee, and Seong No Yoon. 2021. "Application of artificial intelligence-based technologies in the healthcare industry: Opportunities and challenges." *International Journal of Environmental Research and Public Health* 18 (1): 271. https://doi.org/10.3390/ijerph18010271.

Lehane, Elaine, and Geraldine Mccarthy. 2009. "Medication non-adherence-exploring the conceptual mire." *International Journal of Nursing Practice* 15 (1): 25–31. https://doi.org/10.1111/j.1440-172x.2008.01722.x.

Li, Tong, Tong Xia, Huandong Wang, Zhen Tu, Sasu Tarkoma, Zhu Han, and Pan Hui. 2022. "Smartphone app usage analysis: Datasets, methods, and applications." *IEEE Communications Surveys & Tutorials* 24 (2): 937–966. https://doi.org/10.1109/comst.2022.3163176.

Lubitz, Steven A., Anthony Z. Faranesh, Steven J. Atlas, David D. McManus, Daniel E. Singer, Sherry Pagoto, Alexandros Pantelopoulos, and Andrea S. Foulkes. 2021. "Rationale and design of a large population study to validate software for the assessment of atrial fibrillation from data acquired by a consumer tracker or smartwatch: The Fitbit heart study." *American Heart Journal* 238: 16–26.

Mantovani, Fabrizia, Gianluca Castelnuovo, Andrea Gaggioli, and Giuseppe Riva. 2003. "Virtual reality training for health-care professionals." *CyberPsychology & Behavior* 6 (4): 389–395.

Marangunić, Nikola, and Andrina Granić. 2015. "Technology acceptance model: A literature review from 1986 to 2013." *Universal Access in the Information Society* 14 (1): 81–95. https://doi.org/10.1007/s10209-014-0348-1.

Martin, Leslie R., Summer L. Williams, Kelly B. Haskard, and M. Robin Dimatteo. 2005. "The challenge of patient adherence." *Ther Clin Risk Manag* 1 (3): 189–199.

McKinsey Global Institute. 2020. "Prioritizing health; a prescription for prosperity." Accessed 09/08/2022. https://www.mckinsey.com/industries/healthcare-systems-and-services/our-insights/prioritizing-health-a-prescription-for-prosperity.

Moko Smart. 2022. "IoT in healthcare." Accessed 07/08/2022. https://www.mokosmart.com/iot-in-healthcare/.

Mück, Julia E., Barış Ünal, Haider Butt, and Ali K. Yetisen. 2019. "Market and patent analyses of wearables in medicine." *Trends in biotechnology* 37 (6): 563–566.

Nambiar, Ankita R., Nikitha Reddy, and Debojyoti Dutta. 2017. "Connected health: Opportunities and challenges." *2017 IEEE International Conference on Big Data (Big Data)*, 2017.

National Institutes of Health. 2018. "Mobile health: Technology and outcomes in low and middle income countries." Accessed 06/08/2022. https://grants.nih.gov/grants/guide/pa-files/PAR-18-242.html.

Nilsen, Wendy, Santosh Kumar, Albert Shar, Carrie Varoquiers, Tisha Wiley, William T. Riley, Misha Pavel, and Audie A. Atienza. 2012. "Advancing the Science of mHealth." *Journal of Health Communication* 17 (sup1): 5–10. https://doi.org/10.1080/10810730.2012.677394.

O'Leary, Padraig, Noel Carroll, Paul Clarke, and Ita Richardson. 2015. "Untangling the complexity of connected health evaluations." *2015 International Conference on Healthcare Informatics*, 2015.

O'Connor, Siobhan. 2019. "Virtual reality and avatars in health care." *Clinical Nursing Research* 28 (5): 523–528. https://doi.org/10.1177/1054773819845824.

Olivieri, N.F, D. Matsui, C. Hermann, and G. Koren. 1991. "Compliance assessed by the medication event monitoring system." *Archives of Disease in Childhood* 66 (12): 1399–1402. https://doi.org/10.1136/adc.66.12.1399.

Omboni, Stefano. 2019. "Connected health in hypertension management." *Frontiers in Cardiovascular Medicine* 6: 76.

Orruño, Estibalitz, Marie Pierre Gagnon, José Asua, and Anis Ben Abdeljelil. 2011. "Evaluation of teledermatology adoption by health-care professionals using a modified Technology Acceptance Model." *Journal of Telemedicine and Telecare* 17 (6): 303–307. https://doi.org/10.1258/jtt.2011.101101.

Pan, Tingrui, and Yong Xu. 2014. "Mobile medicine: Can emerging mobile technologies enable patient-oriented medicine?" *Annals of Biomedical Engineering* 42(11): 2203–2204. https://doi.org/10.1007/s10439-014-1138-x. https://link.springer.com/content/pdf/10.1007%2Fs10439-014-1138-x.pdf.

Panch, Trishan, Heather Mattie, and Leo Anthony Celi. 2019. "The "inconvenient truth" about AI in healthcare." *NPJ Digital Medicine* 2 (1). https://doi.org/10.1038/s41746-019-0155-4. https://www.nature.com/articles/s41746-019-0155-4.pdf.

Pantelopoulos, Alexandros, Anthony Faranesh, Andreea Milescu, Paige Hosking, Subramaniam Venkatraman, and Conor Heneghan. 2017. "Screening of atrial fibrillation using wrist photoplethysmography from a fitbit tracker." *IPROC* 3 (1): e17. https://doi.org/10.2196/iproc.8447. http://www.iproc.org/2017/1/e17/.

Pare, Guy, Mirou Jaana, and Claude Sicotte. 2007. "Systematic review of home telemonitoring for chronic diseases: The evidence base." *Journal of the American Medical Informatics Association* 14 (3): 269–277. https://doi.org/10.1197/jamia.m2270.

Pérez-Jover, Virtudes, Marina Sala-González, Mercedes Guilabert, and José Joaquín Mira. 2019. "Mobile apps for increasing treatment adherence: Systematic review." *Journal of Medical Internet Research* 21 (6): e12505. https://doi.org/10.2196/12505.

Pipich, Michael G. 2018. *Owning Bipolar: How Patients and Families Can Take Control of Bipolar Disorder*. Citadel Press.

Portnoy, Jay, Morgan Waller, and Tania Elliott. 2020. "Telemedicine in the Era of COVID-19." *The Journal of Allergy and Clinical Immunology: In Practice* 8 (5): 1489–1491. https://doi.org/10.1016/j.jaip.2020.03.008.

Rahimi, Bahlol, Hamed Nadri, Hadi Lotfnezhad Afshar, and Toomas Timpka. 2018. "A systematic review of the technology acceptance model in health informatics." *Applied Clinical Informatics* 09 (03): 604–634. https://doi.org/10.1055/s-0038-1668091.

Ramalingam, Mritha, Ramalingam. Puviarasi, Elanchezhian Chinnavan, Quah Chia Shern, and Mohamad Fadli Zolkipli. 2021. "Alarming assistive technology: An IoT enabled sitting posture monitoring system." *2021 International Conference on Software Engineering & Computer Systems and 4th International Conference on Computational Science and Information Management (ICSECS-ICOCSIM)*, 2021.

Rogers, Anne, and Nicola Mead. 2004. "More than technology and access: Primary care patients' views on the use and non-use of health information in the Internet age." *Health and Social Care in the Community* 12 (2): 102–110. https://doi.org/10.1111/j.0966-0410.2004.00473.x.

Ross Kerr, A. 2020. "Diagnostic adjuncts for oral cavity squamous cell carcinoma and oral potentially malignant disorders." In *Textbook of Oral Cancer: Prevention, Diagnosis and Management*, edited by Saman Warnakulasuriya and John S. Greenspan, pp. 99–117. Cham: Springer International Publishing.

Salisbury, Chris, Alicia O'Cathain, Clare Thomas, Louisa Edwards, Alan A. Montgomery, Sandra Hollinghurst, Shirley Large, Jon Nicholl, Catherine Pope, Anne Rogers, Glyn Lewis, Tom Fahey, Lucy Yardley, Simon Brownsell, Padraig Dixon, Sarah Drabble, Lisa Esmonde, Alexis Foster, Katy Garner, Daisy Gaunt, Kim Horspool, Mei-See Man, Alison Rowsell, and Julia Segar. 2017. "An evidence-based approach to the use of telehealth in long-term health conditions: Development of an intervention and evaluation through pragmatic randomised controlled trials in patients with depression or raised cardiovascular risk." *Programme*

Grants for Applied Research 5 (1): 1–468. https://doi.org/10.3310/pgfar05010.

Schiefer, Lisa M, Gunnar Treff, Franziska Treff, Peter Schmidt, Larissa Schäfer, Josef Niebauer, Kai E. Swenson, Erik R. Swenson, Marc M. Berger, and Mahdi Sareban. 2021. "Validity of peripheral oxygen saturation measurements with the Garmin Fēnix® 5X plus wearable device at 4559 m." *Sensors* 21(19): 6363.

Shi, Lizheng, Jinan Liu, Vivian Fonseca, Philip Walker, Anupama Kalsekar, and Manjiri Pawaskar. 2010. "Correlation between adherence rates measured by MEMS and self-reported questionnaires: A meta-analysis." *Health and Quality of Life Outcomes* 8 (1): 99. https://doi.org/10.1186/1477-7525-8-99. https://hqlo.biomedcentral.com/track/pdf/10.1186/1477-7525-8-99.

Smallman, Melanie. 2022. "Multi Scale Ethics–Why we need a sociological approach to the ethics of AI in healthcare at different scales."

Smith, Daisy, Janaka Lovell, Carolina Weller, Briohny Kennedy, Margaret Winbolt, Carmel Young, and Joseph Ibrahim. 2017. "A systematic review of medication non-adherence in persons with dementia or cognitive impairment." *PLoS One* 12 (2): e0170651. https://doi.org/10.1371/journal.pone.0170651. https://doi.org/10.1371/journal.pone.0170651.

Söderström, Sylvia, and Borgunn Ytterhus. 2010. "The use and non-use of assistive technologies from the world of information and communication technology by visually impaired young people: A walk on the tightrope of peer inclusion." *Disability & Society* 25 (3): 303–315. https://doi.org/10.1080/09687591003701215.

Statista. 2022. "Number of smartphone subscriptions worldwide from 2016 to 2027." Statista. Accessed 04/08/2022.

Strandbygaard, Ulla, Simon Francis Thomsen, and Vibeke Backer. 2010. "A daily SMS reminder increases adherence to asthma treatment: A three-month follow-up study." *Respiratory Medicine* 104 (2): 166–171. https://doi.org/10.1016/j.rmed.2009.10.003. https://doi.org/10.1016/j.rmed.2009.10.003.

Tao, Da, Leiyan Xie, Tieyan Wang, and Tieshan Wang. 2015. "A meta-analysis of the use of electronic reminders for patient adherence to medication in chronic disease care." *Journal of Telemedicine and Telecare* 21 (1): 3–13. https://doi.org/10.1177/1357633x14541041.

Thangam, Dhanabalan, Anil B. Malali, Gopalakrishnan Subramanian, Sumathy Mohan, and Jin Yong Park. 2022. "Internet of things: a smart technology for healthcare industries." In *Healthcare Systems and Health Informatics*, edited by Pawan Singh Mehra, Lalit Mohan Goyal, Arvind Dagur, Anshu Kumar Dwivedi, pp. 3–15. CRC Press, Boca Raton, FL.

Ulrik, Charlotte Suppli, Vibeke Backer, Ulrik Søes-Petersen, Peter Lange, Henrik Harving, and Peter P. Plaschke. 2006. "The patient's perspective: Adherence or non-adherence to asthma controller therapy?" *Journal of Asthma* 43(9): 701–704.

Vermeire, Etienne, Hilary Hearnshaw, Paul Van Royen, and Joke Denekens. 2001. "Patient adherence to treatment: Three decades of research. A comprehensive review." *Journal of Clinical Pharmacy and Therapeutics* 26 (5): 331–342. https://doi.org/10.1046/j.1365-2710.2001.00363.x.

West, Cameron, Fenerty, Feldman, Kaplan, and Davis. 2012. "The effect of reminder systems on patients' adherence to treatment." *Patient Preference and Adherence*: 127. https://doi.org/10.2147/ppa.s26314.

Wongvibulsin, Shannon, Seth S. Martin, Steven R. Steinhubl, and Evan D. Muse. 2019. "Connected health technology for cardiovascular disease prevention and management." *Current Treatment Options in Cardiovascular Medicine* 21 (6). https://doi.org/10.1007/s11936-019-0729-0.

Wootton, R. 2001. "Recent advances: Telemedicine." *BMJ* 323 (7312): 557–560. https://doi.org/10.1136/bmj.323.7312.557.

Wu, Jiandong, Meili Dong, Claudio Rigatto, Yong Liu, and Francis Lin. 2018. "Lab-on-chip technology for chronic disease diagnosis." *NPJ Digital Medicine* 1 (1). https://doi.org/10.1038/s41746-017-0014-0.

Yetisen, Ali K., Juan Leonardo Martinez-Hurtado, Barış Ünal, Ali Khademhosseini, and Haider Butt. 2018. "Wearables in medicine." *Advanced Materials* 30 (33): 1706910. https://doi.org/10.1002/adma.201706910.

14 Emerging Technologies for Tackling Pandemics

Elliot Mbunge
University of Eswatini

John Batani
Botho University

Godfrey Musuka
Columbia University

Itai Chitungo
University of Zimbabwe

Innocent Chingombe
Columbia University

Tafadzwa Dzinamarira
Columbia University
University of Pretoria

Benhildah Muchemwa
University of Eswatini

CONTENTS

DOI: 10.1201/9781003224464-14

14.1   INTRODUCTION

Globally, health systems have been experiencing unprecedented challenges posed by the outbreak of pandemics and consequently increase demand for care in the past centuries. The outbreak of pandemics such as influenza viruses and coronavirus tremendously affect healthcare service delivery and subsequently overburden health systems in the process of alleviating the spread of the disease in societies [1]. For instance, the deadliest influenza viruses such as HCoV-OC43, H1N1, H2N2 and H3N2 caused catastrophic impact globally, leading to over one million deaths. For instance, over one million people died due to HCoV-OC43 reported in 1889s [2]; Spanish flu killed over 50 million people, between 1918 and 1920 [3]; Asian flu killed between 1 and 4 million people in 1957–1958; Hong Kong flu (reported in 1968–1970); severe acute respiratory syndrome (SARS) killed over 774 people, between 2002 and 2004; and influenza A virus subtype H1N1 (Swine flu) reported in 2009–2010 killed over 284,000 people [4]. In addition, over 881 (as of 30 May 2020) and 6.3 million+ (as of May 2022) people succumbed to Middle East respiratory syndrome/MERS-CoV and SARS-CoV-2, respectively [3]. The unexpected public health crisis exacerbated by the outbreak of pandemics continues to overwhelm health systems and consequently retards milestones gained to alleviate other diseases in the past especially when resources are channelled to fight the pandemic.

However, due to several preventive and control measures usually imposed to reduce the spread of the pandemic, different emerging technologies have been adopted to continue providing care while observing social distancing and quarantine guidelines. Such technologies include artificial intelligence, nanotechnology, deep learning, machine learning, robotics, drones, Internet of Things, virtual reality, Internet of Medical Things, geographical information systems, 5G technology, telemedicine, cloud computing, big data, additive manufacturing, blockchain and smart health applications [5–12]. However, these technologies generally encounter different impediments in tackling pandemics exacerbated by various existing infrastructural, funding, human capital, regulatory barriers and unequal distribution of health facilities, among others. The emergence of global pandemics is inevitable in future; therefore, there is a need for a comprehensive review of emerging technologies used for tackling pandemics in the past, while paving research contributions in the present and identifying challenges and emerging research areas that need to be addressed in the future. This study sought to address this research gap by achieving the following objectives:

i. Identify and explain emerging technologies utilized for tackling pandemics in the past while introspecting their implementation challenges and barriers in various healthcare application domains.

ii. Explain the role of emerging technologies in tackling pandemics

iii. Propose recommendations for the effective adoption of emerging technologies to tackle future pandemics.

The remainder of this chapter is structured as follows. Section 2 presents the global overview of pandemics reported in the past. The methodology adopted in this study was comprehensively explained in Section 3. Emerging technologies and their respective functions as well as implementation challenges and limitations are explained in Section 4. Finally, Sections 5 and 6 present the recommendations and conclusion, respectively.

14.2   METHODOLOGY

14.2.1   Study Design

This comprehensive review followed the guidelines and principles of the Preferred Reporting Items for Systematic Reviews and Meta-Analysis (PRISMA) [13] model for screening and selecting relevant papers from various prominent and reputable electronic databases. The systematic review of emerging digital health technologies was conducted in line with guidelines for undertaking systematic reviews in healthcare.

14.2.2   Search Strategy

We searched published articles from various electronic databases such as CINAHL with full text, Embase, IEEE, Springer, Taylor and Francis, PubMed, MEDLINE, ScienceDirect databases, Google Scholar and World Health Organization library databases for relevant studies. The search was guided by the search strategy that used both text word and previous/ current pandemics subject heading searches. Therefore, the literature search was based on the following search terms: *"Emerging technologies"* OR *"Digital technology"* OR *"Digital health"* OR *"information technology"* AND *"tackling pandemics"* OR *"COVID-19"* OR *"SARS-CoV-2"* OR *"coronavirus disease"* OR *"Ebola"* OR *"MERS"* OR *"Swine Flu"*.

14.2.3   Study Selection

Authors screened and selected published research papers from reputable outlets using the following inclusion and exclusion criteria. The selected articles were screened based on the following: titles, abstracts and full text. The selection of relevant studies all met the following inclusion criteria: (1) written in English or have English translations; (2) peer-reviewed; and (3) applied digital technology for tackling health pandemics. Thus, incomplete, opinion pieces, letters to the editors, non-peer-reviewed articles and articles without English translations were excluded from the study.

14.2.4 DATA EXTRACTION

The literature search extracted a total of 300 articles from different electronic databases. Authors screened retrieved papers using the titles and abstracts. A total of 103 papers were removed from the study. In addition, we further removed duplicate papers from a pool of articles and left with 197 papers. Authors further read and assessed full-text papers for eligibility and several papers were discarded, as shown in Figure 14.1. Therefore, 24 papers were considered in this study. The PRISMA steps followed for screening and selecting relevant papers are shown in Figure 14.1.

After selecting relevant papers, data such as the pandemic, emerging technology, name, purpose of the emerging technology in tackling the pandemic, implementation barriers and challenges were captured (see Section 4). Table 14.1 shows the review studies conducted on the application of emerging technologies to tackle various health pandemics.

14.3 RESULTS

Table 14.1 shows that several emerging technologies have been deployed to fight pandemics in different health systems. These technologies include artificial intelligence (deep learning, machine learning), robotics, drones, Internet of Things, virtual reality, Internet of Medical Things, geographical Information systems, 5G technology, telemedicine, blockchain and smart health applications. The following sections provide a detailed explanation of these emerging technologies.

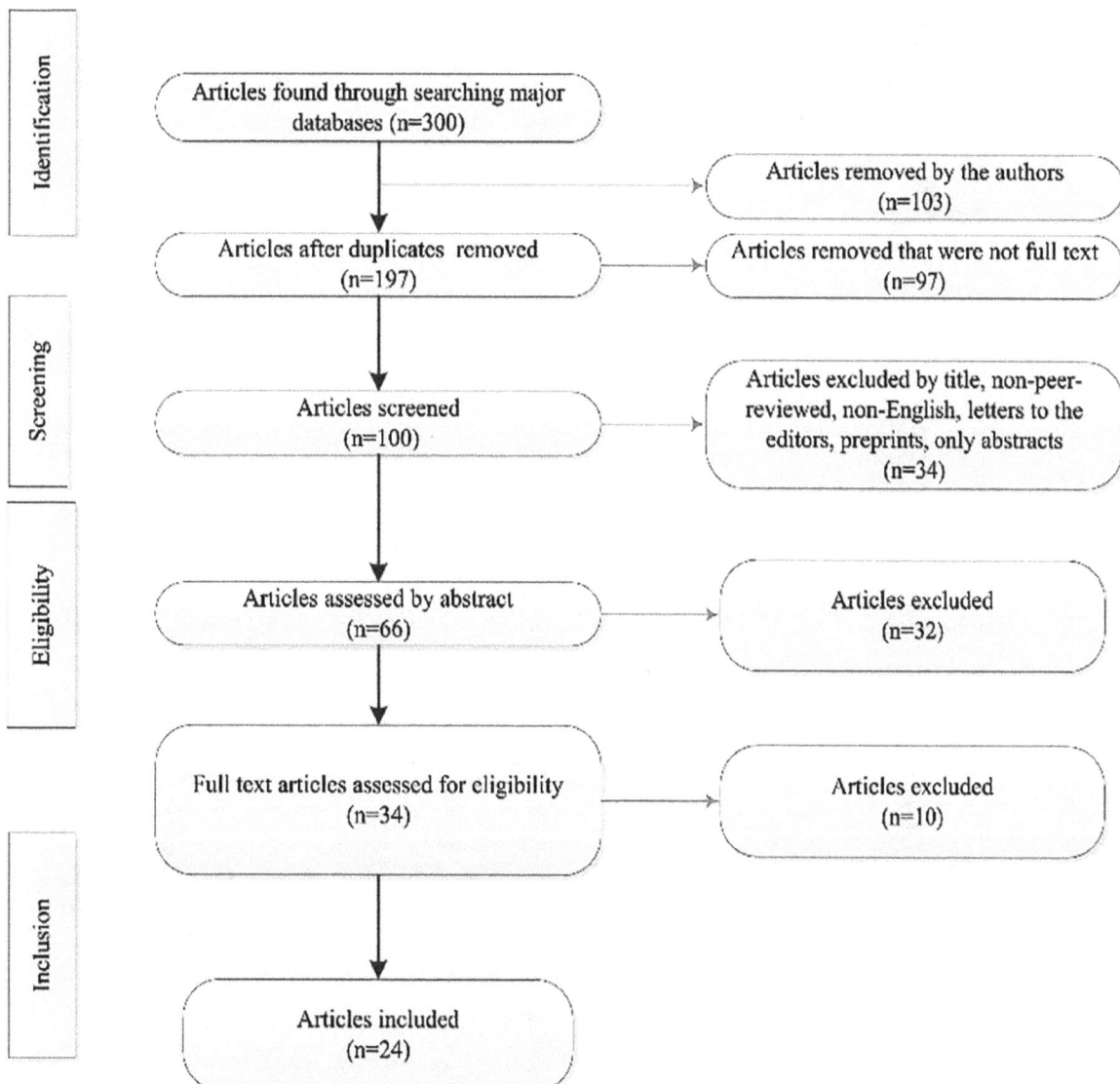

FIGURE 14.1 PRISMA model.

TABLE 14.1

Emerging Technologies for Tackling Pandemics

Emerging Technology	Functions	References
Artificial intelligence	• Genome analysis	[5–7,14–17]
	• Protein structure prediction	
	• Vaccines and drugs development	
	• Detection, diagnosis, treatment and prognosis	
	• Infection identification and tracking	
	• Prediction of cases and mortality rate	
	• Medicines perspective and repurposing	
	• Drug discovery	
	• Curbing the spread of misinformation	
Internet of Things and IOMT	• Real-time data collection using smart wearable devices	[10,18–20]
	• Real-time patient monitoring using sensor-based wearable devices	
	• Enforcing contact tracing and social distancing guidelines	
	• Smart sharing of patients' data	
	• Internet-connected healthcare service delivery and treatment	
	• Screening and surveillance	
	• Real-time tracking	
Telemedicine	• Telehealth consultation	[10,21–30]
	• Tele-rehabilitation	
	• Remote consultation and prevention and control	
	• Support tele-pharmacy and telenursing	
	• Support remote diagnosis through Internet technology	
Blockchain technology	• Improve health data security	[3,8,31]
	• Facilitates integration of verified regional and international data sources	
	• Development of tamper-proof smart apps for contact tracing.	
	• Facilitates disease testing and reporting	
5G Technology	• Better performance in terms of higher speed, lower latency, wider range, increased availability and more reliability	[32,33] [3,17–20,34,35]
	• Support teleservices and telemedicine	
	• Facilitates real-time healthcare service delivery in smart healthcare systems	
	• Support real-time health data exchange and sharing	
	• Support real-time and virtual communication between healthcare professionals and patients	
Geographical information systems	• Mapping infections hotspot zones	[36–39]
	• Spatial modelling of pandemics	
	• Tracking migration patterns of the affected populations	
	• Mapping transmission and assessment risk	

14.3.1 Artificial Intelligence

Artificial intelligence (AI) is an emerging technology inspired by human-like intelligent in machines by incorporating intelligent models such as deep learning and machine learning in developing smart devices and smart applications (apps). AI-based devices have been used to improve healthcare service delivery during health emergencies in different ways including early diagnosis and recognition of the patient's infection, vaccines and drug development, and controlling remediation. Most recently, AI has been successfully used in fighting COVID-19 by developing expert systems for detecting COVID-19 from X-ray images using deep learning, remote monitoring treatment and adherence, maintaining social distancing and quarantine guidelines, tracking individuals using AI-based contact tracing apps [16], predicting cases and mortality rates, remote monitoring of migration patterns, development of robots and drones to help in scaling down the workload of healthcare workers and risk exposure. AI techniques, especially deep learning models, have been significantly used to identify and screen COVID-19 infected patients with high accuracy, analyse COVID-19 genome and detect COVID-19 face masks. However, the development of AI-based expert systems requires numerous and quality datasets for training and testing the models, of which such data is not always available, especially in the early stages of the pandemic. Also, there are various ethical, social and human rights issues associated with the deployment of AI-based tools in public health and sharing of patients' health data.

14.3.2 Internet of Things and Internet of Medical Things

Internet of Things (IoT) involves the interconnectedness of computing and digital devices to share and transmit data over the Internet with minimal human interaction. IoT devices are progressively utilized for predicting, preventing and monitoring health pandemics by connecting digital such as smart thermometers, smart wearable devices, pulse oximeters, smart robots and smart apps. These digital tools can assist health workers by continuously monitoring patients' pathological details and collecting important body parameter readings by using sensors, wireless networks, wearable devices and implantable sensors [40–42]. The continuous monitoring of patients is critical, especially for patients with critical health conditions through telehealth consultation during pandemics [6] and automated treatment. IoT and IoMT devices collect huge data that can be analysed to provide useful insights, monitor patient and predict disease trends and subsequently monitor quarantine and social distancing compliances. Due to the huge data collected by IoMT that requires more storage space, cloud computing [43] becomes imperative as it uses virtualization to store more data that can be accessed ubiquitously. Besides, in times of pandemics, cloud computing enables the continuation of services and daily activities by facilitating communication, collaboration and online services available even in the face of lockdowns [44].

14.3.3 Drones and Robots

The use of drones and autonomous robots can significantly reduce the risk of exposure for health workers and other frontline employees. For instance, during Ebola and COVID-19, drones and robots have been used to deliver medical supplies [6], nasal swabs and food items in quarantine facilities without direct human contact, public announcements, enforcing lockdown rules, crowd surveillance [3], disinfection and sanitizing hotspot areas. Also, robots and drones equipped with cameras and sensors can be used to check an individual's temperature, maintain social distance, remote monitoring and face masking guidelines [8]. Keeping hotspot areas disinfected and sanitized using robots and drones can tremendously help reduce the spread of infection, especially in countries with fragile health systems and poor sanitation conditions [35]. However, drones are vulnerable to GPS jamming and hacking which might affect their implementation. Despite battery life and load capacity constraints, the integration of drones in health systems is generally affected by the lack of a clear drone regulatory framework [3].

14.3.4 5G Technology

5G is among paramount emerging technologies that are paramount in tackling pandemics-related network connectivity challenges by providing a wide range of wireless including enhanced mobile broadband communication [3], ultra-reliable and low-latency communication and massive machine-type communications [45]. The utilization of 5G technology can enable real time communication of interconnected IoT, AI and IoMT devices to exchange and share health-related data. Together with concomitant technologies, 5G technology can revolutionize health systems globally to become robust in providing healthcare services. 5G technology can help improve the smart healthcare system's strong communication infrastructure in relation to better reliability, stability of connections, ultra-massive accessibility, scalability of the network and rapid response flexibility, enabling monitoring and preventive action to be taken in pandemic situations. AI-based digital tools for real-time detection of disease, pandemic monitoring, social distancing as well as quarantine tools require fast Internet speed to effectively monitor the spread of pandemics while providing healthcare services remotely to minimize physical contact between health workers and patients.

14.3.5 Virtual Reality and Augmented Reality

The outbreak of pandemics such as COVID-19 witnessed a rapid increase in virtual reality and augmented reality devices. Such devices create a simulated environment for experimental and laboratory work to provide enhanced interactivity in the virtual space. Virtual reality devices have been used in simulation-based training in medical education through real-time virtual classrooms [6], virtual diagnosis, virtual laboratory simulation and design.

14.3.6 Geographic Information Systems and Smart Apps

Geographic information systems (GIS) can help identify high-risk areas and information dissemination during pandemics [46]. It is a useful tool for examining the spatial distribution of diseases [47]. Identifying high-risk areas and examining the spatial distribution of diseases during pandemics are useful as they inform policymakers and health authorities with information critical for determining and prioritizing resource allocation and interventions. Coupled with other emerging digital technologies like AI, GIS can be used to model the spread of infection in real time [48]. Smart mobile applications play a significant role in several ways. These include data collection, reporting, information provision, contact tracing [49], travel monitoring and vaccination status verification by scanning quick response (QR) codes [50]. However, problems arise when the data collected end up repurposed without the users' knowledge, such as being used for tracking and surveillance for political purposes instead of responding to the pandemic. There are also issues of privacy and security arising from the use of smart apps.

14.3.7 Telemedicine

Telemedicine refers to remote clinical services such as healthcare delivery, diagnosis, consultation and treatment where a healthcare professional utilizes communication infrastructure to deliver care to a patient at a remote site. Telemedicine provides a new means to support and promote long-distance clinical care, education and healthcare, from first response to recovery with low cost and extensive coverage. In the age of increased risk pandemic, outbreaks telemedicine is ideal for the management of communicable diseases. For pandemics such as the current COVID-19 and similar high transmissible infections such as Ebola, the key to slowing the transmission reduced contact through measures such as quarantine and social distancing. During infectious disease outbreaks, telemedicine can enable remote triaging of care and provide rapidly accessible information through technology – such as chatbots. Furthermore, the adoption of telemedicine will help decongest healthcare centres and allow the general population to continue accessing safe and convenient quality care regardless of COVID-19 exposure [27]. The rest of the population, especially the vulnerable and elderly, can continue to access routine care and treatment free of risk of infection through the convenience of telemedicine. Telemedicine can thus address health care inequalities, mental challenges and shortages of personal protective equipment (PPE) during the period of crisis or emergencies as what transpired during the global pandemic.

Telemedicine can provide rapid access to specialists who are not physically available to clients in remote settings who are usually financially constrained and limited in their mobility to access quality healthcare [51]. The system leverages specialists at distant sites to provide virtual emergency consultations and distribute work among subspecialty providers. Under COVID-19 lockdown, telemedicine has the potential to create linkages between ICU experts in developed countries and clinicians attending to a patient in resource-limited settings, thus improving health outcomes. Furthermore, telemedicine can be used to provide on-demand training or remote training [52]. Thus developed countries experienced a surge in the uptake and use of telemedicine amid social distancing and stay-at-home directives. Concerns about both healthcare workers (HCW) and general population safety in combination with urgent regulatory changes have been the main driving forces behind the rapid adoption of telemedicine [53].

Telemedicine offers countries the opportunity to collect and organize patient data for epidemiological surveillance by assisting in identifying and tracking public health issues and illustrating trends [1]. Leveraging app-facilitated screening, triaging and tracing of populations will improve the efficiency and effectiveness of the traditional healthcare setting. Data collected from these platforms make it possible to map disease hotspots and identify populations at most risk, thus allowing for a more coordinated response. While telemedicine is unable to accurately predict natural disasters or infectious pandemics, it surely gives an advantage in preparing for such eventualities and therefore, timeous response. Telemedicine allows for a more integrated approach to patient management and maintaining wellness, and with patients easily accessing their data, there is increased engagement, autonomy, and agency. The technological revolution and increase in the IoT bring both the young and elderly onto digital platforms making it easy to provide upskill the population on health literacy and

provide the required tools to adapt and adopt telemedicine. This will help eliminate barriers and increase access. Thus, telemedicine has the potential to accelerate the implementation and adoption of personalized medicine. Telemedicine's value proposition is its potential to make healthcare more personalized and integrated, thereby improving efficiency, patient and clinician satisfaction and health outcomes [51].

14.4 CHALLENGES ENCOUNTERED IN DEPLOYING EMERGING TECHNOLOGIES FOR TACKLING PANDEMICS

14.4.1 POOR INFRASTRUCTURE

Oftentimes, these technologies tend to be more successful in developed countries than in developing ones due to poor, dilapidated or unavailability of the necessary infrastructure to support their deployment, use and sustenance [54]. Unreliable electricity supply [55] and communication networks affect the deployment and use of these emerging digital technologies. In developing countries, infrastructure is usually often available in urban areas but is usually unavailable in rural areas, or if there, it will mostly be relatively poor.

14.4.2 LACK OF CLEAR GOVERNMENT REGULATORY POLICIES

Technology is moving too quick for most regulatory bodies and governments. Consequently, government regulatory policies often lag behind the technologies. This presents challenges in introducing emerging technologies since there will be a lack of regulatory frameworks guiding their implementation. The critical regulatory frameworks that often lag include device regulation and technology certification, security and patient data protection frameworks [26,56].

14.4.3 ETHICAL ISSUES

Ethical issues are often raised when it comes to digital technologies' use in healthcare, including breaches of security and privacy [57,58], as well as over surveillance and repurposing. Repurposing means using the collected data for other purposes other than what they were collected for, which is an unethical practice.

14.4.4 LACK OF SYSTEMS INTEROPERABILITY

An ecosystem of digital health systems requires sharing data capabilities to fully benefit from data collected by different systems. However, there is a lack of interoperability of the digital systems and tools in healthcare, affecting data sharing and communication by such tools and technologies [29].

14.4.5 LACK OF EQUIPMENT

Low-resource countries are always faced with a lack of equipment. The deployment of emerging digital technologies needs various digital equipment, which may not be readily available. Such equipment includes tablets, computers and sensors, among others. Lack of equipment is inherent in developing countries.

14.4.6 FUNDING INADEQUACY

Funding is central to the deployment of emerging digital technologies, and its inadequacy is the root of other challenges like insufficiency of equipment. Development of digital tools, training of personnel and purchasing of equipment and other resources require funds. However, funding is often a challenge in developing countries [58,59], leading to heavy reliance on donations.

14.4.7 LACK OF TRUST AND SECURITY ISSUES

Security of digital technologies is a real concern among some users and patients who have little trust in such technologies. The deployment of emerging digital technologies is likely to meet with resistance due to perceived security and privacy concerns.

14.5 RECOMMENDATIONS

i. **Poor infrastructure**: The continent should commit to investing in infrastructure to compete in the global space. Governments can engage in public-private partnership and adopt models such as build, operate and transfer (BOT), to modernize infrastructure.

ii. **Lack of policy**: Policy development by individual countries may be hampered by limited technical skills and funding. However, the intergovernmental agencies can help to identify the requisite technical expertise available on the continent to craft and/or advise governments on policy issues relating to various emerging technologies. Individual governments will then adapt these based on socio-cultural demands/needs.

iii. **Ethical issues**: It is imperative to safeguard human dignity by ensuring adopted technology that do not breach citizen privacy and trust. Therefore, emerging technology implementation must be transparent in the operations, secure from data breaches, seek informed consent and adapt to the socio-cultural determinants of the community to guarantee public trust.

iv. **Lack of interoperability**: Legacy issues will hamper the implementation of emerging technologies, hence must be addressed through skills gap training and exposure to technologies from other countries.

v. **Lack of equipment**: In the short to medium term, adequate funding to purchase equipment will increase the use of emerging technologies during pandemics. However, there is need for long-term

sustainable investiment strategy in local high tertiary education institutions to innovate and develop relevant technologies that also address the African idiosyncrasies.

vi. **Funding inadequacy**: The African CDC model of funding has achieved great success with regard to directing funds to fight the COVID-19 pandemic. Regional blocks and African Union are vital to closing funding gaps for investing in strengthening the health system through investing in emerging technology on the continent.

vii. **Lack of trust and security issues**: Increasing user awareness on the operation of technologies, being transparent in its use, seeking user consent, ensuring data security through end-to-end encryption and also guaranteeing non-repurposing of citizen data will help improve user trust and security of emerging technologies.

14.6 CONCLUSION

Emerging technologies such as AI, nanotechnology, deep learning, machine learning, robotics, drones, IoT, virtual reality, sensors, IoMT, GIS, 5G technology, telemedicine, cloud computing, big data, additive manufacturing, blockchain and smart health applications continue bringing unprecedented opportunities for tackling pandemics and improving healthcare service delivering during health emergencies. For instance, during the COVID-19 pandemic amid social distancing and stay-at-home directives spurred by necessity and supported by urgent regulatory changes, emerging technologies such as AI and telemedicine gained traction as an alternative primary mode of detecting COVID-19, development of vaccines and provision of virtual healthcare services. These technologies offer options to meet all patients' needs during pandemics, emergencies and particularly those inaccessible areas, with limited mobility or financially constrained time. While it is not feasible to predict the occurrence of infectious pandemics in the future, history has taught us the need for developing adaptive, intelligent, robust and feasible digital health technologies to prepare for the unexpected public health crisis. The advantage of having emerging technologies integrated into the health system will have a critical role in emergency responses. They can be rapidly deployed facilitating forward triaging and ameliorating pressures on health facilities and frontline workers while ensuring the continued provision of routine care for the general population while the limited risk of transmission of communicable diseases.

REFERENCES

1. Sezgin E, Huang Y, Ramtekkar U, Lin S. Readiness for voice assistants to support healthcare delivery during a health crisis and pandemic. *Npj Digit Med* 2020;3(1):1–4. https://doi.org/10.1038/s41746-020-00332-0.

2. Worobey M, Han GZ, Rambaut A. Genesis and pathogenesis of the 1918 pandemic H1N1 influenza a virus. *Proc Natl Acad Sci U S A* 2014;111:8107–12. https://doi.org/10.1073/PNAS.1324197111/SUPPL_FILE/PNAS.1324197111.SAPP.PDF.

3. Chamola V, Hassija V, Gupta V, Guizani M. A comprehensive review of the COVID-19 pandemic and the role of IoT, Drones, AI, Blockchain, and 5G in managing its impact. *IEEE Access* 2020;8:90225–65. https://doi.org/10.1109/ACCESS.2020.2992341.

4. Ankomah AA, Moa A, Chughtai AA. The long road of pandemic vaccine development to rollout: A systematic review on the lessons learnt from the 2009 H1N1 influenza pandemic. *Am J Infect Control* 2022. https://doi.org/10.1016/J.AJIC.2022.01.026.

5. Rana A, Malik S. A review of computational intelligence technologies for tackling Covid-19 pandemic. *Internet of Things* 2021:223–42. https://doi.org/10.1007/978-3-030-75220-0_11/COVER/.

6. Chandra M, Kumar K, Thakur P, Chattopadhyaya S, Alam F, Kumar S. Digital technologies, healthcare and Covid-19: Insights from developing and emerging nations. *Health Technol (Berl)* 2022;12:547–68. https://doi.org/10.1007/S12553-022-00650-1/TABLES/4.

7. Ghimire A, Thapa S, Jha AK, Kumar A, Kumar A, Adhikari S. AI and IoT solutions for tackling COVID-19 pandemic. *Proc 4th Int Conf Electron Commun Aerosp Technol ICECA 2020*, 2020, pp. 1083–1092. https://doi.org/10.1109/ICECA49313.2020.9297454.

8. Khan H, Kushwah KK, Singh S, Urkude H, Maurya MR, Sadasivuni KK. Smart technologies driven approaches to tackle COVID-19 pandemic: A review. *3 Biotech* 2021;11:1–22. https://doi.org/10.1007/S13205-020-02581-Y/FIGURES/7.

9. Arora N, Banerjee AK, Narasu ML. The role of artificial intelligence in tackling COVID-19. 2020;15:717–24. https://doi.org/10.2217/FVL-2020-0130.

10. Ye J. The role of health technology and informatics in a global public health emergency: Practices and implications from the COVID-19 pandemic. *JMIR Med Inf* 2020;8(-7):e19866. https://doi.org/10.2196/19866.

11. Alharbi A, Md, Rahman A. Review of recent technologies for tackling COVID-19. *SN Comput Sci* 2021;2(6):1–27. https://doi.org/10.1007/S42979-021-00841-Z.

12. Khan SAR, Yu Z, Umar M, Lopes de Sousa Jabbour AB, Mor RS. Tackling post-pandemic challenges with digital technologies: An empirical study. *J Enterp Inf Manag* 2022;35:36–57. https://doi.org/10.1108/JEIM-01-2021-0040/FULL/PDF.

13. Shamseer L, Moher D, Clarke M, Ghersi D, Liberati A, Petticrew M, et al. Preferred reporting items for systematic review and meta-analysis protocols (PRISMA-P) 2015: Elaboration and explanation. *BMJ* 2015;349. https://doi.org/10.1136/BMJ.G7647.

14. Vaishya R, Haleem A, Vaish A, Javaid M. Emerging technologies to combat the COVID-19 pandemic. *J Clin Exp Hepatol* 2020;10:409–11. https://doi.org/10.1016/J.JCEH.2020.04.019.

15. Lalmuanawma S, Hussain J, Chhakchhuak L. Applications of machine learning and artificial intelligence for Covid-19 (SARS-CoV-2) pandemic: A review. *Chaos, Solit Fractals* 2020;139:110059. https://doi.org/10.1016/J.CHAOS.2020.110059.

16. Swayamsiddha S, Prashant K, Shaw D, Mohanty C. The prospective of artificial intelligence in COVID-19 pandemic. *Health Technol (Berl)* 2021;11:1311–20. https://doi.org/10.1007/S12553-021-00601-2/TABLES/1.

17. Shamman AH, Hadi AA, Ramul AR, Abdul Zahra MM, Gheni HM. The artificial intelligence (AI) role for tackling against COVID-19 pandemic. *Mater Today Proc* 2021. https://doi.org/10.1016/J.MATPR.2021.07.357.

18. Singh RP, Javaid M, Haleem A, Suman R. Internet of things (IoT) applications to fight against COVID-19 pandemic. *Diabetes Metab Syndr Clin Res Rev* 2020;14:521–4. https://doi.org/10.1016/J.DSX.2020.04.041.

19. Wang K, Xie S, Rodrigues J. Medical data security of wearable tele-rehabilitation under internet of things. *Internet Things Cyber-Physical Syst* 2022;2:1–11. https://doi.org/10.1016/J.IOTCPS.2022.02.001.

20. Swayamsiddha S, Mohanty C. Application of cognitive internet of medical things for COVID-19 pandemic. *Diabetes Metab Syndr Clin Res Rev* 2020;14:911–5. https://doi.org/10.1016/J.DSX.2020.06.014.

21. Nittari G, Khuman R, Baldoni S, Pallotta G, Battineni G, Sirignano A, et al. Telemedicine practice: Review of the current ethical and legal challenges. *Telemed e-Health* 2020;26:1427–37. https://doi.org/10.1089/TMJ.2019.0158/ASSET/IMAGES/LARGE/TMJ.2019.0158_FIGURE1.JPEG.

22. Mayoka KG, Rwashana AS, Mbarika VW, Isabalija S. A framework for designing sustainable telemedicine information systems in developing countries. *J Syst Inf Technol* 2012;14:200–19. https://doi.org/10.1108/13287261211255329/FULL/PDF.

23. Adepoju P. Africa turns to telemedicine to close mental health gap. *Lancet Digit Heal* 2020;2:e571–2. https://doi.org/10.1016/s2589-7500(20)30252-1.

24. Garattini L, Badinella Martini M, Zanetti M. More room for telemedicine after COVID-19: Lessons for primary care? *Eur J Heal Econ* 2020;22(2):183–6. https://doi.org/10.1007/S10198-020-01248-Y.

25. Mbunge E, Muchemwa B, Batani J. Are we there yet? Unbundling the potential adoption and integration of telemedicine to improve virtual healthcare services in African health systems. *Sensors Int* 2022;3:100152. https://doi.org/10.1016/J.SINTL.2021.100152.

26. Chitungo I, Mhango M, Mbunge E, Dzobo M, Musuka G, Dzinamarira T. Utility of telemedicine in sub-Saharan Africa during the COVID-19 pandemic. A rapid review. *Hum Behav Emerg Technol* 2021. https://doi.org/10.1002/HBE2.297.

27. Portnoy J, Waller M, Elliott T. Telemedicine in the era of COVID-19. *J Allergy Clin Immunol Pract* 2020;8:1489–91. https://doi.org/10.1016/J.JAIP.2020.03.008.

28. Pappot N, Taarnhøj GA, Pappot H. Telemedicine and e-health solutions for COVID-19: Patients' perspective. 2020;26:847–9. https://doi.org/10.1089/TMJ.2020.0099.

29. Bitar H, Alismail S. The role of eHealth, telehealth, and telemedicine for chronic disease patients during COVID-19 pandemic: A rapid systematic review. *Digit Heal* 2021;7. https://doi.org/10.1177/20552076211009396.

30. Ferenczi G (Gyoergy), Mahmood AN, Bergmann RK. Telemedicine pre and post Covid-19: Lessons for commercialisation based on previous use cases. *J Int Soc Telemed EHealth* 2020;8:e8 (1–10). https://doi.org/10.29086/JISFTEH.8.E8.

31. Elsersy M, Sherif A, Darwsih A, Hassanien AE. Digital transformation and emerging technologies for tackling COVID-19 pandemic. *Stud Syst Decis Control* 2021;322:3–19. https://doi.org/10.1007/978-3-030-63307-3_1/COVER/.

32. Siriwardhana Y, De Alwis C, Gur G, Ylianttila M, Liyanage M. The fight against the COVID-19 pandemic with 5G technologies. *IEEE Eng Manag Rev* 2020;48:72–84. https://doi.org/10.1109/EMR.2020.3017451.

33. Siriwardhana Y, Gür G, Ylianttila M, Liyanage M. The role of 5G for digital healthcare against COVID-19 pandemic: Opportunities and challenges. *ICT Express* 2021;7:244–52. https://doi.org/10.1016/J.ICTE.2020.10.002.

34. Alshammari N, Sarker MNI, Kamruzzaman MM, Alruwaili M, Alanazi SA, Raihan ML, et al. Technology-driven 5G enabled e-healthcare system during COVID-19 pandemic. IET Commun 2022;16:449–63. https://doi.org/10.1049/CMU2.12240.

35. Manavi SY, Nekkanti V, Choudhary RS, Jayapandian N. Review on emerging internet of things technologies to fight the COVID-19. *Proc -2020 5th Int Conf Res Comput Intell Commun Networks, ICRCICN 2020*, 2020, pp. 202–208 https://doi.org/10.1109/ICRCICN50933.2020.9296161.

36. Mousavi SH, Zahid SU, Wardak K, Azimi KA, Hosseini SMR, Wafaee M, et al. Mapping the changes on incidence, case fatality rates and recovery proportion of COVID-19 in Afghanistan using geographical information systems. *Arch Med Res* 2020;51:600. https://doi.org/10.1016/J.ARCMED.2020.06.010.

37. Adeyelure TS. Geographical information system readiness assessment framework in the South African private health sector for pre- to post-COVID-19 pandemic management. *Lect Notes Networks Syst* 2022;217:9–14. https://doi.org/10.1007/978-981-16-2102-4_2/COVER/.

38. Arab-Mazar Z, Sah R, Rabaan AA, Dhama K, Rodriguez-Morales AJ. Mapping the incidence of the COVID-19 hotspot in Iran – Implications for travellers. *Travel Med Infect Dis* 2020;34:101630. https://doi.org/10.1016/J.TMAID.2020.101630.

39. Zhou C, Su F, Pei T, Zhang A, Du Y, Luo B, et al. COVID-19: Challenges to GIS with big data. *Geogr Sustain* 2020;1:77–87. https://doi.org/10.1016/J.GEOSUS.2020.03.005.

40. Sun W, Cai Z, Li Y, Liu F, Fang S, Wang G. Security and privacy in the medical internet of things: A review. *Secur Commun Networks* 2018;2018:1–9. https://doi.org/10.1155/2018/5978636.

41. Mbunge E, Muchemwa B, Jiyane S, Batani J. Sensors and healthcare 5.0: Transformative shift in virtual care through emerging digital health technologies. *Glob Heal J* 2021. https://doi.org/10.1016/J.GLOHJ.2021.11.008.

42. Gaobotse G, Mbunge E, Batani J, Muchemwa B. Non-invasive smart implants in healthcare: Redefining healthcare services delivery through sensors and emerging digital health technologies. *Sensors Int* 2022;3:100156. https://doi.org/10.1016/J.SINTL.2022.100156.

43. Tuli S, Tuli S, Wander G, Wander P, Gill SS, Dustdar S, et al. Next generation technologies for smart healthcare: Challenges, vision, model, trends and future directions. *Internet Technol Lett* 2020;3:e145. https://doi.org/10.1002/itl2.145.

44. Singh RP, Haleem A, Javaid M, Kataria R, Singhal S. Cloud computing in solving problems of COVID-19 pandemic. *J Ind Integr Manag* 2021;06:209–19. https://doi.org/10.1142/S2424862221500044.

45. Abubakar AI, Omeke KG, Ozturk M, Hussain S, Imran MA. The role of artificial intelligence driven 5G networks in COVID-19 outbreak: Opportunities, challenges, and

46. Franch-Pardo I, Desjardins MR, Barea-Navarro I, Cerdà A. A review of GIS methodologies to analyze the dynamics of COVID-19 in the second half of 2020. *Trans GIS* 2021;25:2191–239. https://doi.org/10.1111/tgis.12792.

47. Mollalo A, Vahedi B, Rivera KM. GIS-based spatial modeling of COVID-19 incidence rate in the continental United States. *Sci Total Environ* 2020;728:138884. https://doi.org/10.1016/j.scitotenv.2020.138884.

48. Yahya BM, Yahya FS, Thannoun RG. COVID-19 prediction analysis using artificial intelligence procedures and GIS spatial analyst: A case study for Iraq. *Appl Geomatics* 2021;13:481–91. https://doi.org/10.1007/s12518-021-00365-4.

49. Mbunge E, Dzinamarira T, Fashoto SG, Batani J. Emerging technologies and COVID-19 digital vaccination certificates and passports. *Public Heal Pract* 2021;2:100136. https://doi.org/10.1016/j.puhip.2021.100136.

50. Mbunge E, Fashoto SG, Batani J. COVID-19 digital vaccination certificates and digital technologies: Lessons from digital contact tracing apps. *SSRN Electron J* 2021. https://doi.org/10.2139/ssrn.3805803.

51. Poppas A, Rumsfeld JS, Wessler JD. Telehealth is having a moment: Will it last? *J Am Coll Cardiol* 2020;75:2989–91. https://doi.org/10.1016/J.JACC.2020.05.002.

52. Sims JM. Communities of practice: Telemedicine and online medical communities. *Technol Forecast Soc Change* 2018;126:53–63. https://doi.org/10.1016/J.TECHFORE.2016.08.030.

53. Dandachi D, Dang BN, Lucari B, Teti M, Giordano TP. Exploring the attitude of patients with HIV about using telehealth for HIV care. 2020;34:166–72. https://doi.org/10.1089/APC.2019.0261.

54. Wong ZSY, Rigby M. Identifying and addressing digital health risks associated with emergency pandemic response: Problem identification, scoping review, and directions toward evidence-based evaluation. *Int J Med Inform* 2022;157:104639. https://doi.org/10.1016/J.IJMEDINF.2021.104639.

55. Batani J, Dzambo S, Magodi I. Household power optimisation and monitoring system. *Int J Comp. Sci. Bus. Inform.* 2017;17:23–37.

56. Bhaskar S, Nurtazina A, Mittoo S, Banach M, Weissert R. Editorial: Telemedicine during and beyond COVID-19. *Front Public Heal* 2021;0:233. https://doi.org/10.3389/FPUBH.2021.662617.

57. Mbunge E, Fashoto SG, Akinnuwesi B, Metfula A, Simelane S, Ndumiso N. Ethics for integrating emerging technologies to contain COVID-19 in Zimbabwe. *Hum Behav Emerg Technol* 2021. https://doi.org/10.1002/HBE2.277.

58. Furusa SS, Coleman A. Factors influencing e-health implementation by medical doctors in public hospitals in Zimbabwe. *SA J Inf Manag* 2018;20. https://doi.org/10.4102/SAJIM.V20I1.928.

59. Panchal M, Singh S, Rodriguez-Villegas E. Analysis of the factors affecting the adoption and compliance of the NHS COVID-19 mobile application. *MedRxiv* 2021:2021.03.04.21252924. https://doi.org/10.1101/2021.03.04.21252924.

15 Emerging Technologies in Age-Related Therapies

Moe Elbadawi, Corinna Schlosser, Manal Eid Alkahtani,
Simon Gaisford, Abdul W. Basit, and Mine Orlu
UCL School of Pharmacy, University College London

CONTENTS

15.1 CHALLENGES IN GERIATRIC MEDICINES

Recent advances in medicine and public health have increased life expectancy and changed the age distribution of the world's population [1–6]. As such, geriatric medicine is increasingly becoming a field of topical importance, presenting with a unique set of challenges for the pharmaceutical industry to address [7–10]. With advancing age, the acquisition and accumulation of chronic diseases become more likely [11]. Indeed, the majority of people aged 65 or over are diagnosed with multiple diseases requiring treatment with several different medication simultaneously [12–14]. Thus, multimorbidity, which describes the co-existence of two or more chronic health conditions, increases the complexity of therapeutic treatment and is likely to have a negative health outcome [15]. Further, the global health burden associated with multimorbidity is expected to increase significantly as a consequence of the growing number of older people [15].

Multimorbidity consequently leads to patients requiring multiple medications. Polypharmacy describes the use of multiple pharmaceutical treatments [15]. It is evident that multimorbidity is also closely related to polypharmacy as patients often require one or more drugs for the treatment of each condition [15]. Further, polypharmacy is associated with adverse outcomes including mortality, falls, adverse drug reactions, increased length of stay in hospital and readmission to hospital soon after discharge [15]. This is of particular concern to older adults because they may have unique barriers and challenges associated with

the multidrug treatment, including cognitive impairment, functional limitations, or multiple healthcare providers to only name a few [16].

It has been found that patient's subjected to polypharmacy often find it challenging to adhere to their medication, with approximately 50% of older adults not adhering to at least one of their chronic medications [12,16]. However, the adherence of a patient to their drug regimen is crucial for the successful therapeutic outcome [12]. There are different mechanisms by which polypharmacy can lead to non-adherence. The most simple is that the greater number of medicines increases the chances of a patient forgetting to take one or more pills on a daily basis [16]. Additionally, the complexity of the dosing regimen (e.g. frequency), routes of administrations and specific dosing instructions increase with the number of drugs to take [16], while physical constraints, such as impaired vision, poor handgrip strength, loss of fine motor skills and dysphagia, can hinder the intake of medicine; these factors all worsen with age [17], and there is evidence that decreasing the regimen complexity would lead to improved adherence [16].

Fixed-dose combinations (FDC) are drug products which combine at least two active pharmaceutical ingredients in a single dosage form at a fixed-dose ratio [12,18–21]. Combining drugs with similar treatment regimens into a single dose facilitates the treatment plan, improves patient compliance and provides a better therapeutic outcome [12,18]. Simplification of the treatment regimen by FDC might be one solution to improve patient's acceptability and adherence. A meta-analysis from 2007 provides evidence that FDCs can reduce the risk of non-adherence by 26%

[22]. Further, combining different (classes of) therapeutics allows for a reduction in individual doses, thus reducing side effects, while maintaining or even improving the therapeutic outcome [18]. Existing fixed-dose preparation methods include among others tableting, hot-melt co-extrusion, spray-drying and 3D printing [18,23–26]. Current FDCs may help patients to manage on multiple prescriptions; however, they do not always provide desired release characteristics for all drugs within the FDC nor allow for dosing flexibility of the individual components [18,27].

The remaining chapter details examples of emerging technologies that are set to transform the pharmaceutical industry, with a focus on their application in older patients. The chapter concludes with examples of how these technologies have been merged with one another to either accelerate or enhance developments.

15.2   ARTIFICIAL INTELLIGENCE

Artificial intelligence (AI) is a collection of technologies programmed to artificially replicate human tasks, including critical thinking and physical movement [28]. The recent uptake of AI tools is a result of recent progress made in computers and other processors. Training machines to think like humans is not cheap but fortunately what were once considered high-performing computers are now much more accessible and can meet the computational demands to develop AI applications. In fact, of late, AI has demonstrated hyper-intelligence in outperforming humans in tasks, including clinicians in diagnostic tests, and of late has garnered interest in pharmaceutics.

The main technologies in AI are machine learning (ML), natural language processing (NLP) and machine vision (MV), where the combination of at least two results in AI. ML is the ability of machines to establish a pattern in the data being fed [29]. The main applications for ML are prediction and knowledge insight. At the core of ML are statistical-learning algorithms that look to find a correlation between the explanatory variables to the response variables. However, ML is a pipeline, encompassing several stages that involve data pre-processing, model training (where the statistical-learning algorithm is trained) and model evaluation, where the latter is used to assess how well the machine has learnt. The advantages provided by ML over other predictive tools are that it can handle high-dimensional data and large datasets and is compatible with different data formats, including numbers, texts and images. Moreover, ML encompasses learning strategies for both linear and non-linear events, which allows it to learn the most complex of relationships.

ML has been explored for polypharmacy to improve clinical precision. The innumerable combination of medications and other variables is difficult to anticipate with simple observational data and mathematical modelling [30,31]. Hence, ML has been pursued to comprehend the level of complexity. One challenge in this domain is to identify potential drug adverse effects. Fortunately, ML

was found to have a 92% accuracy for identifying medication errors and adverse effects, which was an improvement over conventional rule-based methods [30]. Besides being applied at the clinical level, ML has been investigated to interpret the cause of drug adverse reaction, primarily at the research level, with a view to improving drug development and preventing adverse reactions. Here, ML is paired with cheminformatics, a merger of physical chemistry and informatics, to help identify patterns in drug chemical structure that lead to, for example, drug-drug interaction. The majority of research has successfully identified chemical features that cause drug-drug interaction [32,33], which has inspired research to leverage ML for predicting other drug interactions, such as drug-food and drug-microbiome interactions, to facilitate drug development [34–36].

NLP is another commonly used AI subset technology, where machines are programmed to understand human languages, thereby allowing interactions between humans and machines. NLP is complementary to ML as the majority of information is "hidden" in text, such as electronic health records. The focus here is information retrieval, information extraction, text summarisation and topic modelling [37]. Common applications include predictive text, speech recognition and language translation. Similar to ML, NLP can be used in developing predictive tools or extracting knowledge from a corpus of text and also encompasses a pipeline of stages that involve data pre-processing model training and evaluation. For example, NLP has been used to comprehend abstracts from scientific literature to recognise adverse drug events both at the development and clinical stages [38,39]. These findings demonstrate that NLP can act as an early warning system to unknown adverse events [40,41]. Moreover, NLP has found a unique application in omics, which has seen a recent explosion in data generation. Omics is a collection of disciplines that study the structure and function of biological molecules, such as genes (genomics), proteins (proteomics) and metabolites (metabolomics) [42]. In particular, the data in genomics and proteomics are presented in the form of texts, such as nucleotide sequences (e.g., AAGT) and amino acid sequences (e.g. Ala-Leu-His), where NLP has been used to identify patterns between sequences and function. The majority of pharmaceutics relies heavily on articulating observations as text, and hence NLP possesses potential to further elucidate as well as automate decision making. However, one limitation is the lack of structured corpora that can be seamlessly inputted into NLP models.

MV seeks to replicate human sight, where machines are trained to recognise visual cues. Unsurprisingly, MV also involves a similar pipeline to ML and NLP, where the data is first pre-processed to make it machine-compatible, and then model training is performed. The majority of MV applications in pharmaceutics have been directed towards manufacturing, where MV can recognise and organise delivery systems by a designated category, as well as facilitating other aspects of the pharmaceutical pipeline [43–47]. MV has also been demonstrated to help patients in recognising the correct drug among multiple prescribed medicines, which is discussed in Section 15.1.3.2. It is worth acknowledging that while NLP and MV can be used to recognise text or visual cues, respectively, as trained by the user, integrating ML into either subset technologies allows the machine to extrapolate its recognition into events that the machine has not been trained on. Overall, the aforementioned subset technologies of AI show promise in identifying DDI to help minimise adverse effects.

15.3 EMERGING MANUFACTURING TECHNOLOGIES

Emerging manufacturing technologies are one of the physical entities of digital pharmaceutics that offer precision and on-demand release, and other sought-after features, for healthcare for the elderly [48–52]. The most prominent is additive manufacturing (AM), also referred to as three-dimensional (3D) printing. AM, as the name suggests, manufactures products in an additive manner, layer-by-layer, until the product has been obtained. The main advantages of AM are fast production time, the ability to produce personalised and complex geometries, and high digital resolution for precision fabrication [48]. Furthermore, most, if not all, AM technologies are digitalised, where the pipeline involves digitally designing the product on a computer-aided software, uploading the design to the AM machine and selecting the manufacturing parameters in software. These digitalised features of AM allow it to be paired with other digital technologies, such as medical imaging, and make it possible to integrate manufacturing into the world of Internet of Things (IoT). Extensive research has been undertaken in applying AM to the medical sector, with

applications such as prosthetics, medicines and tissue engineering. Thus, understandably, AM has been heavily implicated in the next industrial revolution, Industry 4.0 [53].

The flexibility of AM to produce different and complex geometries has been exploited to develop a variety of dosage forms, including oro-dispersible films [54,55]. AM has also been successfully demonstrated to produce tablets with braille to help patients with visual impairments recognise the different tablets, again leveraging the flexibility of seamlessly designing different geometries [56]. Moreover, the digital precision of AM has allowed for multiple drugs to be incorporated with high spatial control and temporal drug release into one dosage form – referred to as a "Polypill" [57]. Therefore, despite being a nascent fabrication technology, these early applications of AM highlight the potential of the technology to address existing challenges in geriatric medicine.

15.3.1 ELECTROHYDRODYNAMIC TECHNIQUES

Aside from AM, electrohydrodynamic (EHD) techniques offer solutions to improving medicines for the elderly. Electrospinning and spraying techniques, commonly referred to as electrohydrodynamic techniques, have attracted huge interest in recent years. This is exemplified by performing a search on SciFinder for articles with the topics "electrospinning" and "drug delivery". In 2002, only one published article met these criteria; in 2010, just over 50; and in 2020, over 130 such papers can be found in the literature. The technology is widely applied in drug delivery system for poorly soluble small molecules, biopharmaceuticals (e.g. proteins, DNA, siRNA), cells, tissue scaffolds, and for wound healing [27,58–61]. Table 15.1 enumerates

TABLE 15.1
Examples of EHD Technologies

Product (Manufacturer)	Details	Application	Reference
SpinCare™ (Nanomedic Technologies Ltd.)	Personalised electrospun wound dressing	Portable bedside wound device	[65]
PK Papyrus (Biotronik AG)	Electrospun coating of stent	Covered coronary stent system	[65,66]
ReBOSSIS (Ortho ReBirth Co. Ltd.)	Synthetic cotton-like bone-void/defect filling material consisting of beta-TCP (beta-tricalcium phosphate), bioabsorbable polymer and SiV (silicon containing calcium carbonate)	Bone-void filler	[65,67]
Rivelin® patch (Bionicia and Dermtreat A/S)	Multi-component product designed for unidirectional delivery of a pharmaceutical drug to a mucosal surface	Mucoadhesive drug delivery patch	[65,67]
SurgiCLOT® (St. Theresa Medical, Inc.)	Electrospun dextran fibres to deliver thrombin and fibrinogen	Wound dressing	[65,66]
PATHON	Polyurethan fibres for nitric oxide release	Diabetic foot ulcers and cutaneous leishmaniasis ulcers	[66]
ReDura™ (Medprin Biotech GmbH)	Poly-l-lactic acid fibres	Dural substitute patch	[65]
HealSmart™ (PolyRemedy®, Inc).	Composed of microfibre polymers	Personalised wound care system	[67]
	Dressing composition is adjusted based on the wound's functional needs		[65]
TPP-fibres	Polyurethan/PEG fibres containing tetraphenyporphyrin (TPP)	Chronic leg ulcers	[66]
Bioweb™ (Zeus Industrial Products, Inc.)	PTFE and PU-based nanofibers	Vascular stent covering	[65,67]

examples of notable EHD technologies, some of which have successfully been translated into clinical application or have received FDA approval, demonstrating the potential of EHD for addressing unmet needs [62–64].

Electrospinning is a simple and relatively low-cost processing technique to generate nano to micrometre-sized fibres providing a high level of control over the composition and structure of the fibres produced [58,68]. The process is applicable to a wide range of polymers, which can be loaded with a multitude of compounds such as drugs or pro-drugs [58]. Liquid samples can be electrospun as aqueous solution, organic solution, organic-aqueous solution, polymer melts, nano-emulsion, suspension and nanosuspensions [69]. Fibres are obtained by applying a strong electric field to a pumped solution producing charged liquid jets which narrow and allow for solvent evaporation while travelling towards a grounded collector [60,70,71]. The process generally yields fibres with a narrow size distribution and high API encapsulation efficiency [68]. Melt electrospinning (MES) compared with solution electrospinning (SES) has the advantage of avoiding solvents, thus being a safer and greener method [72–74]. Additionally, fibres produced by MES can provide even faster release profiles compared with SES-produced fibres [72,75]. Comparison of Carvedilol-Eudragit E and Carvedilol-PVP64 fibres prepared by MES and SES show that despite the MES fibres being one to two orders of magnitude larger, they result in faster dissolution [72,75]. MES fibres are collected loosely and thus provide a higher porosity compared with the tightly packed SES fibres, hence providing better accessibility to the dissolution media [75].

Electrospun fibres show great potential for use as an orally dissolving immediate-release dosage form system owing to their high surface area, high porosity and ability to encapsulate high drug loadings [76,77]. Williams et al. [58] suggest that their most promising uses would be for topical delivery as well as for fast-dissolving oral drug delivery systems. Fibres offer additional benefits such as improved wettability, mechanical flexibility and ease of handling, tuneable diameter and structure, as well as functionalisation [78,79]. Thus, the fibres being at the nanoscale but also forming macroscale webs offer simultaneously the advantages of nanotechnologies and those of conventional solid-dosage forms (e.g. easy processing, good API stability, ease of packaging and shipping) [80].

One application where EHD outperforms other fabrication technologies is in fast-dissolving drug delivery systems. Oral administration of therapeutics is commonly the most preferred route as it presents the advantage of high patient compliance and allows self-administration [81]. However, older people may struggle with oral dosage forms due to swallowing difficulties. This is highlighted by a study aiming to identify practical problems faced by older people in their daily use of medication, where 29% of the participants reported they had problems with swallowing certain tablets due to their size [17]. To address this issue, oro-dispersible dosage forms have been developed [82]. Oro-dispersible drug formulations are innovative dosage forms which rapidly disintegrate in the oral cavity forming a fine suspension or solution of the active pharmaceutical ingredient (API) in the saliva generally providing rapid onset of action [77,78,82,83]. Thus, such dosage forms allow ease in the administration of oral medicine and make their use more convenient while maintaining the benefits of solid-dosage forms [82,83]. These formulations do not require intake of liquid and are generally considered first choice in patients with swallowing difficulties or dysphagia which is common in geriatric patients [12,83]. The convenience and ease of use are expected to increase patient's acceptance and treatment adherence [82,83]. These systems exist in the form of granules, tablets, films, wafers and buccal or sublingual films [77,83].

Electrospun fibres have been used to overcome drug hydrophobicity while also stabilising APIs in the amorphous state producing so-called amorphous solid dispersions, which is a common formulation approach [58]. Hydrophobic compounds solubilise more readily when formulated into solid dispersions with highly hydrophilic compounds [58]. In the electrospinning process, the API and polymer are prepared in a single solvent and are thus distributed on a molecular scale in an amorphous state [58,79]. Due to the lack of any lattice energy, the API's free energy is increased when compared with that of the crystalline form, which will result in the improvement of thermodynamic solubility and augment the driving force for dissolution [76]. The fibre's nanostructrues provide a high surface area and thus enhance the dissolution rate of poorly soluble drugs allowing the disintegration of nanofibers within seconds and total drug release within minutes [69,76,84]. During production of the formulation, the polymer should be adapted to the APIs allowing miscibility of the components while providing adequate solubility, glass transition temperature, hygroscopicity and the possibility to form intermolecular interactions [79]. These properties greatly affect production, stability and dissolution [79] (Figure 15.1).

The potential of EHD further extends to addressing the challenges associated with FDCs. Coaxial electrospinning is an approach where the spinneret is adapted to accommodate two or more fluids concentric to each other producing core-shell fibres [58]. Electrohydrodynamic techniques allow multiple fluids to be processed simultaneously, leading to fibre or particle products with highly tuneable properties [85]. The coaxial/multi-axial set up allows for additional control over composition and properties of the material. Indeed, the polymer-drug composition within the core and shell (bi-axial) or the different layers (multi-axial) of the fibres/particles can be modulated [58,59]. For these reasons, EHD techniques have received considerable attention in the production of multi-layered drug delivery systems not only allowing targeted drug delivery but also modulation of the release profile and orderly release of the APIs [27,59,68]. The technique offers the possibility to encapsulate more than one active pharmaceutical ingredient

(a)

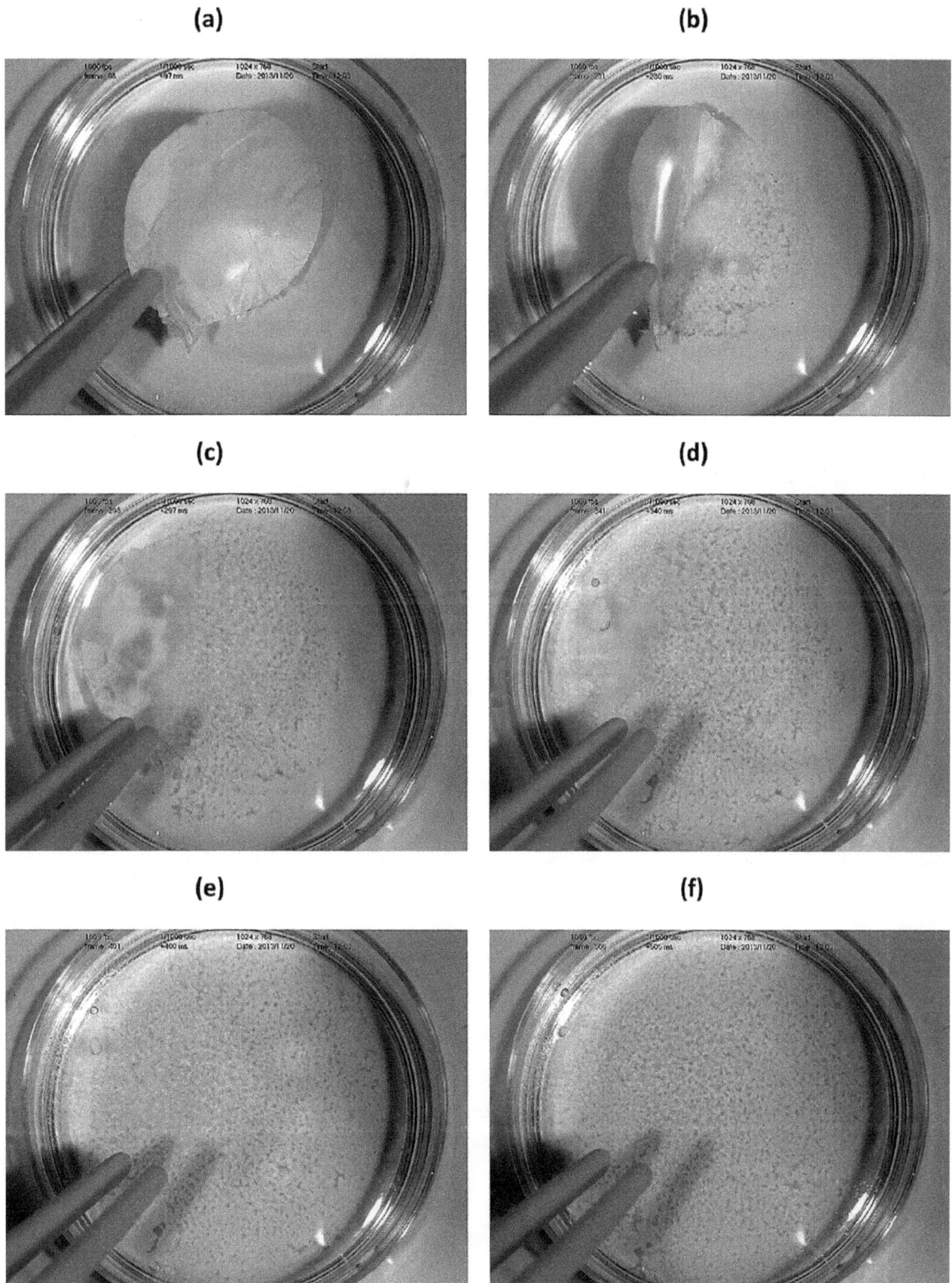

(b)

(c)

(d)

(e)

(f)

FIGURE 15.1 High-speed camera images of the disintegration of paracetamol-caffeine loaded fibres. Images were taken at (a) 0 ms; (b) 133 ms; (c) 200 ms; (d) 243 ms; (e) 303 ms; and (f) 408 ms after the fibres were added to simulated saliva. (Reproduced with permission [77]. Copyright 2014, Elsevier B.V.)

making it an interesting technique for the preparation of fixed-dose combinations [77]. An additional advantage of separating APIs in different layers could be to overcome unintended interactions among them [59]. Incorporation of bitter-blocking agents into the formulation, thus improving the taste averseness associated with certain APIs, is also achievable [77].

Moreover, the application of co- and multi-axial EHD processing can overcome drug-drug interaction and drug instability. Cardiovascular disease is the leading cause of death worldwide which risk can be reduced significantly by adopting a healthy lifestyle and pharmacological treatment [86,87]. Pharmaceutical treatment often consists of one or a combination of classes of therapeutic agents such as antihypertensive drugs (e.g. beta blocker, diuretics and angiotensin-converting enzyme inhibitors), lipid-lowering drugs (statins) and anti-platelet drugs [86]. A combination of therapeutic agents and their dosing regimens can rapidly increase the daily number of pharmaceutical formulations a patient is treated with and simultaneously impact treatment adherence [86]. Several electrospun FDCs for the treatment of cardiovascular disease have been prepared in the literature, e.g., combination of the diuretic spironolactone and calcium channel blocker nifedipine or association of calcium channel blocker amlodipine besylate and angiotensin-converting enzyme inhibitor valsartan [12,87].

In particular, Smeets et al. have prepared an electrospun FDC which overcomes not only physicochemical drug-drug interactions and solubility issues but also achieves the required release profile [86]. The FDC was composed of three APIs, namely atenolol (beta blocker), acetyl salicylic acid (anti-platelet) and lovastatin (lipid-lowering compound). The challenges for this formulation are all associated with lovastatin. Lovastatin (LOV) is a poorly soluble prodrug which is hydrolysed *in vivo* into its active form. However, premature acid catalysed hydrolysis needs to be prevented. Thus, the system will require separation of lovastatin and acetyl salicylic acid (ASA) as well as an enteric coating avoiding drug release in the gastric media. Additionally, it would be favourable to delay its release until the ileum as LOV is strongly metabolised. Core-shell particles were produced by electrospraying where the core contained lovastatin and atenolol and which were coated with the gastro-resistant polymer Eudragit S100 containing ASA. Soluplus was added to the core of the particles to obtain high supersaturation of lovastatin and thus helping with solubility issues. Figure 15.2 illustrates the composition of the different particle layers and the in vitro release profiles of each API obtained with this particle formulation.

Release studies in acidic media show rapid release of ASA and atenolol with 80% and 70% of the drug released after 90 min respectively. However, little to none of the LOV was released. After 90 min, the pH was increased to 7.5 which resulted in the release of LOV (~80% after 3.5 h) as well as a full release of ASA and atenolol (ATE). Despite the surprisingly rapid release of ASA and ATE in acidic media – which was attributed to their aqueous solubility – electrospraying allowed tight control over the composition of the different compartment and addressed all formulation challenges.

15.3.2 MERGING AM WITH EHD

No technology is perfect, and the limitation of one, in some cases, is the advantage of the other; therefore, combining technologies will enhance the properties of the final product as well as mitigating some shortfalls during processing. In the scope of AM and EHD methods, some limitations could be overcome by using hybrid systems. For example, AM develops structures with superior mechanical characteristics that provide the required support for nanofibers/nanoparticles. Several combination strategies were explored

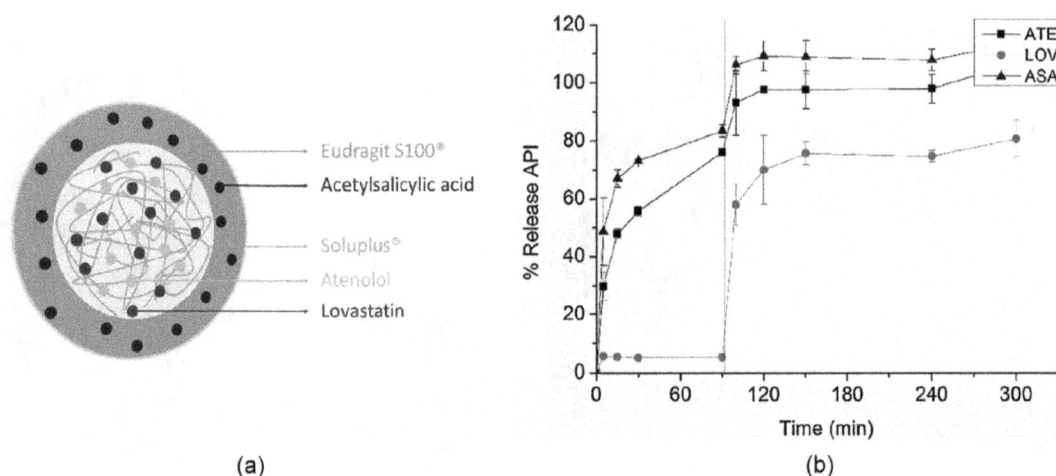

(a)

(b)

FIGURE 15.2 (a) Design of the coated ASD particles. The core exists of Soluplus© (SOL), atenolol (ATE) and lovastatine (LOV), while the coating layer exists of Eudragit S100 © and acetylsalicylic acid (ASA). (b) Dissolution profile of the particles with ATE (squares), LOV (circles) and ASA (triangles). The vertical line indicates the time point of the switch from acidic to neutral medium. (Reproduced with permission [86]. Copyright 2020, Elsevier B.V.)

including electrospinning onto 3D-printed objects such as scaffolds; this allows the production of personalised scaffolds with complex structures and better mechanical properties. However, it is associated with poor distribution of electrospun fibres and lower adhesion on the 3D-printed scaffold. Choi et al. [88] have used this strategy to develop a multi-layer skin substitute for skin wounds, the base consisting of a polycaprolactone (PCL) 3D-printed scaffold and electrospun fibres comprising PCL and keratin. The electrospun fibre's diameter helped in stopping fibroblasts to infiltrate the nanofibrous layer and promoted the formation of distinctive layers of keratinocytes and fibroblasts, while the 3D-printed scaffold provided mechanical support for the skin substitute.

Another strategy for combining AM and EHD processing is using electrospun fibres as inks of AM. This ensures a uniform distribution of nanofibers and better bonding. Electrospun nanofibers containing gelatin/poly lactic-co-glycolic acid (PLGA) were used as bio-ink for an extrusion-based AM to fabricate scaffolds for cartilage regeneration [89]. After electrospinning, the fibre mats were subjected to dehydration and homogenisation processes to yield short fibres, which were dispersed into a viscose solution containing water, hyaluronic acid and polyethylene oxide (PEO) to form the ink.

15.4 WEARABLE TECHNOLOGIES

Another class of emerging technologies anticipated to improve quality of life (QoL) in the elderly are wearable technologies: devices that are worn by users to monitor bodily functions and/or deliver drug. As sensors and actuators are becoming smaller and portable, it is becoming feasible to incorporate such technologies into commonly worn accessories, such as watches that can monitor glucose or heart rhythms in real time. As an alternative to conventional diagnostic tools, such as magnetic resonance imaging (MRI), X-ray or traditional electrocardiogram tests, wearables also provide either non- or minimally invasive diagnosis but with the added benefit of continuous monitoring remote from clinical settings. In other words, wearables provide the opportunity for patients to continue with their daily activities without sacrifice to their health. According to the World Health Organization, approximately 50% of patients with chronic medical conditions follow the prescribed dosing schedule in developed countries [90]. Incidentally, it has come to light that wearable devices encourage patients to take greater interest in their own health, providing an element of empowerment in decision making. In addition to diagnostic, wearable drug delivery platforms have also emerged that provide on-demand and continuous drug release. Prominent examples here include insulin delivery systems.

As an emerging technology, the terminology of wearables is yet to be defined. For sensing capabilities, wearables can monitor body temperature, heart and respiration rhythm, motion and many biological compounds, such as glucose, lactate and oxygen. The sensing modality will vary for the different categories, and while it is outside of the scope of this chapter, readers can find further information in references [91,92]. Briefly, sensors compromise of a transducer to detect/measure the signal; a series of electronics to help process the signal, such as filters and amplifiers; and a communication interface for users to interpret the results. For example, a glucose monitor will have a transducer that facilitates in the detection of glucose, followed by amplifiers and filters to stabilise the detected signal and display it on a monitor. There are additional instrumentation blocks involved, such as data transmission to the relevant clinician, or the signalling of an alarm in case of a drastic signal detected, which will depend on the healthcare application. The following sub-section will detail wearable technologies that are currently being trialled and/or emerging.

15.4.1 WEARABLES FOR MONITORING HEALTH

As mentioned, wearables can be used to monitor a broad range of physical and biological activities. A well-known sensor is the glucose metre that allows diabetics to self-monitor their glucose levels without needing constant attention. Diabetes is becoming one of the most widespread health problems in elderly, with a global prevalence of 123 million in 2017 [93]. Unfortunately, this number is expected to double by 2045. Moreover, elderly patients with the disease inherently possess higher risk of comorbidities, such as cognitive impairment, traumatic fractures and side effects from polypharmacy. Understandably, diabetes in the elderly population is an economic burden. Glucose metres have been widely accessible in the developed world, are portable and easy to use, as well as easy to interpret. One drawback with glucose monitors is that the constant pricking of the finger can be uncomfortable for patients, which could lead to poor patient adherence. Moreover, the patient is constantly required to take measurements, which is challenging for patients with memory loss. Fortunately, wearable glucose monitors have been recently introduced to address these problems that provide continuous glucose monitoring while being minimally invasive. One such platform involves the insertion of a small glucose sensor implanted under the skin, on the back of the upper arm. This platform leverages the ubiquity of smartphones by connecting and providing real-time measurements on the smartphone for users to see. The system records measurements at five-minute intervals and can generate alerts if levels exceed a set threshold. While more costly than traditional glucose metres, the wearable systems are less dependent on compliance and are pain-free.

Aside from wearable glucose monitors, wearables for detecting seizures have also obtained FDA approval. In the US, 1 million adults aged 55 or older have active epilepsy, where a hard fall can have much more severe consequences in comparison to a younger epileptic patient [94]. Thus, wearable devices for detecting seizures can mitigate the effects of the disease [95]. One newly FDA-approved smartwatch automatically contacts carers when it suspects its wearer of having a seizure. Similar to glucose wearables, it also stores data and analyse patient activity to help patients and clinicians to better understand when and why a seizure might come about. In addition to detecting seizures, a wearable smartwatch was also recently cleared to detect falls, using in-build accelerometers and gyroscopes to detect the movement and positioning of an individual. Similarly, the system contacts carers to inform them that the wearer may require assistance.

Wearable devices have evolved to incorporate actuating mechanisms that permit the administration of drug. Again, the diabetes market is leading the field with FDA-approved wearable insulin pumps. For example, Omnipod® provides continuous insulin delivery through a tubeless, waterproof pump. The system can hold up to 200 units of U-100 insulin and offers customisable insulin bolus amounts. The device itself wirelessly connects to smart devices to control and monitor pump activity [96]. Other systems can be paired with wearable glucose monitors to allow for a controlled feedback loop that administers the amount of insulin according to the glucose detected in the body, which consequently achieves a stable glucose value. Such a platform provides its users with independence, but it is worth acknowledging that wearable insulin pumps can be more costly compared with multiple daily injections, which has been noted to prohibit patient uptake of the technology [97].

15.4.2 Wearables for Assisted Living

Wearables can also be employed to help empower patients by allowing them to achieve their daily tasks. One of the best and earliest examples are hearing aids, which allow patients with hearing impairment to engage with their surroundings. Other emerging wearables are assisting patients with chronic visual impairment. For example, patients on multiple medications and who suffer from chronic visual impairment, risk taking the wrong medicine. In addressing this issue, researchers in Taiwan developed a wearable smart-glasses that recognise the medicine. The system comprises an image sensor and a microprocessor that encodes the AI algorithm integrated into sunglasses. It performs the task by providing sound prompts to alert the user if the medicine they are taking is the correct one and if it is being taken at the correct time [98]. Hence, it does not only help patients with visual impairment to adhere to their medical regime but also patients with memory loss. Moreover, taking the wrong medicine may lead to further complications, and hence such technology prevents complications from arising.

15.5 CONCLUDING REMARKS

The challenges related to the monitoring and management of age-related clinical conditions have only recently come to light, as more people are living longer. Understandably, the field remains in its infancy, and there is more to learn.

Fortunately, the advent of AI can help accelerate both discoveries and developments. As discussed in this chapter, AI can interpret large and complex information, and at faster speeds than humans can, with the aim to identify rapidly patterns that would otherwise have been overlooked. Moreover, AI is expected to facilitate pharmaceutical developments, by improving manufacturing processes, including the two discussed herein [99,100].

While AI can establish patterns in data that can provide actionable insight, it is the physical technologies discussed that will enact those decisions. AM offers an unprecedented level of digital precision and complex product design fabrication that can facilitate the development of personalised medicines. EHD, relatively more established than AM, has found application in addressing topical geriatric issues, such as dysphagia and treating wounds. More recently, the two fabrication technologies have been merged together, and their combined benefits are yet to be thoroughly realised. Aside from AM and EHD, emerging materials, referred to as "smart materials", could also improve the development of drug delivery systems [101–103].

Wearables will also play a prominent role in geriatric medicine and will undoubtedly complement the above technologies. AI relies heavily on the quality of the data, which at the moment is intermittent. It is anticipated that the continuous monitoring, whether monitoring patient vital sign or monitoring drug administration, will provide improvements to the quality of the data, and subsequently allow AI to make more informative decisions. Similarly, recent developments have demonstrated that AM can be leveraged to produce wearables with form-fitting features, which can improve wearable compliance [104]. Therefore, it is unsurprising that the emerging technologies discussed in the current chapter have garnered recent attention, and the examples discussed herein highlight their potential in tackling challenges associated with geriatric medicine.

REFERENCES

1. F. Baum, J. Popay, T. Delany-Crowe, T. Freeman, C. Musolino, C. Alvarez-Dardet, V. Ariyaratne, K. Baral, P. Basinga, M. Bassett, D.M. Bishai, M. Chopra, S. Friel, E. Giugliani, H. Hashimoto, J. Macinko, M. McKee, H.T. Nguyen, N. Schaay, O. Solar, S. Thiagarajan, D. Sanders, Punching above their weight: a network to understand broader determinants of increasing life expectancy, *International Journal for Equity in Health*, 17 (2018) 117.
2. S.H. Woolf, H. Schoomaker, Life expectancy and mortality rates in the United States, 1959–2017, *JAMA*, 322 (2019) 1996–2016.
3. H. Chen, L. Hao, C. Yang, B. Yan, Q. Sun, L. Sun, H. Chen, Y. Chen, Understanding the rapid increase in life expectancy in shanghai, China: a population-based retrospective analysis, *BMC Public Health*, 18 (2018) 1–8.
4. E.M. Crimmins, Recent trends and increasing differences in life expectancy present opportunities for multidisciplinary research on aging, *Nature Aging*, 1 (2021) 12–13.
5. C.E. Welsh, F.E. Matthews, C. Jagger, Trends in life expectancy and healthy life years at birth and age 65 in the UK,

2008–2016, and other countries of the EU28: an observational cross-sectional study, *The Lancet Regional Health-Europe*, 2 (2021) 100023.

6. N. Vidra, S. Trias-Llimós, F. Janssen, Impact of obesity on life expectancy among different European countries: secondary analysis of population-level data over the 1975–2012 period, *BMJ Open*, 9 (2019) e028086.

7. R. Patel, B.J. McKinnon, Hearing loss in the elderly, *Clinics in Geriatric Medicine*, 34 (2018) 163–174.

8. K. Iijima, H. Arai, M. Akishita, T. Endo, K. Ogasawara, N. Kashihara, Y.K. Hayashi, W. Yumura, M. Yokode, Y. Ouchi, Toward the development of a vibrant, super-aged society: the future of medicine and society in Japan, *Geriatrics & Gerontology International*, 21 (2021) 601–613.

9. R. Montejano, R. de Miguel, J.I. Bernardino, Older HIV-infected adults: complex patients—comorbidity (I), *European Geriatric Medicine*, 10 (2019) 189–197.

10. A. Cherubini, J. Mateos-Nozal, D.T. Lo, M.A. Paniagua, Geriatric medicine education in Europe and the United States, *Pathy's Principles and Practice of Geriatric Medicine*, 2 (2022) 1585–1593.

11. J. Wahlich, M. Orlu, A. Mair, S. Stegemann, D. van Riet-Nales, Age-related medicine, *Pharmaceutics*, 11 (2019) 172.

12. H. Bukhary, G.R. Williams, M. Orlu, Electrospun fixed dose formulations of amlodipine besylate and valsartan, *International Journal of Pharmaceutics*, 549 (2018) 446–455.

13. J.W. Wastesson, L. Morin, E.C. Tan, K. Johnell, An update on the clinical consequences of polypharmacy in older adults: a narrative review, *Expert Opinion on Drug Safety*, 17 (2018) 1185–1196.

14. M. Khezrian, C.J. McNeil, A.D. Murray, P.K. Myint, An overview of prevalence, determinants and health outcomes of polypharmacy, *Therapeutic Advances in Drug Safety*, 11 (2020) 2042098620933741.

15. N. Masnoon, S. Shakib, L. Kalisch-Ellett, G.E. Caughey, What is polypharmacy? A systematic review of definitions, *BMC Geriatrics*, 17 (2017) 230.

16. Z.A. Marcum, W.F. Gellad, Medication adherence to multidrug regimens, *Clinics in Geriatric Medicine*, 28 (2012) 287–300.

17. K. Notenboom, E. Beers, D.A. van Riet-Nales, T.C.G. Egberts, H.G.M. Leufkens, P.A.F. Jansen, M.L. Bouvy, Practical problems with medication use that older people experience: a qualitative study, *Journal of the American Geriatrics Society*, 62 (2014) 2339–2344.

18. D.-W. Kim, K.Y. Weon, Pharmaceutical application and development of fixed-dose combination: dosage form review, *Journal of Pharmaceutical Investigation*, 51 (2021) 555–570.

19. A.A. Verma, W. Khuu, M. Tadrous, T. Gomes, M.M. Mamdani, Fixed-dose combination antihypertensive medications, adherence, and clinical outcomes: a population-based retrospective cohort study, *PLoS Medicine*, 15 (2018) e1002584.

20. M. Sadia, A. Isreb, I. Abbadi, M. Isreb, D. Aziz, A. Selo, P. Timmins, M.A. Alhnan, From 'fixed dose combinations' to 'a dynamic dose combiner': 3D printed bi-layer antihypertensive tablets, *European Journal of Pharmaceutical Sciences*, 123 (2018) 484–494.

21. M.R. Savona, O. Odenike, P.C. Amrein, D.P. Steensma, A.E. DeZern, L.C. Michaelis, S. Faderl, W. Harb, H. Kantarjian, J. Lowder, An oral fixed-dose combination of decitabine and cedazuridine in myelodysplastic syndromes: a multicentre, open-label, dose-escalation, phase 1 study, *The Lancet Haematology*, 6 (2019) e194–e203.

22. S. Bangalore, G. Kamalakkannan, S. Parkar, F.H. Messerli, Fixed-dose combinations improve medication compliance: a meta-analysis, *The American Journal of Medicine*, 120 (2007) 713–719.

23. J. Kelleher, G. Gilvary, A. Madi, D. Jones, S. Li, Y. Tian, A. Almajaan, Z. Senta-Loys, G. Andrews, A. Healy, A comparative study between hot-melt extrusion and spray-drying for the manufacture of anti-hypertension compatible monolithic fixed-dose combination products, *International Journal of Pharmaceutics*, 545 (2018) 183–196.

24. R. Fernández-García, M. Prada, F. Bolás-Fernández, M.P. Ballesteros, D.R. Serrano, Oral fixed-dose combination pharmaceutical products: industrial manufacturing versus personalized 3D printing, *Pharmaceutical Research*, 37 (2020) 1–22.

25. A. Doty, J. Schroeder, K. Vang, M. Sommerville, M. Taylor, B. Flynn, D. Lechuga-Ballesteros, P. Mack, Drug delivery from an innovative LAMA/LABA co-suspension delivery technology fixed-dose combination MDI: evidence of consistency, robustness, and reliability, *AAPS PharmSciTech*, 19 (2018) 837–844.

26. M. Davis, G. Walker, Recent strategies in spray drying for the enhanced bioavailability of poorly water-soluble drugs, *Journal of Controlled Release*, 269 (2018) 110–127.

27. M. Orlu-Gul, A.A. Topcu, T. Shams, S. Mahalingam, M. Edirisinghe, Novel encapsulation systems and processes for overcoming the challenges of polypharmacy, *Current Opinion in Pharmacology*, 18 (2014) 28–34.

28. J. He, S.L. Baxter, J. Xu, J. Xu, X. Zhou, K. Zhang, The practical implementation of artificial intelligence technologies in medicine, *Nature Medicine*, 25 (2019) 30–36.

29. S. Badillo, B. Banfai, F. Birzele, I.I. Davydov, L. Hutchinson, T. Kam-Thong, J. Siebourg-Polster, B. Steiert, J.D. Zhang, An introduction to machine learning, *Clinical Pharmacology & Therapeutics*, 107 (2020) 871–885.

30. R.S. Mehta, B.D. Kochar, K. Kennelty, M.E. Ernst, A.T. Chan, Emerging approaches to polypharmacy among older adults, *Nature Aging*, 1 (2021) 347–356.

31. M. Zitnik, M. Agrawal, J. Leskovec, Modeling polypharmacy side effects with graph convolutional networks, *Bioinformatics*, 34 (2018) i457–i466.

32. A.V. Zakharov, E.V. Varlamova, A.A. Lagunin, A.V. Dmitriev, E.N. Muratov, D. Fourches, V.E. Kuz'min, V.V. Poroikov, A. Tropsha, M.C. Nicklaus, QSAR modeling and prediction of drug–drug interactions, *Molecular Pharmaceutics*, 13 (2016) 545–556.

33. S. Mei, K. Zhang, A machine learning framework for predicting drug–drug interactions, *Scientific Reports*, 11 (2021) 17619.

34. D. Reker, Y. Shi, A.R. Kirtane, K. Hess, G.J. Zhong, E. Crane, C.-H. Lin, R. Langer, G. Traverso, Machine learning uncovers food- and excipient-drug interactions, *Cell Reports*, 30 (2020) 3710–3716.e3714.

35. F.K.H. Gavins, Z. Fu, M. Elbadawi, A.W. Basit, M.R.D. Rodrigues, M. Orlu, Machine learning predicts the effect of food on orally administered medicines, *International Journal of Pharmaceutics*, 611 (2022) 121329.

36. L.E. McCoubrey, S. Thomaidou, M. Elbadawi, S. Gaisford, M. Orlu, A.W. Basit, Machine learning predicts drug metabolism and bioaccumulation by intestinal microbiota, *Pharmaceutics*, 13 (2021) 2001.

37. K.R. Chowdhary, Natural language processing, in: K.R. Chowdhary (Ed.) *Fundamentals of Artificial Intelligence*, Springer India, New Delhi, 2020, pp. 603–649.

38. M. Herrero-Zazo, I. Segura-Bedmar, P. Martínez, T. Declerck, The DDI corpus: an annotated corpus with pharmacological substances and drug–drug interactions, *Journal of Biomedical Informatics*, 46 (2013) 914–920.

39. Y. Luo, W.K. Thompson, T.M. Herr, Z. Zeng, M.A. Berendsen, S.R. Jonnalagadda, M.B. Carson, J. Starren, Natural language processing for EHR-based pharmacovigilance: a structured review, *Drug Safety*, 40 (2017) 1075–1089.

40. A. Nikfarjam, G.H. Gonzalez, Pattern mining for extraction of mentions of adverse drug reactions from user comments, *AMIA Annual Symposium Proceedings*, 2011 (2011) 1019–1026.

41. A. Wong, J.M. Plasek, S.P. Montecalvo, L. Zhou, Natural language processing and its implications for the future of medication safety: a narrative review of recent advances and challenges, *Pharmacotherapy: The Journal of Human Pharmacology and Drug Therapy*, 38 (2018) 822–841.

42. M. Vailati-Riboni, V. Palombo, J.J. Loor, What are omics sciences? in: B.N. Ametaj (Ed.) *Periparturient Diseases of Dairy Cows: A Systems Biology Approach*, Springer International Publishing, Cham, 2017, pp. 1–7.

43. M. Ficzere, L.A. Mészáros, L. Madarász, M. Novák, Z.K. Nagy, D.L. Galata, Indirect monitoring of ultralow dose API content in continuous wet granulation and tableting by machine vision, *International Journal of Pharmaceutics*, 607 (2021) 121008.

44. L.A. Mészáros, A. Farkas, L. Madarász, R. Bicsár, D.L. Galata, B. Nagy, Z.K. Nagy, UV/VIS imaging-based PAT tool for drug particle size inspection in intact tablets supported by pattern recognition neural networks, *International Journal of Pharmaceutics*, 620 (2022) 121773.

45. S. Floryanzia, P. Ramesh, M. Mills, S. Kulkarni, G. Chen, P. Shah, D. Lavrich, Disintegration testing augmented by computer Vision technology, *International Journal of Pharmaceutics*, 619 (2022) 121668.

46. D.L. Galata, Z. Könyves, B. Nagy, M. Novák, L.A. Mészáros, E. Szabó, A. Farkas, G. Marosi, Z.K. Nagy, Real-time release testing of dissolution based on surrogate models developed by machine learning algorithms using NIR spectra, compression force and particle size distribution as input data, *International Journal of Pharmaceutics*, 597 (2021) 120338.

47. S. Barimani, D. Tomaževič, R. Meier, P. Kleinebudde, 100% visual inspection of tablets produced with continuous direct compression and coating, *International Journal of Pharmaceutics*, 614 (2022) 121465.

48. G. Chen, Y. Xu, P. Chi Lip Kwok, L. Kang, Pharmaceutical applications of 3D printing, *Additive Manufacturing*, 34 (2020) 101209.

49. M. Elbadawi, L.E. McCoubrey, F.K.H. Gavins, J.J. Ong, A. Goyanes, S. Gaisford, A.W. Basit, Disrupting 3D printing of medicines with machine learning, *Trends in Pharmacological Sciences*, 42 (2021) 745–757.

50. M.S. Gupta, T.P. Kumar, R. Davidson, G.R. Kuppu, K. Pathak, D.V. Gowda, Printing Methods in the Production of Orodispersible Films, *AAPS PharmSciTech*, 22 (2021) 129.

51. A.J. Capel, R.P. Rimington, M.P. Lewis, S.D.R. Christie, 3D printing for chemical, pharmaceutical and biological applications, *Nature Reviews Chemistry*, 2 (2018) 422–436.

52. N. Beer, I. Hegger, S. Kaae, M.L. De Bruin, N. Genina, T.L. Alves, J. Hoebert, S. Kälvemark Sporrong, Scenarios for 3D printing of personalized medicines - A case study, *Exploratory Research in Clinical and Social Pharmacy*, 4 (2021) 100073.

53. U.M. Dilberoglu, B. Gharehpapagh, U. Yaman, M. Dolen, The role of additive manufacturing in the era of industry 4.0, *Procedia Manufacturing*, 11 (2017) 545–554.

54. C.S. O'Reilly, M. Elbadawi, N. Desai, S. Gaisford, A.W. Basit, M. Orlu, Machine learning and machine vision accelerate 3D printed orodispersible film development, *Pharmaceutics*, 13 (2021) 2187.

55. M. Elbadawi, D. Nikjoo, T. Gustafsson, S. Gaisford, A.W. Basit, Pressure-assisted microsyringe 3D printing of oral films based on pullulan and hydroxypropyl methylcellulose, *International Journal of Pharmaceutics*, 595 (2021) 120197.

56. A. Awad, A. Yao, S.J. Trenfield, A. Goyanes, S. Gaisford, A.W. Basit, 3D printed tablets (printlets) with braille and moon patterns for visually impaired patients, *Pharmaceutics*, 12 (2020) 172.

57. S.A. Khaled, J.C. Burley, M.R. Alexander, J. Yang, C.J. Roberts, 3D printing of five-in-one dose combination polypill with defined immediate and sustained release profiles, *Journal of Controlled Release*, 217 (2015) 308–314.

58. G.R. Williams, N.P. Chatterton, T. Nazir, D.-G. Yu, L.-M. Zhu, C.J. Branford-White, Electrospun nanofibers in drug delivery: recent developments and perspectives, *Therapeutic Delivery*, 3 (2012) 515–533.

59. D.K. Sahu, G. Ghosh, G. Rath, Chapter 9- Nanofiber: An immerging novel drug delivery system, in: A.T. Azar (Ed.) *Modeling and Control of Drug Delivery Systems*, Academic Press, Cambridge, MA, 2021, pp. 145–152.

60. A. Moreira, D. Lawson, L. Onyekuru, K. Dziemidowicz, U. Angkawinitwong, P.F. Costa, N. Radacsi, G.R. Williams, Protein encapsulation by electrospinning and electrospraying, *Journal of Controlled Release*, 329 (2021) 1172–1197.

61. F. Wang, M. Elbadawi, S.L. Tsilova, S. Gaisford, A.W. Basit, M. Parhizkar, Machine learning to empower electrohydrodynamic processing, *Materials Science and Engineering: C*, 132 (2022) 112553.

62. Nanomedic, Spincare, 2022.

63. BioTronik, PK Papyrus, 2022.

64. O. ReBirth, ReBOSSIS, 2022.

65. Z. Liu, S. Ramakrishna, X. Liu, Electrospinning and emerging healthcare and medicine possibilities, *APL Bioengineering*, 4 (2020) 030901.

66. R.J. Stoddard, A.L. Steger, A.K. Blakney, K.A. Woodrow, In pursuit of functional electrospun materials for clinical applications in humans, *Therapeutic Delivery*, 7 (2016) 387–409.

67. S. Omer, L. Forgách, R. Zelkó, I. Sebe, Scale-up of electrospinning: market overview of products and devices for pharmaceutical and biomedical purposes, *Pharmaceutics*, 13 (2021) 286.

68. S. Labbaf, H. Ghanbar, E. Stride, M. Edirisinghe, Preparation of multilayered polymeric structures using a novel four-needle coaxial electrohydrodynamic device, *Macromolecular Rapid Communications*, 35 (2014) 618–623.

69. S. Kajdič, F. Vrečer, P. Kocbek, Preparation of poloxamer-based nanofibers for enhanced dissolution of carvedilol, *European Journal of Pharmaceutical Sciences*, 117 (2018) 331–340.

70. M. Zamani, M.P. Prabhakaran, S. Ramakrishna, Advances in drug delivery via electrospun and electrosprayed nanomaterials, *International Journal of Nanomedicine*, 8 (2013) 2997.

71. C. Huang, S.J. Soenen, J. Rejman, B. Lucas, K. Braeckmans, J. Demeester, S.C. De Smedt, Stimuli-responsive electrospun fibers and their applications, *Chemical Society Reviews*, 40 (2011) 2417–2434.

72. Z.K. Nagy, A. Balogh, G. Drávavölgyi, J. Ferguson, H. Pataki, B. Vajna, G. Marosi, Solvent-free melt electrospinning for preparation of fast dissolving drug delivery system and comparison with solvent-based electrospun and melt extruded systems, *Journal of Pharmaceutical Sciences*, 102 (2013) 508–517.

73. T.M. Robinson, D.W. Hutmacher, P.D. Dalton, The next frontier in melt electrospinning: taming the jet, *Advanced Functional Materials*, 29 (2019) 1904664.

74. F.M. Wunner, M.-L. Wille, T.G. Noonan, O. Bas, P.D. Dalton, E.M. De-Juan-Pardo, D.W. Hutmacher, Melt electrospinning writing of highly ordered large volume scaffold architectures, *Advanced Materials*, 30 (2018) 1706570.

75. A. Balogh, B. Farkas, K. Faragó, A. Farkas, I. Wagner, I. Van assche, G. Verreck, Z.K. Nagy, G. Marosi, Melt-blown and electrospun drug-loaded polymer fiber mats for dissolution enhancement: a comparative study, *Journal of Pharmaceutical Sciences*, 104 (2015) 1767–1776.

76. T. Vigh, T. Horváthová, A. Balogh, P.L. Sóti, G. Drávavölgyi, Z.K. Nagy, G. Marosi, Polymer-free and polyvinylpyrrolidone-based electrospun solid dosage forms for drug dissolution enhancement, *European Journal of Pharmaceutical Sciences*, 49 (2013) 595–602.

77. U.E. Illangakoon, H. Gill, G.C. Shearman, M. Parhizkar, S. Mahalingam, N.P. Chatterton, G.R. Williams, Fast dissolving paracetamol/caffeine nanofibers prepared by electrospinning, *International Journal of Pharmaceutics*, 477 (2014) 369–379.

78. B. Balusamy, A. Celebioglu, A. Senthamizhan, T. Uyar, Progress in the design and development of "fast-dissolving" electrospun nanofibers based drug delivery systems - A systematic review, *Journal of Controlled Release*, 326 (2020) 482–509.

79. T. Casian, E. Borbás, K. Ilyés, B. Démuth, A. Farkas, Z. Rapi, C. Bogdan, S. Iurian, V. Toma, R. Știufiuc, B. Farkas, A. Balogh, G. Marosi, I. Tomuță, Z.K. Nagy, Electrospun amorphous solid dispersions of meloxicam: influence of polymer type and downstream processing to orodispersible dosage forms, *International Journal of Pharmaceutics*, 569 (2019) 118593.

80. D.-G. Yu, X.-X. Shen, C. Branford-White, K. White, L.-M. Zhu, S.W. Annie Bligh, Oral fast-dissolving drug delivery membranes prepared from electrospun polyvinylpyrrolidone ultrafine fibers, *Nanotechnology*, 20 (2009) 055104.

81. R. Bajracharya, J.G. Song, S.Y. Back, H.-K. Han, Recent advancements in non-invasive formulations for protein drug delivery, *Computational and Structural Biotechnology Journal*, 17 (2019) 1290–1308.

82. F. Cilurzo, U.M. Musazzi, S. Franzé, F. Selmin, P. Minghetti, Orodispersible dosage forms: biopharmaceutical improvements and regulatory requirements, *Drug Discovery Today*, 23 (2018) 251–259.

83. M. Slavkova, J. Breitkreutz, Orodispersible drug formulations for children and elderly, *European Journal of Pharmaceutical Sciences*, 75 (2015) 2–9.

84. M.E. Cam, Y. Zhang, M. Edirisinghe, Electrosprayed microparticles: a novel drug delivery method, *Expert Opinion on Drug Delivery*, 16 (2019) 895–901.

85. H. Bukhary, G.R. Williams, M. Orlu, Fabrication of electrospun levodopa-carbidopa fixed-dose combinations, *Advanced Fiber Materials*, 2 (2020) 194–203.

86. A. Smeets, I.L. Re, C. Clasen, G. Van den Mooter, Fixed dose combinations for cardiovascular treatment via coaxial electrospraying: coated amorphous solid dispersion particles, *International Journal of Pharmaceutics*, 577 (2020) 118949.

87. L. Zhao, M. Orlu, G.R. Williams, Electrospun fixed dose combination fibers for the treatment of cardiovascular disease, *International Journal of Pharmaceutics*, 599 (2021) 120426.

88. W.S. Choi, J.H. Kim, C.B. Ahn, J.H. Lee, Y.J. Kim, K.H. Son, J.W. Lee, Development of a multi-layer skin substitute using human hair keratinic extract-based hybrid 3D printing, *Polymers*, 13 (2021) 2584.

89. H. Chen, Y. Jin, J. Wang, Y. Wang, W. Jiang, H. Dai, S. Pang, L. Lei, J. Ji, B. Wang, Design of smart targeted and responsive drug delivery systems with enhanced antibacterial properties, *Nanoscale*, 1 (2018) 2946–2962.

90. A. Kar, N. Ahamad, M. Dewani, L. Awasthi, R. Patil, R. Banerjee, Wearable and implantable devices for drug delivery: applications and challenges, *Biomaterials*, 283 (2022) 121435.

91. V. Garzón, D.G. Pinacho, R.-H. Bustos, G. Garzón, S. Bustamante, Optical biosensors for therapeutic drug monitoring, *Biosensors*, 9 (2019) 132.

92. T.D. Pollard, J.J. Ong, A. Goyanes, M. Orlu, S. Gaisford, M. Elbadawi, A.W. Basit, Electrochemical biosensors: a nexus for precision medicine, *Drug Discovery Today*, 26 (2021) 69–79.

93. M. Longo, G. Bellastella, M.I. Maiorino, J.J. Meier, K. Esposito, D. Giugliano, Diabetes and aging: from treatment goals to pharmacologic therapy, *Frontiers in Endocrinology*, 10 (2019).

94. CDC, Epilepsy and Seizures in Older Adults, 2021.

95. M. Cella, Ł. Okruszek, M. Lawrence, V. Zarlenga, Z. He, T. Wykes, Using wearable technology to detect the autonomic signature of illness severity in schizophrenia, *Schizophrenia Research*, 195 (2018) 537–542.

96. T.T. Ly, J.E. Layne, L.M. Huyett, D. Nazzaro, J.B. O'Connor, Novel bluetooth-enabled tubeless insulin pump: innovating pump therapy for patients in the digital age, *Journal of Diabetes Science and Technology*, 13 (2019) 20–26.

97. L.R. Curtis, K. Alington, H.L. Partridge, Insulin pumps: are services and health equity undermining technological progression? *Practical Diabetes*, 38 (2021) 27–32.

98. W. Chang, L. Chen, C. Hsu, J. Chen, T. Yang, C. Lin, MedGlasses: a wearable smart-glasses-based drug pill recognition system using deep learning for visually impaired chronic patients, *IEEE Access*, 8 (2020) 17013–17024.

99. M. Elbadawi, B.M. Castro, F. Gavins, J. Ong, S. Gaisford, G. Perez, A.W. Basit, P. Cabalar, A. Goyanes, M3DISEEN: a novel machine learning approach for predicting the 3D printability of medicines, *International Journal of Pharmaceutics*, 590 (2020) 119837.

100. F. Wang, M. Elbadawi, S. Liu Tsilova, S. Gaisford, A.W. Basit, M. Parhizkar, Machine learning predicts electrospray particle size, *Materials & Design*, 219 (2022) 110735.

101. S. Chatterjee, P. Chi-Leung Hui, Review of stimuli-responsive polymers in drug delivery and textile application, *Molecules*, 24 (2019) 2547.

102. M.A. Rahim, N. Jan, S. Khan, H. Shah, A. Madni, A. Khan, A. Jabar, S. Khan, A. Elhissi, Z. Hussain, H.C. Aziz, M. Sohail, M. Khan, H.E. Thu, Recent advancements in stimuli responsive drug delivery platforms for active and passive cancer targeting, *Cancers (Basel)*, 13 (2021) 670.

103. B. Singh, N. Shukla, J. Kim, K. Kim, M.H. Park, Stimuli-responsive nanofibers containing gold nanorods for on-demand drug delivery platforms, *Pharmaceutics*, 13 (2021) 1319.

104. M. Elbadawi, J.J. Ong, T.D. Pollard, S. Gaisford, A.W. Basit, Additive manufacturable materials for electrochemical biosensor electrodes, *Advanced Functional Materials*, 31 (2021) 2006407.

16 Innovative Management of Pharmaceutical Product Design and Manufacturing

Angelo Kenneth Romasanta, Jonathan Wareham,
Laia Pujol Priego, and Gozal Ahmadova
ESADE Business School

CONTENTS

16.1 INTRODUCTION

Innovation in the pharmaceutical industry has been key to addressing the health challenges facing society (Romasanta, van der Sijde, and van Muijlwijk-Koezen 2020; Bianchi et al. 2011). However, developing new drugs and delivering them to patients have become more difficult due to various challenges across all the phases from basic research to marketing (Pammolli, Magazzini, and Riccaboni 2011; Stott 2017; Hess and Rothaermel 2012). To address these challenges, this chapter explores the emerging management practices in pharmaceutical product design and manufacturing.

To be clear, this chapter does not aim to go deep into the specific technologies or process innovations shaping the drug development as these have been covered in the previous chapters in the book. Instead, it focuses on the management practices that are increasingly adopted in the industry to keep up with the complexity and risks involved in developing, designing, and manufacturing innovative pharmaceutical products. This chapter provides a framework to understand these different challenges and to guide decision-makers involved in the process. Industry reports from top consulting firms and academic publications were reviewed, complementing them with the authors' experience being involved in projects in the industry.

This chapter is structured as follows. First, overall trends in the pharmaceutical industry are highlighted, examining the new regulatory, economic, and societal challenges surrounding the development of innovative pharmaceutical products. These trends are foreseen to impact the industry across all its phases from early-stage drug development to downstream supply chains. To address them, emerging management practices were then highlighted, dividing this into internal-oriented and external-oriented management. Concerning internal management, that is, management inside organizational boundaries, the promises of agile methodology in pharmaceuticals were explored. Concerning external management, which refers to organizing inter-organizational collaborations, FAIR data management was featured due to its promise to enable organizations

DOI: 10.1201/9781003224464-16

to fully leverage data from outside sources. For each of these topics, reports from industry insiders and case studies of organizations leveraging these ideas were presented. Having discussed trends in the industry, the implications were investigated for different actors within the life science ecosystem: big pharmaceutical companies, biotech start-ups, and academia.

16.2 CHALLENGES IN THE PHARMACEUTICAL INDUSTRY

Research and development (R&D) in the pharmaceutical industry has faced various challenges in recent years. First, the cost of bringing new drugs to the market has risen (Ranade and Singh 2019; Pammolli, Magazzini, and Riccaboni 2011). Estimates of the cost of developing a new drug range from $161 million to $4.54 billion, with R&D costs increasing over time (Schlander et al. 2021). Second, the types of drugs for development have increased in number and complexity (IQVIA 2019), pressuring companies to adapt quickly to new technologies. Third, firms are expected to have a deeper understanding of their patient needs, away from the old blockbuster paradigm (KPMG 2019). Consequently, regulatory bodies also have different expectations of the benefits and risks associated with new treatments (KPMG 2019). These trends are further investigated in the following.

16.2.1 Increasing Costs in Developing New Drugs

It has been well documented that the R&D productivity in the pharmaceutical industry has been declining (Stott 2017; KPMG 2018). Although the industry is spending more on developing new drugs at around a 3.6% compound annual growth rate, the return on investment has been continually declining, leading to the cost being below the cost of capital currently (KPMG 2018). Many factors are driving this low productivity, including the intensifying competition across companies taking the same classes of drugs through the trials, the increased pressure to be first in the market (Ranade and Singh 2019), and the low-hanging fruit effect where easier drug targets have been already discovered (Scannell et al. 2012).

Even breaking down productivity across different disease areas and phases in clinical trials reveals the same trend. Most disease areas have been declining in clinical development productivity but with some variance across some fields. Success rates can be as high as 32% for rare diseases to less than 10% for neurology and cardiovascular (IQVIA 2021). Across the different phases, the average success rate in clinical trials in the last 10 years is 9.8% (IQVIA 2021). Breaking this down further, phase 1 clinical trials had 56%, phase 2 clinical trials had 38%, and Phase III had a success rate of 66% (IQVIA 2021). In total, the pandemic already considered, productivity is still below its previous levels due to complexity and the increased duration of trials (IQVIA 2021).

16.2.2 Need for Novel Therapies

New technologies and scientific discoveries are enabling companies to address new diseases and explore difficult targets. This has then enabled new types of drugs including cell and gene therapies, antibodies, proteins and peptides, and other forms of biologics to go through the clinical trials (Berggren et al. 2018). As a consequence, it is not surprising that in 2021, 27 of the 50 approved novel drugs are first-in-class (US FDA 2021). Apart from advances in basic research, manufacturing innovations have also led to increased viability of novel therapeutics such as biologics.

These advances have led to changing expectations from the industry in terms of the future treatments to come out of the pipeline. From a survey of life science executives (Ford et al. 2020), customized treatments which enable patients to be treated at the individual level, informed by data, are seen as the biggest trend. However, in this survey, it was also found to be the area that most companies are not prepared for. Another major trend is precision interventions that leverage new technologies like robotics, nanotechnology, tissue engineering, and sensors for accurate treatments. Besides these two, executives also noted the changing focus on prevention and early detection. In this case, vaccines have been seen as crucial to preventing diseases even before they progress (Ford et al. 2020). New-generation drugs are also exploring ways to slow down or halt disease progression after its early identification (IQVIA 2019). Finally, digital therapeutics that modify behavior was also seen as a potential area that can even reduce the need for medications.

16.2.3 Changing Patient and Regulatory Demands

Drug development has also moved away from the one-size-fits-all blockbuster model, where a drug is supposed to cater to a large number of patients (KPMG 2019; 2018). Instead, pharma companies are expected to cater to the individual needs of patients, harnessing advances in proteomics, genomics, and metabolomics. Alongside this, the advances in biomarker technology mean that diseases can be tracked and monitored more accurately as it progresses.

This evolution away from the blockbuster paradigm can be seen in the field of lung cancer oncology. In the 1960s, chemotherapies were used to treat all forms of lung cancer. However, with new diagnostic capabilities, new therapies were created in the 2010s to treat specific variants of the disease. By 2020, different subgroups of patients receive these different therapies as estimated by BCG (Richenberger et al. 2021): ~40% Tecentriq/Avastin, ~30% with Keytruda, 8% (Tagrisso), and 1% (Rozlytrek).

With access to troves of data, companies can map the individual journeys of their patients as they engage with the healthcare system and understand better how they can holistically address their needs. As healthcare moves toward the precision medicine model, companies can provide individualized medicine instead of catering to patient subgroups where a drug is tested to have some benefits (Hartl et al. 2021). Closer relationships between companies and patients are expected as these next-generation therapies may also require longer and continuous follow-up to track progress. Current manufacturing and supply chain processes also have to transform to accommodate these advances (Rambaldini and Fernandez Giove 2018).

This shift has also led to greater attention paid to diseases afflicting only a small population of patients. When a disease has a high unmet need, regulators are more open to fast track the regulatory process and operating based on limited data (Berggren et al. 2018). To illustrate, the number of drug approvals for orphan drugs in the US has increased from merely 156 in the 2000s to 495 in the 2010s (US FDA 2022). A disease is given an orphan status if they are so rare that drug development would not be able to recoup costs without government help.

Summarizing the previous section, the pharmaceutical industry has been facing challenges including increased costs in developing new drugs, shifting expectations for the types of drugs, and changing patient and regulatory demands. Overall, these trends are impacting the industry not only in terms of how drug development is conducted upstream but also in how downstream pharmaceutical supply chains provide value to patients (Rambaldini and Fernandez Giove 2018; Brooks 2014). To overcome these challenges, the industry can experiment with emerging management practices (Figure 16.1).

16.3 INNOVATIVE MANAGEMENT OF PHARMACEUTICALS

16.3.1 INTERNAL MANAGEMENT

Large pharmaceutical companies have a proven track record in bringing new drugs to the market. The challenge however arises when these practices become so ingrained that they hinder these organizations from flexibly pursuing new opportunities. In a study by the consultancy firm Accenture (Philipp and Miinch 2021), they find that most companies have settled on a particular way of organizing their research activities. Instead of experimenting with different approaches, companies tend to share the same characteristics in how they structure innovation.

Summarizing their findings (Philipp and Miinch 2021), first, in terms of leadership structure, companies tend to have a separate leadership across the three major functions of research, development, and medical affairs. Integrated leadership across these three phases has not been common. Second, in terms of governing which projects to prioritize and progress, most firms are focused on the stage-gate model (horizontal governance), with other firms prioritizing projects by therapeutic areas (vertical governance). Mixed models of the two only account for a small number of companies. Third, for the timing of the transition to clinical development, most companies transfer the ownership of the project to the clinical development team only at the proof-of-concept stage, with others doing this before phase 1 clinical trials. Embracing an integrated translational organization is still not common. Fourth, concerning decision-making, most firms follow a centralized global functional steering model, while other firms follow a matrix of functions and therapeutic areas. Enabling local regions to have autonomy toward decentralized decision-making is not typical. Fifth, companies also have similar ways of adopting and diffusing new capabilities to serve therapeutic areas. Most companies build new technological capabilities within global functions and then later share these across each therapeutic area. Companies that build capabilities to address specific therapeutic area needs to account for a smaller number of firms. Finally, regarding the externalization of clinical development, firms tend to be divided into three different ways of working: internally enabling clinical activities, externalizing clinical activities to contract research organizations, and a mixture of the two.

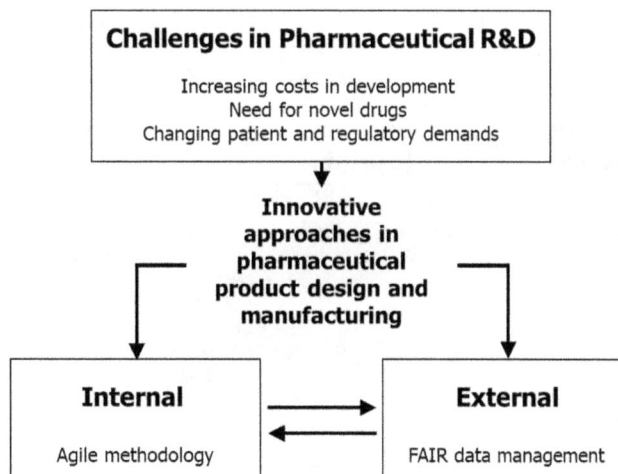

FIGURE 16.1 Innovative management to address challenges in pharmaceutical R&D.

Despite the homogeneity of these extant practices, managers in the industry have increasingly questioned whether these are appropriate in addressing their current challenges. In a survey by the consulting firm McKinsey, 82% of respondents from the pharmaceutical industry anticipate that the industry would need to prioritize agile ways of working (Balz et al. 2021). Although agile originates in software development, interest in applying its principles in the pharmaceutical setting has surged as a potential way to address the slow development timelines, strict requirements, and dependence on cross-functional interactions in drug development (Karaivanov 2020). Since it is common for competing firms with similar projects to be separated only by a few months of progress, agile was also seen as crucial to enable firms to flexibly reprioritize and deploy their resources to respond to rapidly changing conditions (Apple et al. 2019).

16.3.1.1 Agile Methodology

Agile practices have been widely used in software engineering to help coders respond quickly and nimbly to changing demands of customers. Instead of agile being a standard, it refers to a set of principles (Beck et al. 2001):

- **Individuals and interactions over processes and tools**: emphasis on the roles of humans in the process instead of strictly following institutionalized processes and tools
- **Working software over comprehensive documentation**: the focus is on continuously testing the product to get feedback and for improvement
- **Customer collaboration over contract negotiation**: relationships across different stakeholders are prioritized
- **Responding to change over following a plan**: members of the team are informed of the current conditions and challenges to adapt to potential changes

Applied in organizations, agile can be characterized by the following features: Rapid iterations are adopted in development to gain feedback early in the process. Agile teams aim to manifest their knowledge in the form of the minimum viable product – prototypes that incorporate all the critical features in the desired solution without requiring much resources or approval from decision-makers (Ries 2009). These prototypes are presented to stakeholders as soon as possible for early feedback to anticipate critical issues soon as possible. Such feedback is also used to inform the team whether to continue, terminate, or pivot the project into a new direction. Through this emphasis, startups in software development have especially been effective when projects are complex and carry high uncertainties, when the processes are not well-established, and when different stakeholders need to be coordinated (Osmond 2019). Under the broad agile philosophy, different frameworks exist to guide its implementation in organizations – one of the most famous of which is scrum. The typical scrum workflow is illustrated in Figure 16.2 and further elaborated in the following.

Core to scrum are the sprints, which refer to a set amount of time for a team to achieve different tasks. During this period, the team members focus on working on the agreed list of goals (Osmond 2019). In software development, sprints typically last from a week to 4 weeks. The goal is that the sprints are long enough to have substantial progression but short enough to effectively hold the team's attention. Unsurprisingly, if scrum were to be embraced in a pharmaceutical organization, adjustments have to be made to ensure its relevance. Unlike engineering organizations where the science is typically settled, activities in this sector would typically need longer sprints to cope with the high uncertainties and the longer development cycles. As such, initial adoption would require openness to experimentation from the staff, and their leaders as the early stages

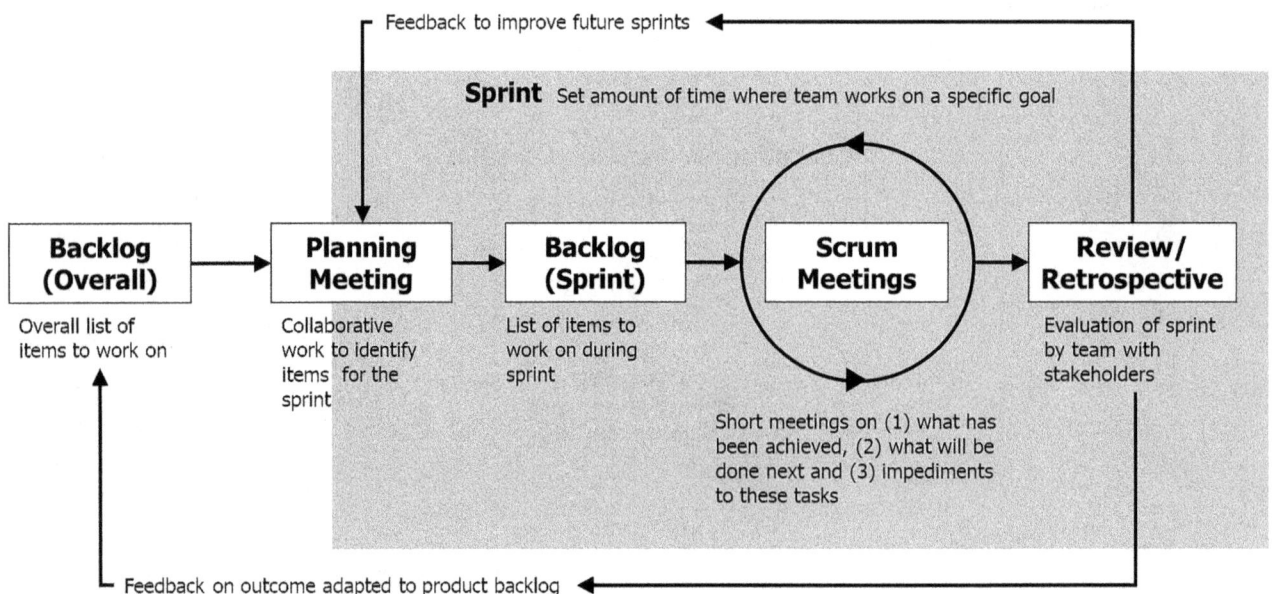

FIGURE 16.2 Simplified scrum workflow. (Adapted from scrum.org.)

are expected to produce inefficiencies and not immediately show beneficial results.

Before each sprint, everyone in the team agrees on the work to be done and the measures to track progress. To do this, a meeting is done with the team members to review the previous task list or create a new task list for the team and its specific members. These meetings discuss what the goals are, how they will be measured, and what the potential next steps are after the sprint.

Another cornerstone of the agile approach is the scrum meetings. These daily meetings, typically held with people standing up, enable the team to update each other about their progress. This tends to be daily with a length of around 15 min. Typically, three things are mentioned by each member: what has been achieved since the last meeting, what is planned to be done, and the challenges along the way. These meetings are meant to be brief, with comments or questions discussed in a more relevant time. The tasks are then checked whether they are completed or needed to be reprioritized. When the Broad Institute introduced agile to its staff, researchers found that the daily meetings are too ineffective and intrusive to their current workflows. Researchers then suggested changing the frequency of the meetings from daily to weekly or biweekly since laboratory results can take longer compared to typical software development (Fiore, West, and Segnalini 2019).

At the end of sprints, teams conduct a review to evaluate the outcomes of the sprint together with stakeholders outside the team. In software development, the result of a sprint is typically referred to as an "increment" based on the idea of improving the product on a small yet frequent and meaningful basis. Apart from the review, a retrospective is also conducted by the team to understand how they can be more effective in future sprints. The team discusses the things that went well and those that can be done differently. Moreover, previous retrospectives are also rechecked whether their changes were implemented (Osmond 2019). These meetings typically include external stakeholders who are not part of the team. As a consequence, this can help solve the problems of siloing of knowledge typical in large organizations. For instance, in an unnamed biotech company featured in Harvard Business Review, it was found that many teams lose 30% of their time just coordinating with each other to prepare the presentation to update key stakeholders (Fiore, West, and Segnalini 2019). The different teams barely collaborated and typically just deferred decisions to their specific heads. This lack of coordination has led to the inability to iterate quickly and pursue better options. Adopting agile methods was then seen as the solution to these inefficiencies.

16.3.1.2 Agile Adoption

To successfully adopt agile then, preparations must be made by organizations in the pharmaceutical industry to ensure relevance. For instance, one challenge would be overcoming the current ways of working that practitioners in the pharmaceutical sector are already accustomed to, with the staff already highly specialized and their routines highly structured toward stability and consistency. Companies then have to overcome the resistance to change in these processes and be more open to a more flexible and less sequential workflow (Apple et al. 2019).

Many organizations might have to redesign their workflows and build routines to support the flexibility and speed in decision-making promised by agile (Freytag 2019; Vaidyanathan et al. 2019). Moreover, since agile methods rely on promoting the sharing of knowledge across siloed units (Alaedini, Ozbas, and Akdemir 2014), staff must be trained to be more comfortable with interacting with other departments. Typically, in agile, cross-functional teams are assembled and disbanded based on a deliverable (Agrawal et al. 2019). Instead of the sequential waterfall model where teams are divided by their expertise, agile teams often are smaller, but they integrate different subject experts from different levels of seniority within the organization (Alaedini, Ozbas, and Akdemir 2014). By ensuring that the team is from the start created for a certain purpose, the members are then brought together based on their potential contribution, with their individual responsibilities clear. Important decisions then are done with the lead of this cross-functional team. With their small size, the consensus is reached quicker, empowering teams to respond quickly to emergent issues. These teams are made the main decision-makers toward their deliverables, with the guidance of the leaders of the firm (Agrawal et al. 2019).

Coordination between the team is promoted through different practices. Visual representations can help members understand their different roles and improve delegation and alignment of tasks across the team (Alaedini, Ozbas, and Akdemir 2014). In this approach called Kanban, boards that are either physical or digital are used to monitor progression. Moreover, incremental improvements are desired in current workflows, to avoid disrupting processes that are already working well. Automated software tools are also available that can guide firms in implementing agile workflows (Karaivanov 2020).

16.3.1.3 Examples of Agile

Due to its promises, agile has started to be experimented with within the research community to address many inefficiencies in the scientific process (Fiore, West, and Segnalini 2019; George and Fridley 2021). It has been adopted by medical device companies (Rottier and Rodrigues 2008), pharmaceutical firms (Vaidyanathan et al. 2019), and even top academic institutes (West 2018).

The Finnish subsidiary of Roche has also experimented with the adoption of scrum (Lilja, Kailanto, and Saanila-Sotamaa 2021). They were motivated to adopt agile to improve customer satisfaction, to be able to test products early for decision-making, and to promote more effective communication across the organization. To kickstart this transformation, the company sent a message to everyone to ensure that the employees know that the method is not being forced on anyone; instead, was being used as a tool to help manage the work overload that people were already experiencing. One big change was improving flexibility by allowing disease area teams to be added and modified flexibly to respond to business needs. The term "enabling structures" was introduced which refers to simple rules that guide people on how they can self-organize. Beyond this common structure, teams can decide the specifics of implementation within their team. Previously, the company culture has emphasized individual work. As such, making the shift to working in self-organized and autonomous teams initially was difficult for the staff to get accustomed to.

The organization started doing 1-month sprints with all the typical processes in scrum, and this led them to make some tweaks to make the method fit the organization. For instance, instead of a regular sprint review where all the stakeholders are invited, this was limited to a maximum of four guests from other teams to make it easier for employees to commit to attending such meetings. The daily scrum was replaced with a weekly meeting. Instead of a finely defined backlog with specific tasks, the company had to structure it more as a general roadmap due to the uncertainties in the process. Despite these challenges and adjustments, the shift to scrum was found to be overall positive. It enabled the company to adapt quickly to the Covid-19 restrictions.

Another company that adopted agile is PTC Therapeutics, a company specializing in rare diseases (Fiore, West, and Segnalini 2019). They saw agile as a way to cope with the quickly growing number and complexity in their drug development pipeline. Before proceeding to adopt agile methods, leaders in the company first had to get the buy-in from employees within the firm. To do this, the company held a workshop to align everyone on why such an initiative had to be undertaken in the first place. Only when it was agreed that adopting these methods were necessary to handle the amount and scale of projects did the company decide to proceed. The main change they enacted was to reorganize into a cross-functional team-based structure composed of basic research, clinical research, and business staff. To coordinate this, the company had "pooling functions" which could mobilize resources to teams when needed so that they can quickly move forward with their activities. These also enable people to move to projects where there is much demand and enable interactions across different departments that do not typically interact due to siloing of information. These teams are led by heads who have a bigger picture of the different projects across the similar therapeutic area to create alignment across the company's different initiatives.

Apart from drug development, agile methods are also being explored in downstream supply chain operations. In a white paper, the logistics firm DHL discussed how they are helping the pharmaceutical supply chain deal with rapidly changing consumer demands and fast technology evolution (Rambaldini and Fernandez Giove 2018). They argue that traditional supply chains had not been able to adequately address these trends given their linear nature with information flowing from one step to another. If supply chains were

more agile, they may be able to address demand fluctuations and respond to unforeseen events (Mehralian, Zarenezhad, and Ghatari 2015). This would mean that companies can quickly scale up or down the supply of materials according to needs, adapt to local capacities with minimal cost, rebalance the network of suppliers easily, make information transparent for planning, and manage issues in product segmentation and complexity (Ebel, Kubik, and Lösch 2012).

Beyond individual companies, even governments have been entertaining the idea of adopting agile to aid the pharmaceutical industry. Particularly, in the field of precision medicine, policymakers have been embracing agile to improve the efficiency of international collaborations and reduce regulatory barriers (Doxzen, Signe, and Bowman 2022). To support this, various agile governance tools have been recommended including policy labs, regulatory sandboxes, crowdsourcing policy, public-private data sharing, and direct representation in governance (Doxzen, Signe, and Bowman 2022).

In summary, this section explores how organizations can align their internal innovation with the agile philosophy to address the increasing complexities in drug development, manufacturing, and supply chains. Complementing this internal-oriented aspect, the following section explores how firms can effectively manage external innovation activities.

16.3.2 External Management

With the pace of advances in the sciences, it is not feasible for firms to have all the expertise in-house. To adapt, firms engage with external sources of innovation. This paradigm of leveraging external sources of knowledge and innovation has often been referred to as open innovation (Schuhmacher et al. 2013; Hughes and Wareham 2010; Xia 2013; Reichman and Simpson 2016; Bianchi et al. 2011; Michelino et al. 2015; Hunter and Stephens 2010; Chesbrough 2003). This embrace of open innovation is not surprising considering the myriad sources of innovation external to the firm. Contract research organizations that specialize in key technologies like AI, 3D printing, and lab-on-chip are enabling firms to further eliminate barriers they face in other areas of clinical development so that they can focus on their main capabilities (KPMG 2018). Outsider technology players like IBM, Microsoft, Google, and Apple are also increasingly entering pharmaceuticals and healthcare such as by increasing sources of data, enabling innovative diagnostics through wearables, virtualizing clinical trials, and facilitating patient enrollment (KPMG 2018).

To successfully collaborate with these external players, companies have to establish proper governance models across parties that may have different or even competing strategic priorities (Besel 2019). For instance, a company may engage in such collaborations to develop a new therapy for commercialization while another firm may be using such collaborations to merely learn from the other party. It is thus critical for firms to have clear alignment on their goals. Moreover, even when companies are aligned, their staff may not give the collaborations the attention needed to guarantee success.

Apart from sourcing the services of other companies, open innovation also emphasizes that companies can deploy their assets externally if they can be leveraged by other groups. AstraZeneca, for example, has collaborations on their high-throughput screening platform, allowing academics to use their large compound library for screening (Murray, Wigglesworth, and Preston 2019).

While open innovation is widely accepted in the industry, these previously acknowledged forms of collaboration have mostly been based on physical, explicit ties among the different parties. However, digital technologies are enabling new ways of collaborating across these different partners beyond these previous models constrained by physical connections. A paradigm championed by academia and policymakers to broaden openness in research is Open Science. Open Science has been defined as "transparent and accessible knowledge that is shared and developed through collaborative networks" (Vicente-Saez and Martinez-Fuentes 2018). This movement values the sharing of data, methods, code, and results to external stakeholders so that they can also evaluate, reuse, and build on these findings. Open Science, especially in drug development, can help organizations gain insights from a wider group of experts, accelerate the pace of feedback toward more robust solutions, and reduce unnecessary duplication of research through reuse (Robertson et al. 2014). One noteworthy example is that of the Pathogen Box, created and distributed by the Medicines for Malaria Venture (Veale 2019). It is a collection of 400 molecules selected for their potential activity against several neglected diseases. These compounds were assembled from successful screenings from companies' drug programs. The use of the compounds was free as long as their results are shared with the research community within 2 years.

A more extreme implementation of Open Science has also diffused through the research community. Open source, which has originated in the software development industry, is also becoming increasingly familiar in science. In a manifesto on Open Source Drug Discovery (OSDD), the following values were highlighted (Todd 2019): (1) All data and ideas are freely shared; (2) anyone may participate at any level; (3) There will be no patents; (4) Suggestions are the best form of criticism; (5) Public discussions are more valuable than private ones; and (6) An Open project is bigger than and is not owned by any lab. Thus, open-source abandon secrecy and intellectual property protection in favor of opening the research cycle for outside scrutiny. This model would mandate that the results, reagents, compounds, software, and clinical trial results are accessible to the public (Han et al. 2018).

The open-source approach has been championed by researchers in tropical or neglected diseases like tuberculosis and malaria (Robertson et al. 2014; Årdal and Røttingen 2012; Bhardwaj et al. 2011), where knowledge sharing is much needed to overcome current barriers. In this open approach, stakeholders from both academia and industry can collaborate and share both positive findings and failed results and integrate them into their process (Bhardwaj et al. 2011). In one project called SysBorg TB, researchers can share data, methods, and algorithms on tuberculosis drug discovery for others to use and build on. The platform was structured in such a way that investigators that build on previous research are asked to give back and also let others build on their work, with the help of clear licensing rules and a reward system within the portal. To ease the concerns related to intellectual property claims, the OSDD model allows for such claims as long as companies give non-exclusive licenses in developing countries.

As seen, the pharmaceutical industry has been increasingly embracing approaches such as Open Innovation, Open Science, and Open Source due to its dependence on external innovation. However, to fully unlock the potential of these external sources of knowledge, organizations also would have to adopt new practices. The following section introduces and highlights the paradigm of FAIR data management to leverage fully external knowledge.

16.3.2.1 FAIR Data Management

The pharmaceutical industry is generating tremendous amounts of data from its scientific and clinical activities. The problem arises when valuable data is generated on large scales never to reused again (Pujol Priego, Wareham, and Romasanta 2022). Without proper preparation, scientists who would have to use them would have to spend more time processing the dataset instead of readily leveraging them to inform their research (Wienken 2021; Romasanta and Wareham 2021). As such, academia, policymakers, and many companies are pushing for data to be FAIR, meaning that data is findable, accessible, interoperable, and reusable (Wilkinson et al. 2016; Wise et al. 2019). These principles are being embraced by a diverse set of stakeholders including academia, industry, funding agencies, and even governmental organizations such as the European Commission, G7, and National Institutes of Health (Wise et al. 2019; Mons et al. 2017). Figure 16.3 summarizes its principles.

FAIR principles act as a guide to facilitate knowledge discovery with an emphasis on machine operability (Schultes and Wittenburg 2019). First, data should be easy to find for both humans and computer systems by enriching data with metadata (data that provides information about other data) and providing persistent identifiers. Second, once found, data should be accessible to be easily downloaded from trusted repositories. Third, data should be interoperable so that they can be used, processed, and analyzed across different computing environments. Finally, data should be reusable through proper descriptions of their provenance and usage rights. Through FAIR, researchers would be able to minimize research duplication and build on top of previous research for new insights. By being able to link data from silos across the organization through common language and clear data standards, organizations can then help bring insights that are hidden in the data (Cerwin, Guenard, and Alosi 2021).

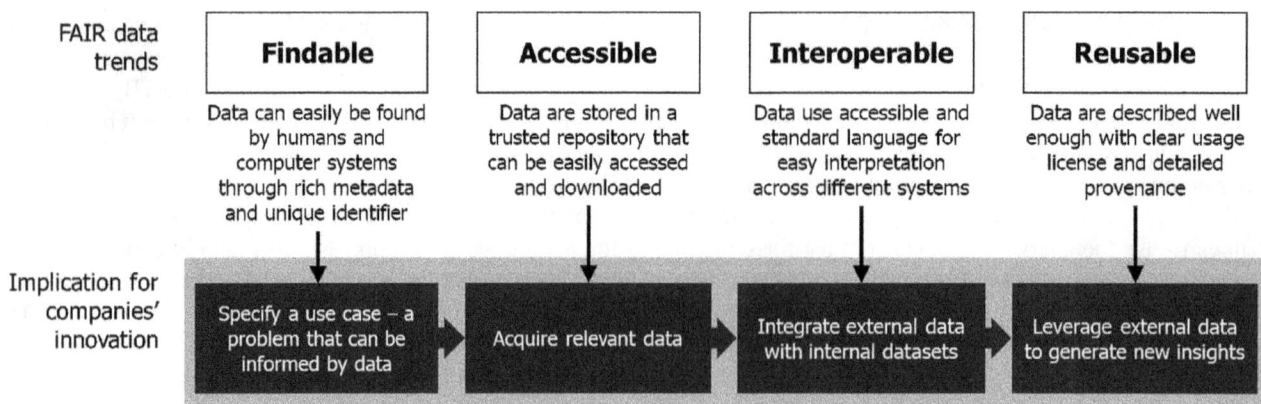

FIGURE 16.3 FAIR data management.

FAIR promises to impact processes across the industry. For instance, in precision and personalized medicine, FAIR allows research centers to pool data together to create prognostic and predictive models of superior accuracy (Vesteghem et al. 2020; Hulsen et al. 2019; Cirillo and Valencia 2019; Parra-Calderón 2019; Blasimme et al. 2018). Downstream, it also aims to help as the industry moves toward a more sustainable pharmaceutical supply chain by improving the coordination and information sharing across different actors within this ecosystem (Ding 2018). Beyond specific phases, FAIR can support innovation across the entire organization by supporting the development of predictive models, generating new insights, and improving decision-making (Wienken 2021; Van Vlijmen et al. 2020).

16.3.2.2 Areas Where FAIR Can Be Leveraged

To unlock these ambitious promises of FAIR, companies would have to implement several practices in their data management and stewardship. Most pharma companies are still at an early stage of FAIR alignment due to various implementation barriers (Harrow et al. 2022; Jacobsen et al. 2020; Alharbi et al. 2021). Apart from the technical skills needed to create and maintain the data infrastructure, it can also be challenging to implement a data-centric culture around the workplace. To be successful then, leaders in an organization need to transmit the value of FAIR to employees through training and education programs. Nonetheless, despite these barriers, the hope is that leveraging FAIR data would enable companies to reuse their assets for new lines of research and generate innovations from the available information.

Traditionally, companies partner with other firms to gain access to data (Besel 2019). For instance, GSK has partnered with 23 and me to access the genetic and phenotypic data of their consenting customers. However, beyond these time-intensive setting up of partnerships, the hope is that FAIR data will enable organizations to tap into other valuable datasets. For instance, one of the most important sources of data is research infrastructures like the European Molecular Biology Laboratory (EMBL) which houses data generated by other researchers that can then be

taken up by other companies to create value. Showing the value of such infrastructure, the economic impact analysis of EMBL-EBI by Beagrie and Houghton (2016) demonstrated the long-term value of biological data in the pharmaceutical industry. A survey of 4509 respondents showed that EMBL-EBI data and services made their research significantly more efficient: at an estimated value of 1 billion per annum worldwide – equivalent to more than 20 times the direct operational cost. Long-term economic value to R&D was estimated to be worth some 920 million annually. To further enable organizations to benefit from their data, EMBL-EBI has further invested to make their data align with FAIR principles (Harrow et al. 2021).

The move toward FAIR would also open research to novel sources of data. Recently, there has been a larger emphasis on real-world evidence (RWE) – atypical sources of data such as health care information from unconventional sources such as electronic health records, billing databases, and disease registries (Sherman, 2016). Such real-world evidence provides increased statistical power, enhanced sensitivity, and specificity, ultimately allowing a smaller sample size, shorter study duration, and real-time feedback for early decision-making (Hartl et al. 2021; David Champagne et al. 2020). For instance, data from smartphones and wearables, if accessible to organizations, sensors can serve as digital biomarkers, giving insights into the physiological, biochemical, and behavioral characteristics of a patient. These digital devices can be powerful sources of data as they can collect objective and longitudinal patient data in a real-life context. Diverse biological data such as motor movement, sleep patterns, speech and executive function can be collected through these devices can be leveraged as objective endpoints with clinical and patient relevance (IQVIA 2021. If they were made to be easily findable, such data can explain, influence, or predict health-related results toward a more comprehensive assessment of therapeutic effectiveness and safety (Babrak et al. 2019).

Crucial to making data much easier to leverage for researchers are the infrastructures referred to as data commons. Data commons collocate data, storage, and computing services together with other web and informatics tools for managing, processing, analyzing, and sharing data across the community (Grossman et al. 2016). With many datasets generated by research institutes publicly available, one challenge may then be integrating these datasets and making use of them for critical insights. To address this problem, data commons harmonize datasets so users can conveniently assemble datasets for new experiments. For example, Open Targets is a partnership and a platform created by an ever-growing group of organizations in bioinformatics, genomics, and pharmaceuticals dedicated to collaboratively establishing links between genetic targets and disease development. The partnership began in 2015 and is the result of the collaboration across the European Bioinformatics Institute unit of the European Molecular Biology Laboratory (EBI-EMBL), the Wellcome Sanger Institute, GSK, Biogen, Takeda, Celgene/Bristol Meyers Squibb, Sanofi, and Pfizer. In order to support the integration and sharing of the experimental data collaboratively generated by its members with other data sources, Open Targets developed an open platform capable of searching, assessing, and integrating a vast quantity of genetic and biological data that are openly shared as data commons for any third-party to reuse (Pujol Priego and Wareham 2019).

An additional example is Therapeutics Data Commons, which enables researchers to access and evaluate machine learning across therapeutics (Huang et al. 2021). This platform provides different tools and datasets such as integrating 66 AI-ready datasets spread across 22 learning tasks. These resources cover the entire process from discovery to safety evaluation, to accelerate the use of machine learning in drug development.

The excitement for FAIR data also comes from the high expectations of the value that AI can provide in the innovation process. Since data is only useful if it can be properly processed and analyzed, computational tools and models play an important role (Mak and Pichika 2019). These technologies including artificial intelligence, cloud computing, augmented/virtual reality (AR/VR), wearables, digital twins, IoT, blockchain, and quantum computing have all been cited as potential solutions to current challenges in pharmaceutical development (Ford et al. 2020).

Current AI-powered solutions have been actively introduced in many areas of the pharmaceutical industry to help in drug discovery such as in organic synthesis and design; scoring synthetic complexity; automation of molecule design; predicting organic reaction outcomes and toxicity; and computer-aided synthesis and retrosynthesis (Vijayan et al. 2022; Federation of Indian Chambers of Commerce and Industry 2019; Paul et al. 2021; Lou and Wu 2020). Recently, one of the most publicized advances in AI use is AlphaFold developed by DeepMind, owned by Alphabet Inc., the parent company of Google. This tool paves the way to creating the most complete and accurate database yet of the predicted human protein structures freely and openly available to the scientific community. A complete understanding of the estimated 20,000+ proteins in the human body is considered one of the most important scientific challenges in the medical and pharmacological sciences.

Other tools are also being developed by different actors. For instance, researchers have released VirtualFlow, an open-source platform for large-scale virtual screening (Gorgulla et al. 2020). This platform was designed in an automated and scalable manner for researchers to easily perform docking processes. Researchers can take VirtualFlow and run it in their cloud computing platforms or computing clusters to screen 1 billion compounds in 2 weeks. Initiatives too from entrepreneurial companies are transforming the industry.

Showing the large expectations for these computational advances, all leading pharmaceutical companies have at least one research collaboration built around AI-related activities (Schuhmacher et al. 2021). For instance, Exscientia, an AI drug discovery firm, has partnered with various large pharmaceutical firms to integrate their approach into their processes. Partnering can also help improve diagnostics and direct treatments. Similarly, Novartis has worked with Microsoft to bolster its AI capabilities in drug discovery and development, including cell and gene therapy, clinical trials, manufacturing, and commercial operations. Through AI-enabled systems, project management within R&D can be improved through monitoring procedures, anticipating risks, accelerating the decision-making process, and enabling automation. These changes are then expected to replace traditional centralized and formalized ways of project management, and thus enable a more comprehensive model that includes technology players, CROs, and project-focused players (Federation of Indian Chambers of Commerce and Industry 2019).

Despite the promise of FAIR, many critical and sensitive data are still not readily accessible for any researcher to use. Nonetheless, new technologies are solving this by enabling companies to collaborate without disclosing sensitive information. For instance, blockchain-based technological solutions have the potential to deliver value to diverse areas of the pharma industry (Schöner et al. 2017). The Machine Learning Ledger Orchestration for Drug Discovery (MELLODDY) project is an EU project bringing together 17 partners across Europe including some of the largest big pharma and tech companies like Nvidia (Burki 2019). The 10 large pharmaceutical companies are bringing in their collection of small molecules together with their biochemical activity. Using blockchain technology, the project is exploring whether it would be possible to train machine learning models across this pooled dataset without leaking the IP-sensitive information of individual partners. The results of which are hoped to predict the potential of compounds for drug development.

16.3.2.3 Examples of FAIR Management

Many notable companies are already adopting FAIR. The big pharma firm Roche has adopted FAIR to take advantage of the clinical trial data they have collected across decades and use them to generate new insights. In 2017, they launched a program to improve data management within the firm by investing in the infrastructure and also implementing a workflow for FAIR. FAIR involved identifying the relevant use case, locating the datasets, confirming usage rights, scoping and defining the process for standardizing data for curation, and publishing data to catalog and evaluate the value of data with scientists. This was applied to four therapeutic areas so that historical clinical data can inform the design of clinical trials and help in decision-making (Pistoia Alliance 2021)

AstraZeneca has also pursued FAIRification, applying it to competitive intelligence, clinical study design, and translational medicine (Harrow et al. 2022). To demonstrate its utility, they conducted a "datathon" to enable their translational medicine team to evaluate the mechanism of lung cancer therapy resistance using clinical data. Through FAIRification of the model they built, they were then able to align the clinical outcomes with other data including subjects, genes, variants, and drugs for newer insights. In the long run, this FAIRification is seen as a good investment since previously solved models can be built upon for new questions.

Beyond established companies adopting FAIR, new forms of companies have sprung up to take advantage of the amount of external data being generated. Taking this to the extreme, companies do not even need to have internal development or production facilities, leading to the emergence of virtual pharmaceutical companies (VPCs). These VPCs can be composed of a small management team coordinating a network of service organizations to conduct development (Forster et al. 2014). One big advantage of collaborating with external partners and leveraging external data is that it can minimize the costs of building infrastructure to start operating. Debiopharm, Puma Biotechnology, Tioga Pharmaceuticals, VDDI Pharmaceuticals, and ReNeuroGen are some examples of such companies (Naylor and Pritchard Jr. 2019a).

Apart from pharmaceutical companies, broader healthcare systems are pursuing FAIR (Queralt-Rosinach et al. 2022). For instance, through the FAIR genomes initiative, the Dutch healthcare system would be able to link and leverage data that have been previously captured and stored across heterogeneous systems and organizations (van der Velde et al. 2022). Through FAIR, researchers and doctors would then be able to easily discover patients with rare diseases, assemble research cohorts based on more refined parameters, and enable the reuse of data for personalized medicine (van der Velde et al. 2022).

16.4   IMPLICATIONS AND CONCLUSIONS

The following section discusses the implications of the previously described advances in agile and FAIR data management in the pharmaceutical industry. These implications are explored for big pharmaceutical firms, small and medium biotech companies, and finally, academic researchers.

16.4.1   Large Pharmaceutical Companies

To cope with the changing demands in drug development, large pharmaceutical firms may have to transform their entire way of doing R&D (Apple et al. 2019; Lubkeman, Kronimus, and Hansen 2021). Should companies want to adopt agile, they must be clear about why they want to apply agile methods in the first place (Berggren et al. 2018). In general, the goal is to increase the pace by which companies develop and update their strategies to become more adaptable to changing conditions.

To support a more fluid R&D, companies can evaluate how their structure contributes to or impedes the speed of decision-making (Apple et al. 2019). As leaders cannot get the full picture of the situation within the company, staff should be able to escalate issues as soon as possible to the corresponding leaders. To respond to emerging needs, companies can also have an adaptive workforce where talent is allocated depending on the need (Philipp and Miinch 2021). Apart from speed, organizational structure can be harnessed these structures to promote interdisciplinary interactions. Companies can experiment with softening the boundaries across therapeutic areas and enable teams to look at disease areas from different perspectives (Ranade and Singh 2019). Often, later stage functions such as chemistry, manufacturing, regulations, and marketing only get caught up in the accumulated knowledge in the project in the later phases (Ranade and Singh 2019). Companies may experiment whether involving them earlier can help foresee and address risks that only surface late in the process. Companies can even try to integrate a team of data translators or external relationship managers and even experiment if they can be involved early in the process (Berggren et al. 2018).

Individual staff members and their teams can be empowered to decide and prioritize which assets can progress internally or be further developed with external partners (Philipp and Miinch 2021). Improving this interface between outside and internal knowledge ensures that external knowledge provides value within the firm (Romasanta, van der Sijde, and de Esch 2021). By reducing their emphasis on certain in-house capabilities, companies can flexibly pursue new technologies or disease areas (Berggren et al. 2018).

In terms of infrastructure, data can be standardized and curated for employees across the entire R&D value chain to inform decisions (Philipp and Miinch 2021; Steinwandter, Borchert, and Herwig 2019). Knowledge management systems can then help staff located across different parts of the value chain to learn quickly from other colleagues.

The adoption of agile and FAIR does not have to be abrupt. Companies have to balance the degree of change required and the potential benefits of such changes as they implement these approaches (Darino et al. 2020). Companies can run pilot tests on specific assets, functions, or platforms and learn from these implementations before rolling them out to the rest of the company (Agrawal et al.

2019). In such trials, leaders of the company have to understand the inefficiencies and redundancies that occur as the unit applies these new ways of working. However, through experience, the company can then learn the specific processes that work and do not work. Nonetheless, the goal also of such engagements is to expose staff to the agile and FAIR mindset. Gradually then, the company can identify other parts of the company that can also adopt such practices.

16.4.2 BIOTECH VENTURES

Smaller companies are now more competitive than ever partly due to the previous two trends mentioned: these smaller companies are leveraging agile methodologies to respond quickly to changing environmental conditions, and these smaller companies are also savvier in applying digital technologies to leverage external sources of knowledge. Illustrating their increased role in pharmaceutical development, smaller ventures have taken a larger proportion of clinical trials conducted, diminishing the share of the top ten largest pharmaceutical firms to now just about 30% (Smietana et al. 2020).

Through strategic partnerships, these smaller ventures do not have to invest much in building their technology infrastructure but focus on developing their core scientific ideas. Instead of owning assets, companies have also embraced the sharing economy model to optimize the allocation of their limited resources. Project teams can rent assets from providers and only pay for the specific resources that they need for their projects. Such transformation is helping companies access new technologies, lower costs, minimize underutilization of resources, and decentralize the handling of complex decisions (KPMG 2018). This shift toward sharing model is accelerated by various digital technologies. For instance, Science Exchange is a biotechnology marketplace where researchers can "source, order, and pay for scientific services from 3500+ trusted service providers–with only one contract" (Science Exchange n.d.). Through its network of CROs, CMOs, and academic labs, researchers can mix and match the different services to run the experiments needed to progress their projects.

Instead of demanding firms to know the entire landscape in development, they can focus on specializing further in their specific niche. Platform companies that harness novel approaches such as cell and gene therapies and mRNA have become central in the industry (Morrison 2016). It is estimated that there are more than 100 platform companies with at least five being valued at more than 1 billion USD (Kiernan and Naylor 2018). A typical platform company would have strong competency in a new technology that can be applied across a range of indications (Naylor and Pritchard Jr 2019b). By applying standardized processes across a class of molecules, the platform companies can continuously improve, adding data for every new molecule going through the platform (Drug Development and Delivery n.d.). Perhaps, the most famous recent example

is Moderna, which is the pioneer in mRNA therapeutics. To capitalize on their platform technology, they have created an internal portfolio of companies that apply their technology to different therapeutic areas such as immuno-oncology, viral and rare diseases. In many cases, they also partner with other larger firms like AstraZeneca, Merck, and Vertex to combine their different expertise (Damiani 2017).

As the industry becomes more interconnected, smaller companies have to be clear on what business model they would like to pursue to create value. Companies typically divide themselves across two models: technology platforms and asset creators (Kiernan and Naylor 2018). Technology platforms provide services to other firms for revenue. In this arrangement, the client absorbs all the risks but owns all the produced assets. On the other hand, asset creators develop their drug candidates. Once they have a potential candidate, they can then decide whether to out-license them to other companies or build in-house capacity for development. Most companies however combine both, serving other companies while also developing their candidates. In such cases, it would be important for these firms to differentiate between their therapeutic drugs and platform technologies and not dilute their efforts across these different competencies (Naylor and Pritchard Jr 2019b).

16.4.3 ACADEMIA

Many academic groups have also started to adopt agile project management (Pirro 2019). Other groups can also experiment with the approach and find the best configuration that would fit their specific needs. As for external relationships, academia is also used to collaborating with industry for technology transfer. However, they are mired by many challenges including resource constraints, legal and administrative complexity, challenges in coordination, goal alignment, interpersonal relations, and scientific gaps (Gersdorf et al. 2019). Across these, the lack of coordination had the strongest negative effect to project success.

Beyond dyadic partnerships, academia should embrace other forms of collaboration. Making their data FAIR so that other organizations can reuse their data would be beneficial (Pujol Priego, Wareham, and Romasanta 2022; Romasanta and Wareham 2021). To take advantage of these; however; researchers need to be well-versed in different tools and schemes including electronic laboratory notebooks, data management plans, online within-team coordination tools, and community feedback portals (Robertson et al. 2014).

Moreover, the complexity of science has necessitated academic organizations to form consortia with industry to address different aspects of a problem. It is thus important that academic organizations participating in these consider different aspects to ensure their success (Simpson and Wilkinson 2020). Partners should agree on the goals of the project and how to measure these goals. With partners working on different parts of the project, there should be an agreement on how results would be shared. An independent party is useful to help manage the knowledge generated and ensure that it is carefully shared with relevant organizations. More importantly, the consortium should establish how they will validate the results by different partners. Partners need to agree on how the project should terminate or progress. If the project shows promise, partners need to align how they should decide to invest further in the project. If the consortium finds an interesting area not originally foreseen, members should think about their procedures for pivoting and redirecting their efforts. Finally, it should be clear who gains the intellectual property rights from the results of the consortia and what would be the procedure for deciding this such as if companies can negotiate or purchase. Since the early phases of drug discovery tend to be explorative while later stages are sequential and process-driven, the organizations had aligned in the management of the consortia to cater to these heterogeneous workflows.

To conclude, developing new pharmaceutical products is challenging due to costs, complexity, and changing stakeholder demands. To address these challenges, organizations from large pharmaceutical companies to new ventures to academia can experiment with new trends in management including agile and FAIR. By adopting these innovative ways of management, both internally and externally, organizations can adapt to the rapidly shifting landscape in the development and manufacture of pharmaceutical products.

REFERENCES

Agrawal, Gaurav, Harriet Keane, Maha Prabhakaran, and Michael Steinmann. 2019. "The Pursuit of Excellence in New-Drug Development." *McKinsey*. https://www.mckinsey.com/industries/life-sciences/our-insights/the-pursuit-of-excellence-in-new-drug-development.

Alaedini, Pedram, Birnur Ozbas, and Fahri Akdemir. 2014. "Agile Drug Development: Lessons from the Software Industry." *Contract Pharma*. https://www.contractpharma.com/issues/2014-10-01/view_features/agile-drug-development-lessons-from-the-software-industry.

Alharbi, Ebtisam, Rigina Skeva, Nick Juty, Caroline Jay, and Carole Goble. 2021. "Exploring the Current Practices, Costs and Benefits of FAIR Implementation in Pharmaceutical Research and Development: A Qualitative Interview Study." *Data Intelligence* 3 (4): 507–27. https://doi.org/10.1162/DINT_A_00109.

Apple, Aliza, Harriet Keane, Rachel Moss, and Valentina Sartori. 2019. "Agile Pharma: Transforming R&D Functions through Agility." *McKinsey*. https://www.mckinsey.com/industries/life-sciences/our-insights/designing-an-agile-transformation-in-pharma-r-and-d.

Årdal, Christine, and John Arne Røttingen. 2012. "Open Source Drug Discovery in Practice: A Case Study." *PLOS Neglected Tropical Diseases* 6 (9): e1827. https://doi.org/10.1371/JOURNAL.PNTD.0001827.

Babrak, Lmar M., Joseph Menetski, Michael Rebhan, Giovanni Nisato, Marc Zinggeler, Noé Brasier, Katja Baerenfaller, Thomas Brenzikofer, Laurenz Baltzer, Christian Vogler, and Leo Gschwind, 2019. "Traditional and digital biomarkers: two worlds apart?." *Digital Biomarkers*, 3 (2): 92–102. https://doi.org/10.1159/000502000

Balz, Michael, Alix Burke, Alberto Montagner, and Michele Tarallo. 2021. "A New Operating Model for Pharma: How the Pandemic Has Influenced Priorities." *McKinsey*.

Beagrie, Neil, and John Houghton. 2016. "The Value and Impact of the European Bioinformatics Institute Full Report."

Beck, Kent, Mike Beedle, Arie van Bennekum, Alistair Cockburn, Ward Cunningham, Martin Fowler, James Grenning, et al. 2001. "Manifesto for Agile Software Development." http://agilemanifesto.org/.

Berggren, Roy, Edd Fleming, Harriet Keane, and Rachel Moss. 2018. "An Agile Pharma R&D Operating Model." *McKinsey*. https://www.mckinsey.com/industries/life-sciences/our-insights/r-and-d-in-the-age-of-agile.

Besel, Andreas. 2019. "Alliances as a Tool to Transform Life Science R&D." *KPMG*. https://home.kpmg/ch/en/blogs/home/posts/2019/10/alliances-as-a-tool-to-transform-life-science.html.

Bhardwaj, Anshu, Vinod Scaria, Gajendra Pal Singh Raghava, Andrew Michael Lynn, Nagasuma Chandra, Sulagna Banerjee, Muthukurussi V. Raghunandanan, et al. 2011. "Open Source Drug Discovery– A New Paradigm of Collaborative Research in Tuberculosis Drug Development." *Tuberculosis* 91 (5): 479–86. https://doi.org/10.1016/J.TUBE.2011.06.004.

Bianchi, Mattia, Alberto Cavaliere, Davide Chiaroni, Federico Frattini, and Vittorio Chiesa. 2011. "Organisational Modes for Open Innovation in the Bio-Pharmaceutical Industry: An Exploratory Analysis." *Technovation* 31 (1): 22–33. https://doi.org/10.1016/j.technovation.2010.03.002.

Blasimme, Alessandro, Marta Fadda, Manuel Schneider, and Effy Vayena. 2018. "Data Sharing for Precision Medicine: Policy Lessons and Future Directions." *Health Affairs* 37 (5): 702–9. https://doi.org/10.1377/HLTHAFF.2017.1558/ASSET/IMAGES/LARGE/FIGUREEX2.JPEG.

Brooks, Kristin. 2014. "Agile Drug Development."

Burki, Talha. 2019. "Pharma Blockchains AI for Drug Development." *The Lancet* 393 (10189): 2382. https://doi.org/10.1016/S0140-6736(19)31401-1.

Cerwin, Erika L., Robert D. Guenard, and Timothy B. Alosi. 2021. "Pioneering a Data Strategy to Support Pharmaceutical Operations and Technology | American Pharmaceutical Review - The Review of American Pharmaceutical Business & Technology." *American Pharmaceutical Review*.

Chesbrough, Henry W. 2003. *Open Innovation: The New Imperative for Creating and Profiting from Technology*. Harvard Business School Press, Boston, MA. https://doi.org/10.1111/j.1467-8691.2008.00502.x.

Cirillo, Davide, and Alfonso Valencia. 2019. "Big Data Analytics for Personalized Medicine." *Current Opinion in Biotechnology* 58 (August): 161–67. https://doi.org/10.1016/J.COPBIO.2019.03.004.

Damiani, Marcello. 2017. "Building The Digital Biotech Company: Why and How Digitization Is Mission-Critical for Moderna."

Darino, Lucia, Aaron De Smet, Umar Husain, and Emily Yueh. 2020. "Reimagining How Life Sciences Work Will Be Done in the next Normal." *McKinsey*. https://www.mckinsey.com/industries/life-sciences/our-insights/reimagining-how-life-sciences-work-will-be-done-in-the-next-normal.

David Champagne, Alex Devereson, Lucy Pérez, and David Saunders. 2020. "How Pharma Companies Are Applying Advanced Analytics to Real-World Evidence Generation." *McKinsey*. https://www.mckinsey.com/industries/life-sciences/our-insights/creating-value-from-next-generation-real-world-evidence.

Ding, Baoyang. 2018. "Pharma Industry 4.0: Literature Review and Research Opportunities in Sustainable Pharmaceutical Supply Chains." *Process Safety and Environmental Protection* 119 (October): 115–30. https://doi.org/10.1016/J. PSEP.2018.06.031.

Doxzen, Kevin W., Landry Signe, and Diana M. Bowman. 2022. "Advancing Precision Medicine through Agile Governance." *Brookings.*

Drug Development and Delivery. n.d. "Special Feature - Platform Technologies: Not Just for Big Pharma." *Drug Development and Delivery.* Accessed March 16, 2022. https://drug-dev. com/special-feature-platform-technologies-not-just-for-big-pharma/.

Ebel, Thomas, Kerstin Kubik, and Martin Lösch. 2012. "Light-Footed Operations: The Virtues of Agility in Volatile Times." *McKinsey.*

Federation of Indian Chambers of Commerce and Industry. 2019. "Use of Artificial Intelligence and Advanced Analytics in Pharmaceuticals."

Fiore, Alessandro Di, Kendra West, and Andrea Segnalini. 2019. "Why Science-Driven Companies Should Use Agile." *Harvard Business Review.*

Ford, Jeff, Alex Blair, Bushra Naaz, and Jessica Overman. 2020. "Biopharma Leaders Prioritize R&D, Technological Transformation, and Global Market Presence." *Deloitte Insights.*

Forster, Simon P., Julia Stegmaier, Rene Spycher, and Stefan Seeger. 2014. "Virtual Pharmaceutical Companies: Collaborating Flexibly in Pharmaceutical Development." *Drug Discovery Today* 19 (3): 348–55. https://doi.org/ 10.1016/J.DRUDIS.2013.11.015.

Freytag, Clemens. 2019. "Building an Agile Pharma Organization 4.0." *LinkedIn.* https://www.linkedin.com/pulse/building-agile-pharma-organization-40-clemens-freytag/.

George, K.C., and Gina Fridley. 2021. "The Pandemic Forced Agile Innovation in Healthcare. Now Make It Stick." *Bain.* https://www.bain.com/insights/the-pandemic-forced-agile-innovation-in-healthcare/.

Gersdorf, Thomas, Vivianna Fang He, Ann Schlesinger, Guido Koch, Dominic Ehrismann, Hans Widmer, and Georg von Krogh. 2019. "Demystifying Industry–Academia Collaboration." *Nature Reviews Drug Discovery* 18 (10): 743–44. https://doi.org/10.1038/d41573-019-00001-2.

Gorgulla, Christoph, Andras Boeszoermenyi, Zi Fu Wang, Patrick D. Fischer, Paul W. Coote, Krishna M. Padmanabha Das, Yehor S. Malets, et al. 2020. "An Open-Source Drug Discovery Platform Enables Ultra-Large Virtual Screens." *Nature* 580 (7805): 663–68. https://doi.org/10.1038/s41586-020-2117-z.

Grossman, Robert L., Allison Heath, Mark Murphy, Maria Patterson, and Walt Wells. 2016. "A Case for Data Commons: Toward Data Science as a Service." *Computing in Science and Engineering.* https://doi.org/10.1109/MCSE.2016.92.

Han, Chanshuai, Mathilde Chaineau, Carol X.Q. Chen, Lenore K. Beitel, and Thomas M. Durcan. 2018. "Open Science Meets Stem Cells: A New Drug Discovery Approach for Neurodegenerative Disorders." *Frontiers in Neuroscience* 12 (FEB): 47. https:// doi.org/10.3389/FNINS.2018.00047/BIBTEX.

Harrow, Ian, Rama Balakrishnan, Hande Küçük McGinty, Tom Plasterer, and Martin Romacker. 2022. "Maximizing Data Value for Biopharma through FAIR and Quality Implementation: FAIR plus Q." *Drug Discovery Today* 27 (5): 1441–47. https://doi.org/10.1016/J.DRUDIS.2022.01.006.

Harrow, Jennifer, John Hancock, Niklas Blomberg, Niklas Blomberg, Søren Brunak, Salvador Capella-Gutierrez,

Christine Durinx, et al. 2021. "ELIXIR-EXCELERATE: Establishing Europe's Data Infrastructure for the Life Science Research of the Future." *The EMBO Journal* 40 (6): e107409. https://doi.org/10.15252/EMBJ.2020107409.

Hartl, Dominik, Valeria de Luca, Anna Kostikova, Jason Laramie, Scott Kennedy, Enrico Ferrero, Richard Siegel, et al. 2021. "Translational Precision Medicine: An Industry Perspective." *Journal of Translational Medicine* 19 (1): 1–14. https://doi.org/10.1186/S12967-021-02910-6.

Hess, Andrew, and Frank T. Rothaermel. 2012. "Intellectual Human Capital and the Emergence of Biotechnology: Trends and Patterns, 1974–2006." *IEEE Transactions on Engineering Management.* https://doi.org/10.1109/TEM.2010.2082550.

Huang, Kexin, Tianfan Fu, Wenhao Gao, Yue Zhao, Yusuf Roohani, Jure Leskovec, Connor W. Coley, Cao Xiao, Jimeng Sun, and Marinka Zitnik. 2021. "Therapeutics Data Commons: Machine Learning Datasets and Tasks for Drug Discovery and Development." *ArXiv*, February. https://doi. org/10.48550/arxiv.2102.09548.

Hughes, Benjamin, and Jonathan Wareham. 2010. "Knowledge Arbitrage in Global Pharma: A Synthetic View of Absorptive Capacity and Open Innovation." *R&D Management* 40 (3): 324–43. https://doi.org/10.1111/j.1467-9310.2010.00594.x.

Hulsen, Tim, Saumya S. Jamuar, Alan R. Moody, Jason H. Karnes, Orsolya Varga, Stine Hedensted, Roberto Spreafico, David A. Hafler, and Eoin F. McKinney. 2019. "From Big Data to Precision Medicine." *Frontiers in Medicine* 6 (MAR): 34. https://doi.org/10.3389/FMED.2019.00034/BIBTEX.

Hunter, Jackie, and Susie Stephens. 2010. "Is Open Innovation the Way Forward for Big Pharma?" *Nature Reviews Drug Discovery* 9 (2): 87–88. https://doi.org/10.1038/nrd3099.

IQVIA. 2019. "The Changing Landscape of Research and Development - IQVIA."

———. 2021. "Global Trends in R&D."

Jacobsen, Annika, Ricardo de Miranda Azevedo, Nick Juty, Dominique Batista, Simon Coles, Ronald Cornet, Mélanie Courtot, et al. 2020. "FAIR Principles: Interpretations and Implementation Considerations." *Data Intelligence.* https:// doi.org/10.1162/dint_r_00024.

Karaivanov, Dimitar. 2020. "The Impact of Agile In The Pharmaceutical Industry." *Electronic Health Reporter.* https://electronichealthreporter. com/the-impact-of-agile-in-the-pharmaceutical-industry/.

Kiernan, Urban, and Stephen Naylor. 2018. "Emerging Paradigm of Integrated Platform Drug Discovery and Development Companies." *Drug Discovery World.* https://www. ddw-online.com/media/32/(4)-emerging-paradigm-of-integrated-platform-drug-discovery.pdf.

KPMG. 2018. "R&D 2030: Reinvent Innovation and Become an R&D Front-Runner by 2030."

———. 2019. "Reshaping the Future of Pharma: Four Critical Capabilities for 2030."

Lilja, Sami, Jarkko Kailanto, and Marita Saanila-Sotamaa. 2021. "From Pyramid to Communities: How Pharma Company Reinvented Themselves Using Scrum." *Agile Alliance.*

Lou, Bowen, and Lynn Wu. 2020. "Artificial Intelligence and Drug Innovation: A Large Scale Examination of the Pharmaceutical Industry." *SSRN Electronic Journal.* https:// doi.org/10.2139/ssrn.3524985.

Lubkeman, Mark, André Kronimus, and Filip Hansen. 2021. "Building Effective Business Development in Pharma." *BCG.* https://www.bcg.com/publications/2021/six-ways-to-build-an-effective-pharmaceutical-business-development-strategy.

Mak, Kit-Kay, and Mallikarjuna Rao Pichika. 2019. "Artificial intelligence in drug development: Present status and future prospects." *Drug Discovery Today* 24 (3): 773–80. https://doi.org/10.1016/j.drudis.2018.11.014.

Mehralian, Gholamhossein, Forouzandeh Zarenezhad, and Ali Rajabzadeh Ghatari. 2015. "Developing a Model for an Agile Supply Chain in Pharmaceutical Industry." *International Journal of Pharmaceutical and Healthcare Marketing* 9 (1): 74–91. https://doi.org/10.1108/IJPHM-09-2013-0050/FULL/PDF.

Michelino, Francesca, Emilia Lamberti, Antonello Cammarano, and Mauro Caputo. 2015. "Open Innovation in the Pharmaceutical Industry: An Empirical Analysis on Context Features, *Internal R & D, and Financial Performances*" 62(3): 421–35.

Mons, Barend, Cameron Neylon, Jan Velterop, Michel Dumontier, Luiz Olavo Bonino Da Silva Santos, and Mark D. Wilkinson. 2017. "Cloudy, Increasingly FAIR; Revisiting the FAIR Data Guiding Principles for the European Open Science Cloud." *Information Services and Use.* https://doi.org/10.3233/ISU-170824.

Morrison, Chris. 2016. "Platform Biotechs Bring Out Pharma's Creative Side." Biopharma Dealmakers 2021, September.

Murray, David, Mark Wigglesworth, and Marian Preston. 2019. "Open Innovation – Collaboration between Academia and the Pharma Industry." *Drug Target Review.* https://www.drugtargetreview.com/article/53093/open-innovation-a-collaboration-between-academia-and-the-pharmaceutical-industry-to-further-leverage-drug-discovery-expertise-and-assets/.

Naylor, Stephen, and Kirkwood A. Pritchard Jr. 2019a. "The Reality of Virtual Pharmaceutical Companies." *Drug Discovery World.* https://www.ddw-online.com/the-reality-of-virtual-pharmaceutical-companies-1320-201908/.

Naylor, Stephen, and Kirkwood A. Pritchard Jr. 2019b. "Integrated Platform Drug Discovery and Development Companies Part II: Comparative Analysis." *Drug Discovery World.* https://www.ddw-online.com/media/32/(6)-integrated-platform.pdf.

Osmond, Neil. 2019. "A Guide to Implementing Agile in Pharma." *PF Media.* https://pf-media.co.uk/in-depth/a-guide-to-implementing-agile-in-pharma/.

Pammolli, Fabio, Laura Magazzini, and Massimo Riccaboni. 2011. "The Productivity Crisis in Pharmaceutical R&D." *Nature Reviews Drug Discovery* 10 (6): 428–38. https://doi.org/10.1038/nrd3405.

Parra-Calderón, Carlos Luis. 2019. "Enhancing Precision Medicine: Sharing and Reusing Data."

Paul, Debleena, Gaurav Sanap, Snehal Shenoy, Dnyaneshwar Kalyane, Kiran Kalia, and Rakesh K. Tekade. 2021. "Artificial Intelligence in Drug Discovery and Development." *Drug Discovery Today* 26 (1): 80. https://doi.org/10.1016/J.DRUDIS.2020.10.010.

Philipp, Marc P., and Katie Miinch. 2021. "The End of the R&D Organization as We Know It – Pharma's Pivot to a Liquid Pipeline Progression Model." *STAT News.* https://www.statnews.com/sponsor/2021/08/20/the-end-of-the-rd-organization-as-we-know-it-pharmas-pivot-to-a-liquid-pipeline-progression-model/.

Pirro, Laura. 2019. "How Agile Project Management Can Work for Your Research." *Nature.* https://doi.org/10.1038/D41586-019-01184-9.

Pistoia Alliance. 2021. "FAIRification of Clinical Trial Data – Roche." *FAIR Toolkit.* https://fairtoolkit.pistoiaalliance.org/use-cases/fairification-of-clinical-trial-data-roche/.

Pujol Priego, Laia, and Jonathan D. Wareham. 2019. "Open Targets: Pre-Competitive Collaborative Research in Life Sciences." 2019 (1): 11674. https://doi.org/10.5465/AMBPP.2019.11674ABSTRACT.

Pujol Priego, Laia, and Jonathan D. Wareham, and Angelo Kenneth S. Romasanta. 2022. "The Puzzle of Sharing Scientific Data." 29 (2): 219–50. https://doi.org/10.1080/13662716.2022.2033178.

Queralt-Rosinach, Núria, Rajaram Kaliyaperumal, César H. Bernabé, Qinqin Long, Simone A. Joosten, Henk Jan van der Wijk, Erik L.A. Flikkenschild, et al. 2022. "Applying the FAIR Principles to Data in a Hospital: Challenges and Opportunities in a Pandemic." *Journal of Biomedical Semantics* 13 (1): 1–19. https://doi.org/10.1186/S13326-022-00263-7.

Rambaldini, Joel, and Cesar Fernandez Giove. 2018. "The Future of Pharma & Medical Device Supply Chains in the USA."

Ranade, Vikram, and Navjot Singh. 2019. "An Engineering Approach to Derisking Pharmaceutical R&D for Novel Modalities | McKinsey." *McKinsey.* https://www.mckinsey.com/industries/life-sciences/our-insights/an-engineering-approach-to-derisking-r-and-d-for-novel-drug-modalities.

Reichman, Melvin, and Peter B. Simpson. 2016. "Open Innovation in Early Drug Discovery: Roadmaps and Roadblocks." *Drug Discovery Today* 21(5): 779–88.

Richenberger, Carina, Yudong Zhang, Jens Grueger, and Srikant Vaidyanathan. 2021. "How Pharma Can Navigate the Shifting Sands of Oncology Treatments." *BCG.* https://www.bcg.com/en-es/publications/2021/navigating-evolving-oncology-treatment-landscape.

Ries, Eric. 2009. "Minimum Viable Product: A Guide." Startup Lessons Learned. http://www.startuplessonslearned.com/2009/08/minimum-viable-product-guide.html.

Robertson, Murray N., Paul M. Ylioja, Alice E. Williamson, Michael Woelfle, Michael Robins, Katrina A. Badiola, Paul Willis, Piero Olliaro, Timothy N.C. Wells, and Matthew H. Todd. 2014. "Open Source Drug Discovery – A Limited Tutorial." *Parasitology* 141 (1): 148–57. https://doi.org/10.1017/S0031182013001121.

Romasanta, Angelo K.S., Peter van der Sijde, and Iwan J.P. de Esch. 2021. "Absorbing Knowledge from an Emerging Field: The Role of Interfacing by Proponents in Big Pharma." *Technovation* 110, 102363. https://doi.org/10.1016/j.technovation.2021.102363.

Romasanta, Angelo Kenneth S., Peter van der Sijde, and Jacqueline van Muijlwijk-Koezen. 2020. "Innovation in Pharmaceutical R&D: Mapping the Research Landscape." *Scientometrics.* https://doi.org/10.1007/s11192-020-03707-y.

Romasanta, Angelo, and Jonathan Wareham. 2021. "Fair Data through a Federated Cloud Infrastructure: Exploring the Science Mesh." In *ECIS 2021 Research-in-Progress Papers*, 14.

Rottier, Pieter Adriaan, and Victor Rodrigues. 2008. "Agile Development in a Medical Device Company." *Proceedings - Agile 2008 Conference*, 218–23. https://doi.org/10.1109/AGILE.2008.52.

Scannell, Jack W., Alex Blanckley, Helen Boldon, and Brian Warrington. 2012. "Diagnosing the Decline in Pharmaceutical R&D Efficiency." *Nature Reviews Drug Discovery* 11 (3): 191–200. https://doi.org/10.1038/nrd3681.

Schlander, Michael, Karla Hernandez-Villafuerte, Chih Yuan Cheng, Jorge Mestre-Ferrandiz, and Michael Baumann. 2021. "How Much Does It Cost to Research and Develop a New Drug? A Systematic Review and Assessment."

PharmacoEconomics 39 (11): 1243–69. https://doi.org/10.1007/S40273-021-01065-Y/TABLES/4.

Schöner, Manuela M., Dimitris Kourouklis, Philipp Sandner, Erick Gonzalez, and Jonas Förster. 2017. "Blockchain Technology in the Pharmaceutical Industry."

Schuhmacher, Alexander, Oliver Gassmann, Markus Hinder, and Michael Kuss. 2021. "The Present and Future of Project Management in Pharmaceutical R&D." *Drug Discovery Today* 26 (1): 1–4. https://doi.org/10.1016/j.drudis.2020.07.020.

Schuhmacher, Alexander, Paul-Georg Germann, Henning Trill, and Oliver Gassmann. 2013. "Models for Open Innovation in the Pharmaceutical Industry." *Drug Discovery Today* 18 (23–24): 1133–37. https://doi.org/10.1016/j.drudis.2013.07.013.

Schultes, Erik, and Peter Wittenburg. 2019. "FAIR Principles and Digital Objects: Accelerating Convergence on a Data Infrastructure." In *Communications in Computer and Information Science*. https://doi.org/10.1007/978-3-030-23584-0_1.

Science Exchange. n.d. "Science Exchange." Accessed March 16, 2022. https://ww2.scienceexchange.com/s/.

Sherman, Rachel E., Steven A. Anderson, Gerald J. Dal Pan, Gerry W. Gray, Thomas Gross, Nina L. Hunter, Lisa LaVange, Danica Marinac-Dabic, Peter W. Marks, Melissa A. Robb, and Jeffrey Shuren, 2016. Real-world evidence—what is it and what can it tell us?. *New England Journal of Medicine*, 375 (23): 2293–2297. https://doi.org/10.1056/NEJMsb1609216.

Simpson, Peter B., and Graeme F. Wilkinson. 2020. "What Makes a Drug Discovery Consortium Successful?" *Nature Reviews Drug Discovery* 19 (11): 737–38. https://doi.org/10.1038/D41573-020-00079-Z.

Smietana, Katarzyna, David Quigley, Bart Van de Vyver, and Martin Møller. 2020. "The Fragmentation of Biopharmaceutical Innovation." *Nature Reviews. Drug Discovery* 19 (1): 17–18. https://doi.org/10.1038/D41573-019-00046-3.

Steinwandter, Valentin, Daniel Borchert, and Christoph Herwig. 2019. "Data Science Tools and Applications on the Way to Pharma 4.0." *Drug Discovery Today* 24 (9): 1795–805. https://doi.org/10.1016/J.DRUDIS.2019.06.005.

Stott, Kelvin. 2017. "Pharma's Broken Business Model - Part 1: An Industry on the Brink of Terminal Decline." https://www.linkedin.com/pulse/pharmas-broken-business-model-industry-brink-terminal-kelvin-stott/.

Todd, Matthew H. 2019. "Six Laws of Open Source Drug Discovery." *ChemMedChem* 14 (21): 1804–9. https://doi.org/10.1002/CMDC.201900565.

US FDA. 2021. "Advancing Health through Innovation New Drug Therapy Approvals 2020."

———. 2022. "Orphan Drug Designations and Approvals." *FDA*. https://www.accessdata.fda.gov/scripts/opdlisting/oopd/listResult.cfm.

Vaidyanathan, Srikant, Grant Freeland, María López, and David Greber. 2019. "Agile Can Work Wonders in Pharma." *BCG*. https://www.bcg.com/publications/2019/agile-work-wonders-pharma.

Veale, Clinton G.L. 2019. "Unpacking the Pathogen Box- An Open Source Tool for Fighting Neglected Tropical Disease." *ChemMedChem* 14 (4): 386–453. https://doi.org/10.1002/CMDC.201800755.

van der Velde, K. Joeri, Gurnoor Singh, Rajaram Kaliyaperumal, XiaoFeng Liao, Sander de Ridder, Susanne Rebers, Hindrik H.D. Kerstens, et al. 2022. "FAIR Genomes Metadata Schema Promoting Next Generation Sequencing Data Reuse in Dutch Healthcare and Research." *Scientific Data* 9 (1): 1–13. https://doi.org/10.1038/s41597-022-01265-x.

Vesteghem, Charles, Rasmus Froberg Brøndum, Mads Sønderkær, Mia Sommer, Alexander Schmitz, Julie Støve Bødker, Karen Dybkær, Tarec Christoffer El-Galaly, and Martin Bøgsted. 2020. "Implementing the FAIR Data Principles in Precision Oncology: Review of Supporting Initiatives." *Briefings in Bioinformatics* 21 (3): 936–45. https://doi.org/10.1093/BIB/BBZ044.

Vicente-Saez, Ruben, and Clara Martinez-Fuentes. 2018. "Open Science Now: A Systematic Literature Review for an Integrated Definition." *Journal of Business Research* 88 (July): 428–36. https://doi.org/10.1016/j.jbusres.2017.12.043.

Vijayan, R. S.K., Jan Kihlberg, Jason B. Cross, and Vasanthanathan Poongavanam. 2022. "Enhancing Preclinical Drug Discovery with Artificial Intelligence." *Drug Discovery Today* 27 (4): 967–84. https://doi.org/10.1016/J.DRUDIS.2021.11.023.

Van Vlijmen, Herman, Albert Mons, Arne Waalkens, Wouter Franke, Arie Baak, Gerbrand Ruiter, Christine Kirkpatrick, et al. 2020. "The Need of Industry to Go FAIR." *Data Intelligence* 2: 276–84. https://doi.org/10.1162/dint_a_00050.

West, Kendra. 2018. "Reinventing Research: Agile in the Academic Laboratory I." *Agile Alliance*. https://www.agilealliance.org/resources/experience-reports/reinventing-research-agile-in-the-academic-laboratory/.

Wienken, Magdalena. 2021. "FAIR Data or the Journey Towards Data-Centricity – Pushing Towards Making Data Discoverable, Accessible, Interoperable, and Reusable – Enlightenbio Blog." *Enlighten Bio*. https://enlightenbio.com/news-and-features/2021/05/06/fair-data-or-the-journey-towards-data-centricity-pushing-towards-making-data-discoverable-accessible-interoperable-and-reusable/.

Wilkinson, Mark D., Michel Dumontier, IJsbrand Jan Aalbersberg, Gabrielle Appleton, Myles Axton, Arie Baak, Niklas Blomberg, et al. 2016. "The FAIR Guiding Principles for Scientific Data Management and Stewardship." *Scientific Data* 3 (1): 160018. https://doi.org/10.1038/sdata.2016.18.

Wise, John, Alexandra Grebe de Barron, Andrea Splendiani, Beeta Balali-Mood, Drashtti Vasant, Eric Little, Gaspare Mellino, et al. 2019. "Implementation and Relevance of FAIR Data Principles in Biopharmaceutical R&D." *Drug Discovery Today* 24 (4): 933–38. https://doi.org/10.1016/j.drudis.2019.01.008.

Xia, Tianjiao. 2013. "Absorptive Capacity and Openness of Small Biopharmaceutical Firms - a European Union-United States Comparison." *R and D Management* 43 (4): 333–51. https://doi.org/10.1111/radm.12017.

Index

Note: **Bold** page numbers refer to tables and *italic* page numbers refer to figures.

For Product Safety Concerns and Information please contact our EU
representative GPSR@taylorandfrancis.com
Taylor & Francis Verlag GmbH, Kaufingerstraße 24, 80331 München, Germany